Lecture Notes in Mathematics

Editors:
J.-M. Morel, Cachan
F. Takens, Groningen
B. Teissier, Paris

Jean-Pierre Antoine · Camillo Trapani

Partial Inner Product Spaces

Theory and Applications

 Springer

Jean-Pierre Antoine
Institut de Physique Théorique
Université catholique de Louvain
2, chemin du Cyclotron
1348 Louvain-la-Neuve
Belgium
jean-pierre.antoine@uclouvain.be

Camillo Trapani
Dipartimento di Matematica ed Applicazioni
Università di Palermo
Via Archirafi, 34
90123 Palermo
Italy
trapani@unipa.it

ISBN: 978-3-642-05135-7   e-ISBN: 978-3-642-05136-4
DOI 10.1007/978-3-642-05136-4
Springer Heidelberg Dordrecht London New York

Lecture Notes in Mathematics ISSN print edition: 0075-8434
                             ISSN electronic edition: 1617-9692

Library of Congress Control Number: 2009941068

Mathematics Subject Classification (2000): 46C50, 46Axx, 46Exx, 46Fxx, 47L60, 47Bxx, 81Qxx, 94A12

Cover design: SPi Publisher Services

Printed on acid-free paper

springer.com

Foreword

This volume has its origin in a longterm collaboration between Alex Grossmann (Marseille) and one of us (JPA), going back to 1967. This has resulted in a whole collection of notes, manuscripts, and joint papers. In particular, a large set of unpublished notes by AG (dubbed the 'skeleton') has proven extremely valuable for writing the book, and we thank him warmly for putting it at our disposal. JPA also thanks the Centre de Physique Théorique (CPT, Marseille) for its hearty hospitality at the time.

Later on, almost thirty years ago, the two authors of this book started to interact (with a strong initial impulse of G. Epifanio, CT's advisor at the time), mostly in the domain of partial operator algebras. This last collaboration has consisted entirely of bilateral visits between Louvain-la-Neuve and Palermo. We thank our home institutions for a constantly warm hospitality, as well as various funding agencies that made it possible, namely, the Commisariat Général aux Relations Internationales de la Communauté Française de Belgique (Belgium), the Direzione Generale per le Relazioni Culturali del Ministero degli Affari Esteri Italiano and the Ministero dell'Università e della Ricerca Scientifica (Italy). In the meantime, we also enjoyed the collaboration of many colleagues and students such as F. (Debacker)-Mathot, J-R. Fontaine, J. Shabani (LLN), G. Epifanio, F. Bagarello, A. Russo, F. Tschinke (Palermo), G. Lassner[†], K-D. Kürsten (Leipzig), W. Karwowski (Wrocław), and A. Inoue (Fukuoka). We thank them all.

Last, but not least, we owe much to our respective wives Nicole and Adriana for their loving support and patience throughout this work.

Contents

Prologue

In the course of their curriculum, physics and mathematics students are usually taught the basics of Hilbert space, including operators of various types. The justification of this choice is twofold. On the mathematical side, Hilbert space is the example of an infinite dimensional topological vector space that more closely resembles the familiar Euclidean space and thus it offers the student a smooth introduction into functional analysis. On the physics side, the fact is simply that Hilbert space is the daily language of quantum theory, so that mastering it is an essential tool for the quantum physicist.

Beyond Hilbert Space

However, after a few years of practice, the former student will discover that the tool in question is actually insufficient. If he is a mathematician, he will notice, for instance, that Fourier transform is more naturally formulated in the space L^1 of integrable functions than in the space L^2 of square integrable functions, since the latter requires a nontrivial limiting procedure. Thus enter Banach spaces. More striking, a close look at most partial differential equations of interest for applications reveals that the interesting solutions are seldom smooth or square integrable. Physically meaningful events correspond to changes of regime, which mean discontinuities and/or distributions. Shock waves are a typical example. Actually this state of affairs was recognized long ago by authors like Leray or Sobolev, whence they introduced the notion of *weak solution*. Thus it is no coincidence that many textbooks on PDEs begin with a thorough study of distribution theory. Famous examples are those of Hörmander [Hör63] or Lions-Magenes [LM68].

As for physics, it is true that the very first mathematically precise formulation of quantum mechanics is that of J. von Neumann [vNe55], in 1933, which by the way yielded also the first exact definition of Hilbert space as we know it. However, a pure Hilbert space formulation of quantum mechanics is both inconvenient and foreign to the daily behavior of most physicists, who stick to the more suggestive version of Dirac [Dir30]. A glance at the textbook of Prugovečki [Pru71] will easily convince the reader.... An additional drawback is the universal character of Hilbert space: all separable Hilbert spaces are isomorphic, but physical systems are not! It would be more logical that the structure of the state space carry some information about the system it describes. In addition, there are many interesting objects that do not find their place in Hilbert space. Plane waves or δ-functions do not belong in L^2, yet they are immensely useful. The same is true of wave functions belonging to the continuous spectrum of the Hamiltonian.

As a matter of fact, all these objects can receive a precise mathematical meaning as distributions or generalized functions, that is, linear functionals over a space of nice test functions. Thus the door opens on Quantum Mechanics beyond Hilbert space [14]. Many different structures have emerged along this line, such the rigged Hilbert spaces (RHS) of Gel'fand et al. [GV64], the equipped Hilbert spaces of Berezanskii [Ber68], the extended Hilbert spaces of Prugovečki [170], the analyticity/trajectory spaces of van Eijndhoven and de Graaf [Eij83, EG85, EG86] or the nested Hilbert spaces (NHS) of Grossmann [114]. Among these, the RHS is the best known and it answers the objections made above to the sole use of Hilbert space. A different approach to its introduction is via the consideration of unbounded operators representing observables, as proposed independently by J. Roberts [171, 172], A. Böhm [47], and one of us (JPA) [8, 9]. We will discuss this approach at length in Chapter 5.

The central topic of this volume, namely *partial inner product spaces (PIP-spaces)*, has its origin in the first meeting between A. Grossmann and JPA, in 1967. Both of us were already working beyond Hilbert space, with NHS for AG and RHS for JPA. We realized that we were in fact basically doing the same thing, using different languages. After many discussions, we were able to extract the quintessence of our respective approaches, namely, the notions of *partial inner product* and *partial inner product space* (PIP-space). A thorough analysis followed, that led to a number of joint publications [12, 13, 17–19], later with W. Karwowski [22, 23]. Students joined in, such as F. Mathot [Mat75], A-M. Nachin [Nac72] and J. Shabani [177]. But gradually interest moved to other subjects, such as algebras of unbounded operators and partial operator algebras, culminating in the monograph by the two of us with A. Inoue [AIT02]. But sometimes PIP-spaces came back on the stage also when considering partial *-algebras. Indeed, in their study on partial *-algebras of distribution kernels, Epifanio and Trapani [77] introduced the notion of *multiplication framework*, to be developed later by Trapani and Tschinke [183] when analysing the multiplication of operators acting in a RHS. A multiplication framework is nothing but a family of intermediate spaces (interspaces) between the smallest space and the largest one of a RHS and these spaces indeed generate a true PIP-space.

But, on the whole, the topic of PIP-spaces remained dormant for a number of years, until one of us (JPA) was drawn back into it by the mathematical considerations of the signal processing community. There, indeed, it is commonplace to exploit families of function or distribution spaces that are indexed by one or several parameters controlling, for instance, regularity or behavior at infinity. Such are the Lebesgue spaces $\{L^p, 1 \leqslant p \leqslant \infty\}$, the Wiener amalgam spaces $W(L^p, \ell^q), 1 \leqslant p, q \leqslant \infty$, the modulation spaces $M_m^{p,q}, 1 \leqslant p, q \leqslant \infty$, the Besov spaces $B_{pq}^s, 1 \leqslant p, q \leqslant \infty$. The interesting point is that individual spaces have little individual value, it is the whole family that counts. Taking into account the duality properties among the various spaces, one concludes that, in all such cases, the underlying structure is that of a PIP-space. In addition, one needs operators that are defined over all spaces of the family, such as translation, modulation or Fourier transform. And the PIP-space formalism yields precisely such a notion of global operator.

Thus it seemed to us that time was ripe for having a second look at the subject and write a synthesis, the result being the present volume.

About the Contents of the Book

The work is organized as follows. We begin by a short introductive chapter, in which we restrict ourselves to the simplest case of a chain or a lattice of Hilbert spaces or Banach spaces. This allows one to get a feeling about the general theory and, in particular, about the machinery of operators on such spaces. The following two chapters are the core of the general theory. It is convenient to divide our study of PIP-spaces into two stages.

In Chapter 1 we consider only the algebraic aspects, focusing of the generation of a PIP-spaces from a so-called *linear compatibility relation* on a vector space V and a partial inner product defined exactly on compatible pairs of vectors. Standard examples are the space ω of *all* complex sequences, with the partial inner product inherited from ℓ^2, whenever defined, and the space $L_{\text{loc}}^1(X, d\mu)$ of all measurable, locally integrable, functions on a measure space (X, μ), with the partial inner product inherited from L^2. The key notion here is that of *assaying subspaces*, particular subspaces of V which are in a sense the building blocks of the whole construction. Given a linear compatibility $\#$ on a vector space V, it turns out that the set of all assaying subspaces, (partially) ordered by inclusion, is a complete involutive lattice denoted by $\mathcal{F}(V, \#)$. This will lead us to another equivalent formulation, in terms of particular coverings of V by families of subspaces. Now the complete lattice $\mathcal{F}(V, \#)$ defined by a given linear compatibility can be recovered from much smaller families of subspaces, called *generating families*. An interesting observation is that, in many cases, including the two standard examples mentioned above, there exists a generating family consisting entirely of Hilbert spaces. The existence of such generating families in crucial for practical applications; indeed they play the same role as a basis of neighborhoods or a basis of open sets does in topology. And in fact, they will naturally lead to the introduction, in Chapter 2, of a reduced structure called an *indexed PIP-space*.

We conclude the chapter with the problem of comparing different compatibilities on the same vector space. *A priori* several order relations may be considered. It turns out that the useful definition is to say that a given compatibility $\#_1$ on V is *coarser* than another one $\#_2$ if, and only if, the complete lattice $\mathcal{F}(V, \#_1)$ is a sublattice of $\mathcal{F}(V, \#_2)$, on which the two involutions coincide (*involutive* sublattice). This concept is useful for the construction of PIP-space structures on a given vector space V. Most vector spaces used in mathematical physics carry a natural (partial) inner product, defined on a suitable domain $\Gamma \subseteq V \times V$. With trivial restrictions on Γ (symmetry, bilinearity), the condition: $f \# g \Leftrightarrow \{f, g\} \in \Gamma$, actually defines a linear compatibility $\#$ on V. Then *all* linear compatibilities which are admissible for that particular inner product are precisely those that are coarser than $\#$, which in turn are determined by all involutive sublattices of $\mathcal{F}(V, \#)$. On the other hand, the problem of refining a given compatibility (and then a given PIP-space structure) admits in general no solution, even less a unique maximal one. However, partial answers to the refinement problem can be given, but some additional structure is needed, namely topological restrictions on individual assaying subsets. This will be the main topic of Chapter 5.

Then, in Chapter 2, we introduce topologies on the assaying subspaces. With a basic nondegeneracy assumption, the latter come as compatible pairs $(V_r, V_{\overline{r}})$, which are dual pairs in the sense of topological vector spaces. This allows one to consider various canonical topologies on these subspaces and explore the consequences of their choice. It turns out that the structure so obtained is extremely rich, but may contain plenty of pathologies. Since the goal of the whole construction is to provide an elementary substitute to the theory of distributions, we are led to consider a particular case, in which all assaying subspaces are of the same type, Hilbert spaces or reflexive Banach spaces. The resulting structure is called an *indexed PIP-space*, of type (H), resp. type (B). However, a further restriction is necessary. Indeed, in such a case, the two spaces of a dual pair $(V_r, V_{\overline{r}})$ are conjugate duals of each other, but we require now, in addition, that each of them is given with an explicit norm, not only a normed topology, and the two norms are supposed to be conjugate to each other also. In that case, we speak of a *lattice of Hilbert spaces (LHS)*, resp. a *lattice of Banach spaces (LBS)*. These are finally the structures that are useful in practice, and plenty of examples will be described in the subsequent chapters.

Chapter 3 is devoted to the other central topic of the book, namely, operators on (indexed) PIP-spaces. As we have seen so far, the basic idea of PIP-spaces is that vectors should not be considered individually, but only in terms of the assaying subspaces V_r, which are the basic units of the structure. Correspondingly, an operator on a PIP-space should be defined in terms of assaying subspaces only, with the proviso that only continuous or bounded operators are allowed. Thus an operator is what we will call a *coherent collection* of continuous operators. Its domain is a nonempty union of assaying subspaces of V and its restriction to each of these is linear and bounded into the target space. In addition, and this is the crucial condition, the operator is *maximal*, in the sense that it has no proper extension satisfying the two conditions above. Requiring this essentially eliminates all the pathologies associated to unbounded operators and their extensions, while at the same time allowing more singular objects.

Once the general definition of operator on a PIP-space is settled, we may turn to various classes, that more or less mimic the standard notions. For instance, regular and totally regular operators, homomorphisms and isomorphisms, unitary operators (with application to group representations), symmetric operators. The last class examplifies what we said above, for it leads to powerful generalizations of various self-adjointness criteria for Hilbert space operators, even for very singular ones (the central topic here is the well-known KLMN theorem). For instance, this technique allows one to treat correctly very singular Schrödinger operators (Hamiltonians with various δ-potentials). In the last section, we turn to another central object, that of a projection operator, and the attending notion of subspace. It turns out that an appropriate definition of PIP-subspace permits to reproduce the familiar Hilbert space situation, namely the bijection between projections and closed subspaces. In addition, this leads to interesting results about finite dimensional subspaces and pre-Hilbert spaces.

Chapter 4 is a collection of concrete examples of PIP-spaces. There are two main classes, spaces of (locally) integrable functions and spaces of sequences. The simplest example of the former is the family of Lebesgue space L^p, $1 < p < \infty$, first over a finite interval (in which case, one gets a chain of Banach spaces), then over $\mathbb{R}$ or $\mathbb{R}^n$, where a genuine lattice is generated. A further generalization is the (wide) class of Köthe function spaces, which contains, among others, most of the spaces of interest in signal processing (see Chapter 8). The other class consists of the Köthe sequence spaces, which incidentally provide most of the pathological situations about topological vector spaces! Next we briefly describe the so-called analyticity/trajectory spaces, which were actually meant as a substitute to distribution theory, better adapted to a rigorous formulation of Dirac's formalism of quantum mechanics. Another class of PIP-spaces concludes the chapter, namely spaces of analytic functions. Starting from the familiar Bargmann space of entire functions [42, 43], we consider first a LBS that generates it (also defined by Bargmann). Then we turn to spaces of functions analytic in a sector. The PIP-space structure we describe, inspired by the work of van Winter in quantum scattering theory [188, 189], leads to a new insight into the latter. In the same way, we present some PIP-space variations around the Bergman or Hardy spaces of functions analytic in a disk.

In Chapter 5, we return to the problem of refinement of PIP-space structures, in particular, the extension from a discrete chain to a continuous one, and similarly for a lattice. When the individual spaces are Banach spaces, we are clearly in the realm of interpolation theory. In the Hilbert case, one can also exploit the spectral theorem of self-adjoint operators. The simplest example is that of the canonical chain generated by the powers of a positive self-adjoint operator in a Hilbert space, where both techniques can be used. The next case is that of a genuine lattice of Hilbert spaces. In both cases, there are infinitely many solutions. Next we explore how a RHS $\mathcal{D} \subset \mathcal{H} \subset \mathcal{D}^\times$ can be refined into a LHS (here $\mathcal{H}$ is a Hilbert space, $\mathcal{D}$ a dense subspace, endowed with a suitable finer topology, and $\mathcal{D}^\times$ its strong conjugate dual). This is an old problem, connected to the proper definition of a multiplication rule for operators on a RHS. The key is the introduction of the so-called *interspaces*, that is, subspaces $\mathcal{E}$ such that $\mathcal{D} \subset \mathcal{E} \subset \mathcal{D}^\times$. This is, of course, strongly reminiscent of Laurent Schwartz's hilbertian subspaces of a topological vector space and corresponding kernels [175]. Indeed, the crucial condition on interspaces may already be found in that paper. As an application, we construct a family of Banach spaces that generalize the well-known Bessel potential or Sobolev spaces. We also discuss the PIP-space structure of distribution spaces. In particular, we review the elegant construction of the so-called Hilbert spaces of type S of Grossmann, which enables one to construct manageable spaces of nontempered distributions.

The next step is the construction of PIP-spaces generated by a family or an algebra of unbounded operators, equivalently, a (compatible) family of quadratic forms on a Hilbert space. In particular, if one starts from the algebra of regular operators on a PIP-space V, one ends up with *two* PIP-space structures on the same vector space. Comparison between the two leads to several situations, from the 'natural' one to a downright pathological one. Examples may be given for all cases, and this might give some hints for a classification of PIP-spaces.

In Chapter 6, we consider the set $\mathrm{Op}(V)$ of all operators on a PIP-space V as a partial *-algebra. This concept, developed at length in the monograph by Antoine-Inoue-Trapani [AIT02], sheds new light on the operators. Of particular interest is the case where the PIP-space V is a RHS. The proper definition of a multiplication scheme in that context has generated some controversies in the literature [135, 136], but the PIP-space point of view eliminates the pathologies unearthed in these papers. In the same vein, we consider also the construction of representations of partial *-algebras, in particular, the Gel'fand–Naĭmark–Segal (GNS) construction suitably generalized to the PIP-space context. As for general partial *-algebras, one has to take account of the fact that the product of two operators is not always defined, which requires replacing positive linear functionals by sesquilinear ones, in particular the so-called *weights*. Clearly this kind of topic implies borrowing ideas and techniques from operator algebras.

The last two chapters are devoted to applications of PIP-spaces. In Chapter 7, we consider applications in mathematical physics, in the next one applications in signal processing. We begin with quantum mechanics. As mentioned at the beginning of this prologue, the insufficient character of a pure Hilbert space formulation led mathematical physicists to introduce the RHS approach, which then generalizes in straightforward way to a PIP-space approach, via the consideration of the observables characterizing a physical system (the so-called *labeled* observables). A different generalization that we quickly mention is that of the analyticity/trajectory spaces. A spectacular application of the RHS point of view is a mathematically correct treatment of very singular interactions (δ-potentials or worse). A case where a PIP-space formulation yields a new insight is that of quantum scattering theory, along the lines developed by van Winter [188, 189]. At play here are the spaces of functions analytic in a sector, described in Chapter 4. Also we obtain a precise link to the dilation analyticity or complex scaling method (CSM), nowadays a workhorse in quantum chemistry. We also make some remarks on the still controversial time-asymmetric quantum mechanics, which is based on an energy-valued RHS. The next topic where PIP-spaces are used since a long time is, of course, quantum field theory. In the axiomatic Wightman formulation, based on (tempered) distributions, a RHS language emerges naturally. Two explicit instances, that we describe in some detail, are the construction of the theory from the so-called Borchers algebra and the Euclidean approach of Nelson. Similarly, a proper definition of *unsmeared* field operators require some sort of PIP-space structure. Another area where PIP-spaces have been exploited is that of Lie group representations, using Nelson's theory of analytic vectors, that we touch briefly for concluding the chapter.

The final Chapter 8 is devoted to applications in signal processing. Namely, we explore in some detail a number of families of function spaces that yield the 'natural' framework for some specific applications. Typically, each class is indexed by two indices, at least. One of them characterizes the local behavior (local growth, smoothness), whereas the other specifies the global behavior, for instance the decay properties at infinity. The first example is that of mixed-norm Lebesgue spaces and Wiener amalgam spaces (the first spaces of this type were introduced by N. Wiener in his study of Tauberian theorems). For instance, the amalgam space $W(L^p, \ell^q)$ consists of functions on $\mathbb{R}$ which are locally in L^p and such that the L^p norms over the intervals $(n, n+1)$ form an ℓ^q sequence. This clearly corresponds to the local vs. global behavior announced above. It turns out that such spaces (and generalizations thereof) provide a natural framework for the time-frequency analysis of signals. The same may be said, *a fortiori*, for the modulation spaces $M_m^{p,q}$, which are defined in terms of the Short-Time Fourier (or Gabor) Transform (m is a weight function and $1 \leqslant p, q \leqslant \infty$). Among these, a special role is played by the space $M_1^{1,1}$, called the Feichtinger algebra and denoted usually by $\mathcal{S}_0$ (it is indeed an algebra both under pointwise multiplication and under convolution). $\mathcal{S}_0$ is a reflexive Banach space and one has indeed $\mathcal{S} \subset \mathcal{S}_0 \subset L^2 \subset \mathcal{S}_0^\times \subset \mathcal{S}^\times$ (thus $\mathcal{S}_0$ and its conjugate dual are interspaces in the Schwartz RHS). In practice, $\mathcal{S}_0$ may often advantageously replace Schwartz's space $\mathcal{S}$, yielding the prototypical Banach Gel'fand triple $\mathcal{S}_0 \subset L^2 \subset \mathcal{S}_0^\times$, which plays an important role in time-frequency analysis.

A second important class is that of Besov spaces, which are intrinsically related to the (discrete) wavelet transform. Typical results concern the specification of a space to which a given function belongs through the decay properties of its wavelet coefficients in an appropriate wavelet basis. Finally, we survey briefly a far reaching generalization of all the preceding spaces, namely, the so-called co-orbit spaces. These spaces are defined in terms of an integrable representation of a suitable Lie group. For instance, the Weyl-Heisenberg group leads to modulation spaces, the affine group of the line yields Besov spaces, $SL(2,\mathbb{R})$ gives Bergman spaces.

For the convenience of the reader, we conclude the volume with two short appendices. The first one (A) gives some indications about the so-called Galois connections (used in Chapter 1), and the second (B) collects some basic facts about (locally convex) topological vector spaces, mostly needed in Chapter 2.

A final word about the presentation. Although a large literature already exists on the subject, we have decided to mention very few papers in the body of the chapters. Instead, each of them concludes with notes that give all the relevant bibliography. We have tried, in particular, to trace most of the results to the original papers. Thus a substantial part of the book consists of a survey of known results, often reformulated in the PIP-space language. This means that, in most cases, we state and comment the relevant results, but skip the proofs, referring instead to the literature. Clearly there are omissions and misrepresentations, due to our own ignorance and prejudices. We take responsibility for this and apologize in advance to those authors whose work we might have mistreated.   ∎

Jean-Pierre Antoine (Louvain-la-Neuve)
Camillo Trapani (Palermo)

Introduction: Lattices of Hilbert or Banach Spaces and Operators on Them

I.1 Motivation

It is a fact that many function spaces that play a central role in analysis come in the form of families, indexed by one or several parameters that characterize the behavior of functions (smoothness, behavior at infinity, ...). The simplest structure is a *chain of Hilbert or (reflexive) Banach spaces*. Let us give two familiar examples.

(i) *The Lebesgue L^p spaces on a finite interval*, e.g. $\mathcal{I} = \{L^p([0,1], dx), 1 \leqslant p \leqslant \infty\}$:

$$L^\infty \subset \ldots \subset L^{\bar{q}} \subset L^{\bar{r}} \subset \ldots \subset L^2 \subset \ldots \subset L^r \subset L^q \subset \ldots \subset L^1, \tag{I.1}$$

where $1 < q < r < 2$. Here L^q and $L^{\bar{q}}$ are dual to each other $(1/q + 1/\bar{q} = 1)$, and similarly $L^r, L^{\bar{r}}$ $(1/r + 1/\bar{r} = 1)$. By the Hölder inequality, the (L^2) inner product

$$\langle f | g \rangle = \int_0^1 \overline{f(x)}\, g(x)\, dx \tag{I.2}$$

is well-defined if $f \in L^q$, $g \in L^{\bar{q}}$. However, it is *not* well-defined for two arbitrary functions $f, g \in L^1$. Take for instance, $f(x) = g(x) = x^{-1/2}: f \in L^1$, but $fg = f^2 \notin L^1$. Thus, on L^1, (I.2) defines only a *partial* inner product. The same result holds for any finite interval of $\mathbb{R}$ instead of $[0,1]$.

(ii) *The chain of Hilbert spaces built on the powers of a positive self-adjoint operator $A \geqslant 1$ in a Hilbert space $\mathcal{H}_0$*. Let $\mathcal{H}_n$ be $D(A^n)$, the domain of A^n, equipped with the graph norm $\|f\|_n = \|A^n f\|$, $f \in D(A^n)$, for $n \in \mathbb{N}$ or $n \in \mathbb{R}^+$, and $\mathcal{H}_{\bar{n}} := \mathcal{H}_{-n} = \mathcal{H}_n^\times$ (conjugate dual):

$$\mathcal{D}^\infty(A) := \bigcap_n \mathcal{H}_n \subset \ldots \subset \mathcal{H}_2 \subset \mathcal{H}_1 \subset \mathcal{H}_0 \subset \mathcal{H}_{\bar{1}} \subset \mathcal{H}_{\bar{2}} \ldots \subset \mathcal{D}_{\bar{\infty}}(A) := \bigcup_n \mathcal{H}_n. \tag{I.3}$$

Note that here the index n may be integer or real, the link between the two cases being established by the spectral theorem for self-adjoint operators.

Here again the inner product of $\mathcal{H}_0$ extends to each pair $\mathcal{H}_n, \mathcal{H}_{\overline{n}}$, but on $\mathcal{D}_{\overline{\infty}}(A)$ it yields only a *partial* inner product. The following examples, all three in $\mathcal{H}_0 = L^2(\mathbb{R}, dx)$ are standard:

- $(A_p f)(x) = (1 + x^2)^{1/2} f(x)$.
- $(A_m f)(x) = (1 - \frac{d^2}{dx^2})^{1/2} f(x)$.
- $(A_{osc} f)(x) = (1 + x^2 - \frac{d^2}{dx^2}) f(x)$.

(The notation is suggested by the operators of position, momentum and harmonic oscillator energy in quantum mechanics, respectively). In the case of A_m, the intermediate spaces are the Bessel potential (or Sobolev) spaces $H^s(\mathbb{R})$, $s \in \mathbb{Z}$ or $\mathbb{R}$. Note that both $\mathcal{D}^\infty(A_p) \cap \mathcal{D}^\infty(A_m)$ and $\mathcal{D}^\infty(A_{osc})$ coincide with the Schwartz space $\mathcal{S}(\mathbb{R})$ of smooth functions of fast decay, and $\mathcal{D}_{\overline{\infty}}(A_{osc})$ with the space $\mathcal{S}^\times(\mathbb{R})$ of tempered distributions.[1]

However, a moment's reflection shows that the total order relation inherent in a chain is in fact an unnecessary restriction, partially ordered structures are sufficient, and indeed necessary in practice. For instance, in order to get a better control on the behavior of individual functions, one may consider the lattice built on the powers of A_p and A_m simultaneously. Then the extreme spaces are still $\mathcal{S}(\mathbb{R})$ and $\mathcal{S}^\times(\mathbb{R})$. Similarly, in the case of several variables, controlling the behavior of a function in each variable separately requires a nonordered set of spaces. This is in fact a statement about tensor products (remember that $L^2(X \times Y) \simeq L^2(X) \otimes L^2(Y)$). Indeed a glance at the work of Palais on chains of Hilbert spaces shows that the tensor product of two chains of Hilbert spaces, $\{\mathcal{H}_n\} \otimes \{\mathcal{K}_m\}$ is naturally a lattice $\{\mathcal{H}_n \otimes \mathcal{K}_m\}$ of Hilbert spaces. For instance, in the example above, for two variables x, y, that would mean considering intermediate Hilbert spaces corresponding to the product of two operators, $\left(A_m(x)\right)^n \left(A_m(y)\right)^m$.

Thus the structure we want to analyze is that of *lattices of Hilbert or Banach spaces*. Many examples are around us, for instance the lattice generated by the spaces $L^p(\mathbb{R}, dx)$, the amalgam spaces $W(L^p, \ell^q)$, the mixed norm spaces $L_m^{p,q}(\mathbb{R}, dx)$, and many more (these spaces will be discussed in detail in Chapters 4 and 8, where the references to original papers will be given). In all these cases, which contain most families of function spaces of interest in analysis and in signal processing, a common structure emerges for the "large" space V, defined as the union of all individual spaces. There is a lattice of Hilbert or reflexive Banach spaces V_r, with an (order-reversing) involution $V_r \leftrightarrow V_{\overline{r}}$, where $V_{\overline{r}} = V_r^\times$ (the space of continuous antilinear functionals on V_r), a central Hilbert space $V_o \simeq V_{\overline{o}}$, and a partial inner product on V that extends the inner product of V_o to pairs of dual spaces $V_r, V_{\overline{r}}$.

Actually, in many cases, it is the family $\{V_r\}$ as a whole that is meaningful, not the individual spaces. The spaces $L^p(\mathbb{R})$ are a good example. Therefore,

---

[1] considered here as continuous *conjugate linear* functionals on $\mathcal{S}$. See the Notes to Chapter 1, Section 1.1.

many operators should be considered globally, for the whole chain or lattice, instead of on individual spaces. For instance, in many spaces of interest in signal processing, this would apply to operators implementing translations $(x \mapsto x - y)$ or dilations $(x \mapsto x/a)$, convolution operators, Fourier transform, etc. In the same spirit, it is often useful to have a *common* basis for the whole family of spaces, such as the Haar basis for the spaces $L^p(\mathbb{R})$, $1 < p < \infty$. Thus we need a notion of operator and basis defined globally for the chain or lattice itself.

The subject matter of the present volume is to present a formalism that answers these questions, namely, the theory of *partial inner product spaces* or PIP-*spaces*. However, before analyzing in detail the general theory, we will concentrate in this introductory chapter on the simple case of lattices of Hilbert or Banach spaces and operators on them. These are indeed the most useful families of spaces for the applications.

I.2 Lattices of Hilbert or Banach Spaces

I.2.1 Definitions

Let thus $\mathcal{J} = \{\mathcal{H}_p,\, p \in J\}$ be a family of Hilbert spaces, partially ordered by inclusion (the index set J has the same order structure). Then $\mathcal{J}$ generates a lattice $\mathcal{I}$, indexed by I, by the operations:

- $\mathcal{H}_{p \wedge q} = \mathcal{H}_p \cap \mathcal{H}_q$, with the projective norm

$$\|f\|_{p \wedge q}^2 = \|f\|_p^2 + \|f\|_q^2\, ; \tag{I.4}$$

- $\mathcal{H}_{p \vee q} = \mathcal{H}_p + \mathcal{H}_q$, the vector sum, with the inductive norm

$$\|f\|_{p \vee q}^2 = \inf_{f = g + h} \left(\|g\|_p^2 + \|h\|_q^2 \right),\ g \in \mathcal{H}_p, f \in \mathcal{H}_q\,. \tag{I.5}$$

It turns out that both $\mathcal{H}_{p \wedge q}$ and $\mathcal{H}_{p \vee q}$ are Hilbert spaces, that is, they are complete with the norms indicated. These statements will be proved, and the corresponding mathematical structure analyzed, in Chapter 2, Section 2.2.

Assume that the original index set J has an involution $q \leftrightarrow \bar{q}$, with $\mathcal{H}_{\bar{q}} = \mathcal{H}_q^\times$ (by an involution, we mean a one-to-one correspondence such that $p \leqslant q$ implies $\bar{q} \leqslant \bar{p}$ and $\bar{\bar{p}} = p$). Then the lattice $\mathcal{I}$ inherits the same duality structure, with $\mathcal{H}_{p \wedge q} \leftrightarrow \mathcal{H}_{\bar{p} \vee \bar{q}}$ (it is then called an *involutive lattice*; a precise definition will be given in Section 1.1). Finally, we assume the family $\mathcal{J}$ contains a unique self-dual space $V_o = V_{\bar{o}}$. The resulting structure is called a *lattice of Hilbert spaces* or LHS.

In addition to the family $\mathcal{I} = \{V_r,\, r \in I\}$, it is convenient to consider the two spaces $V^{\#}$ and V defined as

$$V = \sum_{q \in I} \mathcal{H}_q, \quad V^{\#} = \bigcap_{q \in I} \mathcal{H}_q. \tag{I.6}$$

These two spaces themselves usually do *not* belong to $\mathcal{I}$.

The concept of LHS is closely related to that of nested Hilbert space, which will be discussed in Section 2.4). More important, this construction is the basic structure of interpolation theory.

A similar construction can be performed with a family $\mathcal{J} = \{V_p, \, p \in J\}$ of reflexive Banach spaces, the resulting structure being then a *lattice of Banach spaces* or LBS. In this case, one considers the following norms, which are usual in interpolation theory.

- $V_{p \wedge q} = V_p \cap V_q$, with the *projective* norm

$$\|f\|_{p \wedge q} = \|f\|_p + \|f\|_q\,; \tag{I.7}$$

- $V_{p \vee q} = V_p + V_q$, with the *inductive* norm

$$\|f\|_{p \vee q} = \inf_{f = g + h} \left(\|g\|_p + \|h\|_q \right), g \in V_p, f \in V_q. \tag{I.8}$$

Here too, we assume the family $\mathcal{J}$ contains a unique self-dual space $V_o = V_{\bar{o}}$, which is a Hilbert space.

I.2.2 Partial Inner Product on a LHS/LBS

The basic question is how to generate such structures in a systematic fashion. In order to answer it, we may reformulate it as follows: given a vector space V and two vectors $f, g \in V$, when does their inner product make sense? A way of formalizing the answer is given by the idea of *compatibility*.

Let $\mathcal{I} := \{V_r, \, r \in I\}$ be a LHS or a LBS and $f, g \in V$ two vectors. Then we say that f and g are *compatible*, which we note $f \# g$, if the following relation holds:

$$f \# g \quad \Leftrightarrow \quad \exists \, r \in I \text{ such that } f \in V_r, g \in V_{\bar{r}}. \tag{I.9}$$

Clearly the relation $\#$ is a symmetric binary relation which preserves linearity:

$$f \# g \iff g \# f, \, \forall \, f, g \in V,$$
$$f \# g, f \# h \implies f \# (\alpha g + \beta h), \, \forall \, f, g, h \in V, \forall \, \alpha, \beta \in \mathbb{C}.$$

From now on, we write $\mathcal{I} = (V, \#)$. A formal definition will be given in Section 1.1.

Now we introduce the basic notions of our structure, namely, a partial inner product and a partial inner product space.

A *partial inner product* on $(V, \#)$ is a hermitian form $\langle \cdot | \cdot \rangle$ defined exactly on compatible pairs of vectors, that is, on $\Delta = \left(\bigcup_{V_r \in \mathcal{I}} V_r \times V_{\overline{r}} \right) \cup \left(V^\# \times V \right)$. A *partial inner product space (PIP-space)* is a vector space V equipped with a linear compatibility and a partial inner product.

In general, the partial inner product is not required to be positive definite, but it will be in all the examples given in this chapter. In the present case of a LHS or a LBS $\mathcal{I} = \{V_r, r \in I\}$, this simply means that the partial inner product is defined between elements of two spaces in duality $(V_r, V_{\overline{r}})$; in particular, for $r = o$, it is a Hermitian form on the Hilbert space V_o, which we take as equal to the inner product. Thus, the partial inner product may be seen as the extension of the inner product of V_o to the whole of V, whenever possible. Clearly, with the compatibility (I.9) and this definition of partial inner product, the LHS/LBS $\mathcal{I}$ is a PIP-space, that we henceforth denote as $(V, \#, \langle \cdot | \cdot \rangle)$.

The partial inner product defines a notion of *orthogonality* : $f \perp g$ if and only if $f \# g$ and $\langle f | g \rangle = 0$. Then we say that the PIP-space $(V, \#, \langle \cdot | \cdot \rangle)$ is *nondegenerate* if $(V^\#)^\perp = \{0\}$, that is, if $\langle f | g \rangle = 0$ for all $f \in V^\#$ implies $g = 0$.

In this introductory chapter, we will assume that our PIP-space $(V, \#, \langle \cdot | \cdot \rangle)$ is nondegenerate. This assumption has important topological consequences, that will be explored at length in Chapter 2. In a nutshell, $(V^\#, V)$, like every couple $(V_r, V_{\overline{r}})$, $r \in I$, is a dual pair in the sense of topological vector spaces. Furthermore, $r < s$ implies $V_r \subset V_s$, and the embedding operator $E_{sr} : V_r \to V_s$ is continuous and has dense range. In particular, $V^\#$ is dense in every V_r.

I.2.3 Two Examples of LHS

Let us give two simple examples of LHS, thus of PIP-spaces as well.

(i) Sequence spaces

Let V be the space ω of *all* complex sequences $x = (x_n)$. Consider the following family of weighted Hilbert spaces, which are obviously subspaces of ω:

$$\ell^2(r) = \{(x_n) \in \omega : (x_n/r_n) \in \ell^2, i.e., \sum_{n=1}^{\infty} |x_n|^2 \, r_n^{-2} < \infty\}, \qquad (I.10)$$

where $r = (r_n)$, $r_n > 0$, is a sequence of positive numbers. The family possesses an involution:

$$\ell^2(r) \leftrightarrow \ell^2(\overline{r}) = \ell^2(r)^\times, \text{ where } \overline{r}_n = 1/r_n.$$

In addition, there is a central, self-dual Hilbert space, namely, $\ell^2(1) = \ell^2(\overline{1}) = \ell^2$, where 1 denotes the unit sequence, $r_n = 1$, for all n.

As a matter of fact, the collection $\mathcal{I} := \{\ell^2(r)\}$ of those vector subspaces of ω is an involutive lattice.

(i) $\mathcal{I}$ is a lattice for the following operations:

$$\ell^2(r) \wedge \ell^2(s) = \ell^2(u), \quad \text{where } u_n = \min\{r_n, s_n\},$$
$$\ell^2(r) \vee \ell^2(s) = \ell^2(v), \quad \text{where } v_n = \max\{r_n, s_n\}.$$

Indeed one shows easily that the norms of $\ell^2(u)$ and $\ell^2(v)$ are equivalent, respectively, to the projective and inductive norms defined in (I.4), (I.5) above (proofs will be given in Section 4.3).

(ii) $\mathcal{I}$ is an involutive lattice, with the involution $r \leftrightarrow \overline{r} \equiv (r_n^{-1})$. Indeed:

$$[\ell^2(u)]^{\#} = \ell^2(\overline{u}) = \ell^2(\overline{r}) \vee \ell^2(\overline{s}),$$
$$[\ell^2(v)]^{\#} = \ell^2(\overline{v}) = \ell^2(\overline{r}) \wedge \ell^2(\overline{s}).$$

Actually, $\mathcal{I}$ is a sublattice of $\mathcal{L}(\omega)$, the lattice of all vector subspaces of ω, i.e.,

$$\ell^2(r) \wedge \ell^2(s) = \ell^2(r) \cap \ell^2(s),$$
$$\ell^2(r) \vee \ell^2(s) = \ell^2(r) + \ell^2(s).$$

As for the extreme spaces, it is easy to see that the family $\{\ell^2(r)\}$ generates the space ω of *all* complex sequences, while the intersection is the space φ of all *finite* sequences:

$$\bigcup_{r \in I} \ell^2(r) = \omega, \quad \bigcap_{r \in I} \ell^2(r) = \varphi.$$

Thus, with the partial inner product $\langle x|y \rangle = \sum_{n=1}^{\infty} \overline{x}_n\, y_n$, inherited from $V_o = \ell^2$, the family $\mathcal{I} = \{\ell^2(r)\}$ is a nondegenerate LHS.

(ii) Spaces of locally integrable functions

Instead of sequences, we consider locally integrable functions, i.e., Lebesgue measurable functions, integrable over compact subsets, $f \in L^1_{\text{loc}}(\mathbb{R}, dx)$, and define again weighted Hilbert spaces:

$$L^2(r) = \{f \in L^1_{\text{loc}}(\mathbb{R}, dx) : fr^{-1} \in L^2, i.e. \int_{\mathbb{R}} |f(x)|^2\, r(x)^{-2}\, dx < \infty\}, \quad (\text{I.11})$$

with $r, r^{-1} \in L^2_{\text{loc}}(\mathbb{R}, dx), r(x) > 0$ a.e. The family $\mathcal{I} = \{L^2(r)\}$ has an involution, $L^2(r) \leftrightarrow L^2(\overline{r})$, with $\overline{r} = r^{-1}$, and a central, self-dual Hilbert space, $L^2(\mathbb{R}, dx)$. This is, of course, the continuous analogue of the preceding example. Thus we get exactly the same structure as in (i), namely the family $\mathcal{I} = \{L^2(r)\}$ is an involutive lattice, for the operations:

- infimum: $L^2(p \wedge q) = L^2(p) \wedge L^2(q) = L^2(r)$, $r(x) = \min(p(x), q(x))$;
- supremum: $L^2(p \vee q) = L^2(p) \vee L^2(q) = L^2(s)$, $s(x) = \max(p(x), q(x))$;
- duality: $L^2(p \wedge q) \leftrightarrow L^2(\overline{p} \vee \overline{q})$, $L^2(p \vee q) \leftrightarrow L^2(\overline{p} \wedge \overline{q})$.

Here too, it is easily shown that the lattice $\mathcal{I}$ generates the extreme spaces:

$$\bigcup_{r \in I} L^2(r) = L^1_{\mathrm{loc}}(\mathbb{R}, dx), \quad \bigcap_{r \in I} L^2(r) = L^\infty_c(\mathbb{R}),$$

where $L^\infty_c(\mathbb{R})$ denotes the space of essentially bounded measurable functions of compact support. With the partial inner product inherited from the central space L^2,

$$\langle f | g \rangle = \int_{\mathbb{R}} \overline{f(x)} g(x) \, dx,$$

the family $\mathcal{I} = \{L^2(r)\}$ becomes a nondegenerate LHS. The construction extends trivially to $\mathbb{R}^n$, or to any manifold (X, μ). It may also be done around Fock space, instead of L^2 (see Section 1.1.3, Example (iv)).

I.3 Operators on a LHS/LBS

As follows from the compatibility relation (I.9), the basic idea of LHS/LBS (and, more generally, PIP-spaces, as we shall see in the next chapter) is that vectors should not be considered individually, but only in terms of the subspaces V_r ($r \in I$), the building blocks of the structure. Correspondingly, an operator on such a space should be defined in terms of the defining subspaces only, with the proviso that only *bounded* operators between Hilbert or Banach spaces are allowed. Thus an operator is a *coherent collection* of bounded operators. More precisely,

Definition I.3.1. Given a LHS or LBS $V_I = \{V_r, r \in I\}$, an *operator* on V_I is a map A from a subset $\mathcal{D} \subseteq V$ into V, where

(i) $\mathcal{D}$ is a nonempty union of defining subspaces of V_I;
(ii) for every defining subspace V_q contained in $\mathcal{D}$, there exists a $p \in I$ such that the restriction of A to V_q is linear and continuous into V_p (we denote this restriction by A_{pq});
(iii) A has no proper extension satisfying (i) and (ii), i.e., it is maximal.

According to Condition (iii), the domain $\mathcal{D}$ is called the *natural domain* of A and denoted $\mathcal{D}(A)$.

The linear bounded operator $A_{pq} : V_q \to V_p$ is called a *representative* of A. In terms of the latter, the operator A may be characterized by the set $j(A) = \{(q, p) \in I \times I : A_{pq} \text{ exists}\}$. Thus the operator A may be identified with the collection of its representatives,

$$A \simeq \{A_{pq} : V_q \to V_p : (q, p) \in j(A)\}.$$

We also need the two sets obtained by projecting $j(A)$ on the "coordinate" axes, namely,

$$d(A) = \{q \in I : \text{there is a } p \text{ such that } A_{pq} \text{ exists}\},$$
$$i(A) = \{p \in I : \text{there is a } q \text{ such that } A_{pq} \text{ exists}\}.$$

The following properties are immediate:

- $d(A)$ is an initial subset of I: if $q \in d(A)$ and $q' < q$, then $q' \in d(A)$, and $A_{pq'} = A_{pq} E_{qq'}$, where $E_{qq'}$ is a representative of the unit operator (this is what we mean by a 'coherent' collection).
- $i(A)$ is a final subset of I: if $p \in i(A)$ and $p' > p$, then $p' \in i(A)$ and $A_{p'q} = E_{p'p} A_{pq}$.
- $j(A) \subset d(A) \times i(A)$, with strict inclusion in general.

We denote by $\mathrm{Op}(V_I)$ the set of all operators on V_I. Of course, a similar definition may be given for operators $A : V_I \to Y_K$ between two LHSs or LBSs.

Since $V^\#$ is dense in V_r, for every $r \in I$, an operator may be identified with a sesquilinear form on $V^\# \times V^\#$. Indeed, the restriction of any representative A_{pq} to $V^\# \times V^\#$ is such a form, and all these restrictions coincide (these sesquilinear forms are even separately continuous for appropriate topologies on $V^\#$, see Chapter 3). Equivalently, an operator may be identified with a linear map from $V^\#$ into V (here also continuity may be obtained). But the idea behind the notion of operator is to keep also the *algebraic operations* on operators, namely:

(i) *Adjoint $A^\times$:* every $A \in \mathrm{Op}(V_I)$ has a unique adjoint $A^\times \in \mathrm{Op}(V_I)$, defined by the relation

$$\langle A^\times x | y \rangle = \langle x | Ay \rangle, \text{ for } y \in V_r, r \in d(A), \text{ and } x \in V_{\bar{s}}, s \in i(A),$$

that is, $(A^\times)_{\bar{r}\bar{s}} = (A_{sr})^*$ (usual Hilbert/Banach space adjoint).

It follows that $A^{\times\times} = A$, for every $A \in \mathrm{Op}(V_I)$: no extension is allowed, by the maximality condition (iii) of Definition I.3.1.

(ii) *Partial multiplication:* AB is defined if and only if there is a $q \in i(B) \cap d(A)$, that is, if and only if there is a continuous factorization through some V_q:

$$V_r \xrightarrow{B} V_q \xrightarrow{A} V_s, \quad \text{i.e., } (AB)_{sr} = A_{sq} B_{qr}.$$

It is worth noting that, for a LHS/LBS, the natural domain $\mathcal{D}(A)$ is always a vector subspace of V (this is not true for a general PIP-space). Therefore, $\mathrm{Op}(V_I)$ is a vector space and a *partial *-algebra*.

The concept of PIP-space operator is very simple, yet it is a far reaching generalization of bounded operators. It allows indeed to treat on the same footing all kinds of operators, from bounded ones to very singular ones. By this, we mean the following, loosely speaking. Given $A \in Op(V_I)$, when looked at from the central Hilbert space $V_o = \mathcal{H}$, there are three possibilities:

- if $(o, o) \in j(A)$, i.e., A_{oo} exists, then A corresponds to a bounded operator $V_o \to V_o$;
- if $(o, o) \notin j(A)$, but there is an $r < o$ such that $(r, o) \in j(A)$, i.e., A_{oo} does not exist, but only $A_{or} : V_r \to V_o$, with $r < o$, then A corresponds to an unbounded operator A_{or}, with hilbertian domain containing V_r;
- if $(r, o) \notin j(A)$, for any $r \leqslant o$, i.e., no A_{or} exists, then A is a sesquilinear form on some V_s, $s \leqslant o$, and, as an operator on $\mathcal{H}$, its domain does not contain any V_r (it may be reduced to $\{0\}$): then A corresponds to a singular operator; this happens, for instance, if $(r, s) \in j(A)$ with $r < o < s$, i.e., there exists $A_{sr} : V_r \to V_s$.

Exactly as for Hilbert or Banach spaces, one may define various types of operators between PIP-spaces, in particular LBS/LHS, such that regular operators, orthogonal projections, homomorphisms and isomorphisms, symmetric operators, unitary operators, etc. We will describe those classes in detail in Chapter 3.

In the following chapters, we will extend this discussion to a general PIP-space and operators between two PIP-spaces. A slightly more restrictive structure, called an *indexed PIP-space*, will also be introduced. Many concrete examples will be discussed in detail.

Notes

Section I.1. For unbounded operators, see Reed-Simon I [RS72, Section VIII. 2]. For the spectral theorem, see Kato [Kat76, RS72] or Reed-Simon I [RS72, Section VIII. 1]. The work of Palais may be found in [162, Chap.XIV].

Section I.2. Nested Hilbert spaces were introduced by Grossmann [114].

- For interpolation theory, one may consult the monographs of Bergh-Löfström [BL76] or Triebel [Tri78a].
- Our standard references for topological vector spaces are the monographs of Köthe [Köt69] and Schaefer [Sch71].

Section I.3. Partial *-algebras are studied in detail in the monograph of Antoine-Inoue-Trapani [AIT02].

Chapter 1
General Theory: Algebraic Point of View

It is convenient to divide our study of PIP-spaces into two stages. In the first one, we consider only the algebraic aspects. That is, we explore the structure generated by a linear compatibility relation on a vector space V, as introduced in Section I.2, without any other ingredient. This will lead us to another equivalent formulation, in terms of particular coverings of V by families of subspaces. This first approach, purely algebraic, is the subject matter of the present chapter.

Then, in a second stage, we introduce topologies on the so-called assaying subspaces $\{V_r\}$. Indeed, as already mentioned in Section I.2, assuming the partial inner product to be nondegenerate implies that every matching pair $(V_r, V_{\bar{r}})$ of assaying subspaces is a dual pair in the sense of topological vector spaces. This in turn allows one to consider various canonical topologies on these subspaces and explore the consequences of their choice. These considerations will be developed at length in Chapter 2.

1.1 Linear Compatibility on a Vector Space, Partial Inner Product Space

1.1.1 From Hilbert Spaces to PIP-Spaces

Definition 1.1.1. Let E be a complex vector space. A map $\langle \cdot | \cdot \rangle$ from $E \times E$ into $\mathbb{C}$ is called a positive *inner product* if

(ip$_1$) $\langle \cdot | \cdot \rangle$ is linear in the second argument, i.e., $\langle f | \lambda g + \mu h \rangle = \lambda \langle f | g \rangle + \mu \langle f | h \rangle$, for every $f, g \in E$ and $\lambda, \mu \in \mathbb{C}$.
(ip$_2$) $\langle g | f \rangle = \overline{\langle f | g \rangle}$, for every $f, g \in E$.
(ip$_3$) $\langle f | f \rangle \geqslant 0$, for every $f \in E$ and $\langle f | f \rangle = 0$ implies $f = 0$.

Notice that (ip$_1$) and (ip$_2$) imply that $\langle \cdot | \cdot \rangle$ is conjugate linear in the first variable and also that $\langle f | f \rangle \in \mathbb{R}$, for every $f \in E$.

J.-P. Antoine and C. Trapani, *Partial Inner Product Spaces:*
Theory and Applications, Lecture Notes in Mathematics 1986,
DOI 10.1007/978-3-642-05136-4_1, © Springer-Verlag Berlin Heidelberg 2009

Definition 1.1.2. If $\langle \cdot | \cdot \rangle$ is a positive inner product on E, a natural norm is defined by

$$\|f\| = \langle f | f \rangle^{1/2}, \quad f \in E.$$

The space E, endowed with the topology defined by this norm, is called a *pre-Hilbert space*. If, in addition, $E[\| \cdot \|]$ is complete, then $E[\| \cdot \|]$ is called a *Hilbert space*

Note that one can consider also nonpositive inner product spaces, i.e., spaces for which condition (ip$_3$) does not hold. Some examples will be mentioned below.

Examples 1.1.3. The following examples should be very familiar to the reader.

(i) The space $\mathbb{C}^n$, $n \in \mathbb{N}$, of all n-uples of complex numbers $z = (z_1, z_2, \ldots, z_n)$ is a Hilbert space if the inner product of z and $w = (w_1, w_2, \ldots, w_n)$ is defined by

$$\langle z | w \rangle = \sum_{j=1}^{n} \overline{z_j} w_j.$$

(ii) The space ℓ^2 of all sequences $z = (z_n)$ of complex numbers such that $\sum_{n=1}^{\infty} |z_n|^2 < \infty$ is a Hilbert space with inner product

$$\langle z | w \rangle = \sum_{n=1}^{\infty} \overline{z_n} w_n, \quad w = (w_n) \in \ell^2.$$

(iii) The space $L^2(\mathbb{R})$ of all Lebesgue square integrable functions (modulo the set of all almost everywhere null functions) is a Hilbert space with respect to the inner product

$$\langle f | g \rangle = \int_{\mathbb{R}} \overline{f(x)} g(x) \, dx, \quad f, g \in L^2(\mathbb{R}). \tag{1.1}$$

(iv) The space $C[0, 1]$ of all complex continuous functions on $[0, 1]$ provides a simple example of a pre-Hilbert space which is not Hilbert, i.e., it is not complete. Once more the inner product is defined by

$$\langle f | g \rangle = \int_0^1 \overline{f(x)} g(x) \, dx$$

To see that this space is not complete under the norm defined by the inner product, let us consider the following sequence, for $n > 2$,

$$f_n(x) = \begin{cases} 0, & \text{if } 0 \leqslant x \leqslant \frac{1}{2} - \frac{1}{n}, \\ \frac{n}{2}\left(x - \frac{1}{2}\right) + \frac{1}{2}, & \text{if } \frac{1}{2} - \frac{1}{n} \leqslant x \leqslant \frac{1}{2} + \frac{1}{n}, \\ 1, & \text{if } \frac{1}{2} + \frac{1}{n} \leqslant x \leqslant 1. \end{cases}$$

It is easy to check that

$$\lim_{n \to \infty} \int_0^1 |f_n(x) - f(x)|^2 \, dx = 0,$$

where f is the discontinuous function

$$f(x) = \begin{cases} 0, & \text{if } 0 \leqslant x \leqslant \frac{1}{2}, \\ 1, & \text{if } \frac{1}{2} < x \leqslant 1. \end{cases}$$

Hence (f_n) is a Cauchy sequence in $C[0,1]$, but $f \notin C[0,1]$.

The notion of *partial inner product space* (PIP-space) generalizes that of inner product space. The main difference lies in the fact that in a PIP-space the inner product is defined only for pairs of *compatible* elements.

1.1.2 Linear Compatibility on a Vector Space

In order to deal with partially defined inner products, we need some general properties of the relation "the inner product of g and f is defined." As already mentioned in the Introduction, a convenient way of doing this is through the notion of *compatibility*. Here is a formal definition.

Definition 1.1.4. A *linear compatibility relation* on a vector space V is a symmetric binary relation $f \# g$ which preserves linearity:

$$f \# g \iff g \# f, \ \forall \, f, g \in V,$$
$$f \# g, f \# h \implies f \# (\alpha g + \beta h), \ \forall \, f, g, h \in V, \forall \, \alpha, \beta \in \mathbb{C}.$$

The relation $\#$ is in general neither reflexive nor transitive. As a consequence of this definition, for every subset $S \subset V$, the set $S^{\#} = \{g \in V : g \# f, \forall \, f \in S\}$ is a vector subspace of V and one has

$$S^{\#\#} = (S^{\#})^{\#} \supseteq S, \ S^{\#\#\#} = S^{\#}.$$

Thus one gets the following equivalences:

$$\begin{aligned} f \# g &\iff f \in \{g\}^{\#} \iff \{f\}^{\#\#} \subseteq \{g\}^{\#} \\ &\iff g \in \{f\}^{\#} \iff \{g\}^{\#\#} \subseteq \{f\}^{\#}. \end{aligned} \tag{1.2}$$

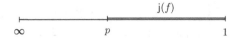

Fig. 1.1 The set $j(f)$ in the case of the scale $\mathcal{I} = \{L^p([0,1], dx), 1 \leqslant p \leqslant \infty\}$

From now on, we will call *assaying subspace* of V a subspace S such that $S^{\#\#} = S$ and denote by $\mathcal{F}(V, \#)$ the family of all assaying subspaces of V, ordered by inclusion. Let F be the isomorphy class of $\mathcal{F}$, that is, $\mathcal{F}$ considered as an abstract partially ordered set.[1] Elements of F will be denoted by $q, r, s, \ldots$, and the corresponding assaying subspaces by $V_q, V_r, V_s, \ldots$. By definition, $r \leqslant s$ if and only if $V_r \subseteq V_s$. We also write $V_{\overline{r}} = V_r^\#$, $r \in F$.

In order to characterize the behavior of individual vectors $f \in V$, it is useful to introduce the following set:

$$j(f) := \{r \in F : f \in V_r\}. \tag{1.3}$$

Intuitively, the larger the set $j(f)$, the "better" the behavior of f (for instance, in terms of regularity or behavior at infinity, see the examples of Section I.1). Clearly, $j(f)$ is a final subset of F: if $r \in j(f)$ and $s \geqslant r$, then $s \in j(f)$. To get a feeling, consider the scale of Lebesgue spaces on $[0,1]$, $\mathcal{I} = \{L^p([0,1], dx), 1 \leqslant p \leqslant \infty\}$. Then, for every $f \in L^1$, the set $j(f)$ is the semi-infinite interval $j(f) = \{q \geqslant 1 : f \in L^q\}$ (the spaces are increasing to the right, with *decreasing* p; it would be more natural to index the spaces by $1/p$, but we follow the tradition). Note that, if we define $p = \sup j(f)$, with $1 \leqslant p \leqslant \infty$, we still have to consider two cases, namely,

(i) $j(f) = [p, 1]$, a closed interval, i.e., $f \in L^p$, but $f \notin L^s, \forall s > p$.
(ii) $j(f) = (p, 1]$, a semi-open interval, i.e., $f \in L^q, \forall q < p$, hence $f \in \bigcap_{q<p} L^q$, but $f \notin L^p$.

This example shows that the set $j(f)$ indeed allows a "fine" control on the behavior of f.

In that language, the relations (1.2) mean that $f \# g$ if and only if there is an index $r \in F$ such that $f \in V_r$, $g \in V_{\overline{r}}$ or, equivalently, $\overline{j}(f) \cap j(g) \neq \emptyset$, where $\overline{j}(f) := \{\overline{r} : r \in j(f)\}$. In other words, vectors should not be considered individually, but only in terms of assaying subspaces, which are the building blocks of the whole structure.

In order to proceed, we need some more terminology.

Definition 1.1.5. A partially ordered set F is a *lattice* if every pair $r, s \in F$ admits a (unique) infimum (greatest lower bound) $r \wedge s \in F$ and a (unique)

---

[1] We recall that a partial order on a set X is a binary relation $\leqslant$ such that $\forall x, y, z \in X$, (i) $x \leqslant x$; (ii) $x \leqslant y$ and $y \leqslant x$ imply $x = y$; and (iii) $x \leqslant y$ and $y \leqslant z$ imply $x \leqslant z$; but two arbitrary elements x, y need not be comparable.

supremum (least upper bound) $r \vee s \in F$. The lattice is said to be *complete* if every family $J \subseteq F$ admits an infimum $\bigwedge_{r_j \in J} r_j \in F$ and a supremum $\bigvee_{r_j \in J} r_j \in F$.

Definition 1.1.6. Let F be a partially ordered set, with order $\leqslant$. An (order reversing) *involution* on F is a bijection $r \mapsto \bar{r}$ of F onto itself, such that

(i) $\bar{\bar{r}} = r$, $\forall\, r \in F$,
(ii) $s \geqslant r$ implies $\bar{s} \leqslant \bar{r}$, $\forall\, r, s \in F$.

In fact, by (i), $s \geqslant r$ if and only if $\bar{s} \leqslant \bar{r}$.

Let now F be a lattice, with lattice operations $\wedge$ and $\vee$. Then, for any involution $r \leftrightarrow \bar{r}$, condition (ii) above is equivalent to

(iii) $\overline{r \vee s} = \bar{r} \wedge \bar{s}$, $\forall\, r, s \in F$.

In the sequel, we will call *involutive lattice* any lattice with an involution. Similarly, a *complete involutive lattice* is a complete lattice F with an involution $r \leftrightarrow \bar{r}$ that verifies the condition

(iii') $\overline{\left(\bigvee_{r_j \in J} r_j \right)} = \bigwedge_{r_j \in J} \bar{r_j}$ for any family $J \subseteq F$.

It should be noticed that condition (iii') holds true for any involution on a lattice, complete or not, whenever the supremum on the l.h.s. exists in the lattice.

If F is a (complete) involutive lattice, a subset $F_1 \subseteq F$ is a (*complete*) *involutive sublattice* of F if it is a (complete) sublattice stable under the involution.

Remark 1.1.7. An involution should not be confused with a complementation: Even if the lattice has a greatest element z and a least element $a = \bar{z}$, one has in general $r \vee \bar{r} \neq z$ and $r \wedge \bar{r} \neq a$.

Now we may go on. Let again $\#$ be a linear compatibility on the vector space V and $\{V_r, r \in F\}$ the corresponding assaying subspaces. Then it is easy to see that the map $S \mapsto S^{\#\#}$ is a closure, in the sense of universal algebra (see Appendix A), so that the assaying subspaces are precisely the "closed" subsets. Therefore one has the following standard result.

Theorem 1.1.1. *Let V be a vector space and $\#$ a linear compatibility on V. The family $\mathcal{F}(V, \#) := \{V_r, r \in F\}$ of all assaying subspaces, ordered by inclusion, is a complete involutive lattice, under the following operations:*

$$\bigwedge_{j \in J} V_j = \bigcap_{j \in J} V_j, \qquad \bigvee_{j \in J} V_j = \left(\sum_{j \in J} V_j \right)^{\#\#}, \qquad \text{for any subset } J \subset F, \quad (1.4)$$

and the involution $V_r \leftrightarrow V_{\bar{r}} = (V_r)^{\#}$. Moreover, the involution is a lattice anti-isomorphism, that is,

$$(V_r \wedge V_s)^{\#} = V_{\overline{r}} \vee V_{\overline{s}}, \quad (V_r \vee V_s)^{\#} = V_{\overline{r}} \wedge V_{\overline{s}}.$$

The smallest element of $\mathcal{F}(V, \#)$ is $V^{\#} = \bigcap_r V_r$ and the greatest element is $V = \sum_r V_r$. By definition, the index set F is also a complete involutive lattice. Denoting $V_{r \wedge s} := V_r \wedge V_s$ and $V_{r \vee s} := V_r \vee V_s$, we have, for instance,

$$(V_{r \wedge s})^{\#} = V_{\overline{r \wedge s}} = V_{\overline{r} \vee \overline{s}} = V_{\overline{r}} \vee V_{\overline{s}}.$$

Complete involutive lattices arise naturally in the theory of Galois connections; a brief summary of the latter is given in Appendix A.

1.1.3 Partial Inner Product Spaces

Now we turn to partial inner product spaces or PIP-spaces. For the sake of completeness we give again the formal definition.

Definition 1.1.8. A *partial inner product* on $(V, \#)$ is a Hermitian form $\langle \cdot | \cdot \rangle$ defined exactly on compatible pairs of vectors. i.e., a mapping associating to every pair of vectors f, g such that $f \# g$, a complex number $\langle f | g \rangle$ in such a way that

(i) $\langle f | g \rangle = \overline{\langle g | f \rangle}$,
(ii) for fixed g, the correspondence $f \mapsto \langle g | f \rangle$ is linear in f.

If f is not compatible with g, the number $\langle f | g \rangle$ is not defined. If *all* pairs of vectors are compatible, then $\langle \cdot | \cdot \rangle$ is a (possibly indefinite) inner product.

Definition 1.1.9. A *partial inner product space* (*PIP-space*) is a vector space V equipped with a linear compatibility and a partial inner product.

Notice that the partial inner product is not required to be positive definite. Nevertheless, the partial inner product clearly defines a notion of *orthogonality* : $f \perp g$ if and only if $f \# g$ and $\langle f | g \rangle = 0$. The latter in turn leads to the important property of nondegeneracy.

Definition 1.1.10. The PIP-space $(V, \#, \langle \cdot | \cdot \rangle)$ is *nondegenerate* if $(V^{\#})^{\perp} = \{0\}$, that is, if $\langle f | g \rangle = 0$ for all $f \in V^{\#}$ implies $g = 0$.

In order to get a feeling, we present here a number of examples of nondegenerate partial inner product spaces and their assaying subspaces (including, of course, the two LHSs discussed in Section I.2.3).

(i) Inner product spaces

Any nondegenerate *inner* product space (not necessarily positive definite), in particular, any pre-Hilbert space. The only assaying subspace of an inner product space V is V itself.

(ii) Sequence spaces

(a) *Arbitrary sequences:* Let V be the space ω of *all* complex sequences $x = (x_n)$ and define on it:

- a compatibility relation by $x \# y \iff \sum_{n=1}^{\infty} |x_n y_n| < \infty$;
- a partial inner product $\langle x|y \rangle = \sum_{n=1}^{\infty} \overline{x_n} y_n$.

Then $\omega^{\#} = \varphi$, the space of finite sequences, and the complete lattice $\mathcal{F}(\omega, \#)$ consists of all Köthe's perfect sequence spaces. They include all ℓ^p-spaces $(1 \leqslant p \leqslant \infty)$, and, more generally, all weighted spaces $\ell^p(r)$, defined by

$$\|x\|_r^{(p)} = \left(\sum_n \left| \frac{x_n}{r_n} \right|^p \right)^{1/p} < \infty, \quad 1 \leqslant p < \infty,$$

with $\{r_n\}$ any sequence of positive numbers. We have $\ell^p(r)^{\#} = \ell^{\overline{p}}(\overline{r})$ where $1/p + 1/\overline{p} = 1$ and $\overline{r}_n = 1/r_n$. If $r_n = 1, \forall n$, we get the familiar chain of ℓ^p-spaces $(1 \leqslant p \leqslant \infty)$:

$$\ell^1 \subset \ldots \subset \ell^p \ldots \subset \ell^2 \ldots \subset \ell^{\overline{p}} \ldots \subset \ell^{\infty} \quad (1 < p < 2).$$

We find here also the weighted Hilbert spaces $\ell^2(r)$ described in Section I.2.3, which constitute a lattice, and indeed a noncomplete sublattice of $\mathcal{F}(\omega, \#)$. There are, of course, many assaying subspaces in ω in addition to the $\ell^p(r)$ (e.g., arbitrary intersections). These (Köthe) sequence spaces will be discussed at length in Section 4.3.

A word of caution concerning duality is in order here. According to the definition of compatibility given above, one has both $(\ell^1)^{\#} = \ell^{\infty}$ and $(\ell^{\infty})^{\#} = \ell^1$, which means that $(\ell^p)^{\#}$ is the so-called α-*dual* or *Köthe dual* of ℓ^p, which might be smaller than the topological dual. Indeed, whereas $(\ell^p)^{\#} = (\ell^p)^{\times} = \ell^{\overline{p}}$ for $1 \leqslant p < \infty$, $(\ell^{\infty})^{\#} = \ell^1 \subsetneqq (\ell^{\infty})^{\times}$ (the latter is not even a sequence space). We will come back to Köthe duality in Section 4.3.

From this example, we can easily construct a partial inner product which is not positive definite. With the same space and the same compatibility, we may change the inner product to $\langle x|y \rangle = \sum_{n=1}^{\infty} \epsilon_n \overline{x_n} y_n$ where ϵ_n is any fixed sequence of nonzero real numbers, for instance $\epsilon_n = \pm 1$. Then we get another PIP-space, with the same assaying subsets as before.

(b) *Sequences of slow increase:* Let $s^{\times}$ be the vector space of all sequences (x_n) of slow increase $(|x_n| \leqslant C(1 + n)^N$ for some $C > 0$ and some $N > 0)$ and equip it with the compatibility and the inner product inherited from ω. Then $V^{\#} = s$ is the space of all sequences (y_n) of fast decrease (i.e., $|y_n| \leqslant C(1 + n)^{-N}$ for some $C > 0$ and all $N > 0)$. The assaying subspaces are precisely the perfect spaces of Köthe that lie between s and $s^{\times} : s \subset V_r \subset s^{\times}$.

(iii) Spaces of locally integrable functions

Consider $V := L^1_{\mathrm{loc}}(X, d\mu)$, the space of all measurable, locally integrable functions on a measure space (X, μ). Define on it

- a compatibility relation by $f \# g \iff \overline{\int_X |f(x)g(x)| \, d\mu} < \infty$
- a partial inner product $\langle f | g \rangle = \int_X \overline{f(x)} g(x) \, d\mu$.

Then $V^\# = L^\infty_c(X, d\mu)$ consists of all essentially bounded measurable functions of compact support. The complete lattice $\mathcal{F}(L^1_{\mathrm{loc}}, \#)$ consists of all Köthe function spaces (Section 4.4). In particular, one may obtain the following result. Given any non-negative function $g \in L^1_{\mathrm{loc}}(X, d\mu)$, one has

$$\{g\}^\# = L^1_{\mathrm{loc}}(X, d\mu) \cap L^1(X, g d\mu)$$
$$\{g\}^{\#\#} = L^\infty_c(X, d\mu) \cap g \, L^\infty(X, d\mu),$$

where $g L^\infty(X, d\mu) := \{gh : h \in L^\infty(X, d\mu)\}$. Here again, typical assaying subspaces are weighted Hilbert spaces $L^2(r)$, which form a noncomplete sublattice of $\mathcal{F}(L^1_{\mathrm{loc}}, \#)$. The PIP-spaces of locally integrable functions will be studied in detail in Section 4.2.

The same structure may be defined on $L^2_{\mathrm{loc}}(X, d\mu)$, the space of all locally square integrable functions. Here $V^\#$ consists of all square integrable functions of compact support.

(iv) Fock space

In this example, we extend the machinery of Fock space (a symmetric tensor algebra over a Hilbert space) to partial inner product spaces. Let (X, μ) be a measure space. Consider, as in the previous example, the space $V^{(1)} = L^1_{\mathrm{loc}}(X, d\mu)$ with

$$f_1 \# g_1 \iff \int_X |f_1 g_1| \, d\mu < \infty, \quad f_1, g_1 \in V^{(1)}.$$

Define $V^{(n)}$ as $V^{(n)} = L^1_{\mathrm{loc}}(X^n, d\mu_n)$, where $X^n := X \times X \dots \times X$ and $d\mu_n := d\mu \times \dots \times d\mu$, with

$$f_n \# g_n \iff \int_{X^n} |f_n g_n| \, d\mu_n < \infty, \quad f_n, g_n \in V^{(n)}.$$

Finally, consider $V := \Gamma(V^{(1)})$, the space of all (i.e., arbitrary) sequences

$$f = \{f_0, \dots, f_n \dots\}, \quad f_0 \in V^{(0)} = \mathbb{C}, f_n \in V^{(n)}.$$

Define compatibility and partial inner product by

$$f \# g \iff \begin{cases} f_n \# g_n, & \text{for all } n = 1, 2, \ldots \\[2mm] \displaystyle\sum_{n=0}^{\infty} \int_{X^n} |f_n g_n| d\mu_n < \infty, \end{cases}$$

$$\langle f | g \rangle = \sum_{n=0}^{\infty} \int_{X^n} \overline{f_n} g_n \, d\mu_n.$$

One verifies immediately that $V^{\#}$ consists of all sequences with finitely many nonzero f_n and $f_n \in (V^{(n)})^{\#} = L_c^{\infty}(X^n, d\mu_n)$.

Assaying subspaces are obtained as before. For each $n = 1, 2, \ldots$, let

$$r^{(n)}(x) := r^{(n)}(x_1, x_2, \ldots, x_n), \quad x_i \in X, \ 1 = 1, 2, \ldots, n,$$

be a positive weight function such that $r^{(n)}, [r^{(n)}]^{-1} \in L_{\text{loc}}^2(X^n, d\mu_n)$. Then, as before, one defines

$$L^2(r^{(n)}) := \{ f \in L_{\text{loc}}^1(X^n, d\mu_n) : \int_{X^n} |f(x)|^2 \, [r^{(n)}]^{-2} \, d\mu_n(x) < \infty \}. \quad (1.5)$$

For each $n = 1, 2, \ldots$, these assaying spaces constitute a LHS with global space $V^{(n)} = L_{\text{loc}}^1(X^n, d\mu_n)$. Taking the direct sum over n, we obtain a LHS in the Fock space $V = \Gamma(V^{(1)})$.

In the same way as the central Hilbert space of $V^{(1)} = L_{\text{loc}}^1(X, d\mu)$ is $L^2(X, d\mu)$, that of $V = \Gamma(V^{(1)})$ is the usual Fock space $\mathcal{H}_F = \Gamma(L^2)$.

(v) Spaces of entire functions

Consider the chain of Hilbert spaces $\mathcal{F}^\rho$ $(-\infty < \rho < \infty)$, where $\mathcal{F}^\rho$ consists of all those entire functions $f(z) \in \mathbb{C}^n$, such that

$$\int_{\mathbb{C}^n} |f(z)|^2 (1 + |z|^2)^\rho \, e^{-|z|^2} d\nu(z) < \infty.$$

Here $d\nu(z) := \prod_{k=1}^{n} dx_k dy_k$, $z_k = x_k + iy_k$, is the Lebesgue measure on $\mathbb{C}^n$. This is indeed a chain, since $\mathfrak{F}^\rho \subset \mathfrak{F}^\sigma$ whenever $\rho > \sigma$. Then one defines Bargmann's spaces as $\mathfrak{E} = \bigcap_{\rho \in \mathbb{R}} \mathfrak{F}^\rho$ and $\mathfrak{E}^{\times} = \bigcup_{\rho \in \mathbb{R}} \mathfrak{F}^\rho$, the space on continuous conjugate linear functionals on $\mathfrak{E}$.

Define compatibility between two elements of $\mathcal{E}^{\times}$ to mean

$$f \# g \iff \int_{\mathbb{C}^n} |f(z) \, g(z)| e^{-|z|^2} d\nu(z) < \infty.$$

Then, with the partial inner product

$$\langle f|g \rangle := \int_{\mathbb{C}^n} \overline{f(z)}\, g(z) e^{-|z|^2}\, d\nu(z),$$

the space $\mathcal{E}^\times$ becomes a PIP-space and every $\mathcal{F}^\rho$ is an assaying subspace. PIP-spaces of analytic functions, including $\mathcal{E}^\times$, are discussed in Section 4.6.

(vi) Spaces of operators

Consider the space $\mathfrak{B}(\mathcal{H})$ of bounded operators in a Hilbert space $\mathcal{H}$. Define compatibility of $A \in \mathfrak{B}(\mathcal{H})$ and $B \in \mathfrak{B}(\mathcal{H})$ as the requirement that A^*B be of trace class. Set $\langle A|B \rangle = \mathrm{tr}(A^*B)$. Then $V^\#$ consists of all operators of trace class. Among the assaying subspaces, one finds the ideals of compact operators such as $\mathcal{C}^p$ and their generalizations. Note that $\{\mathcal{C}^p, 1 < p < \infty\}$ is a chain of Banach spaces, isomorphic to the chain $\{\ell^p,\ 1 < p < \infty\}$.

From these examples, we learn that $\mathcal{F}(V, \#)$ is a huge lattice (it is complete!) and that assaying subspaces may be complicated, such as Fréchet spaces, nonmetrizable spaces, etc. This situation suggests to choose a sublattice $\mathcal{I} \subset \mathcal{F}$, indexed by I, such that

(i) $\mathcal{I}$ is *generating*, by which we mean:

$$f \# g \ \Leftrightarrow\ \exists\, r \in I \text{ such that } f \in V_r, g \in V_{\bar{r}}; \qquad (1.6)$$

(ii) every $V_r, r \in I$, is a Hilbert space or a reflexive Banach space;
(iii) there is a unique self-dual, Hilbert, assaying subspace $V_o = V_{\bar{o}}$.

This is exactly what we have done in the two examples of LHS of weighted Hilbert spaces. The same operation will lead to the notion of *indexed* PIP-*spaces*, that will be studied systematically in Chapter 2.

1.2 Orthocomplemented Subspaces of a PIP-Space

Let V be a vector space with a linear compatibility relation $\#$. Given a direct sum decomposition $V = V_1 \oplus V_2$, we can ask whether $\#$ can be reconstructed from its restriction to V_1, V_2 and the knowledge of compatible pairs $f_1 \in V_1, f_2 \in V_2$.

In order to answer that question, we introduce the more or less standard notion of reduction.

Definition 1.2.1. A direct sum decomposition $V = V_1 \oplus V_2$ *reduces the linear compatibility* $\#$ if the condition $f \# g$ ($f \in V, g \in V$) is equivalent to the four conditions

$$f_1 \# g_1,\ f_2 \# g_2,\ f_1 \# g_2,\ f_2 \# g_1\ (f = f_1 + f_2,\ g = g_1 + g_2, f_1, g_1 \in V_1,\ f_2, g_2 \in V_2),$$

i.e., if # can be expressed in terms of components. The compatibility relation # is *absolutely* reduced if every vector in V_1 is compatible with every vector in V_2. The condition $f\#g$ is then equivalent to $f_1\#g_1$ and $f_2\#g_2$; the spaces V_1 and V_2 are decoupled from each other.

Example 1.2.2. Starting with two vector spaces V_1, V_2 with linear compatibility relations $\#_1, \#_2$, one can easily endow $V = V_1 \oplus V_2$ with compatibility relations that are reduced by the direct sum decomposition. It is enough to introduce any linear compatibility *between* V_1 and V_2 (i.e., a linear symmetric relation $f_1 \#_{12} g_2$) and declare $f\#g$ if and only if $f_1\#_1 g_1, f_2\#_2 g_2, f_1\#_{12} g_2, f_2\#_{12} g_1$).

Example 1.2.3. To get a direct sum decomposition that does *not* reduce a given compatibility, take $V = V_1 \oplus V_2$, with V_1 one-dimensional and spanned by a vector f that is not compatible with itself.

Proposition 1.2.4. *Let V be a vector space with a linear compatibility relation #. Let*

$$V = V_1 \oplus V_2 \qquad (1.7)$$

be a direct sum decomposition of V, and P_1, P_2 the corresponding (algebraic) projections. Then the following four conditions are equivalent:

(i) The compatibility relation # is reduced by the direct sum decomposition (1.7);
(ii) $\{P_1f\}^\# \supseteq \{f\}^\#$ for every $f \in V$;
(iii) $\{P_2f\}^\# \supseteq \{f\}^\#$ for every $f \in V$;
(iv) $\{f\}^\# = \{P_1f\}^\# \cap \{P_2f\}^\#$ for every $f \in V$.

The proof is straightforward and will be omitted.

The compatibility relation # is absolutely reduced by the decomposition (1.7) if and only if $P_1f\#P_1g$ implies $P_1f\#g$ and $f\#P_1g$; the same relations follow then for P_2. When this happens, we shall say that P_1 and P_2 are *absolute*. If, say, $V_1 \subseteq V^\#$, then P_1 and P_2 are absolute.

Now, mimicking the familiar Hilbert space situation, we define the notion of orthocomplemented subspace of a PIP-space. Notice that we do *not* assume that the partial inner product is positive.

Definition 1.2.5. Let V be a nondegenerate partial inner product space and W a vector subspace of V. We shall say that W is *orthocomplemented* if there exists a vector subspace $Z \subseteq V$ such that

(i) $W \cap Z = \{0\}, W + Z = V$,
(ii) the direct sum decomposition $V = W \oplus Z$ reduces the compatibility relation in V,
(iii) if $f \in W, g \in Z$ and $f\#g$, then $\langle f|g \rangle = 0$.

The goal of this definition is, obviously, to recover the standard Hilbert space situation, namely, the one-to-one correspondence between orthogonal projections and (orthocomplemented) subspaces. This will be obtained indeed, but for that purpose we need both the definition of operators in a PIP-space and some topological tools as well. The detailed analysis will be made in Chapter 3, Section 3.4.

1.3 Involutive Covering of a Vector Space Vs. Linear Compatibility

In this section, we shall show the equivalence between the notion of linear compatibility on a vector space and another one, in a sense more "global", namely, that of *involutive covering* of a vector space. This means, a covering family of vector subspaces, stable under finite intersection and equipped with a natural involution, which makes it into a lattice. More precisely, an involutive covering $\mathcal{I}$ of V uniquely defines a linear compatibility $\#$ on V, such that the complete involutive lattice of assaying subspaces $\mathcal{F}(V, \#)$ is the lattice completion of $\mathcal{I}$. Conversely, given V and $\#$, $\mathcal{F}(V, \#)$ is an involutive covering of V.

Given a vector space V, we consider families of subspaces of V, that is, subsets of $\mathcal{L}(V)$, the set of all vector subspaces of V, ordered by inclusion. Let $\mathcal{I} = \{V_r\}_{r \in I} \subset \mathcal{L}(V)$ be such a subset.

Definition 1.3.1. The family $\mathcal{I} = \{V_r, r \in I\} \subset \mathcal{L}(V)$ is called an *involutive covering* of V if:

(i) $\mathcal{I}$ is a covering of V : $\bigcup_{r \in I} V_r = V$;
(ii) $\mathcal{I}$ is stable under finite intersection;
(iii) $\mathcal{I}$ carries an involution $V_r \leftrightarrow V_{\overline{r}}$.

Remarks: (1) It would be mathematically more correct to say that V is the algebraic inductive limit of the family $\mathcal{I} = \{V_r, r \in I\}$, that is, the quotient of $\cup_{r \in I} V_r$ by an equivalence relation, but the present formulation is simpler and sufficient for our purposes. Of course one has also $V = \sum_{r \in I} V_r$.

(2) As we will see below, the property (ii) guarantees that the notion of involutive covering is equivalent to that of linear compatibility, and thus allows the construction of PIP-spaces.

Proposition 1.3.2. *Any involutive covering of V is an involutive lattice, with respect to the given involution and the following lattice operations:*

$$V_r \wedge V_s = V_r \cap V_s$$
$$V_r \vee V_s = V_u, \quad where \ V_{\overline{u}} = V_{\overline{r}} \cap V_{\overline{s}}$$

Proof. Simple verification that $\vee$ defines indeed a supremum.   ∎

Notice that, in general, one has $V_r \vee V_s \neq V_r + V_s$, i.e., $\mathcal{I}$ need not be a sublattice of $\mathcal{L}(V)$: The infimum $\wedge$ is always given by set intersection, but the supremum $\vee$ is determined by the involution.

In Section 1.1, we have introduced the notion of linear compatibility on a vector space. As we shall see now, this concept is equivalent to that of involutive covering. The first result is a simple reformulation of Theorem 1.1.1

Theorem 1.3.3. *Let V be a vector space with a linear compatibility $\#$. Then the family $\mathcal{F}(V, \#)$ of all assaying subspaces is an involutive covering of V, with the involution $V_r \leftrightarrow V_{\bar{r}} = (V_r)^{\#}$.*

Conversely, if we start with an involutive covering $\mathcal{I}$ of V, we can associate to it a linear compatibility such that $\mathcal{F}(V, \#)$ is the lattice completion of $\mathcal{I}$.

Theorem 1.3.4. *Let V be a vector space with an involutive covering $\mathcal{I} = \{V_r, r \in I\}$. Consider in the Cartesian product $V \times V$ the subset $\Delta = \bigcup_{V_r \in \mathcal{I}} V_r \times V_{\bar{r}}$ and define $f \# g$ to mean $(f, g) \in \Delta$. Then:*

(i) $\#$ is a linear compatibility relation;
(ii) $\mathcal{I}$ is an involutive sublattice of $\mathcal{F}(V, \#)$;
(iii) Every element of $\mathcal{F}(V, \#)$ is an intersection of elements of the form $\left(\bigcap_{V_r \in \mathcal{I}'} V_r \right)^{\#}$ where $\mathcal{I}' \subseteq \mathcal{I}$, i.e., $\mathcal{F}(V, \#)$ is the complete involutive lattice generated by $\mathcal{I}$ through unrestricted lattice operations.

Proof. (i) is easily verified. In order to prove (ii), we first notice that, for every $f \in V$, one has

$$\{f\}^{\#} = \bigcup_{V_q \in \mathcal{I}, V_q \ni f} V_{\bar{q}} = \sum_{V_q \in \mathcal{I}, V_q \ni f} V_{\bar{q}}, \tag{1.8}$$

since it is a vector subspace of V.

Next we show that $(V_r)^{\#} = V_{\bar{r}}$ for every $V_r \in \mathcal{I}$. One has obviously $V_{\bar{r}} \subset (V_r)^{\#}$. In order to prove the inclusion in the other direction, let $g \in (V_r)^{\#} = \bigcap_{f \in V_r} \{f\}^{\#}$. That is, given any fixed $h \in V_r$, there exists a $V_q \in \mathcal{I}$, depending on h ($q = q(h)$), such that $h \in V_q$ and $g \in V_{\bar{q}}$. *A fortiori* we have

$$h \in V_{p(h)} := V_{q(h)} \cap V_r \subseteq V_r,$$
$$g \in V_{\overline{p(h)}} \supseteq V_{\overline{q(h)}},$$

and $V_{p(h)} \in \mathcal{I}$, the lattice being stable under intersection. Hence,

$$\bigvee_{h \in V_r} V_{p(h)} \subseteq V_r.$$

On the other hand, every $h \in V_r$ is contained in some $V_{p(h)}$, so,

$$V_r \subseteq \bigcup_{h \in V_r} V_{p(h)} \subseteq \bigvee_{h \in V_r} V_{p(h)}.$$

Thus $\bigvee_{h \in V_r} V_{p(h)}$ exists in $\mathcal{I}$ and equals V_r, This implies that condition (iii') of Definition 1.1.6 holds true:

$$V_{\overline{r}} = \bigcap_{h \in V_r} V_{\overline{p(h)}}.$$

Since $g \in V_{\overline{p(h)}}$ for every $h \in V$, it follows that $g \in V_{\overline{r}}$.

The equality $V_{\overline{r}} = (V_r)^{\#}$ means that all subspaces $V_r \in \mathcal{I}$ are assaying subspaces and that the involution $V_r \leftrightarrow V_{\overline{r}}$ coincides with the involution $V_r \leftrightarrow (V_r)^{\#}$. This proves (ii).

It remains to prove (iii). By definition, $\{h\}^{\#\#}$ is the smallest assaying subspace containing h, thus

$$\{h\}^{\#\#} = \bigcap_{V_r \in \mathcal{I}, V_r \ni h} V_r \tag{1.9}$$

(this follows also from (1.8)). To conclude, it suffices to notice that an arbitrary element $V_q \in \mathcal{F}(V, \#)$ can be written as

$$V_q = \bigcap_{h \in V_{\overline{q}}} \{h\}^{\#} = \bigcap_{h \in V_{\overline{q}}} \left(\bigcap_{V_r \in \mathcal{I}, V_r \ni h} V_r \right)^{\#}. \qquad \blacksquare$$

Remark 1.3.5. Of course, if the set $\mathcal{I}$ is finite, $\mathcal{F}(V, \#) = \mathcal{I}$. This case, although trivial in the present context, is important for applications. Indeed it covers already such concepts as *rigged Hilbert spaces* or *equipped Hilbert spaces*, which are widely used in practice (see for instance the quantum mechanical applications described in Section 7.1).

Theorem 1.3.4 makes contact between the abstract formulation of Section 1.1 and the more concrete "constructive" approach developed previously for particular cases such as chains (or scales) of Hilbert or Banach spaces, nested Hilbert spaces, or rigged Hilbert spaces. In all of these situations, the starting point is a family of vector spaces $\{V_r\}$ which form an involutive lattice (or a scale), and V is defined as $\cup_r V_r$ (technically, algebraic inductive limit). Theorem 1.3.4 shows that the two approaches are, in fact, equivalent. In other words, it makes clear that the concept of partial inner product space is a genuine generalization of all these particular structures.

It also follows that all we need is an involutive covering of V, the complete lattice $\mathcal{F}(V, \#)$ is then uniquely determined. In most cases, the lattice completion will remain implicit, for $\mathcal{F}(V, \#)$ is uncomfortably large. (This is just like a topology: One can seldom exhibit explicitly "all" open sets!). But in certain cases, an arbitrary assaying subspace may be described, as will be clear from the following examples.

1.4 Generating Families of Assaying Subspaces

We have seen in Section 1.3 that an involutive covering on a vector space V uniquely determines a linear compatibility on V. Actually even less is needed: Any family of subsets of V that allows to reconstruct the compatibility will suffice. Considering the fact that the full lattice $\mathcal{F}$ is in general much too large for practical purposes, this result will be most useful for applications.

Definition 1.4.1. A family $\mathcal{I}_o$ of subsets of V is called *generating* for the linear compatibility # if, given any compatible pair $f, g \in V$, there exists $S \in \mathcal{I}_o$ such that $f \in S^{\#}$ and $g \in S^{\#\#}$.

Notice that elements of $\mathcal{I}_o$ need not be vector subspaces of V, the involution automatically generates vector subspaces. But the generating character is not lost in the process. Indeed, if we define the three following families of subspaces:

$$\mathcal{I}_o^{\#} = \{S^{\#} : S \in \mathcal{I}_o\}, \quad \mathcal{I}_o^{\#\#} = \{S^{\#\#} : S \in \mathcal{I}_o\}, \quad \mathcal{I} = \mathcal{I}_o^{\#} \cup \mathcal{I}_o^{\#\#},$$

we have (the proof is straightforward):

Proposition 1.4.2. *Given a generating family $\mathcal{I}_o$ of subsets of V, let $\mathcal{J}$ be any one of $\mathcal{I}_o^{\#}, \mathcal{I}_o^{\#\#}$, and $\mathcal{I}$. Then:*

(i) $\mathcal{J}$ is a subset of $\mathcal{F}(V, \#)$.
(ii) $\mathcal{J}$ is generating and covers $V : \bigcup_{S \in \mathcal{J}} S = V$.
(iii) $V^{\#} = \cap_{s \in \mathcal{J}} S$.

The lesson is clear: If $\mathcal{I}_o$ is generating and $\mathcal{I}_1 \supset \mathcal{I}_o$, then $\mathcal{I}_1$ is also generating. This allows one to enlarge a generating subset by "closing" it with respect to some algebraic operation, involution (which gives $\mathcal{I}$), lattice operations, or both. In the latter case, we get the involutive sublattice of $\mathcal{F}(V, \#)$ generated by $\mathcal{I}_o$, which is an involutive covering. Anyway there is no loss of generality in assuming a generating family to contain only assaying subspaces (although for some applications it might be more convenient not to do so).

Remark: Definition 1.4.1 may also be phrased as follows. A family $\mathcal{I}$ of assaying subspaces is generating if and only if relation (1.7) holds for any assaying subspace V_r or, equivalently, relation (1.9) holds for any $h \in V$.

The main reason for introducing generating families of assaying subspaces in $\mathcal{F}(V, \#)$ is topological. Namely, arbitrary assaying subspaces, equipped with their canonical topologies (in fact, we will always choose the Mackey topology, see Chapter 2), may be difficult to handle, whereas it is easy in most cases to find *homogeneous* generating families, that is generating families $\mathcal{I}$ whose elements are all of the same type, such as Hilbert spaces or reflexive Banach spaces. Two examples are given by the standard weighted Hilbert spaces described in Section 1.1.

1.4.1 Example: Hilbert Spaces of Sequences

We take again the space ω of all sequences and show that $\mathcal{I} = \{\ell^2(r)\}$ is a generating subset of $\mathcal{F}(\omega, \#)$ consisting of Hilbert spaces only. Let $f, g \in \omega$ with $f \# g$, i.e., $\sum_{n=1}^{\infty} |f_n g_n| < \infty$. Define a partition of $n \in N$ into four disjoint subsets as follows :

$$N_1 := \{n \in \mathbb{N} : f_n \neq 0, \ g_n \neq 0\},$$
$$N_2 := \{n \in \mathbb{N} : f_n = 0, \ g_n \neq 0\},$$
$$N_3 := \{n \in \mathbb{N} : f_n \neq 0, \ g_n = 0\},$$
$$N_4 := \{n \in \mathbb{N} : f_n = g_n = 0\}.$$

All four subsets may be finite or infinite. Let $s = (s_n)$ be an arbitrary square integrable sequence, with nonzero elements. Then define a sequence of weights, $r = (r_n)$ as follows:

- for $n \in N_1$, $r_n = |f_n|^{1/2} \, |g_n|^{-1/2}$,
- for $n \in N_2$, $r_n = |s_n| \, |g_n|^{-1}$,
- for $n \in N_3$, $r_n = |f_n| \, |s_n|^{-1}$,
- for $n \in N_4$, $r_n > 0$, arbitrary

Then $f \in \ell^2(r), g \in \ell^2(r^{-1})$; indeed

$$\sum_{n=1}^{\infty} |f_n|^2 r_n^{-2} = \sum_{n \in N_1 \cup N_2} |f_n|^2 r_n^{-2} = \sum_{n \in N_1} |f_n||g_n| + \sum_{n \in N_2} |s_n|^2 < \infty,$$

$$\sum_{n=1}^{\infty} |g_n|^2 r_n^2 = \sum_{h \in N_1 \cup N_2} |g_n|^2 r_n^2 = \sum_{n \in N_1} |f_n||g_n| + \sum_{n \in N_2} |s_n|^2 < \infty.$$

1.4.2 Example: Locally Integrable Functions

Let now $V = L^1_{\mathrm{loc}}(\mathbb{R}^n, dx)$, the space of all Lebesgue locally integrable functions on $\mathbb{R}^n$. With the compatibility.

$$f \# g \iff \int_{\mathbb{R}^n} |f(x)g(x)| \, dx < \infty$$

one has, as we know, $V^{\#} = L_c^{\infty}(\mathbb{R}^n, dx)$, the essentially bounded functions of compact support. Then we claim that the family of weighted Hilbert spaces $\{L^2(r)\}$ is a generating sublattice of $\mathcal{F}(V, \#)$. The statement is proved in Proposition 4.2.1. Additional examples of the same type are discussed in Section 4.2.2.

1.4.3 Two Counterexamples

Not all "natural" families are generating, however. Here we give two counterexamples.

(i) The chain of sequence spaces $\mathcal{I}(\ell^p) := \{\ell^p, \ 1 < p < \infty\}$ is a LBS. The duality $\ell^p \leftrightarrow \ell^{\bar{p}}$ $(1/p + 1/\bar{p} = 1)$ coincides with the compatibility inherited from ω. However, the family $\mathcal{I}(\ell^p)$ is *not* generating in ℓ^∞. Indeed we have:

$$f = (f_n), \text{ where } f_n = (\log n)^{-1}, \text{ belongs to } \ell^\infty, \text{ but not to } \ell^{\infty-} := \cup_{p<\infty} \ell^p,$$

$$g = (g_n), \text{ where } g_n = (n \log n)^{-1}, \text{ belongs to } \ell^{1+} := \cap_{1<p\leqslant\infty} \ell^p, \text{ but not to } \ell^1$$

and yet

$$\sum_{n=1}^{\infty} |f_n g_n| = \sum_{n=1}^{\infty} n^{-1} (\log n)^{-2} < \infty,$$

that is, $f \# g$. This example carries over immediately to the ideals $\mathcal{C}^p(1 \leqslant p \leqslant \infty)$ of compact operators in a Hilbert space.

(ii) The family $\{\ell^2(r^{(k)}), k \in \mathbb{Z}\}$, where $r_n^{(k)} = n^k$, is not generating in the space $s^\times$ of sequences of polynomial growth, for the same compatibility. Take indeed:

$$f_n = n^l \text{ and } g_n = n^{-l-1}(\log n)^{-1-\delta} \ (\delta > 0).$$

Then

$$\sum_{n=1}^{\infty} |f_n g_n| = \sum_{n=1}^{\infty} n^{-1}(\log n)^{-1-\delta} < \infty,$$

but there is no $k \in \mathbb{Z}$ such that $f \in \ell^2(r^{(k)})$ and $g \in \ell^2(\overline{r^{(k)}})$, since one has:

$$\sum_{n=1}^{\infty} |f_n|^2 (r_n^{(k)})^{-2} = \sum_{n=1}^{\infty} n^{2l-2k} < \infty, \qquad\qquad \text{if and only if } k > l + \tfrac{1}{2},$$

$$\sum_{n=1}^{\infty} |g_n|^2 (r_n^{(k)})^{2} = \sum_{n=1}^{\infty} n^{2k-2l-2}(\log n)^{-2-2\delta} < \infty, \text{ if and only if } k \leqslant l + \tfrac{1}{2}.$$

It is interesting to notice that further enlarging the family of weights does not improve the situation, as long as it remains totally ordered: For instance, if one uses $\{r^{(k,j)}\}$ with $r_n^{(k,j)} = n^k (\log n)^j$, a similar counterexample can be obtained. The moral is that, whenever the compatibility $\#$ is given by absolute convergence of a series or an integral, no totally ordered subset of $\mathcal{F}(V, \#)$ will be generating: Two vectors f, g may be compatible, not because one is "good" and the other is "bad", but because there are cancellations between f_n and g_n. An extreme example is that of two wildly increasing sequences, f and g, such that $f_{2m} = 0$ and $g_{2m+1} = 0$ for all m. Simpler yet, the complete lattice generated by a totally ordered subset will be totally ordered, i.e., a chain, while $\mathcal{F}(V, \#)$ is not (compare Section 4.1).

1.5 Comparison of Compatibility Relations on a Vector Space

One of the main reasons to consider partial inner product spaces is the possibility they offer to control very singular operators. For this purpose the question frequently arises as to whether a given compatibility relation is "fine" enough for a particular operator, or, on the contrary, too fine in that it leads to an unmanageable lattice of assaying subspaces. What we need is obviously a way of comparing different compatibility relations on the same vector space.

Let $\mathrm{LC}(V)$ denote the set of all linear compatibility relations on V. It turns out that a good order on $\mathrm{LC}(V)$ must satisfy the two following conditions: If $\#_1$ is finer than $\#_2$, then $\#_2$–compatible vectors should be $\#_1$–compatible and $\mathcal{F}(V, \#_2)$ should be a subset of $\mathcal{F}(V, \#_1)$. However, these requirements are not sufficient for a comparison of compatibilities. It is essential that the involution in $\mathcal{F}(V, \#_2)$ be the restriction of the involution in $\mathcal{F}(V, \#_1)$. This condition leads to the next definition.

Definition 1.5.1. We shall say that $\#_1$ is *finer* than $\#_2$, or that $\#_2$ is *coarser* than $\#_1$, which we note $\#_2 \trianglelefteq \#_1$, if $\mathcal{F}(V, \#_2)$ is an involutive sublattice of $\mathcal{F}(V, \#_1)$.

A simple criterion for comparing two compatibilities on V is given by the following result.

Proposition 1.5.2. $\#_1$ *is finer than* $\#_2$ *if and only if*

$$S^{\#_2 \#_2} = S^{\#_2 \#_1} \tag{1.10}$$

for every subset $S \subseteq V$.

Proof. We know that $A \subseteq V$ belongs to $\mathcal{F}(V, \#_1)$ if and only if $A = B^{\#_1 \#_1}$ for some $B \subseteq V$ or, equivalently, $A = C^{\#_1}$ for some $C \subseteq V$. Thus (1.10) means that every element $S^{\#_2 \#_2}$ of $\mathcal{F}(V, \#_2)$ belongs to $\mathcal{F}(V, \#_1)$. Furthermore, for any $D = S^{\#_2} \in \mathcal{F}(V, \#_2)$, one has $D^{\#_2} = D^{\#_1}$. This proves that $\mathcal{F}(V, \#_2)$ is an involutive sublattice of $\mathcal{F}(V, \#_1)$ if and only if (1.10) holds for every $S \subseteq V$. ∎

Corollary 1.5.3. *If $\#_2 \trianglelefteq \#_1$, then $S^{\#_2} \subseteq S^{\#_1}$ and $S^{\#_2\#_2} \supseteq S^{\#_1\#_1}$ for every subset $S \subseteq V$. In particular, $f\#_2 g$ implies $f\#_1 g$ for any $f, g \in V$.*

Proof: From (1.10), one gets, for any $S \subseteq V$

$$S^{\#_2} = S^{\#_2\#_2\#_2} = S^{\#_2\#_2\#_1} \subseteq S^{\#_1},$$

where the inclusion follows from $S \subseteq S^{\#_2\#_2}$. In particular for $S = \{f\}$ this gives $\{f\}^{\#_2} \subseteq \{f\}^{\#_1}$. Furthermore, using (1.10) again, we have

$$S^{\#_2\#_2} = S^{\#_2\#_1} \subseteq S^{\#_1\#_1}.$$
∎

Remark 1.5.4. Given two linear compatibility relations $\#_1$ and $\#_2$ on a vector space V, it is tempting to use the conclusion of Corollary 1.5.3 as a definition of an order on $\mathrm{LC}(V)$. Thus one could say that $\#_1$ is *weakly finer* than $\#_2$ if $\#_1$ has more compatible pairs; equivalently, $f\#_2 g$ implies $f\#_1 g$ for any pair $f, g \in V$, or $S^{\#_2} \subseteq S^{\#_1}$, for any subset $S \subseteq V$. This relation is a partial order on $\mathrm{LC}(V)$. Actually, $\mathrm{LC}(V)$ is even a complete lattice for this order, since it is stable under arbitrary intersections and has a greatest element $\widehat{\#}$. Here, intersection is defined as follows: $S^{\wedge_{i \in I} \#_i} = \cap_{i \in I} S^{\#_i}$ for an arbitrary subset $S \subseteq V$. The greatest element is the compatibility $\widehat{\#}$ for which every pair of vectors is a compatible pair:

$$f \widehat{\#} g, \, \forall f, g \in V \quad \text{or} \quad \{f\}^{\widehat{\#}} = V, \, \forall f \in V.$$

The lattice $\mathcal{F}(V, \widehat{\#})$ has only one element, namely V itself. Thus $\mathcal{F}(V, \#)$ is *not* a subset of $\mathcal{F}(V, \widehat{\#})$, for any $\# \in \mathrm{LC}(V)$; more generally, a weakly finer compatibility does not lead to more assaying subspaces. As a consequence, this ordering of $\mathrm{LC}(V)$, although standard in the context of Galois connections (see Appendix A), is useless for our purposes.

The order defined on $\mathrm{LC}(V)$ by the week comparability is only partial. Namely, we can exhibit an example of two compatibilities which are not weakly comparable.

Example 1.5.5. Let $\mathcal{E}^\times$ be Bargmann's space of functions introduced in Section 1.1.3, Example (v) (in dimension 1). Define the following compatibilities on $\mathcal{E}^\times$:

$$f \#_1 g \quad \text{if and only if} \quad \int_{\mathbb{C}} |f(z)g(z)| \, e^{-|z|^2} d\nu(z) < \infty \, ,$$

$$f \#_2 g \quad \text{if and only if} \quad \sum_{n=0}^{\infty} n! \, |a_n b_n| < \infty \, ,$$

where $f(z) = \sum_{n=0}^{\infty} a_n z^n$ and $g(z) = \sum_{n=0}^{\infty} b_n z^n$. Let now $f(z) = \exp(\frac{1}{2}z^2)$ and $g_1(z) = \exp(\frac{i}{2}z^2)$. Then f and g_1 are $\#_1$–compatible, but they are not $\#_2$–compatible. On the other hand, if $g_2(z) = z \exp(\frac{1}{2}z^2)$, then f and g_2 are (trivially) $\#_2$–compatible and they are not $\#_1$–compatible. Hence $\#_1$ and $\#_2$ are not comparable in the weak sense.

Let us come back to the set $\mathrm{LC}(V)$. First, the relation $\trianglelefteq$ is a partial order on it, as can be checked immediately. Then $\mathrm{LC}(V)$ is inductively ordered, i.e., every totally ordered subset $\{\#_n, n \in J\}$ has an upper bound $\#_\infty$, defined as follows:

$$f \#_\infty g \quad \text{if and only if there exists } n \in J \text{ such that } f \#_n g.$$

Equivalently, $S^{\#_\infty} = \cup_{n \in J} S^{\#_n}$, for any $S \subseteq V$. Therefore Zorn's lemma applies: Every element of $\mathrm{LC}(V)$ is majorized by a maximal element. However, there is *no* greatest element. The only possible candidate would be $\widehat{\#}$, for which $S^{\widehat{\#}} = V$ for every $S \subseteq V$. But (1.10) implies that $\widehat{\#}$ cannot be finer than any $\# \in \mathrm{LC}(V)$, except itself ! In fact, $\mathrm{LC}(V)$ is in general neither directed to the left, nor to the right, *a fortiori* it is not a lattice.

1.5.1 *Coarsening*

Let again $\#$ be a linear compatibility on V and let $\mathcal{I}$ be a *generating* subset of $\mathcal{F}(V, \#)$. As was remarked after Proposition 1.4.2, we may always assume that $\mathcal{I}$ is an involutive sublattice of $\mathcal{F}(V, \#)$, i.e., an involutive covering of V. Moreover, by Theorem 1.3.4, $\mathcal{F}(V, \#)$ is the lattice completion of $\mathcal{I}$.

What happens now if we start from a sublattice $\mathcal{I}$ cofinal to $\mathcal{F}(V, \#)$, but *not* generating ? (Given a partially ordered set F, a subset $K \subset F$ is *cofinal* to F if, for any element $x \in F$, there is an element $k \in K$ such that $x \leqslant k$.) Again by Theorem 1.3.4, we can associate to $\mathcal{I}$ a new compatibility relation $\#_I$, such that $\mathcal{I}$ generates the complete involutive lattice $\mathcal{F}(V, \#_I)$. Then it is easy to see that $\mathcal{F}(V, \#_I)$ is a complete involutive sublattice of $\mathcal{F}(V, \#)$; in other words, $\#_I \triangleleft \#$. This result in fact gives a description of all compatibility relations on V coarser than a given one $\#$, as the following easy theorem states.

Theorem 1.5.6. *Let V and $\#$ be as usual.*

(a) Let $\mathcal{I}$ be a cofinal involutive sublattice of $\mathcal{F}(V, \#)$. Then the compatibility $\#_I$ determined by $\mathcal{I}$ is coarser than $\#$, that is, $\#_I \trianglelefteq \#$.

(b) Conversely, if $\#_1 \trianglelefteq \#$, then there exists a sublattice $\mathcal{I} \subseteq \mathcal{F}(V, \#)$ cofinal to $\mathcal{F}(V, \#)$ and stable under the involution, such that $\#_{\mathcal{I}} = \#_1$.

Thus, given $\#$, a compatibility relation on V, coarser than $\#$, is the same thing as a complete involutive sublattice of $\mathcal{F}(V, \#)$. The set of all of these is stable under intersection and it has a greatest element, namely $\mathcal{F}(V, \#)$, hence this set is itself a complete lattice, contrary to $\mathrm{LC}(V)$. It has also a smallest element $\mathcal{F}(V, \#_0)$ consisting exactly of $V^{\#}$ and V, and corresponding to the trivial compatibility relation $\#_0$ defined as

$$f \#_0 g \text{ if and only if at least one of them belongs to } V^{\#}.$$

In a sense, this trivial compatibility relation corresponds, in our language, to the standard situation of the theory of distributions: Only two kinds of objects are available, test functions $(V^{\#})$ and distributions (V).

Let us give a few examples of comparable compatibility relations.

Example 1.5.7. As pointed out at the end of Section 1.4.3, no totally ordered subset will be generating for a compatibility which is defined by absolute convergence of a series or an integral. Bargmann's space $\mathcal{E}^{\times}$ illustrates this point beautifully. Three compatibilities arise naturally on $\mathcal{E}^{\times}$: $\#_1$ and $\#_2$, as defined in Example 1.5.5, and $\#_3$ defined by the chain $\{\mathcal{F}^{\rho}, \rho \in \mathbb{R}\}$ described in Section 1.1, Example (v). Then it can be checked easily that $\#_3$ is strictly coarser that both $\#_1$ and $\#_2$: Neither the pair f, g_1 nor the pair f, g_2 are $\#_3$-compatible. Indeed, both f and g_1 belong to $\mathcal{F}^{\rho}$ if and only if $\rho < 0$, and $g_2 \in \mathcal{F}^{\rho}$ if and only if $\rho < -1$.

The main application of Theorem 1.5.6 is to the construction of partial inner product spaces. Given a vector space V, let $\Upsilon \subseteq V \times V$ be a domain such that:

(i) Υ is symmetric: $(f, g) \in \Upsilon$ if and only if $(g, f) \in \Upsilon$.
(ii) Υ is "partially linear": For every $f \in V$, the set $\{g : (f, g) \in \Upsilon\}$ is a vector subspace of V.

Then a Hermitian form defined on the domain Υ is a partial inner product in the sense of Definition 1.1.8. So, a partial inner product uniquely defines a linear compatibility relation $\#$:

$$f \# g \text{ if and only if } (f, g) \in \Upsilon.$$

Typical examples are all those partial inner products whose domain of definition is defined by the absolute convergence of a series or an integral. Such are, for instance, the "natural" inner products on ω, $L^1_{\mathrm{loc}}(X, d\mu)$ or $\mathcal{E}^{\times}$ obtained by extension of the inner product of a Hilbert space.

However, quite often the complete lattice $\mathcal{F}(V, \#)$ generated by $\#$ is too large, and thus one is led, for practical purposes, to consider a coarser compatibility relation on V. The point of Theorem 1.5.6 is that, first, one knows all possible candidates, and second, each of them can be used as a domain for the initial partial inner product. In particular, if the latter is nondegenerate, i.e., $(V^{\#})^{\perp} = \{0\}$, it will remain so, whichever coarser compatibility $\#_1$ one chooses, including the trivial one, $\#_0$, for which $V^{\#_0} = V^{\#_1} = V^{\#}$.

1.5.2 Refining

Theorem 1.5.6 solves the problem of coarsening a given compatibility relation on V. In practice, the converse problem will often arise, namely, how to refine a given compatibility. Here, however, there is no canonical solution.

There is one case where a solution can be found, namely when the compatibility is given in terms of an involutive covering of V and there is an explicit finer covering. Typically, a partial inner product is introduced, which has a bigger "natural" domain. However, even in that case, uniqueness is not guaranteed. Once again, Bargmann's space $\mathcal{E}^{\times}$ gives a counterexample. Let us start with the compatibility $\#_3$ defined by the involutive covering $\{\mathcal{F}^{\rho}, \rho \in \mathbb{R}\}$ and introduce, as in Example 1.5.5, the usual inner product

$$\langle f | g \rangle = \int \overline{f(z)}\, g(z)\, e^{-|z|^2/2} d\nu(z)\,.$$

The latter is obviously defined whenever the integral converges absolutely. This leads to the compatibility $\#_1$, finer than the compatibility $\#_3$. Now if $f \#_3 g$, then the inner product is also given by the expression

$$\langle f | g \rangle = \sum_{n=1}^{\infty} n!\, \overline{a_n b_n}\,,$$

which is defined whenever the series converges absolutely, leading to $\#_2 \trianglerighteq \#_3$. However, we have seen that $\#_1$ and $\#_2$ are not comparable, although they are both finer than $\#_3$.

Apart from that situation, very little can be said about the problem of refinement. As a first step, one might try to increase the number of compatible pairs. Equivalently, one can try to extend the domain of the linear forms $\phi_f := \langle f | \cdot \rangle$, initially defined on $\{f\}^{\#}$, and continuous in the Mackey topology $\tau(\{f\}^{\#}, \{f\}^{\#\#})$ (see the Appendix B). This, however, requires topological considerations, but explicit examples show that the resulting compatibilities are not comparable with the original one, because the involution will be modified. We must conclude that the problem of refinement has in general no solution.

Yet there are cases in which partial results can be obtained, namely one can enlarge the set of *explicit* assaying subspaces. For instance, suppose the compatibility on V is given in terms of an involutive family $\mathcal{I} = \{V_r\}$ of subspaces, which is *not* stable under intersection (thus not a chain). Assume, in addition, that every V_r is a reflexive Banach space (in particular, a Hilbert space). Then it can be shown that every element of the lattice generated by $\mathcal{I}$ is again a reflexive Banach space, as explained in Section I.2. Other examples are the explicit completion of a reflexive chain of Banach spaces ('nonstandard' completion), or the refinement by interpolation of a scale of Hilbert spaces. Both constructions will be described in Chapter 4.

Notice that, in all cases, essential use is made of topological properties of assaying subspaces. These are determined entirely by the partial inner product, which defines the duality between pairs $V_r, V_{\bar{r}}$; the compatibility alone no longer suffices. So the next step in our analysis is to study systematically the topological structure of PIP-spaces. This will be done in the next chapter.

Notes for Chapter 1

Section 1.1. Most of this chapter is borrowed from [12]. The standard result on which Theorem 1.1.1 is based is quoted in Appendix A). For further information, see Birkhoff [Bir66].

- For indefinite inner product spaces, see the monograph of Bognar [Bog74].
- For further information on the Köthe's perfect sequence spaces, in particular the notion of α-dual or Köthe dual of a sequence space, we refer the reader to [Köt69, §30].
- The space of sequences of slow increase is usually taken as s', but we prefer to use $s^\times$, since this allows *linear* embeddings $s \hookrightarrow \ell^2 \hookrightarrow s^\times$. The same remark applies to the space $\mathcal{S}^\times$ of tempered distributions [Sch57] and Bargmann's space $\mathcal{E}^\times$ (see below). The so-called Hermite realization of $\mathcal{S}^\times$ associates to every $f \in \mathcal{S}^\times$ a sequence of slow increase, namely the sequence of coefficients $\langle h_k | f \rangle$, where the h_k are normalized Hermite functions. This realization has been emphasized by Simon [179, Sim71].
- The assaying subspaces of the PIP-space $L^1_{\text{loc}}(X, d\mu)$ have been described by Dieudonné [73] and by Goes-Welland [107]. In particular, the explicit structure of $\{g\}^\#$ and $\{g\}^{\#\#}$ are given in [107].
- The extension to partial inner product spaces of the Fock space formalism, a symmetric tensor algebra over a Hilbert space, was first introduced by Grossmann [117]. See also Antoine-Grossmann [18]. This can serve as a basis for the use of PIP-spaces in quantum field theory. This topic will be revisited in Chapter 7, Section 7.3.
- Bargmann's space is usually taken as $\mathcal{E}'$, which consists of all continuous *linear* functionals on $\mathcal{E}$. As usual in the context of PIP-spaces, we prefer to use $\mathcal{E}^\times$, the space of continuous *conjugate linear* functionals on $\mathcal{E}$, since this allows a *linear* embedding $\mathcal{E} \hookrightarrow \mathcal{E}^\times$.
- For the ideals $\mathcal{C}^p$ of compact operators and their generalizations, we refer to [RS72, Section VI.6] and [RS75, App. to Sec. IX.4]. A description of

all assaying subspaces can be extracted from the books of Gohberg-Krein [GK69], Schatten [Sch70] or Oostenbrink [Oos73].

Section 1.3.

- The notion of algebraic inductive limit is described in Grossmann [114], in the context of nested Hilbert spaces (see also Section 2.4.1, Example (iv)).
- For chains (or scales) of Hilbert or Banach spaces, see Krein-Petunin [132] or Palais [162]; for nested Hilbert spaces, see Grossmann [114]; for equipped Hilbert spaces, see Berezanskii [Ber68]; for rigged Hilbert spaces, see Gel'fand-Vilenkin [GV64].

Section 1.4.

- Generating subsets of $\mathcal{F}(V, \#)$ are called *rich* in [12].
- The summations made in the two counterexamples may be performed with the use of Gradshteyn-Ryzhik [GR65].

Chapter 2
General Theory: Topological Aspects

In Chapter 1, we have analyzed the structure of PIP-spaces from the algebraic point of view only, (i.e., the compatibility relation). Here we will discuss primarily the topological structure given by the partial inner product itself. The aim is to tighten the definitions so as to eliminate as many pathologies as possible. The picture that emerges is reassuringly simple: Only two types of PIP-spaces seem sufficiently regular to have any practical use, namely lattices of Hilbert spaces (LHS) or Banach spaces (LBS), that we have introduced briefly in the Introduction.

Our standard reference on locally convex topological vector spaces (LCS) will be the textbook of Köthe [Köt69]. In addition, for the convenience of the reader, we have collected in Appendix B most of the necessary, but not so familiar, notions needed in the text. Notice that we diverge from [Köt69] for the notation. Namely, for a given dual pair $\langle E, F \rangle$, we denote the weak topology on E by $\sigma(E, F)$, its Mackey topology by $\tau(E, F)$, and its strong topology by $\beta(E, F)$ (see Appendix B).

2.1 Topologies on Dual Pairs of Assaying Subspaces

Let $(V, \#, \langle \cdot | \cdot \rangle)$ be a nondegenerate PIP-space. In this section we will focus our attention to a single dual pair $\langle V_r, V_{\overline{r}} \rangle$ of assaying subspaces. (In fact, we should speak of a *conjugate dual* pair, but it makes no difference; see Schwartz [175] for a full discussion). What are the possible topologies on $V_r, V_{\overline{r}}$?

From the compatibility (or lattice) point of view, there is perfect symmetry between V_r and $V_{\overline{r}}$. In any natural scheme, this feature should be preserved at the topological level as well. This means that the topology $t(V_r)$ of V_r must be such that the dual of V_r is precisely $V_{\overline{r}}$, that is, $t(V_r)$ is a topology of the dual pair $\langle V_r, V_{\overline{r}} \rangle$. Therefore $t(V_r)$ must be finer that the weak topology and coarser than the Mackey topology:

$$\sigma(V_r, V_{\overline{r}}) \prec t(V_r) \prec \tau(V_r, V_{\overline{r}}).$$

J.-P. Antoine and C. Trapani, *Partial Inner Product Spaces:*
Theory and Applications, Lecture Notes in Mathematics 1986,
DOI 10.1007/978-3-642-05136-4_2, © Springer-Verlag Berlin Heidelberg 2009

Assumption: From now on, we will assume that every assaying subspace V_r (including $V^{\#}$ and V) carries the finest topology of the dual pair, i.e., the Mackey topology $\tau(V_r, V_{\overline{r}})$.

When needed, we will write $V_r|_{\tau}$ or even $V_r[\tau(V_r, V_{\overline{r}})]$ if a danger of confusion arises. From this choice (perfectly symmetric with respect to $V_r, V_{\overline{r}}$), it follows that

(i) whenever $V_p \subset V_q$, the injection $E_{qp} : V_p \to V_q$ is continuous and has dense range;

(ii) $V^{\#}$ is dense in every V_r, and every V_r is dense in V.

However, that is not sufficient for eliminating all pathologies. For practical purposes, indeed, the Mackey topology $\tau(V_r, V_{\overline{r}})$ is rather awkward, or at least unfamiliar, unless it coincides with a norm or a metric topology. If a locally convex space $E[\mathsf{t}]$ with topology t is metrizable, then t coincides with the canonical Mackey topology on E, i.e., $\mathsf{t} = \tau(E, E')$ (but not necessarily with the strong topology $\beta(E, E')$ if $E[\mathsf{t}]$ is not complete). Let us give two examples to emphasize the point.

(i) Take the dual pair $\langle \varphi, \ell^2 \rangle$, with respect to the ℓ^2 inner product (as usual, φ denotes the space of all finite sequences); then $\tau(\varphi, \ell^2)$ is the ℓ^2-norm topology on φ, but it is coarser than $\beta(\varphi, \ell^2)$, whereas $\tau(\ell^2, \varphi)$ is not a norm topology (otherwise both spaces would be complete) [Köt69, Sec. 21.5].

(ii) Take $\langle \ell^1, \ell^\infty \rangle$. Then $\tau(\ell^1, \ell^\infty)$ coincides with $\beta(\ell^1, \ell^\infty)$ and the ℓ^1-norm topology, but $\tau(\ell^\infty, \ell^1)$ is weaker than the ℓ^∞-norm topology, and, indeed, is not metrizable.

The origin of the difficulty is clear: $\varphi[\tau(\varphi, \ell^2)]$ is a noncomplete normed space, whereas $\ell^1[\tau(\ell^1, \ell^\infty)]$ is a nonreflexive Banach space. Such pathologies are avoided if the dual pair $\langle V_r, V_{\overline{r}} \rangle$ is *reflexive*, i.e., if the dual of $V_r|_{\beta}$ coincides with $V_{\overline{r}}$ and vice versa. This is indeed equivalent with either (hence both) $V_r|_{\tau}$ or $V_{\overline{r}}|_{\tau}$ being reflexive (a locally convex space E is called *reflexive* if it coincides (topologically) with the strong dual of its strong dual; see Appendix B for further details, as well as for other, equivalent, characterizations of reflexive dual pairs). In addition, each space of a reflexive dual pair is quasi-complete (i.e., closed bounded sets are complete) for the weak, the Mackey and the strong topology (the last two in fact coincide). Typical instances of reflexive dual pairs are the following:

(i) V_r is a Hilbert space; so is then $V_{\overline{r}}$;

(ii) V_r is a *reflexive* Banach space; so is then $V_{\overline{r}}$;

(iii) V_r is a *reflexive* Fréchet space; $V_{\overline{r}}$ is then a reflexive complete (DF)-space;

(iv) V_r is a Montel space; so is then $V_{\overline{r}}$.

As we shall see in the sequel, these cases cover already most spaces of practical interest, in particular all spaces of distributions. Indeed, the Schwartz space of $\mathcal{S}(\mathbb{R})$ of smooth functions of fast decay is a reflexive Fréchet space and the space $\mathcal{D}(\mathbb{R})$ of smooth test functions of compact support is a Montel space.

Actually cases (i) and (ii) play a special role in the theory, for they have particularly nice properties; we will study them systematically in Section 2.2 below. What distinguishes them from the others is metrizability. Indeed:

Proposition 2.1.1. *Let $\langle V_r, V_{\bar{r}} \rangle$ be a reflexive dual pair. Then $V_r|_\tau$ and $V_{\bar{r}}|_\tau$ are reflexive Banach spaces if and only if they are both metrizable.*

Proof. The "only if" part if obvious. Let $V_r|_\tau$ and $V_{\bar{r}}|_\tau$ be metrizable. By reflexivity, they are strong duals of each other, since $\tau(V_r, V_{\bar{r}}) = \beta(V_r, V_{\bar{r}})$ and $\tau(V_{\bar{r}}, V_r) = \beta(V_{\bar{r}}, V_r)$. Hence they are both normable, since the strong dual of a metrizable locally convex space can only be metrizable if both are normable. Finally V_r and $V_{\bar{r}}$ are both normed spaces and reflexive, thus they are both Banach spaces. ∎

We conclude this section with a small proposition that will be useful in the sequel. The easy proof is left to the reader.

Proposition 2.1.2. *Let $\langle V_r, V_{\bar{r}} \rangle$ be a dual pair in V. If L is a vector subspace of V_r, then its closure for $\tau(V_r, V_{\bar{r}})$ is $(L^\perp \cap V_{\bar{r}})^\perp \cap V_r$. In particular, L is dense in V_r if, and only if, $L^\perp \cap V_{\bar{r}} = \{0\}$.*

2.2 Interplay Between Topological and Lattice Properties: The Banach Case

First, we restrict our attention to pairs of Banach spaces. Let $\langle V_p, V_{\bar{p}} \rangle$, $\langle V_q, V_{\bar{q}} \rangle$ be two reflexive dual pairs of assaying subspaces, consisting of (reflexive) Banach spaces. What can be said about the pair $\langle V_{p \wedge q}, V_{\bar{p} \vee \bar{q}} \rangle$? By definition, $V_{p \wedge q}$ is the vector space $V_p \cap V_q$, and $V_{\bar{p} \vee \bar{q}}$ is $(V_{\bar{p}} + V_{\bar{q}})^{\#\#}$, which *a priori* could be larger that $V_{\bar{p}} + V_{\bar{q}}$ (usually they coincide, see Proposition 2.3.2 below). In order to clarify the situation, we will introduce two auxiliary spaces, using a standard construction from interpolation theory. The resulting structure is the one described (without proof) in Section I.2.1.

Let X_a, X_b be two Banach spaces, with norms $\| \cdot \|_a, \| \cdot \|_b$, respectively, and assume they are both continuously embedded in a Hausdorff LCS X (two such Banach spaces are called an *interpolation couple*). The direct sum $X_a \oplus X_b$ of X_a, X_b can be made into a Banach space with the norm

$$\|(f, g)\| = \|f\|_a + \|g\|_b \ (f \in X_a, g \in X_b).$$

We consider the subspace $X_{[a,b]}$ of $X_a \oplus X_b$ which consists of all pairs of the form $(f, -f)$ for some $f \in X_a \cap X_b$. This subspace is obviously isomorphic to $X_a \cap X_b$ and it is closed in $X_a \oplus X_b$. With the induced topology, we will denote $X_{[a,b]}$ by $(X_a \cap X_b)_{\text{proj}}$. Indeed, the induced topology is precisely the projective limit of the two norm (= Mackey) topologies on X_a, X_b. Thus $(X_a \cap X_b)_{\text{proj}}$ is again a Banach space, with norm:

$$\|f\|_{[a,b]} := \|f\|_a + \|f\|_b \qquad (f \in X_a \cap X_b). \tag{2.1}$$

Next we define the quotient $X_{(a,b)} := (X_a \oplus X_b)/X_{[a,b]}$. As a vector space $X_{(a,b)}$ is isomorphic to the vector sum $X_a + X_b$. Equipped with the quotient topology, $X_{(a,b)}$ will be denoted by $(X_a + X_b)_{\text{ind}}$, for it is precisely the inductive limit of X_a, X_b with respect to the identity mappings $X_a \to X_a + X_b, X_b \to X_a + X_b$. It is again a Banach space, with norm

$$\|f\|_{(a,b)} := \inf_{f=g+h} (\|g\|_a + \|h\|_b), \tag{2.2}$$

where the infimum is taken over all possible decompositions $f = g + h, g \in X_a, h \in X_b$; such decompositions are nonunique as soon as $X_a \cap X_b \neq \{0\}$ (the expression (2.2) indeed defines a norm, as a consequence of the continuous embedding of X_a, X_b into X).

Proposition 2.2.1. *Let X_a, X_b be two Banach spaces, both continuously embedded in a Hausdorff LCS X. Then:*

(i) *The two spaces $(X_a \cap X_b)_{\text{proj}}$ and $(X_a + X_b)_{\text{ind}}$ are Banach spaces and the following inclusions hold, where $\hookrightarrow$ denotes a continuous injection:*

$$(X_a \cap X_b)_{\text{proj}} \hookrightarrow \left\{ \begin{matrix} X_a \\ X_b \end{matrix} \right\} \hookrightarrow (X_a + X_b)_{\text{ind}}.$$

(ii) *The norms $\|.\|_a$ and $\|.\|_b$ are consistent on $X_a \cap X_b$: if $\{f_n\} \in X_a \cap X_b$ is Cauchy in both norms and $f_n \to 0$ in X_a, then $f_n \to 0$ in X_b also, and vice-versa.*

Proof. Part (i) is clear from the discussion above. As for (ii), let $X_{ab|a}(\text{resp.}X_{ab|b})$ be the image of $X_a \cap X_b$ in $X_a(\text{resp. } X_b)$ under the identity map. Denote by E_{ba} the identity map $X_{ab|a} \to X_{ab|b}$ The graph of E_{ba} is exactly the set $X_{[a,b]}$, which is closed in $X_a \oplus X_b$; thus E_{ba} is a closed map, which means precisely that the norms $\|.\|_a$ and $\|.\|_b$ are consistent on $X_a \cap X_b$.   ∎

Remarks:

(i) If X_a and X_b are reflexive, so are $(X_a \cap X_b)_{\text{proj}}$ and $(X_a + X_b)_{\text{ind}}$.
(ii) If X_a and X_b are Hilbert spaces, the same construction goes through, using squared norms everywhere (see also Section I.2).
(iii) The construction above and Proposition 2.2.1 remain valid if X_a, X_b are assumed to be Fréchet spaces.

(iv) It is shown in [RS75, Appendix to IX. 4] that the norms $\|.\|_a$ and $\|.\|_b$ are consistent on $X_a \cap X_b$ if and only if the expression $\|.\|_{(a,b)}$ is a norm on $X_a \cap X_b$ and the identity map on $X_a \cap X_b$ extends to continuous injections of $\widehat{X}_a$, resp. $\widehat{X}_b$, into $\widehat{X}_{(a,b)}$, where these three spaces are defined as the completion of $X_a \cap X_b$ under $\|.\|_a$, $\|.\|_b$ and $\|.\|_{(a,b)}$, respectively. When $X_a \cap X_b$ is dense in X_a and in X_b, then $\widehat{X}_a = X_a, \widehat{X}_b = X_b$, and $\widehat{X}_{(a,b)} = (X_a + X_b)_{\text{ind}}$.

Let $X_{\overline{a}}, X_{\overline{b}}$ be the conjugate duals of X_a, X_b respectively. We assume now that $X_a \cap X_b$ is dense in X_a and in X_b. It follows that $X_{\overline{a}}$ and $X_{\overline{b}}$ can be embedded into $(X_a \cap X_b)^{\times}_{\text{proj}}$, i.e., $\{X_{\overline{a}}, X_{\overline{b}}\}$ is also an interpolation couple. Then we have:

Lemma 2.2.2. *Let X_a, X_b be as above and assume that $X_a \cap X_b$ is dense in X_a and in X_b. Then:*

(i) The conjugate dual of $(X_a \cap X_b)_{\text{proj}}$ is $X_{\overline{a}} + X_{\overline{b}}$.
(ii) The conjugate dual of $(X_a + X_b)_{\text{ind}}$ is $X_{\overline{a}} \cap X_{\overline{b}}$.

Proof. The proof of both assertions results from the following two observations:

(i) The conjugate dual of $X_a \oplus X_b$ is $X_{\overline{a}} \oplus X_{\overline{b}}$.
(ii) The closed subspace $X_{[\overline{a},\overline{b}]}$ of $X_{\overline{a}} \oplus X_{\overline{b}}$ is isomorphic, as a vector space, to the orthogonal complement of $X_{[a,b]}$ in $X_{\overline{a}} \oplus X_{\overline{b}}$, the isomorphism being

$$J = \begin{pmatrix} 1 & 0 \\ 0 & -1 \end{pmatrix}.$$

∎

We return now to the nondegenerate PIP-space V and the two pairs of Banach assaying subspaces $\langle V_p, V_{\overline{p}} \rangle$, $\langle V_q, V_{\overline{q}} \rangle$ (they are necessarily reflexive, by Proposition 2.1.1, but the whole discussion that follows is independent of this fact). All four assaying subspaces are continuously embedded into V, and $V_p \cap V_q$ is dense in V_p and V_q. Hence $\{V_p, V_q\}$ and $\{V_{\overline{p}}, V_{\overline{q}}\}$ are interpolation couples and the construction above goes through. Thus, we get the following scheme (and the corresponding one for the duals, taking Lemma 2.2.2 into account), where all injections are continuous and have dense range:

$$V_{p \wedge q}|_\tau \hookrightarrow (V_p \cap V_q)_{\text{proj}} \hookrightarrow \left\{ \begin{matrix} V_p \\ V_q \end{matrix} \right\} \hookrightarrow (V_p + V_q)_{\text{ind}} \hookrightarrow V_{p \vee q}|_\tau.$$

It will prove useful to introduce the following conditions for the couple V_p, V_q:

- (proj) $V_{p \wedge q}|_\tau \simeq (V_p \cap V_q)_{\text{proj}}$ (isomorphism of topological vector spaces).
- (add) $V_{\overline{p} \vee \overline{q}} = V_{\overline{p}} + V_{\overline{q}}$ (as vector spaces).
- (ind) $V_{\overline{p} \vee \overline{q}}|_\tau \simeq (V_{\overline{p}} + V_{\overline{q}})_{\text{ind}}$ (isomorphism of topological vector spaces).

Proposition 2.2.3. *Let V_p, V_q be two Banach assaying subspaces. Then all three conditions* (proj), (add), (ind) *are equivalent.*

Proof. Consider first the equivalence (add) $\Leftrightarrow$ (ind). Let $V_{\overline{p} \vee \overline{q}} = V_{\overline{p}} + V_{\overline{q}}$. Then we have $\langle V_{p \wedge q}, V_{\overline{p} \vee \overline{q}} \rangle = \langle V_p \cap V_q, V_{\overline{p}} + V_{\overline{q}} \rangle$. Since Mackey topologies are inherited both by direct sums and by quotients, it follows that $\tau(V_{\overline{p}} + V_{\overline{q}}, V_p \cap V_q)$ is the quotient topology inherited from $V_{\overline{p}} \oplus V_{\overline{q}}$, that is, the inductive topology on $V_{\overline{p}} + V_{\overline{q}}$. This means that $V_{\overline{p} \vee \overline{q}}|_\tau \simeq (V_{\overline{p}} + V_{\overline{q}})_{\mathrm{ind}}$. The converse assertion is obvious.

As for the implication (proj) $\Rightarrow$ (add), we note that the dual of $V_{p \wedge q}|_\tau$ is, by definition, $V_{\overline{p} \vee \overline{q}}$. Then the result follows from Lemma 2.2.2.

Finally we show that (add) implies (proj). V_p and V_q being Banach spaces, so is $V_p \cap V_q$ under the projective topology. Therefore it carries its Mackey topology, namely, $\tau(V_p \cap V_q, (V_p \cap V_q)^\times_{\mathrm{proj}}) = \tau(V_p \cap V_q, V_{\overline{p}} + V_{\overline{q}}))$ which is $\tau(V_{p \wedge q}, V_{\overline{p} \vee \overline{q}})$ by the condition (add).   ∎

Remark: Proposition 2.2.3 remains true if V_p, V_q are only assumed to be Fréchet spaces. Actually reflexivity has *not* been assumed, although it will hold in most cases. Note also that, although the three conditions above are equivalent in this case, they are *not* necessarily satisfied.

2.3 Interplay Between Topological and Lattice Properties: The General Case

Let now $\langle V_p, V_{\overline{p}} \rangle, \langle V_q, V_{\overline{q}} \rangle$ be two arbitrary dual pairs of assaying subspaces. What can one say about the pair $\langle V_{p \wedge q}, V_{\overline{p} \vee \overline{q}} \rangle$?

We proceed exactly as in the Banach case, constructing first the auxiliary pair $\langle V_p \cap V_q, V_{\overline{p}} + V_{\overline{q}} \rangle$ with the projective and inductive topologies, respectively. The dual pair $\langle V_p \oplus V_q, V_{\overline{p}} \oplus V_{\overline{q}} \rangle$ is again reflexive. The subspace $V_{[p,q]} := V_p \cap V_q$ is closed in $V_p \oplus V_q$, as before. With the induced topology, we denote it again by $(V_p \cap V_q)_{\mathrm{proj}}$ for the same reason. Similarly, $(V_{\overline{p}} + V_{\overline{q}})_{\mathrm{ind}}$ is the quotient of $V_{\overline{p}} \oplus V_{\overline{q}}$ by its closed subspace $V_{\overline{p}} \cap V_{\overline{q}}$, with the quotient, i.e., inductive, topology. Thus, we have again the following continuous embeddings:

$$V_{p \wedge q}|_\tau \hookrightarrow (V_p \cap V_q)_{\mathrm{proj}},$$

$$(V_{\overline{p}} + V_{\overline{q}})_{\mathrm{ind}} \hookrightarrow V_{\overline{p} \vee \overline{q}}|_\tau.$$

We can now repeat part, but not all, of Proposition 2.2.3, namely:

Proposition 2.3.1. *Let $\langle V_p, V_{\overline{p}} \rangle, \langle V_q, V_{\overline{q}} \rangle$ be two arbitrary dual pairs of assaying subspaces. Then, the following implications are true:*

$$(\mathrm{proj}) \Rightarrow (\mathrm{add}) \Leftrightarrow (\mathrm{ind}).$$

The difference with the Banach case lies in the fact that Mackey topologies are inherited by direct sums and by quotients, hence by inductive limits, but they are *not* inherited by subspaces (unless these are metrizable or dense), hence they are in general not inherited by projective limits. In other words, the implication (add) $\Rightarrow$ (proj) holds in the general case if $(V_p \cap V_q)_{\mathrm{proj}}$ is metrizable, and then all three conditions (proj), (add), and (ind) are equivalent, but this need not always be the case.

Condition (add) is particularly interesting in the theory of operators, sketched in Section I.3 and fully developed in Chapter 3. In a nutshell, if (add) holds for sufficiently many assaying subspaces (elements of an involutive covering, in the terminology of Section 1.3), then the domain of every operator is a vector subspace of V, which need not be true in general. By Proposition 2.3.1, it suffices to require that (proj) hold. A convenient criterion is the following.

Proposition 2.3.2. *Let V_p, V_q and $V_{p \wedge q}|_\tau$ be complete metrizable (i.e., Fréchet) spaces. Then $V_{p \wedge q}|_\tau$ is isomorphic to $(V_p \cap V_q)_{\mathrm{proj}}$, that is, (proj) holds.*

Proof. As vector spaces, $V_{p \wedge q} = V_p \cap V_q$. Hence $V_p \cap V_q$ carries two distinct topologies for which it is metrizable and complete, namely, its Mackey topology (by assumption) and the projective topology. The latter is coarser; hence they are equivalent [Köt69, Sec. 15.12(7)]. ∎

Corollary 2.3.3. *Let V_p, V_q be reflexive Banach spaces, and assume that $V_{p \wedge q}|_\tau$ is complete and metrizable. Then $\langle V_{p \wedge q}, V_{\overline{p} \vee \overline{q}} \rangle = \langle V_p \cap V_q, V_{\overline{p}} + V_{\overline{q}} \rangle$ is a reflexive dual pair of Banach spaces.*

Next we assume the two pairs $\langle V_p, V_{\overline{p}} \rangle, \langle V_q, V_{\overline{q}} \rangle$ to be reflexive and look at the auxiliary pair $\langle V_p \cap V_q, V_{\overline{p}} + V_{\overline{q}} \rangle$. We note that $(V_p \cap V_q)_{\mathrm{proj}}$ is semi-reflexive, as a closed subspace of the reflexive space $V_p \oplus V_q$ [Köt69, Sec. 23.3]. Consequently its *strong* dual $(V_{\overline{p}} + V_{\overline{q}})_\beta$ is barreled (see Appendix B). Hence we get the following picture (all topologies refer to the dual pair $\langle V_p \cap V_q, V_{\overline{p}} + V_{\overline{q}} \rangle$):

$$(V_p \cap V_q)_\beta \hookrightarrow (V_p \cap V_q)_\tau \hookrightarrow (V_p \cap V_q)_{\mathrm{proj}}$$

$$(V_{\overline{p}} + V_{\overline{q}})_\beta = (V_{\overline{p}} + V_{\overline{q}})_{\mathrm{ind}} = (V_{\overline{p}} + V_{\overline{q}})_\tau$$

In general nothing more can be said. It is quite possible that $(V_p \cap V_q)_\tau$ be semi-reflexive but not reflexive, and $V_{\overline{p}} + V_{\overline{q}}$ not even semi-reflexive [Köt69, Sec. 23.6]. We will exhibit an example below. Of course this cannot happen if V_p, V_q are Fréchet spaces.

As for the pair $\langle V_{p \wedge q}, V_{\overline{p} \vee \overline{q}} \rangle$, no general conclusion can be drawn, since the right-hand side depends explicitly (as a vector space) on the compatibility and cannot be characterized *a priori* when condition (add) fails. This again suggests that the structure of PIP-space that we have used so far is too general.

What about the pair $\langle V^\#, V \rangle$ itself ? Here one more piece of information is available, namely, a (generalized) condition (add), since we have

$$V^\# = \bigcap_{V_r \in \mathcal{F}} V_r, \qquad V = \sum_{V_r \in \mathcal{F}} V_r$$

Each V_r is, as usual, assumed to carry its Mackey topology $\tau(V_r, V_{\overline{r}})$. Then $V^\#$ carries three natural topologies: the strong topology $\beta(V^\#, V)$, the Mackey topology $\tau(V^\#, V)$ and the projective limit of all the $\tau(V_r, V_{\overline{r}})$ defined exactly as in the previous section. All three are in general distinct, but the last two give the same dual, namely, V itself, whereas $V^\#|_\beta$ could have a larger dual. V also carries three natural topologies, namely, $\beta(V, V^\#), \tau(V, V^\#)$, and the inductive limit of the $\tau(V_r, V_{\overline{r}})$ but the last two always coincide since Mackey topologies are inherited by inductive limits. Thus the general picture is the following (with the same notation as above):

$$V^\#|_\beta \hookrightarrow V^\#|_\tau \hookrightarrow V^\#|_{\text{proj}}. \tag{2.3}$$

$$V|_\beta \hookrightarrow V|_{\text{ind}} = V|_\tau. \tag{2.4}$$

Of course, this by no means implies that the pair $\langle V^\#, V \rangle$ be reflexive, since $V|_\tau$ need not even be semi-reflexive. We will come back to these problems in Chapter 5.

We will now conclude this section with an example, taken again from [Köt69, Sec. 30.4], which illustrates how bad the situation can be in general.

Example 2.3.4. Let, as usual, $\omega = \Pi_{i=1}^\infty \mathbb{C}_{(i)}$ be the space of all complex sequences, with the product topology, and $\varphi = \sum_{i=1}^\infty \mathbb{C}_{(i)}$ the space of all finite sequences, with the direct sum topology. Then $\langle \omega, \varphi \rangle$ is a reflexive dual pair, where ω is Fréchet, φ is a complete (DF)-space, and both are Montel spaces. Then one considers the space of arbitrary double sequences $(a_{ij}), \omega\omega = \Pi_{i=1}^\infty \omega_{(i)}$; one defines similarly $\omega\varphi = \Pi_{i=1}^\infty \varphi_{(i)}, \varphi\omega = \sum_{i=1}^\infty \omega_{(i)}$ and $\varphi\varphi = \sum_{i=1}^\infty \varphi_{(i)}$. Exactly as ω, the space $\omega\omega$ carries a natural PIP-space structure, namely, $(a_{ij})\#(b_{ij}) \Leftrightarrow \sum_{i,j} |a_{ij}b_{ij}| < \infty$, for which $(\omega\omega)^\# = \varphi\varphi, (\omega\varphi)^\# = \varphi\omega, (\varphi\omega)^\# = \omega\varphi, (\varphi\varphi)^\# = \omega\omega$. The intersection $\varphi\omega \cap \omega\varphi$ coincides with $\varphi\varphi$ as a vector space, hence $(\varphi\omega \cap \omega\varphi)^\# = \omega\omega$. On the other hand, $(\varphi\omega)^\# + (\omega\varphi)^\# = \omega\varphi + \varphi\omega \neq \omega\omega$. Thus condition (add) fails. Condition (proj) fails also: on $\varphi\varphi$ the Mackey topology $\tau(\varphi\varphi, \omega\omega) = \beta(\varphi\varphi, \omega\omega)$ is strictly finer than the projective topology induced by $\varphi\omega \oplus \omega\varphi$. Finally it can be seen that $(\varphi\omega \cap \omega\varphi)_{\text{proj}}$ is semi-reflexive as closed subspace of the reflexive space $\varphi\omega \oplus \omega\varphi$, but not reflexive and not barreled (hence not Montel), whereas $(\omega\varphi + \varphi\omega)_{\text{ind}}$ is barreled but not semi-reflexive, *a fortiori* not Montel [Köt69, Sec. 31.5].

The lesson of the example is clear. The PIP-space ω (or $\omega\omega$, since they are isomorphic) contains a generating lattice $\mathcal{I}$ of Hilbert spaces, the family of all weighted ℓ^2-spaces described before, for which all conditions (proj), (add), (ind) hold (Example 1.4.1). But it also contains bad assaying subspaces $\varphi\omega$ and $\omega\varphi$ for which all three conditions fail and the regularity properties are lost. So why not exclude such pathological assaying subspaces and concentrate instead on the nice generating lattice $\mathcal{I}$? The important point is that nothing is lost in this restriction, since the compatibility is fully recovered from $\mathcal{I}$, by the very definition. In fact, concentrating on a fixed generating family is exactly like describing a topology in terms of a fixed, convenient, basis of neighborhoods instead of considering *explicitly* arbitrary open sets. These considerations motivate the next section.

2.4 Indexed PIP-Spaces

The lesson of Section 2.3 is that no general conclusion can be reached as soon as we go beyond the Banach case, for too many pathologies are possible in the general setup. There is a way out, however. As the examples indeed show, the complete lattice $\mathcal{F}$ of *all* assaying subspaces is in general extremely large. Fortunately, as shown in Chapter 1, the whole structure can be reconstructed from a fairly small subset of $\mathcal{F}$, namely, a generating involutive sublattice $\mathcal{I}$ of $\mathcal{F}(V, \#)$. On such a family of subspaces, it makes sense to impose additional restrictions, typically to contain only topological vector spaces of the same type, as in a LHS/LBS. We will now systematize this idea as a way of eliminating pathologies such as those of the example above.

Definition 2.4.1. By an *indexed partial inner product space*, we mean a triple $(V, \mathcal{I}, \langle\cdot|\cdot\rangle)$, where V is a vector space, $\mathcal{I}$ an involutive covering of V and $\langle\cdot|\cdot\rangle$ a Hermitian form defined on those pairs of vectors of V that are compatible for the associated compatibility $\#_{\mathcal{I}}$, namely,

$$ f \#_{\mathcal{I}} g \iff \exists\, V_r \in \mathcal{I} \text{ such that } (f, g) \in V_r \times V_{\bar{r}}. $$

Equivalently, an indexed PIP-space consists of a PIP-space $(V, \#, \langle\cdot|\cdot\rangle)$ together with a fixed generating involutive sublattice $\mathcal{I}$ of $\mathcal{F}(V, \#)$.

For convenience, we will denote the indexed PIP-space $(V, \mathcal{I}, \langle\cdot|\cdot\rangle)$ simply as V_I where I is the isomorphy class of $\mathcal{I}$, i.e., $\mathcal{I}$ considered as an abstract partially ordered set. Thus $\mathcal{I} = \{V_r, r \in I\}$.

   The only difference between an indexed PIP-space and a generating family in the sense of Definition 1.4.1 is the fact that here the generating family is required to be an involutive sublattice of the lattice of all assaying subspaces. Yet the two concepts are closely related. Given an indexed PIP-space V_I, it defines a unique PIP-space, namely, $(V, \#_{\mathcal{I}}, \langle\cdot|\cdot\rangle)$, with $\mathcal{F}(V, \#_{\mathcal{I}})$ the lattice completion of $\mathcal{I}$. And a PIP-space is a particular indexed PIP-space for which $\mathcal{I}$ happens to be a *complete* involutive lattice.

Remark 2.4.2. (1) Although an indexed PIP-space generates a unique PIP-space, the converse is not true. Let # be a compatibility on V. Then each generating involutive sublattice of $\mathcal{F}(V, \#)$ defines an indexed PIP-space that generates the same PIP-space. Actually, the set of *all* involutive sublattices of $\mathcal{F}$ is itself a complete lattice for the following operations:

$$\mathcal{I}_1 \wedge \mathcal{I}_2 = \mathcal{I}_1 \cap \mathcal{I}_2\,,$$
$$\mathcal{I}_1 \vee \mathcal{I}_2 = \text{sublattice generated by } \mathcal{I}_1 \text{ and } \mathcal{I}_2\,.$$

If $\mathcal{I}_1$ and $\mathcal{I}_2$ are distinct and both generating, $\mathcal{I}_1 \wedge \mathcal{I}_2$ might be nongenerating, even empty [see Remark (2) below], but $\mathcal{I}_1 \vee \mathcal{I}_2$ is *a fortiori* generating, hence corresponds to another indexed PIP-space, that still generates the same PIP-space.

(2) In general an involutive covering of V need not contain the extreme elements of $\mathcal{F}$, namely, $V^{\#}$ and V. These, however, are always implicitly present, since they can be recovered from $\mathcal{F} = \{V_r, r \in I\}$:

$$V^{\#} = \bigcap_{r \in I} V_r, \qquad V = \sum_{r \in I} V_r \tag{2.5}$$

Thus it may happen that the intersection $\mathcal{I}_1 \cap \mathcal{I}_2$ of two generating sublattices $\mathcal{I}_1$ and $\mathcal{I}_2$ is empty, that is, they have no common element, although all elements of $\mathcal{I}_1$ and $\mathcal{I}_2$ contain $V^{\#}$ and verify Eq. (2.5).

(3) The discussion so far in this section is purely algebraic. However, if we assume that every defining subspace $V_r \in \mathcal{I}$ carries a locally convex topology t_r, it seems natural to require that the latter coincide with the Mackey topology $\tau(V_r, V_{\overline{r}})$, in order to be consistent with the involution of $\mathcal{I}$.

(4) As for PIP-spaces, the definition on an indexed PIP-space does *not* require that the partial inner product $\langle \cdot | \cdot \rangle$ be nondegenerate, this is an independent assumption. Although this property is needed for the applications, it is by no means automatic. It might happen that $V^{\#}$ is not dense in every V_r, for instance. Also not every involutive covering $\mathcal{I}$ of V may give rise to a nondegenerate PIP-space, it could be that no nondegenerate partial inner product is compatible with $\mathcal{I}$. We will give examples of this pathology in Section 5.4.

The case of interest is when a given involutive covering $\mathcal{I}$ consists of topological spaces of the same type. However, again the symmetry $r \leftrightarrow \overline{r}$ implies that the only case where this happens is that of Hilbert spaces or reflexive Banach spaces (a space and its dual cannot be both metrizable, unless there are both Banach spaces). Then relations (2.5) will imply better properties for $V^{\#}$ and V, equipped with their Mackey or projective, resp. inductive, topology. Similarly, we will then be able to improve the results of Section 2.3 about a given couple V_p, V_q ($p, q \in I$).

For that purpose it is useful to introduce some additional terminology.

Definition 2.4.3. An indexed PIP-space V_I is said to be:

(i) *additive*, if condition (add) holds throughout I:

$$V_{p \vee q} = V_p + V_q, \ \forall p, q \in I,$$

that is, the supremum in $\mathcal{I}$ coincides with that of $\mathcal{L}(V)$, i.e., $\mathcal{I}$ is a sublattice of $\mathcal{L}(V)$, since the two infimums already coincide.

(ii) *projective* or *tight*, if condition (proj) holds throughout I,

$$V_{p \wedge q}|_\tau \simeq (V_p \cap V_q)_{\mathrm{proj}}, \ \forall p, q \in I;$$

(iii) *reflexive*, if $\langle V_p, V_{\overline{p}} \rangle$ is a reflexive dual pair (for their initial topologies), for every $p \in I$.

As we know already by Proposition 2.3.1, a projective indexed PIP-space is always additive. Next we draw some easy consequences of reflexivity.

Proposition 2.4.4. *Let V_I be a reflexive indexed PIP-space. Then $V^{\#}$, with either its projective topology or its Mackey topology, is semi-reflexive and weakly quasi-complete, and V is barreled.*

Proof. Every V_r is reflexive, *a fortiori* semi-reflexive. $V^{\#}|_{\mathrm{proj}}$ is semi-reflexive as projective limit of semi-reflexive spaces [Köt69, Sec. 23.3], thus $V^{\#}|_\tau$ is also semi-reflexive; semi-reflexivity is equivalent to weak quasi-completeness and also to the (common) Mackey dual $V|_\tau$ being barreled: $V|_{\mathrm{ind}} = V|_\tau = V|_\beta$. ∎

Notice that reflexivity of V_I is not sufficient to imply that $\langle V^{\#}, V \rangle$ be a reflexive dual pair, for $V^{\#}|_\tau$ could still be semi-reflexive and nonreflexive. What is missing is $V^{\#}|_\tau$ being barreled, i.e., $V^{\#}|_\tau = V^{\#}|_\beta$. This difficulty is clearly avoided whenever $V^{\#}$ is metrizable.

Proposition 2.4.5. *Let V_I be reflexive and $V^{\#}|_{\mathrm{proj}}$ be metrizable. Then $\langle V^{\#}, V \rangle$ is a reflexive dual pair, with $V^{\#}$ a Fréchet space.*

Proof. The assumptions imply that $V^{\#}|_{\mathrm{proj}}$ is both semi-reflexive and metrizable. That is possible only if it is complete, hence a Fréchet space. A semi-reflexive Fréchet space is necessarily reflexive, and so is its strong dual. ∎

For the case described in Proposition 2.4.5, all three topologies of Eq. (2.3) coincide on $V^{\#}$, and similarly for V, Eq. (2.4); in addition, both spaces are complete. A typical and important example (see Section I.1) is the couple of Schwartz spaces $\langle \mathcal{S}, \mathcal{S}^{\times} \rangle$. However, from the fact that $\langle V^{\#}, V \rangle$ is a reflexive dual pair, we can conclude only that $V^{\#}|_\tau$ is barreled, i.e., $V^{\#}|_\tau = V^{\#}|_\beta$; the projective topology on $V^{\#}$ could still be coarser. Such was the case in the reflexive pair $\langle \varphi \varphi, \omega \omega \rangle$ in Example 2.3.4. Also $V^{\#}|_\tau$ and $V|_\tau$ are then quasi-complete, but not necessarily complete. Fortunately, quasi-completeness of

$V|_\tau$ is sufficient for the two arguments where a completeness result is needed: The existence of a central Hilbert space, to be discussed in Section 2.5, and the identification of the algebra of regular operators (these will be defined in Chapter 3) with an algebra of unbounded operators in that Hilbert space, discussed in Section 3.3.1.

Thus reflexivity of an indexed PIP-space is not sufficient by itself. Actually, as we have already noticed above, the discussion of Sections 2.1 and 2.2 shows that, among reflexive dual pairs, those consisting of reflexive Banach spaces (in particular, Hilbert spaces) are the only ones that are really compatible with the lattice structure, if one insists that all the defining subspaces of $\mathcal{I}$ be of the same type. Hence it is worthwhile to give a separate name to the corresponding indexed PIP-spaces. Of course, there is nothing new here, we have simply recovered in the end the notions of LHS and LBS introduced in the Introduction. However, at this point we have to make an important remark about terminology.

If X is a Banach space and the norm is replaced by an equivalent norm, then the norm on the conjugate dual $X^\times$ is also changed only to within equivalence. In other words, if X is the underlying topological space of a Banach space, then $X^\times$ is a well-defined locally convex vector space (LCS). In that case, we will say that X is a *complete normable LCS*. A Banach space always refers to a specific norm. In the same way, we say that a LCS X is *hilbertian* if it is the underlying LCS of a Hilbert space. Equivalently, a hilbertian space is an equivalence class of Hilbert spaces. By contrast, when we speak of a *Hilbert* space H, we mean a hilbertian space with a specific norm and inner product, the (conjugate) dual $H^\times$ carrying the dual (or conjugate) norm.

First we consider only the underlying topological spaces and introduce the corresponding structures.

Definition 2.4.6. A reflexive indexed PIP-space V_I is said to be of *type* (B) if every $V_r, r \in I$, is a reflexive complete normable space; of *type* (H), if every $V_r, r \in I$ is a hilbertian space.

We emphasize that requiring the family $V_r, r \in I$ to be a *lattice* of complete normable or hilbertian spaces implicitly means that the intersection $V_p \cap V_q$ of two elements of $\mathcal{I}$ must carry its projective topology. Therefore Definition 2.4.6 implies the following.

Proposition 2.4.7. *Let V_I be an indexed PIP-space of type (B) or (H). Then,*

(i) V_I *is projective, hence additive, and each pair $V_p, V_q, p, q \in I$, is an interpolation couple.*

(ii) *If, in addition, I is countable, $V^\#|_{\text{proj}}$ is metrizable and $\langle V^\#, V \rangle$ is a reflexive dual pair.*

Proof. Part (i) results from Propositions 2.2.3 and 2.3.2. As for (ii), it follows from the fact that the projective limit of a countable family of complete normable spaces is a Fréchet space, and Proposition 2.4.5.   ∎

We will conclude this section with a some examples of indexed PIP-spaces of type (H) or (B). Many more will be studied in Chapters 4 and 8. Actually, every concrete example is given by specifying the Hilbert or Banach spaces that constitute them, that is, with explicit norms and inner products. Thus one has to verify that (1) the norm $\| \cdot \|_{\bar{r}}$ on $V_{\bar{r}} = V_r^\times$ is the conjugate of the norm $\| \cdot \|_r$ on V_r; and (2) the norm $\| \cdot \|_{p \wedge q}$ on $V_p \cap V_q$ is equivalent to the projective norm (I.4), resp. (I.7) . Then (1) and (2) imply that the norm $\|\cdot\|_{p \vee q}$ on $V_p + V_q$ is equivalent to the inductive norm (I.5), resp. (I.8). This happens, for instance, in the case of the weighted spaces $\ell^2(r)$ in ω or $L^2(r)$ in L^1_{loc}.

Thus we define more restricted concepts.

Definition 2.4.8.(1) A *lattice of Banach spaces (LBS)* is a nondegenerate indexed PIP-space of type (B) V_I, with positive definite partial inner product, such that

  (b$_1$) each element V_r of $\mathcal{I}$ is a Banach space with norm $\| \cdot \|_r$;
  (b$_2$) $\mathcal{I}$ contains a unique self-dual element $V_o = V_{\bar{o}}$.
  (b$_3$) for every $V_r \in \mathcal{I}$, the norm $\| \cdot \|_{\bar{r}}$ on $V_{\bar{r}} = V_r^\times$ is the conjugate of the norm $\| \cdot \|_r$ on V_r.
  (b$_4$) for every pair $V_p, V_q \in \mathcal{I}$, the norm $\| \cdot \|_{p \wedge q}$ on $V_p \cap V_q$ is equivalent to the projective norm.

(2) A *lattice of Hilbert spaces (LHS)* is a LBS such that

  (h) each element V_r of $\mathcal{I}$ is a Hilbert space with respect to the inner product $\langle \cdot | \cdot \rangle_r$.

(3) In particular, a LHS is called a *chain of Hilbert spaces* if the order of $\mathcal{I}$ is total. The chain is *discrete* if $\mathcal{I}$ is countable and *continuous* if $\mathcal{I}$ has the same cardinality as $\mathbb{R}$.

In other words, a LHS is a nondegenerate indexed PIP-space of type (H) that verifies the five conditions (b$_1$)-(b$_4$) and (h) (the transition from LHS to indexed PIP-space of type (H) can be done in the language of categories, using the so-called forgetful functor; the same thing applies to the Banach case). Notice that the projective norm on an intersection $V_p \cap V_q$ is usually defined as (I.4) in the LHS case (so as to have a Hilbert norm), and as (I.7) in the LBS case.

It is worth noting that, for a LBS, the condition of positive definiteness implies that $V_o = V_{\bar{o}}$ is a Hilbert space, on which the norm $\|\cdot\|_o$ coincides both with the conjugate norm and with the so-called pip-norm $\|f\| := \langle f | f \rangle^{1/2}$. This will result from Proposition 2.5.4 below.

Another situation may arise, however. Namely, there are cases where only a generating family of subspaces of V_I is given explicitly, so that the lattice has to be (re)constructed 'by hand'. Examples will be given in Chapter 4 (the lattice generated by the spaces $L^p(\mathbb{R})$, e.g.) and Chapter 5 (the LHS generated by a family of operators or quadratic forms). In such a case, reflexivity implies that the norm on $V_{\overline{r}} = V_r^\times$ must be the conjugate of the norm on V_r. Then, one has to give an explicit norm on every intersection and direct sum. The simplest solution, that we adopt henceforth, is to require that the norm on $V_p \cap V_q$ be the projective norm (I.4) or (I.7). Then $V_{p\vee q} = V_p + V_q$ automatically carries the inductive norm (I.5), resp. (I.8). This choice has the further advantage of recovering the initial PIP-space, if it was known beforehand.

In that case, we may also speak of a lattice of *Hilbert spaces* or LHS. Otherwise, we call the structure a lattice of *hilbertian spaces*. Similar considerations apply, of course, in the Banach case.

2.4.1 Examples: Indexed PIP-Spaces of Type (H)

(i) Chains of hilbertian or Hilbert spaces

The concept of a chain of hilbertian spaces obviously fits in here: A chain $\{V_k, k \in I = \mathbb{Z} \text{ or } \mathbb{R}\}$ of Hilbert spaces, such that $V_k \subset V_l$ for $k > l$ with continuous injection, V_{-k} being the conjugate dual of V_k, a central Hilbert space $V_0 = V_0^\times$ and the requirement that $\bigcap_{k \in I} V_k$ be dense in every V_k. Typical examples of this structure are the Sobolev spaces, the space $s^\times$ of slowly increasing sequences and the space $\mathcal{S}^\times$ of tempered distributions, for instance in Bargmann's realization $\mathcal{E}^\times = \lim \text{ind}_{\rho \in \mathbb{R}} \mathcal{F}^\rho$ (actually each of these spaces is in fact a LHS, since explicit norms are provided). In the continuous case ($k \in \mathbb{R}$), the projective topology on $V^\#$ may clearly be defined by a cofinal countable subset of $\mathbb{R}$, such as $\mathbb{Z}$, so that $V^\#|_{\text{proj}}$ is still metrizable and Proposition 2.4.5 applies. These chains, as well as the corresponding PIP-spaces, have been discussed in Chapter 1.

(ii) Sequence spaces: Weighted ℓ^2-spaces

Of course, we recover here the familiar lattice of weighted ℓ^2-spaces $\{\ell^2(r)\}$, described in Section I.2.3 (i). All conditions are manifestly satisfied, and V_I is a LHS, in particular it is projective and additive.

(iii) Spaces of locally integrable functions: Weighted L^2-spaces

A similar statement holds true in $L^1_{\text{loc}}(X, \mu)$ for the family

$$\mathcal{I} = \{L^2(r) : r \text{ is } \mu\text{-measurable and a.e. positive, } r \text{ and } r^{-1} \in L^2_{\text{loc}}(X, d\mu).\}$$

This LHS is described in Section I.2.3 (ii).

An interesting subspace of the preceding space is the LHS V_γ generated by the weight functions $r_\alpha(x) = \exp \alpha x$, for $-\gamma \leqslant \alpha \leqslant \gamma \, (\gamma > 0)$. Then all the spaces of the lattice may be obtained by interpolation from $L^2(r_{\pm\gamma})$, and moreover, the extreme spaces are themselves Hilbert spaces, namely

$$V_\gamma^\# = L^2(\mathbb{R}, e^{-\gamma x} dx) \cap L^2(\mathbb{R}, e^{\gamma x} dx) \simeq L^2(\mathbb{R}, e^{\gamma |x|} dx)$$
$$V_\gamma = L^2(\mathbb{R}, e^{-\gamma x} dx) + L^2(\mathbb{R}, e^{\gamma x} dx) \simeq L^2(\mathbb{R}, e^{-\gamma |x|} dx).$$

This LHS plays an interesting role in scattering theory (see Section 7.2).

(iv) Nested Hilbert spaces

For the sake of completeness, it is worth reproducing here the original definition of Grossmann [114]. Consider the triple $V_I = [V_r; E_{sr}; I]$, where

(1) I is an index set, directed to the right, with an order-reversing involution $r \leftrightarrow \bar{r}$ and a self-dual element $o = \bar{o}$.
(2) For each $r \in I$, V_r is a Hilbert space.
(3) For every $p \in I$ and every $q \geq p$, E_{qp} is an injective linear mapping (called a *nesting*) from V_p into V_q, such that E_{pp} is the identity on V_p for every $p \in I$ and $E_{qp} = E_{qr} E_{rp}$ whenever $q \geq r \geq p$.
(4) V_I is the algebraic inductive limit of the family $\{V_r, r \in I\}$ with respect to the nestings E_{qp}; this means the following: in the disjoint union $\bigcup_{r \in I} V_r$, define an equivalence relation by writing $f_p \sim f_r$, $(f_p \in V_p, f_r \in V_r)$ if there is an $s \geq p, r$ such that $E_{sp} f_p = E_{sr} f_r$; then the set of equivalence classes is a vector space, noted V_I. One notes E_{Ir} the natural embedding of V_r into V_I.

Then the triple $V_I = [V_r; E_{sr}; I]$ is called a *nested Hilbert space (NHS)* if the following two conditions are satisfied:

(nh₁) V_I is stable under intersection: for any two $r, q \in I$, there is a $p \leq r, q$ such that $E_{Ip} V_p = E_{Ir} V_r \cap E_{Iq} V_q$;
(nh₂) For every $r \in I$, there exists a (unique) unitary map $u_{r\bar{r}}$ from V_r onto $V_{\bar{r}}$ such that $u_{oo} = 1$ and $(E_{sr})_{rs}^* = u_{r\bar{r}} E_{\bar{r}\bar{s}} u_{\bar{s}s}$.

Condition (nh₁) guarantees the lattice property (proj), hence also (add), whereas Condition (nh₂) states that $V_{\bar{r}}$ is the conjugate dual of V_r, with the dual norm, $u_{r\bar{r}}$ being the unitary (Riesz) operator establishing the duality between the two.

Clearly, every LHS is a nested Hilbert space, but the latter notion is slightly more general. Indeed, in a LHS, each embedding E_{qp}, $p \leq q$, is the identity operator and the $\{p, q\}$ representative of the identity operator on V. In a NHS, instead, the embeddings E_{qp} need not reduce to the identity. To give an example, take the lattice $\{\ell^2(r)\}$ of weighted ℓ^2 spaces, described in Section I.2.3 and, for $p \leq q$, define $(E_{qp}x)_n = \frac{q_n}{p_n} x_n$. Then these operators E_{qp} satisfy the conditions (3) above and are thus nestings (they are all bijective and have norm $\|E_{qp}\| \geq 1$). Thus they define a NHS, distinct from the usual one where all nestings are the identity. Another example is given in Section 4.6.4.

2.4.2 Examples: Indexed PIP-Spaces of Type (B)

(i) Chains of Banach spaces

Reflexive chains of Banach spaces and their (nonstandard) completion yield nice examples of indexed PIP-spaces of type (B). They will be discussed at length in Section 4.1.3.

(ii) Lattices of Banach spaces

A large supply of lattices of reflexive Banach spaces is available. We will discuss in detail the chain of Lebesgue spaces $L^p([0,1], dx)$ (Section 4.1.1), the lattice generated by the Lebesgue spaces $L^p(\mathbb{R}, dx)$ (Section 4.1.2), the Köthe sequence spaces, that generalize the familiar ℓ^p spaces (Section 4.3), the Köthe function spaces, which generalize the L^p spaces (Section 4.4), and various families of spaces of interest in signal processing (Sections 8.2 to 8.4).

2.5 The Central Hilbert Space

In most examples, the (indexed) PIP-space V has two additional properties: (i) The partial inner product is positive definite; and (ii) There exists a central, self-dual Hilbert space. In this section, we will assume the positivity condition (i) and see to what extent it implies (ii). First, it follows from (i) that $(V^{\#}, \| \cdot \|)$ is a pre-Hilbert space for the norm defined by the partial inner product, called the pip-norm, $\|f\| := \langle f|f \rangle^{1/2}$ (the same is true for every assaying subspace V_r such that $V_r \subseteq V_{\overline{r}}$). As such it admits a unique completion $\mathcal{H}$, but *a priori* there is no reason why $\mathcal{H}$ could be identified with a subspace of V, and even less that it should be an assaying subspace. Conditions ensuring these two properties will be given below, for a general (indexed) PIP-space. Here also, these conditions are satisfied automatically for spaces of type (B) and (H), which shows once again that these are the most useful structures for practical applications.

The argument is based on a standard result of L. Schwartz, namely, Hilbert subspaces of V correspond one-to-one to the so-called positive kernels [175]. In this terminology, a *kernel* is a weakly continuous (hence Mackey continuous) linear map from $V^{\#}$ into V, i.e., the restriction to $V^{\#}$ of an operator on V (see Section I.3; a full analysis is given in Chapter 3). A *Hermitian kernel* is the restriction of a symmetric operator $(A = A^{\times})$. A *positive kernel* corresponds to a positive operator. The first result concerns the existence of the central Hilbert space $\mathcal{H}$.

Proposition 2.5.1. *Let V be a nondegenerate (indexed) PIP-space with positive definite partial inner product. Assume there exists an assaying subspace*

V_r such that $V_r \subseteq V_{\overline{r}}$ and $V_{\overline{r}}$ is quasi-complete in its Mackey topology. Then, the completion $\mathcal{H}$ of $V^{\#}$ in the pip-norm is a dense subspace of V, and we have

$$V^{\#} \subseteq V_r \subseteq \mathcal{H} \subseteq V_{\overline{r}} \subseteq V$$

Proof. The spaces V_r and $V_{\overline{r}}$ are dual to each other and $V_{\overline{r}}$ is quasi-complete. By assumption the identity map $I : V_r \to V_{\overline{r}}$ is a positive kernel. Hence, by Proposition 10 and Section 10 of [175], it corresponds to a unique Hilbert subspace $\mathcal{H}$ of $V_{\overline{r}}$ which is the completion of the pre-Hilbert space $(V_r, \| \cdot \|)$ and is dense in $V_{\overline{r}}$. A fortiori, we get $V^{\#} \subseteq V_r \subseteq \mathcal{H} \subseteq V_{\overline{r}} \subseteq V$, and the statement follows.   ■

Before going further, let us mention a few cases where the proposition applies.

(1) $\langle V^{\#}, V \rangle$ is a reflexive dual pair, which implies that V is Mackey-quasi-complete. Typical examples are PIP-spaces of distributions, such as $V = \mathcal{S}^{\times}$ or $\mathcal{D}^{\times}$.

(2) V_I is reflexive, in particular of type (B) or (H): given any assaying subspace V_p $(p \in I)$, the pair $\langle V_{p \wedge \overline{p}}, V_{p \vee \overline{p}} \rangle$ is reflexive and verifies the conditions of Proposition 2.5.1. Plenty of examples have been given.

(3) The following counterexample is instructive. Let (X, μ) be a measure space such that $\mu(X) = \infty$ and μ has no atoms. Then the spaces $L^p(X, d\mu)$ are not comparable to each other, but the family $\{L^1 \cap L^p, 1 \leqslant p \leqslant \infty\}$ is a chain. With the compatibility inherited from L^2, we get $(L^1 \cap L^p)^{\#} = L^1 \cap L^q$ and $(L^1 \cap L^2)^{\#} = L^1 \cap L^2$. However none of them is Mackey-quasi-complete, nor even sequentially complete. Indeed, on $L^1 \cap L^p$ the topology $\tau(L^1 \cap L^p, L^1 \cap L^q)$ is strictly coarser than $\tau(L^1 \cap L^p, L^q)$; but this is just the L^p-norm topology, for which $L^1 \cap L^p$ is not complete, its completion being L^p. Let $(f_n) \in L^1 \cap L^p$ be a sequence that converges, in the L^p norm, to an element $f \in L^p$ which does not belong to $L^1 \cap L^p$. Thus (f_n) converges also for the coarser topology $\tau(L^1 \cap L^p, L^1 \cap L^q)$. Now, if $L^1 \cap L^p$ were sequentially complete for that topology, the limit f of (f_n) would belong to $L^1 \cap L^p$, contrary to the assumption. Hence $(L^1 \cap L^p)[\tau(L^1 \cap L^p, L^1 \cap L^q)]$ is not sequentially complete, *a fortiori* not quasi-complete or complete. Indeed, the completion of $V^{\#} = L^1 \cap L^{\infty}$ in the pip-norm, i.e., the L^2-norm, is L^2, which is not contained in $V = L^1$.

Assume now $\mathcal{H}$ exists as a dense subspace of V. When is it assaying? For answering that question, it is useful to consider the set Λ of all self-compatible vectors:

$$\Lambda = \{f \in V : f \# f\}.$$

Let now $f \in \Lambda$. This means, there exists $r \in I$ such that $f \in V_r \cap V_{\overline{r}}$. Since $V_{r \wedge \overline{r}}$ is a pre-Hilbert space with respect to the pip-norm, we have $V_{r \wedge \overline{r}} \subseteq \mathcal{H}$, and therefore $\Lambda \subseteq \mathcal{H}$. To go further, we need one more assumption.

Proposition 2.5.2. *Let V_I be a nondegenerate, positive definite, indexed* PIP-*space, Λ the subset of self-compatible vectors, $\mathcal{H}$ the completion of $V^{\#}$*

in the pip-norm. Assume that, for every $r \in I$, the assaying subspace $V_{r \vee \bar{r}}$ is quasi-complete in its Mackey topology. Then one has

$$\Lambda \subseteq \mathcal{H} \subseteq \Lambda^{\#}. \tag{2.6}$$

Proof. From the discussion above, we have (as sets)

$$\Lambda = \bigcup_{r \in I} V_{r \wedge \bar{r}}$$

and therefore

$$\Lambda^{\#} = \bigcap_{r \in I} V_{r \vee \bar{r}}.$$

Now, for every $r \in I$, one has $V_{r \wedge \bar{r}} \subseteq V_{r \vee \bar{r}}$ and $V_{r \vee \bar{r}}$ is Mackey-quasi-complete by assumption. Hence, by Proposition 2.5.1, we have

$$V_{r \wedge \bar{r}} \subseteq \mathcal{H} \subseteq V_{r \vee \bar{r}}.$$

Thus we get

$$\Lambda = \bigcup_{r \in I} V_{r \wedge \bar{r}} \subseteq \mathcal{H} \subseteq \Lambda^{\#} = \bigcap_{r \in I} V_{r \vee \bar{r}}. \qquad \blacksquare$$

Corollary 2.5.3. *Under the assumptions of Proposition 2.5.2, Λ is an assaying subspace: $\Lambda = \Lambda^{\#\#} \in \mathcal{F}(V, \#)$*

Proof. Let S be any family of self-compatible vectors, i.e., $S \subseteq S^{\#}$. Then $S \subseteq S^{\#\#} \subseteq S^{\#}$. This implies $S^{\#\#} \subseteq \Lambda$. The result follows by taking $S = \Lambda$. $\qquad \blacksquare$

The corollary implies in particular that Λ is a vector subspace of V, and so $\Lambda = \sum_{r \in I} V_{r \wedge \bar{r}}$. Also any self-compatible assaying subspace, no matter whether it belongs to $\mathcal{I}$ or not, is a pre-Hilbert space with respect to the pip-norm. Thus Λ is also the largest pre-Hilbert subspace of V.

Proposition 2.5.4. *Let $V_I, \Lambda, \mathcal{H}$ be as in Proposition 2.5.2. Then the following inclusions hold:*

$$\Lambda = \mathcal{H}^{\#} \subseteq \mathcal{H} \subseteq \Lambda^{\#} = \mathcal{H}^{\#\#} \tag{2.7}$$

Proof. From Proposition 2.5.2 and the involution, we get $\Lambda = \Lambda^{\#\#} \subseteq \mathcal{H}^{\#}$. Next we show that $\mathcal{H}^{\#} \subseteq \mathcal{H}$. Indeed $\langle \mathcal{H}^{\#}, \mathcal{H} \rangle$ is a dual pair. The Mackey topology $\tau(\mathcal{H}^{\#}, \mathcal{H})$ induces on the dense subspace $V^{\#}$ the topology $\tau(V^{\#}, \mathcal{H})$, which coincides with the one given by the pip-norm. Since $\mathcal{H}$ is the completion of $V^{\#}$ in the latter topology, it follows that $\mathcal{H}^{\#} \subseteq \mathcal{H}$. So $\mathcal{H}^{\#}$ is a self-compatible assaying subspace and therefore $\mathcal{H}^{\#} \subseteq \Lambda$. The statement follows. $\qquad \blacksquare$

Here again, the result applies if V_I is reflexive, for then every $V_{r \vee \bar{r}}$ is reflexive, hence τ-quasi-complete. It applies in particular if V_I is of type (B) or (H).

If all three spaces in Eq. (2.7) coincide, we will say that V_I has a *central Hilbert space*. This happens if any of the following equivalent conditions is satisfied:

(i) Λ is self-dual: $\Lambda = \Lambda^{\#}$;
(ii) I contains a self-dual element o, i.e., $V_o = V_o^{\#}$ (for then $V_o \subseteq \Lambda \subseteq \Lambda^{\#} \subseteq V_o^{\#} = V_o$);
(iii) Λ is complete in the pip-norm.

There are three main cases of interest for applications:

(1) *A LHS*: In each case, the central Hilbert space $\mathcal{H}$ is the natural one, which, so to speak, generates the PIP-space by extension of the natural inner product to a larger domain, e.g., ℓ^2 for $\{\ell^2(r)\}$, L^2 for $\{L^2(r)\}$, etc.
(2) *A LBS*: The structure obtained consists of a lattice of reflexive Banach spaces, the duality $V_r \leftrightarrow V_{\bar{r}}$ being taken with respect to the inner product of $\mathcal{H}$. Examples are $\mathcal{H} = L^2[0,1]$ for the chain $\{L^p[0,1]\}$ or $\mathcal{H} = L^2(X, d\mu)$ for the lattice of Köthe spaces (see Section 4.4).
(3) *A reflexive indexed PIP-space*: The cases of interest are essentially those spaces for which $\mathcal{I}$ is a lattice of type (B) or (H) together with the extreme elements $V^{\#}$ and V, where $\langle V^{\#}, V \rangle$ is a reflexive dual pair. Such are, for instance, rigged Hilbert spaces or, more generally, PIP-spaces of distributions, like $\mathcal{S}^{\times}$ or Bargmann's space $\mathcal{E}^{\times}$.

However, in the general case, the inclusions in (2.7) are strict and $\mathcal{H}$ is not assaying. Examples can be given readily, e.g., simply by omitting the central element in any lattice I of the previous type, but they tend to be artificial.

In fact, the missing central Hilbert space may often be reconstructed by hand. This will be done below for the case where V_I is additive. The key is the following observation. $\mathcal{H}^{\#} \subsetneq \mathcal{H}$ means the given compatibility is too coarse, it does not admit sufficiently many compatible pairs. Thus adding $\mathcal{H}$ to the assaying subspaces amounts to refining the compatibility.

Proposition 2.5.5. *Let V_I be a positive PIP-space for which every $V_{r \vee \bar{r}}, r \in I$ is quasi-complete, but $\mathcal{H}$ is not assaying. Assume V_I is additive. Then V carries a finer PIP-space structure for which $\mathcal{H}$ is assaying and thus a central Hilbert space.*

Proof. V_I additive means than $\mathcal{I}$ is an involutive sublattice of the lattice $\mathcal{L}(V)$ of all vector subspaces of V. Define $\mathcal{I}_1$ as the sublattice of $\mathcal{L}(V)$ generated by $\mathcal{I}$ and $\mathcal{H}$. The lattice $\mathcal{I}_1$ consists of elements of $\mathcal{I}$, plus $\mathcal{H}$ itself and additional elements of the form $V_p \cap \mathcal{H}, V_p + \mathcal{H}, (V_p + \mathcal{H}) \cap V_q$, etc. An involution $\#_1$ can be defined on $\mathcal{I}_1$ as follows:

- for $V_p \in \mathcal{I}$, $V_p^{\#_1} = V_p^{\#} = V_{\bar{p}}$,
- $\mathcal{H}^{\#_1} = \mathcal{H}$,

Fig. 2.1 The involutive lattice $\mathcal{I}$ generated by a single $\ell^2(r)$ space

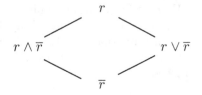

Fig. 2.2 The involutive lattice generated by $\mathcal{I}$ and ℓ^2

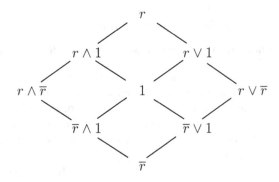

- $(V_p \cap \mathcal{H})^{\#_1} = V_{\bar{p}} + \mathcal{H}$,
- $(V_p + \mathcal{H})^{\#_1} = V_{\bar{r}} \cap \mathcal{H}$, and so on.

By this construction, $\#_1$ is a lattice anti-isomorphism on $\mathcal{I}_1$ and so $\mathcal{I}_1$ is an involutive sublattice of $\mathcal{L}(V)$, and an involutive covering of V. Thus $\#_1$ defines a linear compatibility on V, finer than $\#$. Obviously the new indexed PIP-space V_I has $\mathcal{H}$ as central Hilbert space.   ∎

It is instructive to consider a simple example of the construction just described. Let $\ell^2(r)$ be a weighted ℓ^2-space and $\mathcal{I}$ be the four-element involutive lattice it generates. Denoting each space $\ell^2(p)$ simply by its weight p, we obtain the picture shown in Fig. 2.1 (smaller spaces stand on the left).

Following the general construction, the space $\mathcal{H} = \ell^2 \equiv \ell^2(1)$ is obtained as the completion of $\ell^2(r \wedge \bar{r})$ in the pip-norm. The lattice $\mathcal{I}_1$ generated by $\mathcal{I}$ and ℓ^2 has nine elements and is described in Fig. 2.2.

This is indeed an involutive lattice. One verifies easily, for instance, the following relations: $r \wedge (\bar{r} \vee 1) = r \wedge 1$ and $\overline{r \wedge (\bar{r} \vee 1)} = \bar{r} \vee (r \wedge 1) = \bar{r} \vee 1$.

The crucial fact for the construction of Proposition 2.5.5 is additivity: Since $\mathcal{I}$ and $\mathcal{H}$ are then embedded in the lattice $\mathcal{L}(V)$, the supremum in $\mathcal{I}_1$ is defined independently of the involution. If additivity fails, there is no obvious way of enlarging $\mathcal{I}$ since both the supremum and the involution have to be defined on $\mathcal{I}_1$. This is consistent with the discussion given at the end of Chapter 1: compatibilities may always be coarsened, but not always refined.

As a final remark, we notice that the construction always works for nondegenerate indexed PIP-spaces of type (B) or (H), provided explicit norms are given for every assaying subspace. Each of these can be embedded canonically in a LBS or a LHS, respectively. Thus one may as well assume the existence of the central Hilbert space $\mathcal{H} = \mathcal{H}^{\#}$ from the beginning. Here again we see how simple this class of indexed PIP-spaces is.

Notes for Chapter 2

Section 2.1. Most of this chapter is based on the work of Antoine [13].

Section 2.2.

- The notation $V_{[p,q]}$ and $V_{(\overline{p},\overline{q})}$ is borrowed from Grossmann [114].
- For more information on interpolation theory, we refer to [BL76, Tri78a].
- The interpolation technique for Hilbert spaces, with squared norms throughout, is due to L. Schwartz [175].

Section 2.3. The construction of the auxiliary pair $\langle V_p \cap V_q, V_{\overline{p}} + V_{\overline{q}} \rangle$, with the projective and inductive topologies, respectively, is due to Goulaouic [110].

- Pathologies similar to those described after Eqs. (2.3), (2.3) have been noticed by Friedrich and Lassner, in their study of rigged Hilbert spaces generated by algebras of unbounded operators [100].

Section 2.4.

- The notion of indexed PIP-space was introduced in Antoine [13].
- The distinction between hilbertian spaces and Hilbert spaces is due to Palais [162].
- If V_I is an indexed PIP-space of type (H), with I countable, then V is a *countably Hilbert space*, in the terminology of Gel'fand-Vilenkin [GV64]. This particular case has been discussed by Antoine-Karwowski [22]. For V_I of type (H), with I arbitrary, $V^{\#}$ is a *quasi-Hilbert space*, in the sense of Hirschfeld [124].
- The structure of LBS has been called a *Dirac space* by Grossmann [unpublished seminar notes], see also Antoine [10].
- The notion of a chain of Banach spaces has been developed by Krein and Petunin [132], whereas Palais [162] introduces and discusses chains of hilbertian spaces.
- Köthe function spaces have been introduced (and given their name) by Luxemburg and Zaanen [Zaa61]. We shall discuss them in detail in Section 4.4.

Section 2.5. The results of this section generalize those obtained previously in Antoine-Karwowski [22] for the case where V is a countably Hilbert space, in the sense of Gel'fand [GV64].

- L. Schwartz [175] call these subspaces hilbertian subspaces of V, but we prefer to call them Hilbert subspaces, in order to avoid the confusion with the terminology of Palais that we have adopted. Indeed the subspaces of Schwartz carry an explicit norm.
- The set Λ of all self-compatible vectors was introduced by Popowicz [167] and further studied in Antoine-Karwowski [22].

Chapter 3
Operators on PIP-Spaces and Indexed PIP-Spaces

3.1 Operators on PIP-Spaces

3.1.1 Basic Definitions

As already mentioned, the basic idea of PIP-spaces is that vectors should not be considered individually, but only in terms of the subspaces V_r ($r \in F$), the building blocks of the structure. Correspondingly, an operator on a PIP-space should be defined in terms of assaying subspaces only, with the proviso that only continuous or bounded operators are allowed. Thus an operator is a *coherent collection* of continuous operators. We recall that in a nondegenerate PIP-space, every assaying subspace V_r carries its Mackey topology $\tau(V_r, V_{\bar{r}})$ and thus its dual is $V_{\bar{r}}$. This applies in particular to $V^{\#}$ and V itself. For simplicity, a continuous linear map between two PIP-spaces $\alpha : X \to Y$ will always mean a linear map α continuous for the respective Mackey topologies of X and Y.

For the sake of generality, it is convenient to define directly operators from one PIP-space into another one.

Definition 3.1.1. Let V and Y be two nondegenerate PIP-spaces and A a map from a subset $\mathcal{D} \subseteq V$ into Y. The map A is called an *operator* if

(i) $\mathcal{D}$ is a nonempty union of assaying subspaces of V.
(ii) The restriction of A to any assaying subspace V_r contained in $\mathcal{D}$ is linear and continuous from V_r into Y.
(iii) A has no proper extension satisfying (i) and (ii), i.e., it is maximal.

Here, a proper extension of A satisfying (i) and (ii) would be a map A' defined on a union of assaying subspaces $\mathcal{D}' \supset \mathcal{D}$, coinciding with A on $\mathcal{D}$, linear and continuous on every assaying subspace in its domain. Thus the set $\mathcal{D}$ may be called the *natural* domain of the operator A and will be denoted by $\mathcal{D}(A)$. We write $\mathrm{Op}(V, Y)$ for the set of all operators with domain in V and range in Y. When $V = Y$, we note simply $\mathrm{Op}(V)$.

Given any $A \in \mathrm{Op}(V, Y)$, its restriction to $V^{\#}$ is continuous. Conversely, we have

J.-P. Antoine and C. Trapani, *Partial Inner Product Spaces:*
Theory and Applications, Lecture Notes in Mathematics 1986,
DOI 10.1007/978-3-642-05136-4_3, © Springer-Verlag Berlin Heidelberg 2009

Proposition 3.1.2. *Given any continuous linear map α from $V^{\#}$ into Y, there exists one and only one $A \in \mathrm{Op}(V, Y)$ having α as restriction to $V^{\#}$.*

Proof. (a) An extension of α will mean a map satisfying (i) and (ii), but not necessarily (iii). The family of all extensions of α carries a natural partial order (by inclusion of domains). It is easy to see that it satisfies the conditions of Zorn's lemma, and so has a maximal element. This proves the existence part.

(b) Let $A_1 \in \mathrm{Op}(V, Y)$ and $A_2 \in \mathrm{Op}(V, Y)$ have the same restriction to $V^{\#}$. Notice that A_1 and A_2 coincide not only on $V^{\#}$, but on $\mathcal{D}(A_1) \cap \mathcal{D}(A_2)$. Indeed, on each $V_r \subseteq \mathcal{D}(A_1) \cap \mathcal{D}(A_2)$, the restrictions of A_1 and A_2 are continuous, and $V^{\#}$ is dense on V_r. Next, we see that $\mathcal{D}(A_1) = \mathcal{D}(A_2)$; otherwise we could define an operator A' with domain $\mathcal{D}' = \mathcal{D}(A_1) \cup \mathcal{D}(A_2)$, equal to A_i on $\mathcal{D}(A_i)$ $(i = 1, 2)$. This A' would be a proper extension of A_1 and A_2, contradicting maximality. ∎

Corollary 3.1.3. $\mathrm{Op}(V, Y)$ *is isomorphic, as a vector space, to the space of all linear continuous maps from $V^{\#}$ to Y.*

Thus $\mathrm{Op}(V, Y)$ is a vector space, operators can always be added. We shall see that they can also be multiplied, but not always, as was already hinted in the Introduction. Notice, however, that the natural domain $\mathcal{D}(A)$ of an operator A need *not* be a vector subspace of V, although it will be the case in most situations of interest.

3.1.2 Adjoint Operator

Let V and Y be nondegenerate partial inner product spaces. Let $(V_r, V_{\bar{r}})$ be a dual pair in V and $(Y_u, Y_{\bar{u}})$ a dual pair in Y. Let α be an arbitrary linear map from V_r into Y_u (no continuity assumed). It is clear that the sesquilinear form $b(y, v) = \langle y | \alpha v \rangle_Y$ $(y \in Y_{\bar{u}}, v \in V_r)$ is continuous in y for fixed v. Its continuity in the other argument is equivalent to the continuity of α. Other equivalent continuity properties are listed in Section B.3.

We shall need mostly the following special case, which gives a convenient criterion for continuity on assaying subspaces.

Lemma 3.1.4. *Let $r \in F(V)$ and $u \in F(Y)$. Let α be a linear map from V_r into Y_u, and β a linear map from $Y_{\bar{u}}$ into $V_{\bar{r}}$, such that*

$$\langle y | \alpha v \rangle_Y = \langle \beta y | v \rangle_V, \quad \text{for all } y \in Y_{\bar{u}} \text{ and all } v \in V_r.$$

Then α and β are both continuous for the respective Mackey topologies.

As an application, we prove

Proposition 3.1.5. *Let A be a linear map of V into V, such that*

(i) A improves behavior, i.e., $\{Af\}^{\#} \supseteq \{f\}^{\#}$, for every $f \in V$.
(ii) If f and g are compatible, then $\langle g|Af \rangle = \langle Ag|f \rangle$.

    Then, for every $r \in F(V)$, one has $AV_r \subseteq V_r$, and the restriction of A to V_r is $\tau(V_r, V_{\bar{r}})$-continuous.

Proof. By the definition of compatibility, condition (i) holds true if and only if A maps every V_r into itself. Furthermore, $\langle g|Af \rangle = \langle Ag|f \rangle$ for all $f \in V_r, g \in V_{\bar{r}}$, so that Lemma 3.1.4 applies. ∎

We are now ready to define the adjoint of an arbitrary operator $A \in \mathrm{Op}(V, Y)$. We have seen above that $A \in \mathrm{Op}(V, Y)$ is fully determined by its restriction to $V^{\#}$. This restriction is continuous from $V^{\#}$ into Y and so gives rise to a separately continuous sesquilinear form $b(y, v) = \langle y|Av \rangle_Y$, $y \in Y^{\#}$, $v \in V^{\#}$. This form determines a (unique) map β from $Y^{\#}$ into V, defined by $b(y, v) = \langle \beta y|v \rangle_V$. Furthermore, β is continuous from $Y^{\#}$ into V. Consequently, β has a unique natural extension which we shall call the *adjoint* of A and denote by $A^{\times}$. Thus we have

$$b(y, v) = \langle y|Av \rangle_Y = \langle A^{\times}y|v \rangle_V, \quad y \in Y^{\#}, \ v \in V^{\#}. \tag{3.1}$$

The two correspondences $b \leftrightarrow A \leftrightarrow A^{\times}$ given by (3.1) are bijections between the vector space $\mathfrak{B}(Y^{\#}, V^{\#})$ of all separately continuous sesquilinear forms on $Y^{\#} \times V^{\#}$, $\mathrm{Op}(V, Y)$, and $\mathrm{Op}(Y, V)$, respectively.
    To summarize:

Theorem 3.1.6. *$\mathrm{Op}(V, Y)$ has a natural structure of vector space. The conjugate linear correspondence $A \mapsto A^{\times}$ is a bijection between $\mathrm{Op}(V, Y)$ and $\mathrm{Op}(Y, V)$. One has $A^{\times\times} = A$ for every $A \in \mathrm{Op}(V, Y)$. For $V = Y$, in particular, the correspondence $A \leftrightarrow A^{\times}$ is an involution in $\mathrm{Op}(V)$.*

To say it simply: In adding operators and in taking adjoints, one may proceed algebraically without any special precautions. In particular, we emphasize the relation $A^{\times\times} = A$, $\forall A \in \mathrm{Op}(V, Y)$: no extension is allowed, because of the maximality condition (iii) in Definition 3.1.1. Altogether, this definition and that of adjoint operator imply that operators between PIP-spaces behave (to a certain extent, at least) like bounded operators between Hilbert spaces.

3.1.3 Representatives and Operator Multiplication

We have seen that the vector space structure of $\mathrm{Op}(V, Y)$ is identical to that of the space $\mathfrak{B}(V^{\#}, Y^{\#})$ of sesquilinear forms and to that of a space of continuous linear maps from $V^{\#}$ into Y. One may wonder, then, whether there is any point in considering the natural extensions beyond $V^{\#}$. The answer is yes, if one is interested in studying products of operators, as we shall see now.

Let A belong to $\mathrm{Op}(V, Y)$. As shown by the examples given in the Introduction, the properties of A are conveniently described by the set $\mathrm{j}(A)$ of all pairs $r \in F(V)$, $u \in F(Y)$ such that A maps V_r continuously into Y_u. For such a pair (r, u), we denote by $A_{ur} : V_r \to Y_u$ the restriction of A to V_r by and call it the $\{r, u\}$-*representative* of A. Thus $\mathrm{j}(A)$ is the set of all pairs $\{r, u\}$ for which a representative A_{ur} exists:

$$\mathrm{j}(A) = \{(r, u) \in F(V) \times F(Y) : A_{ur} \text{ exists}\}.$$

If $(r, u) \in \mathrm{j}(A)$, then the representative A_{ur} is uniquely defined. If A_{ur} is any continuous linear map from V_r to Y_u, then there exists a unique $A \in \mathrm{Op}(V, Y)$ having A_{ur} as $\{r, u\}$-representative. This A can be defined by considering A_{ur} as a map from $V^{\#}$ to Y and then extending it to its natural domain. Thus the operator A may be identified with the collection of its representatives,

$$A \simeq \{A_{ur} : V_r \to Y_u : (r, u) \in \mathrm{j}(A)\}.$$

Let $V_{r'}$ and V_r be assaying subspaces of V. If $V_r \subseteq V_{r'}$, denote by $E_{r'r}$ the natural embedding operator from V_r into $V_{r'}$, i.e., the $\{r, r'\}$-representative of the identity $1 \in \mathrm{Op}(V)$. This $E_{r'r}$ is continuous, has dense range, and it is manifestly injective.

If A maps continuously V_r into Y_u, it also maps continuously any smaller $V_{r'} \subseteq V_r$ into any larger $Y_{u'} \supseteq Y_u$. Consequently, it is convenient to introduce in the Cartesian product $F(V) \times F(Y)$ a partial order by

$$(r', u') \geqslant (r, u) \quad \text{if and only if} \quad r' \leqslant r \text{ and } u' \geqslant u \tag{3.2}$$

By the continuity and injectivity of the natural embeddings $E_{rr'}$ and $E_{u'u}$ between assaying subspaces, one has immediately

Proposition 3.1.7. *Let $(r, u) \in \mathrm{j}(A)$, and let (r', u') be a successor of (r, u) with respect to the partial order (3.2). Then:*

    (i) $\mathrm{j}(A)$ is a final subset of $F(V) \times F(Y)$, i.e., (r', u') also belongs to $\mathrm{j}(A)$, and $A_{u'r'} = E_{u'u}A_{ur}E_{rr'}$ is the $\{r', u'\}$-representative of A.
    (ii) If A_{ur} is injective, so is $A_{u'r'}$.
    (iii) If A_{ur} has dense range, so has $A_{u'r'}$.

It will be convenient to call $A_{u'r'}$ a successor of A_{ur}. The relation $A_{u'r'} = E_{u'u}A_{ur}E_{rr'}$ between A_{ur} and any of its successors exemplify what we mean by a *coherent collection* of continuous operators. Property (i) implies that $\mathrm{j}(A)$ is always connected.

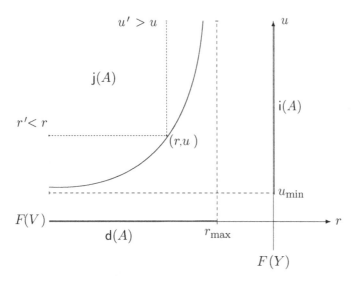

Fig. 3.1 Characterization of the operator $A : V \to Y$, in the case of two chains

We also need the two sets obtained by projecting $j(A)$ on the "coordinate" axes, namely,

$$d(A) := \mathrm{pr}_1 j(A) = \{r \in F(V) : \text{there is a } u \in F(Y) \text{ such that } A_{ur} \text{ exists}\},$$
$$i(A) := \mathrm{pr}_2 j(A) = \{u \in F(Y) : \text{there is a } r \in F(V) \text{ such that } A_{ur} \text{ exists}\}.$$

The following properties are immediate:

- $d(A)$ is an initial subset of $F(V)$: if $r \in d(A)$ and $r' < r$, then $r' \in d(A)$, and $A_{ur'} = A_{ur} E_{rr'}$, where $E_{rr'}$ is a representative of the unit operator.
- $i(A)$ is a final subset of $F(Y)$: if $u \in i(A)$ and $u' > u$, then $u' \in i(A)$ and $A_{u'r} = E_{u'u} A_{ur}$.
- $j(A) \subset d(A) \times i(A)$, with strict inclusion in general.

In order to get a feeling about this machinery of representatives of operators, we present in Fig. 3.1 the case of two chains, i.e., $F(V) \sim \mathbb{R}$ and $F(Y) \sim \mathbb{R}$. Given a pair $(r, u) \in j(A)$, the quarter plane above and on the left of that point (r, u) contains all successors of (r, u) with respect to the partial order $\leqslant$ defined in (3.2). Thus $j(A)$ is a union of such quarter planes. As for its projections, either $d(A) = F(V) \sim \mathbb{R}$ or the domain $d(A)$ has a maximal element $r_{\max} = \vee_{r \in d(A)} r$. Similarly, either $i(A) = F(Y) \sim \mathbb{R}$ or the range $i(A)$ has a minimal element $u_{\min} = \wedge_{u \in i(A)} q$. When both $r_{\max}$ and $u_{\min}$ exist (or are 'finite'), then $j(A)$ is contained in the quarter plane delimited by $r \leqslant r_{\max}$ and $u \geqslant u_{\min}$. In this way one sees that the larger $j(A)$, the 'better' the operator A (think, for instance, of the Sobolev spaces).

As for the adjoint $A^\times$ of an operator $A \in \mathrm{Op}(V, Y)$, one has

$$\langle A^\times y | x \rangle = \langle y | Ax \rangle, \text{ for } x \in V_r, \ r \in \mathsf{d}(A) \text{ and } y \in Y_{\overline{u}}, \ u \in \mathsf{i}(A),$$

that is, $(A^\times)_{\overline{ru}} = (A_{ur})'$, where $(A_{ur})' : Y_{\overline{u}} \to V_{\overline{r}}$ is the adjoint map of A_{ur} (see Section B.3). In other words, $\mathsf{j}(A^\times) = \mathsf{j}^\times(A) := \{(\overline{u}, \overline{r}) : (r, u) \in \mathsf{j}(A)\} \subset F(Y) \times F(V)$. If $V = Y$, $\mathsf{j}(A^\times)$ is obtained by reflecting $\mathsf{j}(A)$ with respect to the anti-diagonal $\{(r, \overline{r}), r \in F(V)\}$. In particular, if $(r, \overline{r}) \in \mathsf{j}(A)$, then $(r, \overline{r}) \in \mathsf{j}(A^\times)$ as well.

We shall now state the conditions under which a product BA is defined. The main point (just as in the definition of partial inner product which is, in fact, a special case) is that the "goodness" of one multiplicand can compensate for the "badness" of the other.

Let V, W, and Y be nondegenerate partial inner product spaces (some, or all, may coincide). Let $A \in \mathrm{Op}(V, W)$ and $B \in \mathrm{Op}(W, Y)$. We say that the product BA is defined if and only if there exist $r \in F(V), t \in F(W), u \in F(Y)$ such that $(r, t) \in \mathsf{j}(A)$ and $(t, u) \in \mathsf{j}(B)$. Then $B_{ut}A_{tr}$ is a continuous map from V_r into Y_u. It is the $\{r, u\}$-representative of a unique element of $BA \in \mathrm{Op}(V, Y)$, called the product of A and B. In other words, BA is defined if and only if there is a $t \in \mathsf{i}(A) \cap \mathsf{d}(B)$, that is, if and only if there is continuous factorization through some W_t:

$$V_r \overset{A}{\to} W_t \overset{B}{\to} Y_u, \quad \text{i.e.} \quad (BA)_{ur} = B_{ut}A_{tr}.$$

If BA is defined, then $A^\times B^\times$ is also defined, and equal to $(BA)^\times \in \mathrm{Op}(Y, V)$.

Similar definitions hold for products of more than two operators. For instance, the product CBA of $A \in \mathrm{Op}(V, W), B \in \mathrm{Op}(W, Y)$, and $C \in \mathrm{Op}(Y, Z)$ is defined whenever there exists a "chain" $\{r, t, v, u\}$ with $(r, t) \in \mathsf{j}(A), (t, v) \in \mathsf{j}(B)$ and $(v, u) \in \mathsf{j}(C)$. Thus one has

$$(CBA)_{ur} = C_{uv}B_{vt}A_{tr}.$$

This definition implies that the multiplication of operators is *associative* (which justifies the notation CBA). Note, however, that the existence of CBA does *not* follow from the existence of CB and of BA.

3.1.4 Examples

(i) Regular linear functionals

Let V be arbitrary. Notice that $\mathbb{C}$, with the obvious inner product $\langle \xi | \eta \rangle = \overline{\xi}\eta$ is an inner product space. Hence both $\mathrm{Op}(\mathbb{C}, V)$ and $\mathrm{Op}(V, \mathbb{C})$ are well defined and anti-isomorphic to each other. Elements of $\mathrm{Op}(V, \mathbb{C})$ will be called *regular linear functionals* on V. They are given precisely by the functionals

$$\langle g| : f \mapsto \langle g|f\rangle \qquad (f \in \{g\}^{\#}).$$

Indeed, for every $f \in V$, define a map $|f\rangle : \mathbb{C} \to V$ by $|f\rangle : \xi \mapsto \xi f$. Clearly, the correspondence $f \mapsto |f\rangle$ is a linear bijection between V and $\mathrm{Op}(\mathbb{C}, V)$. Then, the adjoint of $|f\rangle$ is $\langle f|$ as defined above.

As a result, V has naturally a self-dual structure. In this respect, PIP-spaces genuinely generalize Hilbert spaces, in contrast with schemes such as rigged Hilbert spaces. At this point we see that the algebraic structure of PIP-spaces is extremely simple, once one agrees to take as input the domain of the partial inner product and chooses Mackey topologies on every dual pair that shows up.

(ii) Dyadics

Using the fact that $\langle g| \in \mathrm{Op}(V, \mathbb{C})$ is defined exactly on $\{g\}^{\#}$, one verifies that:

- the product $((\langle g|)(|f\rangle))$ is defined if and only if $g \# f$, and equals $\langle g|f\rangle$.
- the product $|f\rangle\langle g|$ is always defined; it is an element of $\mathrm{Op}(V)$, called a *dyadic*. Its action is given by:

$$|f\rangle\langle g| (h) = \langle g|h\rangle f \quad (h \in \{g\}^{\#}).$$

The adjoint of $|f\rangle\langle g|$ is $|g\rangle\langle f|$. One constructs in the same way operators between different spaces and finite linear combinations of dyadics.

(iii) Matrix elements

The notation introduced above coincides precisely with the familiar Dirac notation of Quantum Mechanics. One can continue in the same vein and define matrix elements of an operator $A \in \mathrm{Op}(V)$: The matrix element $\langle f|A|h\rangle$ is defined whenever $\mathrm{j}(f) \times \bar{\mathrm{j}}(h) \cap \mathrm{j}(A)$ is nonempty. However, one has to be careful, there might be pairs of vectors h, f such that $\langle f|Ah\rangle$ is defined, but $\langle f|A|h\rangle$ is not. This may happen, for instance, if Ah is defined and $Ah = 0$. We see here at work the definition of the product of three operators.

(iv) Hilbert space operators

If V is an *inner* product space, then an operator of $\mathrm{Op}(V)$ is a $\tau(V, V)$-continuous linear map. In particular, if V is a Hilbert space, $\mathrm{Op}(V)$ consists of all bounded operators in V.

(v) Operators in sequence spaces

Consider again ω, the space of all sequences of complex numbers. To every $A \in \mathrm{Op}(\omega)$ there corresponds the infinite matrix $a_{nm} = \langle e_n | A e_m \rangle$ where $e_1 = \{1, 0 \ldots\}, e_2 = \{0, 1, 0, \ldots\}$. Conversely, an *arbitrary* infinite matrix $\{a_{nm}\}$ gives rise to a unique $A \in \mathrm{Op}(\omega)$. Its natural domain is the union of all Köthe sequence spaces V_r which have the property that all the row vectors $\{a_{1n}\}, \{a_{2n}\}, \ldots$ of the matrix belong to $V_{\overline{r}}$ (see Section 4.3).

(vi) Operators on tempered distributions

Let $V = \mathcal{S}^\times$ (tempered distributions). Elements of $\mathrm{Op}(\mathcal{S}^\times)$ are defined by tempered kernels. It should be interesting to study the corresponding natural domains.

If $\mathcal{S}^\times$ is realized as $s^\times$, then the elements of $\mathrm{Op}(s^\times)$ are given by "tempered matrices" satisfying either one of the equivalent conditions

$$|a_{mn}| \leqslant C(1+m)^N (1+n)^{N'},$$
$$|a_{mn}| \leqslant C'(1+m+n)^N,$$

for some $C, C', N, N' > 0$.

(vii) Operators on Fock space

On Fock space $V = \Gamma(V^{(1)})$, where $V^{(1)} = L^1_{\mathrm{loc}}(X, d\mu)$, defined in Section 1.1.3, Example (iv), we consider four kinds of operators:

(1) The operator $P = \bigoplus_{n=0}^\infty P_n$, where P_n is defined by $(P_n f)_k = \delta_{nk} f_n$. This is clearly an absolute projection, with range $0 \oplus \ldots \oplus V^{(n)} \oplus 0 \ldots$.

(2) The symmetrizer $S = \bigoplus_{n=0}^\infty S_n$, where

$$(S_n f_n)(k_1 \ldots k_n) = \frac{1}{n!} \sum_\pi f_n(k_{\pi 1}, \ldots, k_{\pi n}),$$

the sum extending over all $n!$ permutations $\pi = \begin{pmatrix} 1 & 2 & \ldots & n \\ \pi 1 & \pi 2 & \ldots & \pi n \end{pmatrix}$. Then $S \in \mathrm{Op}(V)$, $S^\times = S$. Its natural domain is all of V. However, S is *not* an orthogonal projection. Indeed $S_n f_n$ and g_n are not necessarily compatible, whenever f_n and g_n are, since the integral $\int |S_n f_n| \, |g_n| \, d\mu_n$ need not converge, even if $\int |f_n g_n| \, d\mu_n$ does.

(3) The number operator N, defined by $(Nf)_n = nf_n$. One defines similarly $\varphi(N) = \bigoplus_n \varphi(n) 1_n$ for an arbitrary sequence $\{\varphi(n)\}$ of real numbers (1_n denotes the identity operator on V^n). Here, too, $\varphi(N)$ is defined on all of V, and $\varphi(N)^\times = \varphi(N)$.

(4) The creation and annihilation operators $C^\dagger(f)$, resp. $C(f)$, $f \in V$, defined by

$$[C^\dagger(f)g]_n = \sum_{j+l=n} f_j \otimes g_l,$$

$$(f_j \otimes g_l)(k_1, \ldots, k_j, k_{j+1}, \ldots, k_{j+l}) = f_j(k_1, \ldots, k_j) g_l(k_{j+1}, \ldots, k_{j+l}),$$

$$C(f) = [C^\dagger(f)]^\times.$$

Again, $C^\dagger(f) \in \mathrm{Op}(V)$ is everywhere defined, but $C(f)$ is *not* unless
$f \in V^\#$. Notice that $C^\dagger(f)$ is linear in f, but $C(f)$ is antilinear.

It follows that the boson creation operator

$$a^\dagger(f) = N^{1/2} S C^\dagger(f) S$$

belongs to $\mathrm{Op}(V)$ and has V as natural domain. We may also introduce
the boson annihilation operator $a(f) = [a^\dagger(\overline{f})]^\times$, and the free field operator
$A(f) = a^\dagger(f) + a(\overline{f})$.[1]

We will come back to quantum field theory in these terms in Section 7.3.
We will show in Section 7.3.4, in particular, how the PIP-space approach
allows one to define (nonsmeared) free fields.

3.2 Operators on Indexed PIP-Spaces

In Section 3.1 we have defined operators on PIP-spaces. That definition can
be adapted in an obvious fashion to indexed PIP-spaces, as defined in Section
2.4. For the convenience of the reader, we state the new definition in full.

Definition 3.2.1. Let V_I and Y_K be two nondegenerate indexed PIP-spaces
and let A be a map from a subset $\mathcal{D}(A) \subset V$ into Y. We say that A is an
operator if:

(i) $\mathcal{D}(A)$ is a nonempty union of elements of $\mathcal{I}$:

$$\mathcal{D}(A) = \bigcup_{r \in \mathsf{d}(A)} V_r, \quad \mathsf{d}(A) \subset I.$$

(ii) For every $r \in \mathsf{d}(A)$, there exists $u \in K$ such that the restriction of A to
V_r is a continuous linear map into Y_u.
(iii) A has no proper extension satisfying (i) and (ii).

---

[1] The usual definition is rather $\widetilde{A}(f) = a^\dagger(f) + a(f)$. However, since $a(f)$ is *antilinear* in
f, the definition given here guarantees that $A(f)$ is linear in f, a property required in the
axiomatic approach sketched in Section 7.3.1.

It is worth noting that, if we consider the compatibilities defined by I and K, that is, we consider the corresponding PIP-spaces, then this definition reduces to the previous one, Definition 3.1.1.

Exactly as before, we denote by $\mathrm{Op}(V_I, Y_K)$ the set of all such operators. For a given $A \in \mathrm{Op}(V_I, Y_K)$, $\mathrm{j}(A)$ is the set of all pairs $(r, u) \in I \times K$ such that A maps V_r linearly and continuously into Y_u. The domain of A is $\mathrm{d}(A) = \{r \in I : \exists\, u \in K \text{ such that } (r, u) \in \mathrm{j}(A)\}$, its range is $\mathrm{i}(A) = \{u \in K : \exists\, r \in I$ such that $(r, u) \in \mathrm{j}(A)\}$. For each $(r, u) \in \mathrm{j}(A)$, the continuous linear map $A_{ur} : V_r \to Y_u$ is the corresponding *representative* of A. Thus the whole machinery of representatives can be developed, as was done in Section 3.1.3. The only difference is that here the extreme spaces $V^\#, Y$ are omitted, since they usually do not belong to $\mathcal{I}$, resp. $\mathcal{K}$. This does not change anything: since the embeddings $V^\#|_\tau \to V_r$ and $Y_u \to Y|_\tau$ are continuous for every $r \in I, u \in K$, the extreme representative $A : V^\# \to Y$ always exists.

In this section we will study the lattice properties of the three sets $\mathrm{j}(A), \mathrm{d}(A), \mathrm{i}(A)$, for a given $A \in \mathrm{Op}(V_I, Y_K)$. First we consider $I \times K$. Given any two involutive lattices L and L', their Cartesian product carries a natural partial order, already defined in (3.2):

$$(x, x') \geqslant (y, y') \quad \text{if and only if} \quad x \leqslant y \quad \text{and} \quad x' \geqslant y'.$$

With that order, $L \times L'$ is in fact an involutive lattice with respect to the following operations:

$$(x, x') \wedge (y, y') = (x \vee y, x' \wedge y'),$$
$$(x, x') \vee (y, y') = (x \wedge y, x' \vee y'),$$
$$\overline{(x, x')} = (\overline{x}, \overline{x}').$$

Thus, given two indexed PIP-spaces V_I, Y_K, the product $I \times K$ is an involutive lattice. From the definition we conclude immediately:

Lemma 3.2.2. *Let $A \in \mathrm{Op}(V_I, Y_K)$. Then:*

(i) $\mathrm{j}(A)$ is a final subset of $I \times K$.
(ii) $\mathrm{d}(A)$ is an initial subset of I.
(iii) $\mathrm{i}(A)$ is a final subset of K.

It follows that $\mathrm{j}(A)$ and $\mathrm{i}(A)$ are always $\vee$-stable and directed to the right, whereas $\mathrm{d}(A)$ is always $\wedge$-stable and directed to the left. This property of $\mathrm{j}(A)$ is to be contrasted with the behavior of the set $\mathrm{j}(f) = \{r \in I : f \in V_r\} \subset I$ for a given $f \in V$. Indeed, $\mathrm{j}(f)$ is always a sublattice of I, in particular it is always $\wedge$-stable, whereas $\mathrm{j}(A)$ is not. Indeed, let A map V_p into Y_u and V_q into Y_v continuously, that is (p, u) and (q, v) are in $\mathrm{j}(A)$; this of course does *not* imply that A maps $V_{p \vee q}$ into $Y_{u \wedge v}$, which would mean $(p, u) \wedge (q, v) \in \mathrm{j}(A))$. On the contrary, (p, u) and (q, v) in $\mathrm{j}(A)$ implies $(p \wedge q, u \vee v) = (p, u) \vee (q, v) \in \mathrm{j}(A)$. In fact $\mathrm{j}(A)$ is *never* a sublattice of $I \times K$, even if the only assaying subspaces are $V^{\#}$ and V. Indeed let A map continuously $V^{\#}$ into itself, and V into itself. Then *a fortiori*, it maps $V^{\#} \wedge V = V^{\#}$ into $V^{\#} \vee V = V$ but not the converse! Yet $\mathsf{j}(A)$ characterizes the behavior of A, as $\mathsf{j}(f)$ does for f: the bigger $\mathsf{j}(A)$, the better behaved the operator A.

For the domain $\mathsf{d}(A)$ and the range $\mathsf{i}(A)$, however, the situation can be improved. According to Definition 3.2.1, the operator A is defined on the set $\mathcal{D}(A) = \cup_{r \in \mathsf{d}(A)} V_r$ and such a union of vector subspaces need not be a vector subspace of V. A sufficient condition is that $\mathsf{d}(A)$ be directed to the right, *a fortiori* that $\mathsf{d}(A)$ be $\vee$-stable. Let indeed $f, g \in \mathcal{D}(A)$, with $f \in V_p, g \in V_q, p, q \in \mathsf{d}(A)$. If $\mathsf{d}(A)$ is directed to the right, p and q have a common successor $r \in \mathsf{d}(A)$, i.e., $V_p \subseteq V_r$ and $V_q \subseteq V_r$. Hence $V_p + V_q \subseteq V_r$, or $\lambda f + \mu g \in V_r \subset \mathcal{D}(A), \forall \lambda, \mu \in \mathbb{C}$.

It is of course desirable that *every* operator on the PIP-space have a vector subspace as domain of definition. A sufficient condition is found easily with the results of Section 2.4. Let $A \in Op(V_I, Y_K)$ and $(p, u), (q, v) \in \mathsf{j}(A)$. *A fortiori* $(p, u \vee v)$ and $(q, u \vee v)$ belong to $\mathsf{j}(A)$, that is, A maps both V_p and V_q continuously into $Y_{u \vee v}$. Thus it can be extended by linearity to a *continuous* map from their inductive limit $(V_p + V_q)_{\mathrm{ind}}$ into $Y_{u \vee v}$. Since $(V_p + V_q)$ need not be assaying, we cannot conclude, unless of course $V_p + V_q = V_{p \vee q}$. In a similar fashion, $(p \wedge q, u)$ and $(p \wedge q, v)$ belong to $\mathsf{j}(A)$. Thus A maps $V_{p \wedge q}$ continuously into Y_u and Y_v, hence also into their projective limit $(Y_u \cap Y_v)_{\mathrm{proj}}$. Again, since the Mackey topology on $Y_u \cap Y_v$ might be strictly finer than the projective topology, this does not imply that $A : V_{p \wedge q} \to Y_{u \wedge v}$ is continuous, unless the two topologies coincide on $Y_u \cap Y_v$. So we have proved:

Proposition 3.2.3. *Let V_I, Y_K be two indexed PIP-spaces and A any operator from V_I into Y_K. Then:*

(i) *If V_I is additive, the domain $\mathsf{d}(A)$ is a sublattice of I; in particular, the set $\mathcal{D}(A)$ is a vector subspace of V.*

(ii) *If Y_K is projective, the range $\mathsf{i}(A)$ is a sublattice of K.*

Corollary 3.2.4. *Let V_I be of type (B) or (H), and Y_K arbitrary. Then the domain of definition $\mathcal{D}(A)$ of any operator A from V_I into Y_K is a vector subspace of V.*

Remark 3.2.5. Let again $(p, u), (q, v) \in \mathsf{j}(A)$. Then V_I additive implies $(p \vee q, u \vee v) \in \mathsf{j}(A)$ and Y_K projective implies $(p \wedge q, u \wedge v) \in \mathsf{j}(A)$, but this does not mean that $\mathsf{j}(A)$ is a sublattice of $I \times K$. Assume now that all four pairs $(p, u), (p, v), (q, u), (q, v)$ belong to $\mathsf{j}(A)$. If V_I is additive and Y_K is projective, it follows that A maps $V_{p \vee q} = (V_p + V_q)_{\mathrm{ind}}$ continuously into $Y_{u \wedge v} = (Y_u \cap Y_v)_{\mathrm{proj}}$, i.e., $(p \vee q, u \wedge v) \in \mathsf{j}(A)$.) Of course, this still does not imply that $\mathsf{j}(A)$ be $\wedge$-stable and a sublattice. What this argument does give, in fact, is a three-line proof of Proposition 4.4 of [114]!

Remark 3.2.6. Consider now operators from an indexed PIP-space V_I into itself. Assume V_I to be additive. According to the preceding discussion, this is true if V_I is projective, in particular, if it is a LHS or a LBS. Then, the domain $\mathcal{D}(A)$ of any operator $A \in \mathrm{Op}(V_I)$ is a vector subspace of V, the usual rule of distributivity is valid and $\mathrm{Op}(V_I)$ is equipped with a partial multiplication. This implies that $\mathrm{Op}(V_I)$ is a *partial *-algebra*. This aspect of operator theory will be discussed in Chapter 5. A full treatment may be found in our monograph [AIT02].

3.3 Special Classes of Operators on PIP-Spaces

In a Hilbert space, one considers various classes of operators, besides the bounded ones. We shall do the same in this section for operators between two PIP-spaces. Throughout the section, V_I and Y_K denote two nondegenerate, positive definite PIP-spaces, with central Hilbert space. Actually, by V_I we mean either a general PIP-space or an indexed PIP-space. Indeed, a general PIP-space is just an indexed PIP-space V_I for which $I = F(V_I)$ and $\{V_r, r \in I\}$ is the set of *all* assaying subspaces.

3.3.1 Regular Operators

Definition 3.3.1. An operator $A \in \mathrm{Op}(V_I, Y_K)$ is called *regular* if $\mathsf{d}(A) = I$ and $\mathsf{i}(A) = K$ or, equivalently, if $A : V^\# \to Y^\#$ and $A : V \to Y$ continuously for the respective Mackey topologies.

This notion depends only on the pairs $(V^\#, V)$ and $(Y^\#, Y)$, *not* on the particular compatibilities on them. Accordingly, the set of all regular operators from V to Y is denoted by $\mathrm{Reg}(V, Y)$. Thus a regular operator may be multiplied both on the left and on the right by an arbitrary operator.

Of particular interest is the case $V = Y$, then we write simply $\mathrm{Reg}(V)$ for the set of all regular operators of V onto itself. In this case A is regular if, and only if, $A^\times$ is regular. Clearly the set $\mathrm{Reg}(V)$ is a *-algebra.

We give three examples.

(1) If $V = \omega, V^\# = \varphi$, then $\mathrm{Op}(\omega)$ consists of arbitrary infinite matrices, whereas $\mathrm{Reg}(\omega)$ consists of infinite matrices with a finite number of nonzero entries on each row and on each column.
(2) If $V = \mathcal{S}^\times, V^\# = \mathcal{S}$, then $\mathrm{Op}(\mathcal{S}^\times)$ consists of arbitrary tempered kernels, while $\mathrm{Reg}(\mathcal{S}^\times)$ contains those kernels that can be extended to $\mathcal{S}^\times$ and map $\mathcal{S}$ into itself. A nice example is the Fourier transform.

(3) If $V = \mathcal{E}^\times$, Bargmann's space of entire functions, regular operators are called *properly bounded*. Such are, for instance, multiplication by z and derivation $\frac{d}{dz}$.

Notice that, in each case, the algebra $\mathrm{Reg}(V)$ is nonabelian, but it contains abelian subalgebras: the set of all diagonal matrices for ω, the algebra $\mathfrak{P}[x]$ of all polynomials in x and the algebra $\mathfrak{P}\left[\frac{d}{dx}\right]$ in the case of $\mathcal{S}^\times$, the algebras $\mathfrak{P}[z]$ and $\mathfrak{P}\left[\frac{d}{dz}\right]$ for $\mathcal{E}^\times$.

The interest of the regular operators is that the set $\mathrm{Reg}(V)$ may be identified with a *-algebra of closable operators, called an O*-algebra. This is defined as follows. Let $\mathcal{D}$ be a dense subspace of a Hilbert space $\mathcal{H}$. Let A be a (closable) operator in $\mathcal{H}$ with domain $\mathcal{D}$, such that its adjoint A^* has a domain containing $\mathcal{D}$ and $A\mathcal{D} \subseteq \mathcal{D}$, $A^*\mathcal{D} \subseteq \mathcal{D}$. Equivalently, the restrictions to $\mathcal{D}$ of A and A^*, $A\!\upharpoonright\!\mathcal{D}$ and $A^\dagger := A^*\!\upharpoonright\!\mathcal{D}$, are $\tau(\mathcal{D}, \mathcal{D})$-continuous. The set of all such operators is a *-algebra, denoted by $\mathcal{L}^\dagger(\mathcal{D})$ and an O*-algebra is defined as a *-subalgebra of $\mathcal{L}^\dagger(\mathcal{D})$. In this terminology, we have the following result.

Proposition 3.3.2. *Let V be a positive definite, nondegenerate PIP-space. Then:*

(i) *$\mathrm{Reg}(V)$ is isomorphic to an O*-algebra on $V^\#$;*
(ii) *if V is quasi-complete for its Mackey topology $\tau(V, V^\#)$, $\mathrm{Reg}(V)$ is isomorphic to $\mathcal{L}^\dagger(V^\#)$.*

Proof. Let $A \in \mathrm{Reg}(V)$. Both A and $A^\times$ are continuous in $V[\tau(V, V^\#)]$ and both leave $V^\#$ invariant. Hence their restrictions to $V^\#$ are continuous in the topology induced by $V[\tau(V, V^\#)]$, which is $\tau(V^\#, V^\#)$ since $V^\#$ is dense in V (see Remark 3.3.3). This proves (i). Conversely, let $B \in \mathcal{L}^\dagger(V^\#)$, i.e., B and $B^\dagger = B^*\!\upharpoonright V^\#$ map $V^\#$ into itself, continuously for the topology $\tau(V^\#, V^\#)$ induced by V. This implies that $V \in \mathrm{Op}(V)$ and $B^\dagger$ is the restriction to $V^\#$ of the PIP-space adjoint $B^\times$. Since V is quasi-complete for its Mackey topology $\tau(V, V^\#)$ and $V^\#$ is dense in V, both B and $B^\dagger$ can be extended by continuity to the quasi-completion of $V^\#$, that is, V. Hence B is regular. ∎

Remark 3.3.3. In general, the Mackey topology is *not* inherited by subspaces, but it is for a *dense* subspace. The argument runs as follows. Let H be a linear subspace of a locally convex space E and π be the canonical mapping of E' onto $E'/H^\perp$. Then it is known that the topology $\tau(H, E'/H^\perp)$ is equal to the topology induced by $\tau(E, E')$ if and only if every absolutely convex weakly compact subset of $E'/H^\perp$ is the π-image of an absolutely convex weakly compact subset of E'. Now, if H is dense in E, $H^\perp = \{0\}$ and π is the identity, so that the condition is satisfied.

The condition of Mackey-quasi-completeness is actually very mild and satisfied in almost all useful examples, the only exceptions being quite pathological. It is, of course, trivially satisfied if V is Mackey-complete, for instance if it is a (possibly non-reflexive) Banach or Fréchet space, or the dual of a Fréchet space. It is also satisfied if V is reflexive (in that case, V may fail to be Mackey-complete). Thus, for one or another reason, the following examples are covered: distribution spaces $(\mathcal{S}^\times, \mathcal{E}^\times)$; sequence spaces $(\omega, s^\times)$; the Banach chains $\{\ell^p\}, \{\mathcal{C}^p\}\{L^p[0,1]\}, 1 \leqslant p \leqslant \infty$; the LBS generated by $\{L^p(\mathbb{R})\}, 1 \leqslant p \leqslant \infty$; the spaces $\{L^p_{\mathrm{loc}}(\mathbb{R})\}, 1 \leqslant p \leqslant \infty$, which are reflexive for $1 < p < \infty$ only. Most of these examples are discussed in detail in Chapter 4.

3.3.2 Morphisms: Homomorphisms, Isomorphisms, and all that

Like in any category of mathematical structures (the word 'category' may even be taken in its technical sense, but we will not pursue this line), it is necessary to have a notion of subobject, which in turn is uniquely determined by *morphisms*. A PIP-space is a vector space with a linear compatibility, so intuitively a morphism between two PIP-spaces should be a linear mapping that preserves compatibility. This leads to a formal definition (PIP-subspaces will be discussed in Section 3.4 below).

Definition 3.3.4. An operator $A \in \mathrm{Op}(V_I, Y_K)$ is called a *homomorphism* if

(i) $\mathrm{j}(A) = I \times K$ and $\mathrm{j}(A^\times) = K \times I$;
(ii) $f \#_I g$ implies $Af \#_K Ag$.

In words, for every $r \in I$, there exists $u \in K$ such that $(r, u) \in \mathrm{j}(A)$ and $(\overline{r}, \overline{u}) \in \mathrm{j}(A)$, and for every $u \in K$, there exists $r \in I$ with the same property. The definition may also be rephrased as follows: $A : V_I \to Y_K$ is a homomorphism if

$$\mathrm{pr}_1(\mathrm{j}(A) \cap \overline{\mathrm{j}(A)}) = I \quad \text{and} \quad \mathrm{pr}_2(\mathrm{j}(A) \cap \overline{\mathrm{j}(A)}) = K, \tag{3.3}$$

where $\overline{\mathrm{j}(A)} = \{(\overline{r}, \overline{u}) : (r, u) \in \mathrm{j}(A)\}$.

We denote by $\mathrm{Hom}(V_I, Y_K)$ the set of all homomorphisms from V_I into Y_K and by $\mathrm{Hom}(V_I)$ those from V_I into itself. The following properties are immediate.

Proposition 3.3.5. *Let $V_I, Y_K, \ldots$ be PIP-spaces. Then:*

(i) *Every homomorphism is regular.*
(ii) *$A \in \mathrm{Hom}(V_I, Y_K)$ if and only if $A^\times \in \mathrm{Hom}(Y_K, V_I)$.*
(iii) *The product of any number of homomorphisms (between successive PIP-spaces) is defined and is a homomorphism.*
(iv) *If B_0 and B_{n+1} are arbitrary operators and if $A_1, \ldots, A_n$ are homomorphisms, then the product of $n + 2$ factors $B_0 A_1 \ldots A_n B_{n+1}$ is defined.*
(v) *If $A \in \mathrm{Hom}(V_I, Y_K)$, then $\mathrm{j}(A^\times A)$ contains the diagonal of $I \times I$ and $\mathrm{j}(AA^\times)$ contains the diagonal of $K \times K$.*

The two conditions in (3.3) are independent. Take for instance a dyadic $A = |f\rangle\langle g| \in \mathrm{Op}(V_I, Y_K)$, where $g \in V^{\#}$ and $f \in Y, f \notin Y^{\#}$. This operator satisfies the first condition in (3.3), but not the second one, hence it is not a homomorphism.

In order to verify that a given map is a homomorphism, it is sometimes useful to apply the following result.

Lemma 3.3.6. *Let γ be a map from I onto K, which intertwines the involutions in I and K, that is,*

$$\overline{(\gamma r)} = \gamma \overline{r}, \ \forall r \in I.$$

If $A \in \mathrm{Op}(V_I, Y_K)$ is such that $(r, \gamma r) \in \mathrm{j}(A)$ for every $r \in I$, then $A \in \mathrm{Hom}(V_I, Y_K)$.

Proof. Let $r \in I$ be arbitrary. Then $(r, \gamma r) \in \mathrm{j}(A)$ by assumption. Consequently $(\overline{r}, \overline{\gamma r}) = (\overline{r}, \gamma \overline{r}) \in \overline{\mathrm{j}(A)}$. Since $(\overline{r}, \gamma \overline{r})$ is of the form $(s, \gamma s)$, it belongs also to $\mathrm{j}(A)$. So $(r, \gamma r) \in \mathrm{j}(A) \cap \overline{\mathrm{j}(A)}$. ∎

Incidentally, if the surjectivity ("onto") of γ is dropped, then A satisfies only the first condition in (3.3) and may fail to satisfy the second one. One may also ask whether any homomorphism A admits a map γ satisfying the conditions of Lemma 3.3.6. Does γ have to be order-preserving? Both questions are open.

A special case of Lemma 3.3.6, γ being the identity, is the following. If $A \in \mathrm{Op}(V_I)$ is such that $\mathrm{j}(A)$ contains the diagonal of $I \times I$, then $A \in \mathrm{Hom}(V_I)$. The converse is not true in general. (Such operators are called totally regular, see Section 3.3.3 below).

A scalar multiple of a homomorphism is a homomorphism. However, the sum of two homomorphisms may fail to be one. Of course, if A_1 and A_2 in $\mathrm{Hom}(V_I)$ are such that both $\mathrm{j}(A_1)$ and $\mathrm{j}(A_2)$ contain the diagonal of $I \times I$, then their sum does, too; consequently it belongs to $\mathrm{Hom}(V_I)$. This happens, in particular, for projections, that we shall study in Section 3.4.1.

Next we turn to monomorphisms. The definition is the same as for general categories.

Definition 3.3.7. Let $M \in \mathrm{Hom}(W_L, Y_K)$. Then M is called a *monomorphism* if $MA = MB$ implies $A = B$, for any two elements of $A, B \in \mathrm{Hom}(V_I, W_L)$, where V_I is any PIP-space.

In order to characterize monomorphisms, we need the following result.

Proposition 3.3.8. *Let $A, B \in \mathrm{Op}(V_I, W_L)$, $M \in \mathrm{Op}(W_L, Y_K)$. Assume that*

(i) MA and MB are defined and $MA = MB$.
(ii) Every representative of M is injective.

Then $A = B$.

Proof. Clearly $M(A - B)$ is defined. This means that there exists at least one triple $r \in I, t \in L, u \in K$ such that $(r,t) \in j(A - B), (t, u) \in j(M)$. Then $M_{ut}(A - B)_{tr} = (M(A - B))_{ur} = 0$. Since M_{ut} is injective, it follows that $(A - B)_{tr} = 0$, i.e., that $A - B = 0$.   ∎

Proposition 3.3.9. *If every representative of $M \in \mathrm{Hom}(W_L, Y_K)$ is injective, then M is a monomorphism.*

Proof. Let $A, B \in \mathrm{Hom}(V_I, W_L)$ be two arbitrary morphisms into W_L. Since MA and MB are defined, the result follows from Proposition 3.3.8.   ∎

The converse of Proposition 3.3.9 is open: Does there exist monomorphisms with at least one noninjective representative?

Examples 3.3.10. Typical examples of monomorphisms are the inclusion maps resulting from the restriction of a support. For instance (see Examples 3.4.15):

(1) Take the sequence space ω and define the two subsets:

$$\omega_e := \{x \in \omega : x_{2n+1} = 0, \forall n = 0, 1, 2, \ldots\},$$
$$\omega_o := \{x \in \omega : x_{2n} = 0, \forall n = 1, 2, \ldots\}.$$

Both subsets are PIP-spaces with the operations inherited from ω, and actually, PIP-subspaces of ω, as will be defined in Section 3.4. Then the injection of either of them into ω is a monomorphism, defined as follows:

$$(M_e x)_n = \begin{cases} x_n, & \text{if } n = 2k, \\ 0, & \text{if } n = 2k + 1, \end{cases} \quad x \in \omega_e,$$

and similarly for $M_o : \omega_o \to \omega$.

(2) Take $L^1_{\mathrm{loc}}(X, d\mu)$, the space of locally integrable functions on a measure space (X, μ) (see Section 1.1.3 (iii)). Let Ω be a measurable subset of X and Ω' its complement, both of nonzero measure, and construct the space $L^1_{\mathrm{loc}}(\Omega, d\mu)$, which is a PIP-subspace of $L^1_{\mathrm{loc}}(X, d\mu)$. Given $f \in L^1_{\mathrm{loc}}(X, d\mu)$, define $f^{(\Omega)} = f\chi_\Omega$, where χ_Ω is the characteristic function of χ_Ω. Then we obtain an injection monomorphism $M^{(\Omega)} : L^1_{\mathrm{loc}}(\Omega, d\mu) \to L^1_{\mathrm{loc}}(X, d\mu)$ as follows:

$$(M^{(\Omega)} f^{(\Omega)})(x) = \begin{cases} f^{(\Omega)}(x), & \text{if } x \in \Omega, \\ 0, & \text{if } x \notin \Omega, \end{cases} \quad f^{(\Omega)} \in L^1_{\mathrm{loc}}(\Omega, d\mu).$$

If we consider the lattice of weighted Hilbert spaces $\{L^2(r)\}$ in this PIP-space, then the correspondence $r \leftrightarrow r^{(\Omega)} = r\chi_\Omega$ is a bijection between the corresponding involutive lattices.

Finally, we define the isomorphisms of PIP-spaces.

Definition 3.3.11. An operator $A \in \mathrm{Op}(V_I, , Y_K)$ is an *isomorphism* if $A \in \mathrm{Hom}(V_I, Y_K)$ and there is a homomorphism $B \in \mathrm{Hom}(Y_K, V_I)$ such that $BA = 1_V, AB = 1_Y$, the identity operators on V, Y, respectively.

Note that every isomorphism is a monomorphism as well.

We shall say that a representative A_{ur} is *invertible* if it is bijective and has a continuous inverse. The second condition is automatically satisfied if V_r and Y_q are Fréchet (in particular Banach or Hilbert) spaces. Any successor of an invertible representative is injective and has dense range. An invertible representative has in general no predecessors.

If $A \in \mathrm{Op}(V, Y)$ has an invertible representative, then there exists a $B \in \mathrm{Op}(Y, V)$ such that AB and BA are defined, $AB = 1_Y$, and $BA = 1_V$, but this *does* not imply that A is a isomorphism, because it may fail to be a homomorphism. This does not exclude either the possibility of A having a nontrivial null-space.

To give an example, take $V = L^2_{\mathrm{loc}}(\mathbb{R}^n, dx)$, the space of locally square integrable functions (see Section 1.1.3 (iii)), with $n \geqslant 2$. Let R be the map $(Rf)(x) = f(\rho^{-1}x)$, where $\rho \in \mathrm{SO}(n)$ is an orthogonal transformation of $\mathbb{R}^n$. Then R is an isomorphism of $L^2_{\mathrm{loc}}(\mathbb{R}^n, dx)$ onto itself, $R^\times$ is one, too, but $\mathrm{j}(R)$ does not contain the diagonal of $I \times I$. For instance, the assaying subspace $V_r = L^2(r), r^\pm \in L^\infty$, is invariant under R only if the weight function r is rotation invariant. Note that, in addition, R is unitary, in the sense that both $R^\times R$ and $RR^\times$ are defined and equal 1_V (see Section 3.3.4 below).

3.3.3 Totally Regular Operators and *-Algebras of Operators

In many applications of PIP-spaces, mainly in quantum theories, it is essential to have at one's disposal the notion of projection operator. The notion of orthogonal projection in a PIP-space, already introduced in Section 1.2, will be studied in detail in Section 3.4 below. It turns out that the Hilbert space definition of a projection extends to PIP-spaces, with similar properties, including the one-to-one correspondence with appropriate subspaces. But, *a priori*, one does not know whether a PIP-space V_I possesses sufficiently many projections. In a Hilbert space context, one way of ensuring this is to exhibit a von Neumann algebra, since these are always generated by their projections. Thus, given a PIP-space V_I, we are looking whether $\mathrm{Op}(V_I)$ contains a subset that can be identified with a von Neumann algebra, and first with a *-algebra of bounded operators, in the Hilbert space sense. We proceed in several steps.

Definition 3.3.12. An operator $A \in \mathrm{Op}(V_I)$ is called *totally regular* if $\mathrm{j}(A)$ contains the diagonal of $I \times I$, i.e., A_{rr} exists for every $r \in I$ or A maps every V_r into itself continuously.

Proposition 3.1.5 gives a handy criterion for the total regularity of an operator. As we have seen in Proposition 3.3.5, and also in the previous example, an arbitrary homomorphism $A \in \mathrm{Hom}(V_I)$ need not be totally regular, but both $A^\times A$ and $AA^\times$ are. We denote by $\mathfrak{A}(V_I)$ the set of all totally regular operators from V_I into itself. Clearly $\mathfrak{A}(V_I)$ is a *-algebra.

In order to proceed towards the definition of a von Neumann algebra of operators, we restrict ourselves to the LHS case and denote by $\mathcal{H}_r, r \in I$, the (Hilbert) assaying subspaces. Thus, given $A \in \mathfrak{A}(V_I)$, each representative A_{rr} belongs to the von Neumann algebra $\mathcal{B}(\mathcal{H}_r)$ and the norm $\|A_{rr}\|_{\mathcal{B}(\mathcal{H}_r)}$ is finite. This means that one may identify A with the family $\{A_{rr}, r \in I\}$ of its diagonal representatives. This family is an element of $\prod_{r \in I} \mathcal{B}(\mathcal{H}_r)$, which is an algebra of bounded operators acting in the Hilbert space

$$\bigoplus_{r \in I} \mathcal{H}_r = \{f = (f_r)_{r \in I} : f_r \in \mathcal{H}_r, \ \forall \, r \in I, \ \text{and} \ \sum_{r \in I} \|f_r\|_r^2 < \infty\}.$$

The norm of $\prod_{r \in I} \mathcal{B}(\mathcal{H}_r)$ is that of $\mathcal{B}(\bigoplus_{r \in I} \mathcal{H}_r)$, namely,

$$\|A\|_I := \sup_{r \in I} \|A_{rr}\|_r,$$

and $\prod_{r \in I} \mathcal{B}(\mathcal{H}_r)$ is a C*-algebra with respect to that norm and a von Neumann algebra, called the product von Neumann algebra.

Next we define $\mathfrak{B}(V_I) := \{A \in \mathfrak{A}(V_I) : \|A\|_I < \infty\}$. The key result is then:

Proposition 3.3.13. $\mathfrak{B}(V_I)$ *is a Banach algebra for the norm* $\|\cdot\|_I$.

Proof. The proof proceeds by lifting operators on $\bigoplus_{r \in I} \mathcal{H}_r$ to operators in the LHS V_I. First define, for each $r \in I$, the projection $P_r : \bigoplus_{r \in I} \mathcal{H}_r \to \mathcal{H}_r$, which is an operator in $\mathcal{B}(\bigoplus_{r \in I} \mathcal{H}_r)$, obtained by combining the natural projection $p_r : \bigoplus_{r \in I} \mathcal{H}_r \to \mathcal{H}_r$ with the injection $j_r : \mathcal{H}_r \to \bigoplus_{r \in I} \mathcal{H}_r$, thus $P_r = j_r p_r$. We denote by $\mathcal{P} := \{P_r, r \in I\}$ the set of these projections.

Similarly, for every pair $r, s \in I, s \geqslant r$, we denote by $E^{(s,r)} = j_s E_{sr} p_r$ the embedding of $\mathcal{H}_r$ into $\mathcal{H}_s$, both considered as subspaces of $\bigoplus_{r \in I} \mathcal{H}_r$. We call $\mathcal{E} := \{E^{(s,r)}, r, s \in I, s \geqslant r\}$ the set of these embedding operators.

Now, in $\mathcal{B}(\bigoplus_{r \in I} \mathcal{H}_r)$, the elements of $\prod_{r \in I} \mathcal{B}(\mathcal{H}_r)$ are precisely the operators which commute with every projection P_r, thus $\prod_{r \in I} \mathcal{B}(\mathcal{H}_r) = \mathcal{P}'$. Next, the elements of $A \in \mathfrak{B}(V_I)$ are operators in the LHS, thus they must satisfy the coherence conditions $E_{sr} A_{rr} = A_{ss} E_{sr}$ for every $r, s \in I, s \geqslant r$. Lifted into $\mathcal{B}(\bigoplus_{r \in I} \mathcal{H}_r)$, these conditions read as $E^{(s,r)} A = A E^{(s,r)}, r, s \in I, s \geqslant r$. In other words, $\mathfrak{B}(V_I) = (\mathcal{P} \cup \mathcal{E})'$. Hence, being the commutant of the set $\mathcal{P} \cup \mathcal{E}$ in $\mathcal{B}(\bigoplus_{r \in I} \mathcal{H}_r)$, $\mathfrak{B}(V_I)$ is closed in norm, thus it is a Banach algebra.   ∎

Remark 3.3.14. Suppose we put on $\mathcal{H}_r$ a norm equivalent to the original one (taking $\mathcal{H}_r$ as a hilbertian space):

$$A_r\|f_r\|_r' \leqslant \|f_r\|_r' \leqslant B_r\|f_r\|_r'. \tag{3.4}$$

Then the bounded operators on $\mathcal{H}_r$ remain the same, but *not* necessarily the Hilbert space $\bigoplus_{r\in I} \mathcal{H}_r$, unless the constants A_r, B_r may be taken as independent of r. *A fortiori*, the algebra $\mathfrak{B}(V_I)$ may change. This is why we have to restrict ourselves to a LHS.

The last step is to try and find within $\mathfrak{B}(V_I)$ a von Neumann algebra. But here there is an obstacle, linked to the involution. True, $\mathfrak{B}(V_I)$ is *-invariant, but it is not automatically a C*-algebra, nor a von Neumann algebra, since it is the commutant of the non-self-adjoint subset $\mathcal{P} \cup \mathcal{E}$ in $\mathcal{B}(\bigoplus_{r\in I} \mathcal{H}_r)$. On the latter, the natural adjoint of an operator $A \sim \{A_{rr}\}$ is given by the family $A^\star \sim \{(A_{rr})_{rr}^\star\}$ of the Hilbert space adjoints in individual spaces $\mathcal{H}_r$. Now, for a given $A \in \mathfrak{B}(V_I)$, this adjoint $A^\star$ does *not* belong to $\mathfrak{B}(V_I)$ in general, since its representatives need not satisfy the coherence condition of LHS operators.

In order to proceed, we have to assume that $\mathcal{H}_r$ and $\mathcal{H}_{\bar{r}}$ are dual Hilbert spaces, that is, the two norms are dual to each other. In that case, we can introduce the unitary Riesz operator $u_{\bar{r}r} : \mathcal{H}_r \to \mathcal{H}_{\bar{r}}$ expressing the conjugate duality between the two spaces. In other words, we must assume that V_I is a LHS. Under this restriction, one can show that $\mathfrak{B}(V_I)$ is a C*-algebra if, and only if, the two involutions coincide on it: $A^\times = A^\star, \forall A \in \mathfrak{B}(V_I)$. In terms of representatives, this condition reads:

(c)   For every $r \in I, (A^\times)_{rr} = (A_{rr})_{rr}^\star$, or, equivalently, $A_{rr} = u_{r\bar{r}}A_{\bar{r}\bar{r}}u_{\bar{r}r}$,

Using this condition, we may state the final result. Define the subset of $\mathfrak{B}(V_I)$ on which the two involutions coincide, namely,

$$\mathfrak{C}(V_I) := \{A \in \mathfrak{B}(V_I) : \forall r \in I, \, A_{rr} = u_{r\bar{r}}A_{\bar{r}\bar{r}}u_{\bar{r}r}\}.$$

Proposition 3.3.15. *If V_I is a LHS, then $\mathfrak{C}(V_I)$ is a C*-algebra and a von Neumann algebra.*

Proof. On $\bigoplus_{r\in I} \mathcal{H}_r$, define an operator U as follows: $U : f = (f_r)_{r\in I} \mapsto (u_{r\bar{r}}f_{\bar{r}})_{r\in I}$. Then Condition (c) may be rewritten, in $\mathcal{B}(\bigoplus_{r\in I} \mathcal{H}_r)$, as $AU = UA$. Thus we have, in that space, $\mathfrak{C}(V_I) = (\mathcal{P} \cup \mathcal{E} \cup \{U\})'$.

Next, from the relation $(E^{(s,r)})^\star = UE^{(\bar{r},\bar{s})}U, r, s \in I, s \geqslant r$, we conclude that every operator that commutes with U and $\mathcal{E}$ also commutes with $\mathcal{E}^\star$. Therefore we have $\mathfrak{C}(V_I) = (\mathcal{P} \cup \mathcal{E} \cup \{U\})' = (\mathcal{P} \cup \mathcal{E} \cup \mathcal{E}^\star)'$. Since $\mathcal{P} = \mathcal{P}^\star$, it follows that $\mathfrak{C}(V_I)$ is the commutant of a self-adjoint subset of $\mathcal{B}(\bigoplus_{r\in I} \mathcal{H}_r)$ and, therefore, $\mathfrak{C}(V_I)$ is a von Neumann algebra. Since the two involutions coincide on $\mathfrak{C}(V_I)$, the latter is, *a fortiori*, a C*-algebra.

Remark: It follows from the discussion above that $\mathfrak{C}(V_I) = \mathfrak{B}(V_I) \cap \mathfrak{B}(V_I)^\star$.

Of course, if the LHS is trivial, consisting of a single Hilbert space $\mathcal{H}_o$, we get $\mathfrak{A}(V_I) = \mathfrak{B}(V_I) = \mathfrak{C}(V_I) = \mathcal{B}(\mathcal{H}_o)$. In order to get a nontrivial example, take the LHS $V_I := \{\ell^2(r)\}$ in ω, introduced in Section I.2.3. Operators on ω are represented by infinite matrices. Then one can prove the following results.

- $A \in \mathfrak{A}(V_I)$ if A is the sum of a finite matrix and a diagonal matrix with bounded coefficients.
- $A \in \mathfrak{B}(V_I)$ if A is a diagonal matrix with bounded coefficients.
- $\mathfrak{C}(V_I) = \mathfrak{B}(V_I) = \{A_{mn} = \lambda_m \delta_{mn}, \sup_m |\lambda_m| < \infty\}$, which is an abelian von Neumann algebra.

Notice that the proof of the first statement rests on a not well-known result concerning the characterization of bounded operators on ℓ^2. For further details and other examples, we refer to the original paper [70]. The outcome is that the three alternatives: $\mathfrak{A} = \mathfrak{B}$ or $\mathfrak{A} \neq \mathfrak{B}$, $\mathfrak{B} = \mathfrak{C}$ or $\mathfrak{B} \neq \mathfrak{C}$, $\mathfrak{C}$ abelian or not, are mutually independent, so that eight different types of LHS may be constructed.

There is a whole class of examples where the algebra $\mathrm{Reg}(V_I)$ is trivial, in the sense that it contains only bounded operators and furthermore $\mathrm{Reg} = \mathfrak{A} = \mathfrak{B}$. Take first the scale $\{\ell^p, 1 \leqslant p \leqslant \infty\}$ (Section 4.1.3). For every $S \in \mathrm{Reg}(\ell^\infty)$, one has $S : \ell^1 \to \ell^1$ and $S : \ell^\infty \to \ell^\infty$, continuously for the respective weak and Mackey topologies, and also for the strong (i.e., norm) topology on ℓ^∞, since $\ell^\infty[\tau(\ell^\infty, \ell^1)]$ is semi-reflexive. Then, by the Riesz-Thorin interpolation theorem, $S : \ell^p \to \ell^p$ continuously for every p and $\mathrm{Reg}(\ell^\infty) = \mathfrak{A} = \mathfrak{B}$. The same result holds true for the chain $\{\mathcal{C}^p, 1 \leqslant p \leqslant \infty\}$ of ideals of compact operators on a Hilbert space, as well as for the whole lattice generated by the family $\{L^p(X, d\mu), 1 \leqslant p \leqslant \infty\}$, where (X, μ) is any measure space (see Section 4.1.2). More generally, for any scale of Banach spaces which has the normal interpolation property.

Remark: From the construction, it is clear that the *-algebra $\mathfrak{A}(V_I)$ may be defined for an arbitrary PIP-space. The *-algebra $\mathfrak{B}(V_I)$ may be defined for a LBS also, but the *-algebra $\mathfrak{C}(V_I)$ makes sense for a LHS only.

3.3.4 Unitary Operators and Group Representations

In a Hilbert space, a unitary operator is the same thing as an (isometric) isomorphism (i.e., a bijection that, together with its adjoint, preserves all inner products), but the two notions differ for a general PIP-space.

Definition 3.3.16. An operator $U \in \mathrm{Op}(V_I, , Y_K)$ is *unitary* if $U^\times U$ and $UU^\times$ are defined and $U^\times U = 1_V, UU^\times = 1_Y$, the identity operators on V, Y, respectively.

We emphasize that a unitary operator need *not* be a homomorphism, in fact it is a rather weak notion, and it is insufficient for group representations. Indeed, given a group G and a PIP-space V_I, a unitary representation of G into V_I should be a homomorphism $g \mapsto U(g)$ from G into some class of unitary operators on V_I, that is, one should have $U(g)U(g') = U(gg')$ and $U(g)^\times = U(g^{-1})$ for all $g, g' \in G$. Thus each $U(g)$ should be at least a regular operator. However, in all instances of a group representation in some topological vector space, the operators $U(g)$ always preserve the structure of the representation space. Therefore, in the present case, $U(g)$ should map compatible vectors into compatible vectors. Accordingly, we introduce:

Definition 3.3.17. Let G be a group and V_I a PIP-space. A *unitary representation* of G into V_I is a homomorphism of G into the unitary isomorphisms of V_I.

Given such a unitary representation U of G into V_I, where the latter has the central Hilbert space $\mathcal{H}_0$, consider the representative $U_{00}(g)$ of $U(g)$ in $\mathcal{H}_0$ Then $g \mapsto U_{00}(g)$ is a unitary representation of G into $\mathcal{H}_0$, in the usual sense. In the rest of this section, we will discuss various aspects of the relation between U and U_{00}, leaving aside, however, the continuity properties of $U(g)$ as a function of $g \in G$.

The first question one may ask is the following. Given a strongly continuous unitary representation U_{00} of a Lie group G in $\mathcal{H}_0$, can one build a PIP-space V_I, with $\mathcal{H}_0$ its central Hilbert space, such that U_{00} extends to a unitary representation U into V_I? The solution of this problem is well-known from the theory of C^∞-vectors. We will describe it in Section 7.4.

The second question is, in a sense, the converse of the first one: Given a PIP-space V_I, with central Hilbert space $\mathcal{H}_0$ and a unitary representation U_{00} of G in $\mathcal{H}_0$, under what conditions does U_{00} extend to a unitary representation U in V_I?

First we have to extend $U_{00}(g)$ as a linear map from $\mathcal{H}_0$ to the whole of V. This may often be done by inspection (see the examples below). Otherwise, assume that, for a given $q \in I$, one can find $p \in I$ such that the restriction of $U_{00}(g)$ maps $V^\#$ into V_p, continuously in the topology induced on $V^\#$ by V_q. Then $U_{00}(g)$ extends by continuity to the whole V_q, that is, to a $\{q, p\}$-representative $U_{pq}(g)$. If this can be done for every $q \in I$, then $U_{00}(g)$ extends to the whole of V. In that case, we have the following easy result.

Proposition 3.3.18. Let $U_{00}(g), \mathcal{H}_0, V_I$ be as above. Assume that, for every $g \in G$, $U_{00}(g)$ can be extended to a map $U(g) : V \to V$, such that $U(g) : V^\# \to V^\#$. Then:

(i) $g \mapsto U(g)$ is a representation of G by unitary regular operators on V_I.
(ii) for all $g \in G$, $U(g)V^\#$ is dense in every $V_p, p \in I$.
(iii) Every representative $U_{pq}(g)$ is injective.

Once we have a representation by unitary regular operators, it is easy to see whether each $U(g)$ is a homomorphism. For instance, if V_I is a LHS or a

LBS, a sufficient condition is that, for every $p \in I$, there exists a $q \in I$ such that the representative $U_{pq}(g)$ is *surjective* (a continuous surjection between Banach spaces has a continuous inverse and this, combined with unitarity, shows that $U(g)$ is a morphism). This happens, in particular, if $U(g)$ is *totally regular* for every $g \in G$.

Examples 3.3.19. (1) Let V_I be the scale built on the powers of the operator (Hamiltonian) $H = -\Delta + v(\mathbf{r})$ on $L^2(\mathbb{R}^3, d\mathbf{x})$, where Δ is the Laplacian on $\mathbb{R}^3$ and v is a (nice) rotation invariant potential. The system admits as symmetry group (at least, see Section 7.1.2) $G = \mathrm{SO}(3)$ and the representation U_{00} is the natural representation of $\mathrm{SO}(3)$ in $L^2(\mathbb{R}^3)$:

$$[U_{00}(\rho)\psi](\mathbf{x}) = \psi\left(\rho^{-1}\mathbf{x}\right), \ \rho \in \mathrm{SO}(3).$$

Then U_{00} extends to a unitary representation U by totally regular isomorphisms of V_I and angular momentum decompositions, corresponding to irreducible representations of $\mathrm{SO}(3)$, extend to V_I as well. In addition, this is a good setting also for representations of the Lie algebra $\mathfrak{so}(3)$, as mentioned above.

The same analysis applies to internal symmetries of the physical system, if any, and discrete symmetries, such as space or time reflections. Notice that the representation U is totally regular, but this need not be the case. For instance, if the potential v is not rotation invariant, U will no longer be totally regular, although it is still an isomorphism.

(2) Consider the usual unitary representation of the Lorentz group $\mathrm{SO}_o(1,3)$ generated by pointwise transformations of the positive mass hyperboloid $X := \{k \in \mathbb{R}^4 : k^2 = k_0^2 - \mathbf{k}^3 = m^2 \geqslant 0, k_0 \geqslant 0\}$. Let $\mathcal{H}_0 = L^2(X, d\mu)$, where $d\mu(k) = k_0^{-1} d^3\mathbf{k}$ is the Lorentz-invariant positive measure on X. Then, for $f \in \mathcal{H}_0$, define

$$[U_{00}(\Lambda)f](k) = f(\Lambda^{-1}k), \ \Lambda \in \mathrm{SO}_o(1,3).$$

Take now $V = L^1_{\mathrm{loc}}(X, d\mu)$, as in Section 1.1.3, Example (iii) (see also Section 4.2 below), with its usual PIP-space structure, namely, the compatibility $f \# g \Leftrightarrow \int_X |f(k)g(k)| d\mu(k) < \infty$ and the partial inner product

$$\langle f|g \rangle = \int_X \overline{f(k)}g(k)d\mu(k).$$

Then $V^{\#} = L^{\infty}_c(X, d\mu)$ and $\mathcal{H}_0 = L^2(X, d\mu)$. For any $\Lambda \in \mathrm{SO}_o(1,3)$, the transformation $U(\Lambda) : f(k) \mapsto f(\Lambda^{-1}k)$ can be extended to the whole of V and $U(\Lambda)V^{\#} \subseteq V^{\#}$. Furthermore, both $U(\Lambda)$ and $U(\Lambda)^{\times} = U(\Lambda^{-1})$ preserve compatibility, because of the invariance of the measure. Thus U is a unitary representation of $\mathrm{SO}_o(1,3)$ into V. Notice that, here too, the representation is not totally regular.

The next natural question is, how does the standard Hilbert space theory of direct sum decompositions of unitary representations generalize to PIP-spaces?

In order to investigate this problem, we need first an appropriate definition of irreducibility. The choice is obvious, since we have a good definition of subspace, namely the *orthocomplemented* subspaces introduced in Section 1.2. Indeed, as we shall see in the next section, this is the "categorical" definition of subspace, in the following sense. A subspace W of a PIP-space V is orthocomplemented, and called a PIP-*subspace*, if and only if it is the range of an *orthogonal projection*, that is, a homomorphism $P \in \mathrm{Hom}(V)$ such that $P^2 = P^\times = P$:

$$W = W \oplus W^\perp, \quad \text{with} \quad W = PV, W^\perp = (1 - P)V.$$

Notice that the elements of $W, W^\perp$ need not be compatible. If they are, their inner product vanishes.

Definition 3.3.20. A unitary group representation U in a PIP-space V is *irreducible* if V contains no proper U-invariant PIP-subspace.

Proposition 3.3.21. *Any reducible unitary representation U in a PIP-space V is completely reducible, that is, if W is a U-invariant PIP-subspace, its orthocomplement $W^\perp$ is also U-invariant.*

Proof. The proof is almost identical to the corresponding one in a Hilbert space. To say that $W = PV$ is a U-invariant subspace means that $U(g)P = PU(g)P$, $\forall g \in G$. Pick any element $h \in V^\#$. Then one has, for every $g \in G$ and any $f \in V$:

$$\begin{aligned}
\langle Ph | U(g)(1 - P)f \rangle &= \langle U(g)^\times Ph | (1 - P)f \rangle \\
&= \langle U(g^{-1})Ph | (1 - P)f \rangle \\
&= \langle PU(g^{-1})Ph | (1 - P)f \rangle \\
&= 0.
\end{aligned}$$

Indeed, since both P and $U(g)^{-1} = U(g^{-1})$ are homomorphisms, they leave $V^\#$ invariant, so that $PU(g^{-1})Ph \in V^\#$. Thus both sides of each inner product are compatible, hence orthogonal. Therefore,

$$U(g)(1 - P)f \in (PV^\#)^\perp = (V^\# \cup W)^\perp = W^\perp, \quad \text{for any } (1 - P)f \in W^\perp$$

(see Remark 3.4.5 below).   ■

So far, the theory is identical to the usual one, except for one point. We do not know if V possesses many PIP-subspaces or, equivalently, many orthogonal projections. As in a Hilbert space, abundance of projections is the key to a good theory of direct sum decompositions of unitary representations. Given a unitary representation T in a Hilbert space $\mathcal{H}$, its reducibility is best studied by looking at its commutant $T' := \{B \in \mathcal{B}(\mathcal{H}) : BT(g) = T(g)B, \forall g \in G\}$. This set is a von Neumann algebra, hence it is generated by its projections. Thus, for a unitary representation U in a PIP-space V, the question is whether its commutant $\mathcal{U}'$ (to be defined properly) can be interpreted as a von Neumann algebra, or contains one. This is one of the motivations for introducing the von Neumann algebra $\mathfrak{C}(V_I)$.

3.3.5 Symmetric Operators and Self-Adjointness

In many applications, it is essential to show that a given symmetric operator H in a Hilbert space $\mathcal{H}$ is self-adjoint,[2] since this allows one to apply the powerful spectral theory or to generate one-parameter unitary groups (via Stone's theorem and its generalizations). This is, for instance, the crucial question in quantum mechanics, where H would be the Hamiltonian of the system under consideration. In this section we will translate these problems into a PIP-space context, which seems to be their natural setting. We assume throughout that V is a nondegenerate, positive definite PIP-space with a central Hilbert space $V_o = V_{\bar{o}}$. This could be, in particular, a LBS or a LHS.

Since every operator $A \in \mathrm{Op}(V_I)$ satisfies the condition $A^{\times\times} = A$, there is no room for extensions. Thus the only notion we need is that of symmetry, in the PIP-space sense, namely,

Definition 3.3.22. An operator $A \in \mathrm{Op}(V_I)$ is *symmetric* if $A = A^\times$.

According to the discussion in Section 3.1.3, the set $\mathrm{j}(A)$ of a symmetric operator is symmetric with respect to the anti-diagonal $\{(r, \bar{r}), r \in I\}$. In particular, since it is connected, it contains the initial segment $\{(r, \bar{r}), r \leqslant r_{\max}\}$. But, in fact, we can say much more.

Proposition 3.3.23. *Let V_I be a nondegenerate, positive definite PIP-space with a central Hilbert space $V_o = V_{\bar{o}}$ and let $A = A^\times \in \mathrm{Op}(V_I)$ be a symmetric operator.*

(i) *Assume that V_r is a Banach space comparable to V_o and that $(r, r) \in \mathrm{j}(A)$. Then $(o, o) \in \mathrm{j}(A)$.*

(ii) *Let V_I be projective. Assume that V_r is a Banach space and that $(r, r) \in \mathrm{j}(A)$. Then $(o, o) \in \mathrm{j}(A)$.*

Proof. (i) Let $r < o$. Since $V_r \hookrightarrow V_o$, the result follows from Theorem 2.1 of Lassner [138]: the norm $\|\cdot\|_r$ is stronger that the pip-norm and both A_{rr} and $A_{\bar{r}\bar{r}}$ are continuous, hence A_{oo} is bounded. If $r > o$, then $\bar{r} < o$ and the same reasoning applies.

---

[2] In the sequel we will assume that the reader is familiar with the standard theory of Hilbert space operators. In case of doubt, one may consult Reed & Simon's treatise [RS72, RS75].

(ii) Since $(r, r) \in j(A)$, we have also $(\bar{r}, \bar{r}) \in j(A)$. Since V_I is projective, A maps $V_{r \wedge \bar{r}}$ continuously into itself. Since $V_{r \vee \bar{r}}$ is Banach, hence Mackey-quasi-complete, Proposition 2.5.1 implies that $V_{r \wedge \bar{r}} \hookrightarrow V_o \hookrightarrow V_{r \vee \bar{r}}$. Then A maps each space continuously into itself and the result follows from (i). ∎

The proposition applies, of course, in the cases of interest for applications, namely, a LBS or a LHS.

Corollary 3.3.24. *Let V_I be a LBS or a LHS and let $A = A^\times \in \mathrm{Op}(V_I)$. Then, $(r, r) \in j(A)$, for some $r \in I$, implies $(o, o) \in j(A)$.*

Let V_I be a PIP-space with self-dual element $V_o = V_{\bar{o}}$, which is a Hilbert space. Given a symmetric operator $A \in \mathrm{Op}(V_I)$, it is natural to consider the subspace

$$D(A_o) = \{f \in V_o : Af \in V_o\}$$

and define $A_o f = Af$, for every $f \in D(A_o)$. Then A_o is a symmetric operator in the Hilbert space V_o, with domain $D(A_o)$. Thus, if $D(A_o)$ is dense in V_o (it could be $\{0\}$), it makes sense to ask for the existence of self-adjoint extensions of A_o (if A_o itself is not self-adjoint). The standard technique is to use quadratic forms (the Friedrichs extension) or von Neumann's theory of self-adjoint extensions.

However, there is another possibility. Namely, given a symmetric operator A in $\mathrm{Op}(V_I)$, it is natural to ask directly whether A has *restrictions* that are self-adjoint in V_o. According to Theorem 3.3.27 below, this happens provided some suitable representative of A is invertible (see Section 3.3.2). Since this is essentially the KLMN theorem,[3] we recall first the Hilbert space version of the theorem. The framework is a scale of three Hilbert spaces,

$$\mathcal{H}_1 \hookrightarrow \mathcal{H}_0 \hookrightarrow \mathcal{H}_{\bar{1}}, \tag{3.5}$$

where the embeddings are continuous and have dense range, and $\mathcal{H}_{\bar{1}}$ is the conjugate dual of $\mathcal{H}_1$. Such is, for instance, the central triplet of the scale (I.3). Then the theorem reads as follows.

Theorem 3.3.25 (KLMN theorem). *Let A_o be a continuous map from $\mathcal{H}_1$ into $\mathcal{H}_{\bar{1}}$, such that $A_o - \lambda$ is a bijection for some $\lambda \in \mathbb{R}$ and that $\langle f | A_o g \rangle = \langle A_o f | g \rangle$, $\forall f, g \in \mathcal{H}_1$. Then there is a unique self-adjoint operator A in $\mathcal{H}_0$ with domain $D(A) \subset \mathcal{H}_1$ and $\langle f | Ag \rangle = \langle f | A_o g \rangle$, $\forall g \in D(A)$ and $f \in \mathcal{H}_1$.*

Proof. We can assume $\lambda = 0$ without loss of generality. Define $D(A) := \{g \in \mathcal{H}_1 : Ag \in \mathcal{H}_0\}$ and $A = A_o \upharpoonright D(A)$. Consider the operator $A_o^{-1} \upharpoonright \mathcal{H}_0$, which is well-defined since A_o is a bijection. Take any $g_1, g_2 \in \mathcal{H}_0$ and let $g_i = A_o f_i, i = 1, 2$. Then:

$$\langle g_1 | A_o^{-1} g_2 \rangle = \langle A_o f_1 | f_2 \rangle = \langle f_1 | A_o f_2 \rangle = \langle A_o^{-1} g_1 | g_2 \rangle.$$

---

[3] KLMN stands for Kato, Lax, Lions, Milgram, Nelson.

Thus $A_o^{-1} \restriction \mathcal{H}$ is a symmetric, everywhere defined operator, hence it is bounded and self-adjoint. Therefore its inverse, which is A, is self-adjoint. Uniqueness is obvious. Indeed, if $\tilde{A}$ is another self-adjoint operator satisfying the assumptions, A is an extension of it, hence they must coincide.

■

Before stating the generalized version of the KLMN theorem, we give a useful lemma.

Lemma 3.3.26. *Let $R = R^\times \in \operatorname{Op}(V_I)$ be symmetric. Assume that R has an injective representative R_{rs}, with dense range, from a "big" V_s into a "small" V_r (that is, $V_r \subseteq V_o \subseteq V_s$). Then the representative R_{oo} has a selfadjoint inverse.*

Proof. Indeed, R_{oo} is a successor of R_{rs}, hence it is injective and has dense range, by Proposition 3.1.7. Furthermore, it is bounded and self-adjoint as operator in the Hilbert space V_o. So its inverse is selfadjoint on $R_{oo}V_o$.   ■

In order to avoid confusion in the theorem, we denote by A the PIP-space operator and by A its self-adjoint restriction in V_o.

Theorem 3.3.27 (Generalized KLMN theorem # 1). *Let V_I be a nondegenerate, positive definite PIP-space with a central Hilbert space $V_o = V_{\bar{o}}$ and let $A = A^\times \in \operatorname{Op}(V_I)$ be a symmetric operator. Assume there exists a $\lambda \in \mathbb{R}$ such that $A - \lambda$ has an invertible representative $A_{sr} - \lambda E_{sr}$ from a "small" V_r onto a "big" V_s (i.e., $V_r \subseteq V_o \subseteq V_s$). Then there exists a unique restriction of A_{sr} to a selfadjoint operator A in the Hilbert space V_o. The number λ does not belong to the spectrum of A. The domain of A is obtained by eliminating from V_r exactly the vectors f that are mapped by A_{sr} beyond V_o (i.e., satisfy $A_{sr}f \notin V_o$) The resolvent $(\mathsf{A} - \lambda)^{-1}$ is compact (trace class, etc.) if and only if the natural embedding $E_{sr} : V_r \to V_s$ is compact (trace class, etc.).*

Proof. Define $R_{rs} : V_s \to V_r$ as the inverse of the invertible representative $A_{sr} - \lambda E_{sr}$. Then $R_{oo} = E_{or}R_{rs}E_{so}$ is a restriction of R_{rs}. By Lemma 3.3.26, R_{oo} has a selfadjoint inverse $(R_{oo})^{-1} = \mathsf{A} - \lambda$, which is a restriction of $A_{sr} - \lambda E_{sr}$. Thus A_{sr} has a self-adjoint restriction to V_o. Since $R_{rs} = (A_{sr} - \lambda E_{sr})^{-1}$ is bounded, so is $R_{oo} = (\mathsf{A} - \lambda)^{-1}$, thus λ does not belong to the spectrum of A. The rest is obvious.   ■

It should be clear by now that Theorem 3.3.27 provides the natural setting for the KLMN theorem (even for the proof). Actually, one can generalize the theorem further, using the full PIP-space machinery. Indeed, if V_I is projective (thus, in particular, for a LBS or a LHS), we can drop the assumption that the two spaces V_r and V_r are comparable to V_o, as shown in the next theorem.

Theorem 3.3.28 (Generalized KLMN theorem # 2). *Let V_I be a projective, nondegenerate, positive definite PIP-space with a central Hilbert space $V_o = V_{\bar{o}}$ and $A = A^\times \in \mathrm{Op}(V_I)$. Assume there exists a $\lambda \in \mathbb{R}$ such that $A - \lambda$ has an invertible representative $A_{sr} - \lambda E_{sr} : V_r \to V_s$, where V_r is a Banach space and $V_r \subseteq V_s$. Then the conclusions of Theorem 3.3.27 hold true.*

In order to prove this theorem, we need two lemmas. First we notice that, for any operator in a projective PIP-space, the conditions $(p, u), (q, v) \in j(A)$ imply $(p \wedge q, u \wedge v), (p \vee q, u \vee v) \in j(A)$, by Proposition 3.2.3.

Lemma 3.3.29. *Let V_I and $A = A^\times \in \mathrm{Op}(V_I)$ be as in Theorem 3.3.28. Assume that $(p, u), (q, v) \in j(A)$ and that A_{up} and A_{vq} are both bijective with continuous inverse. Then so are $A_{u \wedge v, p \wedge q}$ and $A_{u \vee v, p \vee q}$.*

Proof. We proceed in four steps.

(i) If A_{up} or A_{vq} is injective, so is $A_{u \wedge v, p \wedge q}$. Indeed, let $f \in V_{p \wedge q}$ be such that $Af = 0$ and take A_{up} injective. Thus,

$$0 = \|Af\|_{u \wedge v} = \|Af\|_u + \|Af\|_v.$$

This implies $\|Af\|_u = 0$, hence $A_{up} f_p = 0$ and thus $f_p = 0$, since A_{up} is injective. This means that $f = 0$. The case A_{vq} injective is analogous. (In the case of Hilbert spaces, one may use squared norms).

(ii) We show now that $A_{u \wedge v, p \wedge q}$ is surjective. Let $g \in V_{u \wedge v} = V_u \cap V_v$. Since $g \in V_u$ and A_{up} is surjective, there exists an $f^{(1)} \in V_p$ such that $g = Af^{(1)}$. Similarly, there exists an $f^{(2)} \in V_q$ such that $g = Af^{(2)}$. Since A_{up} and A_{vq} are both injective, hence invertible, and their inverses are continuous, we may write

$$f_p^{(1)} = (A_{up})^{-1} g_u = (A^{-1}g)_p,$$
$$f_q^{(2)} = (A_{vq})^{-1} g_v = (A^{-1}g)_q,$$

where the continuity of the inverses guarantees that $(A_{up})^{-1}$ and $(A_{vq})^{-1}$ are representatives of the *same* operator A^{-1}. Embedding everything into $V_{p \vee q}$, we get $f_{p \vee q}^{(1)} = f_{p \vee q}^{(2)} = (A^{-1}g)_{p \vee q}$ and thus $f^{(1)} = f^{(2)} \in V_{p \wedge q}$.

(iii) Next we show that $A_{u \vee v, p \vee q}$ is injective. Let $f \in V_{p \vee q}$ be such that $Af = 0$ in $V_{u \vee v}$. However, noting that $0 \in V_{u \wedge v}$, we have in fact $Af \in V_{u \wedge v}$. Since we know by (ii) that $A_{u \wedge v, p \wedge q}$ is bijective, $Af = 0$ implies that $f \in V_{p \wedge q}$ and $f_{p \wedge q} = 0$, hence $f = 0$.

(iv) Finally, we show that $A_{u \vee v, p \vee q}$ is surjective. Let $f \in V_{u \vee v}$. This means, there exist (non unique) $g \in V_u$ and $h \in V_v$ with $f = g + h$. Since A_{up} and A_{vq} are both surjective, there exist $g' \in V_p$ and $h' \in V_q$ such that $g = Ag'$ and $h = Ah'$. Hence, $f = g + h = A(g' + h') = Af'$ with $f' \in V_{p \vee q}$. $\blacksquare$

Note that the continuity assumption for the inverse operators $(A_{up})^{-1}$, $(A_{vq})^{-1}$ is automatically satisfied if the assaying subspaces are Banach or Hilbert spaces, by the closed graph theorem.

Lemma 3.3.30. *If A_{up} and A_{vq} have dense range, then both $A_{u \wedge v, p \wedge q}$ and $A_{u \vee v, p \vee q}$ have dense range.*

Proof. (i) We consider first $A_{u \vee v, p \vee q}$. Let $k \in V_{\overline{u} \wedge \overline{v}}$ be such that $\langle k | Af \rangle = 0$, for all $f \in V_{p \vee q}$. Writing $f = g + h$, with $g \in V_p$, $h \in V_q$, we get

$$\langle k | Ag \rangle + \langle k | Ah \rangle = 0. \tag{3.6}$$

On the other hand (see Section 2.2), $k \in V_{\overline{u} \wedge \overline{v}}$ may be identified with the pair $(k, -k) \in V_{\overline{u}} \oplus V_{\overline{v}}$ and $Af \in V_{u \vee v}$ may be identified with (the equivalence class of) $(Ag, Ah) \in V_u \oplus V_v$, that is, $Af \equiv (Ag + l, Ah + l)$, $l \in V_{u \wedge v}$ (see the proof of Lemma 2.2.2). By these identifications, we may write (here $\langle\!\langle \cdot | \cdot \rangle\!\rangle$ denotes the pairing between $V_{\overline{u}} \oplus V_{\overline{v}}$ and $V_u \oplus V_v$)

$$0 = \langle k | Af \rangle = \langle\!\langle (k, -k) | (Ag, Ah) \rangle\!\rangle$$
$$= \langle k | Ag \rangle - \langle k | Ah \rangle.$$

Comparing with (3.6), we get $\langle k | Ag \rangle = \langle k | Ah \rangle = 0$. Since A_{up} and A_{vq} have dense range, this implies $k_{\overline{u}} = k_{\overline{v}} = 0$, hence $k = 0$.

(ii) Take now $A_{u \wedge v, p \wedge q}$. Let $k \in V_{\overline{u} \vee \overline{v}}$ be such that $\langle k | Af \rangle = 0$, for all $f \in V_{p \wedge q}$. This time, we identify $Af \in V_{u \wedge v}$ with the pair $(Af, -Af) \in V_u \oplus V_v$ and $k = g + h$, with $g \in V_{\overline{u}}, h \in V_{\overline{v}}$ with $(g, h) \in V_{\overline{u}} \oplus V_{\overline{v}}$ (here too, the decomposition is not unique). Then we have

$$0 = \langle k | Af \rangle = \langle g | Af \rangle + \langle h | Af \rangle.$$

On the other hand,

$$0 = \langle k | Af \rangle = \langle\!\langle (g, h) | (Af, -Af) \rangle\!\rangle$$
$$= \langle g | Af \rangle - \langle h | Af \rangle.$$

Thus again $\langle g | Af \rangle = \langle h | Af \rangle = 0$. Since A_{up} has dense range, so has $A_{u, p \wedge q}$, by the nondegeneracy of the partial inner product, hence $g_{\overline{u}} = 0$. In the same way, we get $h_{\overline{v}} = 0$, thus finally $k = g + h = 0$. $\blacksquare$

Proof of Theorem 3.3.28. (i) By assumption, $(A - \lambda)_{sr}$ is invertible, hence its inverse $R_{rs} := (A_{sr} - \lambda E_{sr})^{-1}$ is continuous, injective and has dense range. This defines a symmetric operator $R = R^\times \in \mathrm{Op}(V_I)$. Since $r \leqslant s$, we may consider $R_{rr} = R_{rs}E_{sr}$. Being a successor of an invertible operator, R_{rr} is continuous, injective, with dense range. Since $R = R^\times$, the representative $R_{\bar{r}\bar{r}}$ exists and has the same properties. By Lemmas 3.3.29 and 3.3.30, the representatives $R_{r\wedge\bar{r},r\wedge\bar{r}}$, $R_{r\vee\bar{r},r\vee\bar{r}}$ exist also, they are injective and have dense range. Since V_r is a Banach space, Proposition 3.3.23 (ii) tells us that $V_{r\wedge\bar{r}} \hookrightarrow V_o \hookrightarrow V_{r\vee\bar{r}}$ and that $(o,o) \in \mathrm{j}(R)$. Then R_{oo} is a self-adjoint operator in V_o.

(ii) R_{oo} is injective. Indeed, take $f \in V_o$ such that $Rf = 0$. Then $f \in V_{r\vee\bar{r}}$ and we have

$$\|Rf\|_{r\vee\bar{r}} \leqslant c\|Rf\|_o = 0,$$

thus $R_{r\vee\bar{r},r\vee\bar{r}}f_{r\vee\bar{r}} = 0$. Since $R_{r\vee\bar{r},r\vee\bar{r}}$ is injective, this implies $f = 0$.

(iii) R_{oo} has dense range. Indeed, assume that $\langle Rf|g \rangle = 0$ for all $f \in V_o$. Consider, in particular, $f \in V_{r\wedge\bar{r}}$. then, $\langle R_{r\wedge\bar{r},r\wedge\bar{r}}f_{r\wedge\bar{r}}|g_{r\vee\bar{r}} \rangle = 0$ for all $g \in V_{r\vee\bar{r}}$. But we know that $R_{r\wedge\bar{r},r\wedge\bar{r}}$ has dense range, thus $g = 0$.

Since R_{oo} is injective and has dense range, its inverse $A - \lambda := (R_{oo})^{-1}$, thus also A itself, is defined on a dense domain and is self-adjoint. The rest is as in the proof of Theorem 3.3.27. ∎

Theorem 3.3.28 may be generalized further, for instance, by postulating the existence of two suitable representatives of $A - \lambda$, or assuming that the PIP-space V_I satisfies particular interpolation properties, but the main interest rests with the cases of a LBS or a LHS, and the results stated here are quite sufficient for the applications. Hence, we will refrain from going further in this direction.

There remains the question of the existence of an invertible representative, needed in Theorems 3.3.27 and 3.3.28. If V_r and V_s are Hilbert spaces, we can use the following trivial criterion: If W is unitary from V_r onto V_s, then all bounded operators B in the ball $\|W - B\|_{sr} < 1$ (bound norm with respect to $\|\cdot\|_r$ and $\|\cdot\|_s$) are invertible. This leads to a different formulation of the basic problem in perturbation theory of self-adjoint operators, namely: Given symmetric operators T and U, does $T + U$ admit selfadjoint restrictions and what are their spectral properties? The following proposition answers the first question in the proper PIP-space language.

Proposition 3.3.31. *Let $T = T^\times$ and $U = U^\times$ be two symmetric operators in $\mathrm{Op}(V_I)$. Assume that there exists a $V_r \subseteq V_o$ and a $V_s \supseteq V_o$, such that $T_{sr} - \lambda E_{sr}$ is unitary from V_r onto V_s and that $\|U\|_{sr} < 1$ (bound norm with respect to $\|\cdot\|_r$ and $\|\cdot\|_s$). Then $A = T + U$ has a unique self-adjoint restriction, and all the other conclusions of Theorem 3.3.27 hold.*

Of course, Theorem 3.3.28 admits an identical treatment.

Proposition 3.3.32. *Let V_I be as in Theorem 3.3.28 and let $T = T^\times$ and $U = U^\times$ be two symmetric operators in $\mathrm{Op}(V_I)$. Assume there are assaying* *subspaces $V_r \subseteq V_s$, with V_r a Banach space, such that $(T - \lambda)_{sr} : V_r \to V_s$ is unitary and $\|U\|_{sr} < 1$. Then $A = T + U$ has a unique self-adjoint restriction to V_o, and all the other conclusions of Theorem 3.3.27 hold.*

The proof is immediate, since $A - \lambda$ satisfies the assumptions of Theorem 3.3.28. Of course, in practice, it is not always obvious to check that $\|U\|_{sr} < 1$ for the *same* pair r, s that is adequate for T. However, it is sufficient to ask that $\|U\|_{uv} < 1$ for some $u < s$, $v > r$. In particular, u and v need not be comparable, although r and s are.

The above results can be made entirely explicit if T is given as a multiplication operator by a function $t(p)$ on a measure space and if U is given as a kernel $U(p, p')$ on the same space. An example which covers the Schrödinger equation in momentum representation will be discussed in Section 7.1.3.

If V_I happens to be the scale of Hilbert spaces built on the powers of a self-adjoint operator, then the well-known *commutator theorem* gives a sufficient condition for the symmetric operator acting on V_o to have a self-adjoint extension. This theorem will be conveniently cast into the PIP-space language below.

More precisely, we consider the Hilbert scale built around the self-adjoint operator $N^{1/2} > 1$. That is, for each $n \in \mathbb{N}$, we let $\mathcal{H}_n = Q(N^n) := D(N^{n/2})$, equipped with the norm $\|f\|_n = \|N^{n/2}f\|$, $f \in D(N^{n/2})$, and $\mathcal{H}_{\overline{n}} := \mathcal{H}_{-n} = \mathcal{H}_n^{\times}$, its conjugate dual:

$$\ldots \mathcal{H}_n \subset \ldots \subset \mathcal{H}_2 \subset \mathcal{H}_1 \subset \mathcal{H}_0 \ \subset \mathcal{H}_{\overline{1}} \subset \mathcal{H}_{\overline{2}} \subset \ldots \subset \mathcal{H}_{\overline{n}} \ldots \quad (3.7)$$

Since $\mathcal{H}_{\overline{n}}$ is isomorphic to $Q(N^{-n}) := D(N^{-n/2})$, equipped with the norm $\|f\|_n = \|N^{-n/2}f\|$, we will identify the two, thus we get the LHS $V_I = \{\mathcal{H}_n, n \in \mathbb{Z}\}$.

The Hilbert space operator N determines uniquely a PIP-space operator $N \in \mathrm{Op}(V_I)$. This N is symmetric and $(n + 2, n) \in \mathrm{j}(N)$, $\forall n \in \mathbb{Z}$. Let now $A = A^{\times} \in \mathrm{Op}(V_I)$ be another symmetric operator. Thus $\mathrm{j}(A)$ contains the initial segment $\{(n, \overline{n}), n \geqslant n_{\min}\}$ (beware, the order is inverse here !). For any such n, we can define the commutator $C := [N, A] \in \mathrm{Op}(V_I)$ as follows:

$$Cf = [N, A]f = N(Af) - A(Nf), \quad f \in \mathcal{H}_{n+2},$$

that is (for better readability, we write $A_{p,q}$ instead of A_{pq} for the representatives of the various operators),

$$(Cf)_{\overline{n+2}} = N_{\overline{n+2},\overline{n}} A_{\overline{n},n} E_{n,(n+2)} f_{n+2} - E_{\overline{n+2},\overline{n}} A_{\overline{n},n} N_{n,(n+2)} f_{n+2}.$$

Indeed, since $(n + 2, n) \in \mathrm{j}(N)$, $(n, \overline{n}) \in \mathrm{j}(A)$ and $(\overline{n}, \overline{n+2}) \in \mathrm{j}(N)$, the two triple products are well-defined and we have $(n + 2, \overline{n+2}) \in \mathrm{j}(C)$.

Using this language, we can state the PIP-space version of Nelson's commutator theorem (already the original theorem has a distinctly PIP-space flavor).

Theorem 3.3.33 (Generalized commutator theorem). *Let* $N \geqslant 1$ *be a self-adjoint operator in the Hilbert space* $\mathcal{H}_0$, $V_I := \{\mathcal{H}_n, n \in \mathbb{Z}\}$ *the scale (3.7) built around* $N^{1/2}$ *and* $N \in \mathrm{Op}(V_I)$ *the corresponding* PIP*-space operator. Let* $A = A^\times \in \mathrm{Op}(V_I)$ *with* $(1, \bar{1}) \in j(A)$, *and assume, in addition, that the commutator* $C := [N, A]$ *is such that* $(1, \bar{1}) \in j(C)$ *as well. Then:*

(i) *The operator* $A_{\bar{1}1}$ *admit a unique restriction to a symmetric operator* A *on* $\mathcal{H}_0$, *with dense domain* $D(\mathsf{A}) = \{f \in \mathcal{H}_1 : Af \in \mathcal{H}_0\}$.
(ii) $D(\mathsf{N}) \subset D(\mathsf{A})$ *and, for all* $f \in D(\mathsf{N})$, *one has* $\|\mathsf{A}f\| \leqslant c\|\mathsf{N}f\|$.
(iii) A *is essentially self-adjoint on any core for* N.

It is important to notice that the condition $(1, \bar{1}) \in j(C)$ is crucial, without it, the domain $D(\mathsf{A})$ need not be dense, it might even be $\{0\}$.

It is well-known that the classical KLMN theorem may also conveniently be formulated in terms of quadratic forms. The space $\mathcal{H}_1$ in (3.5) may be taken as the form domain $Q(A) := D(A^{1/2})$ of some positive self-adjoint operator A in $\mathcal{H}_0$, with the form norm $\|f\|_1^2 = \langle f|Af\rangle + \|f\|^2$. This suggests a reformulation of the KLMN theorem in terms of quadratic forms.

Theorem 3.3.34 (KLMN theorem, quadratic form variant). *Let* A *be a positive self-adjoint operator in a Hilbert space and suppose that* $\beta(f, g)$ *is a symmetric quadratic form on the form domain* $Q(A)$ *such that*

$$|\beta(f, f)| \leqslant a(f, Af) + b(f, f), \quad \forall f \in Q(A)$$

for some $a < 1$ *and* $b \in \mathbb{R}$. *Then there exists a unique self-adjoint operator* C *with* $Q(C) = Q(A)$ *and*

$$(f, Cg) = (f, Ag) + \beta(f, g), \quad \forall f, g \in Q(C).$$

C *is bounded below by* $-b$.

In the language of quadratic forms, the triplet (3.5) may be obtained as follows. Let B_{00} a bounded, positive, injective operator in $\mathcal{H}_0$. Define $\mathcal{H}_{\bar{1}}$ as the completion of $\mathcal{H}_0$ with respect to the norm $\|f\|_{\bar{1}} = \|B_{00}^{1/2}f\|_0 = \langle f|B_{00}f\rangle^{1/2}$. Let $\mathcal{H}_1$ be the range of $(B_{00})^{1/2} : \mathcal{H}_1 = (B_{00})^{1/2}\mathcal{H}_0$. It is a Hilbert space with respect to the scalar product $\langle f|g\rangle_1 = \langle B_{00}^{-1/2}f|B_{00}^{-1/2}g\rangle_0$, i.e., $\mathcal{H}_1 = Q(B_{00}^{-1})$, with the form norm. Then $\mathcal{H}^{(B)} = \mathcal{H}_{\bar{1}}$ becomes a nondegenerate PIP-space if $f\#g$ means: either both f and g belong to $\mathcal{H}_0$, or at least one of them belongs to $\mathcal{H}_1$. In the first case, we define $\langle f|g\rangle = \langle f|g\rangle_0$. In the second case, $\langle f|g\rangle = \lim_{n\to\infty}\langle B_{00}^{1/2}f_n|B_{00}^{-1/2}g\rangle_0$ where $\|f_n - f\|_{\bar{1}} \to 0, f_n \in \mathcal{H}_0$. The assaying subspaces of $\mathcal{H}^{(B)}$ are exactly $\mathcal{H}_1$, $\mathcal{H}_0$ and $\mathcal{H}_{\bar{1}}$.

Then the generalized KLMN theorem may be rephrased in terms of quadratic forms as follows.

Proposition 3.3.35. *Let* q *be a densely defined, closed, symmetric quadratic form on a Hilbert space* $\mathcal{H}_0$, *such that* $q+\lambda$ *is strictly positive for some* $\lambda \in \mathbb{R}$.

Equipped with the inner product $\langle f|g\rangle_1 = \lambda\langle f|g\rangle_0 + q(f,g)$, the domain of q is a Hilbert space, that we call $\mathcal{H}_1$. Consider in $\mathcal{H}_0$ the bounded positive injective operator $B_{00} = E_{01}(E_{01})'_{10}$ where E_{01} is the natural embedding of $\mathcal{H}_1$ into $\mathcal{H}_0$ and $(E_{01})'_{10} : \mathcal{H}_0 \to \mathcal{H}_1$ is its adjoint (in the sense of bounded operators between Hilbert spaces). Consider the partial inner product space $\mathcal{H}^{(B)}$, defined above. Let H be the element of $\mathrm{Op}(\mathcal{H}^{(B)})$ defined by $H_{\bar{1}1} - \lambda E_{\bar{1}1} = (B_{1\bar{1}})^{-1}$. Then the self-adjoint restriction of $H_{\bar{1}1}$ is the self-adjoint operator associated to the form q.

Proof. The operator B_{00} is bounded and injective, hence we may construct the triplet (3.5) as above. By construction, the operator $B \in \mathrm{Op}(\mathcal{H}^{(B)})$ which has B_{00} as $\{0,0\}$-representative has a unitary representative $B_{1\bar{1}}$ from $\mathcal{H}_{\bar{1}}$ onto $\mathcal{H}_1$. Thus $(B_{1\bar{1}})^{-1}$ is well-defined, and so is $H_{\bar{1}1}$. According to Theorem 3.3.27, it has a unique self-adjoint restriction.   ■

These considerations will play an important role in applications of the PIP-space formalism in quantum mechanics, for instance, in the rigorous treatment of singular interactions (δ- or δ'-potentials in various configurations). We will discuss these applications in Section 7.1.

3.4 Orthogonal Projections and Orthocomplemented Subspaces

3.4.1 Orthogonal Projections

The notion of orthogonal projection was introduced in Section 1.2, on the purely algebraic level, then exploited more thoroughly in Section 3.4.2 in the context of direct sum decompositions of group representations. As pointed there, we aim at recovering the Hilbert space correspondence between projection operators and orthocomplemented subspaces, the latter being the proper notion of PIP-subspaces. We emphasize that the partial inner product is *not* required to be positive definite, but, of course, it is supposed to be nondegenerate. This justifies the following definition.

Definition 3.4.1. An *orthogonal projection* on a nondegenerate PIP-space V (not necessarily positive definite) is a homomorphism $P \in \mathrm{Hom}(V)$ such that $P^2 = P^\times = P$. A similar definition applies in the case of a LBS or a LHS V_I.

It follows immediately from the definition that an orthogonal projection is totally regular, i.e., $\mathrm{j}(P)$ contains the diagonal $I \times I$, or still that P leaves every assaying subspace invariant. Equivalently, P is an orthogonal projection if P is an idempotent operator (that is, $P^2 = P$) such that $\{Pf\}^\# \supseteq \{f\}^\#$ for every $f \in V$ and $\langle g|Pf\rangle = \langle Pg|f\rangle$ whenever $f \# g$ (see Proposition 1.2.4). We denote by $\mathrm{Proj}(V)$ the set of all orthogonal projections in V and similarly for $\mathrm{Proj}(V_I)$.

In the case of a LHS, a more restricted concept is that of a *totally orthogonal* projection. By this, we mean an orthogonal projection P such that each representative $P_{rr}, r \in I$, is an orthogonal projection in the Hilbert space V_r, thus orthogonal with respect to the inner product $\langle \cdot | \cdot \rangle_r$. Examples will be encountered in Section 5.5.4.

Proposition 3.4.2. *Let V be a nondegenerate partial inner product space, and P_1, P_2 orthogonal projections in V. Then*

- (i) *1 (the identity) and 0 (the zero map) belong to $\mathrm{Proj}(V)$.*
- (ii) *If $P_1 P_2 = P_2 P_1$, then $P_1 P_2 \in \mathrm{Proj}(V)$.*
- (iii) *If $P_1 P_2 = P_2 P_1 = 0$, then $P_1 + P_2 \in \mathrm{Proj}(V)$.*
- (iv) *In particular, $1 - P \in \mathrm{Proj}(V)$ for every $P \in \mathrm{Proj}(V)$.*

The proof is an immediate verification.

Example 3.4.3. If $f \in V^{\#}$ and $\langle f | f \rangle = 1$, then $P_f = |f\rangle\langle f| \in \mathrm{Proj}(V)$. By (iii) above, this generalizes to a finite sum of projections corresponding to a family of mutually orthogonal, normalized vectors. Additional examples of orthogonal projections will be discussed below.

Proposition 3.4.4. *Let $P \in \mathrm{Proj}\, V$, and $Q = 1 - P$. Then, for every $r \in F(V)$,*

- (i) *V_r decomposes into a (topological) direct sum $V_r = PV_r \oplus QV_r$.*
- (ii) *The representatives P_{rr}, Q_{rr} are related by $\mathrm{Ker}\, P_{rr} = \mathrm{Ran}\, Q_{rr}$.*
- (iii) *PV_r and $PV_{\overline{r}}$ are a dual pair, and so are QV_r and $QV_{\overline{r}}$.*
- (iv) *One has $(PV_r)^{\perp} \cap V_{\overline{r}} = QV_{\overline{r}}$.*
- (v) *If $r \leqslant s$, then PV_r is dense in PV_s for the topology $\tau(V_s, V_{\overline{s}})$.*

Proof. (i) and (ii) follow immediately from $PQ = 0$ and $PV_r \subseteq V_r$, $QV_r \subseteq V_r$ (continuously). In order to prove (iii), notice first that PV_r and $PV_{\overline{r}}$, are compatible. Now take $f \in PV_r$ and $f \in (PV_{\overline{r}})^{\perp}$. The first relation shows that f is orthogonal to $QV_{\overline{r}}$. Since it is also orthogonal to $PV_{\overline{r}}$, it is orthogonal to $V_{\overline{r}}$ and is consequently zero. In order to prove (iv), notice that

$$(PV_r)^{\perp} \cap V_{\overline{r}} \subseteq [(PV_r)^{\perp} \cap PV_{\overline{r}}] \cup [(PV_r)^{\perp} \cap QV_{\overline{r}}]$$
$$= (PV_r)^{\perp} \cap QV_{\overline{r}} = QV_{\overline{r}}.$$

The inclusion in the other direction is immediate.

To prove (v), we compute the closure of PV_r in $V_s[\tau(V_s, V_{\overline{s}})]$ using Proposition 2.1.2. We have

$$\mathrm{clos}(PV_r) = ((PV_r)^{\perp} \cap V_{\overline{s}})^{\perp} \cap V_s = ((PV_r)^{\perp} \cap V_{\overline{r}} \cap V_{\overline{s}})^{\perp} \cap V_s$$
$$= (QV_{\overline{r}} \cap V_{\overline{s}})^{\perp} \cap V_s = (QV_{\overline{s}})^{\perp} \cap V_s = PV_s. \qquad \blacksquare$$

Remark 3.4.5. As a special case, we find the useful result:

$$\operatorname{Ker} P = \operatorname{Ran} Q = (PV^{\#})^{\perp} = (W \cap V^{\#})^{\perp}.$$

Given a subspace W of V, we may consider on it the inherited compatibility $\#_W$ and the inherited partial inner product $\langle \cdot | \cdot \rangle_W$ defined as follows. For any $f, g \in W$,

- $f \#_W g$ whenever $f \# g$,
- $\langle f | g \rangle_W = \langle f | g \rangle$.

Corollary 3.4.6. *If W is the range of an orthogonal projection, then it is a nondegenerate partial inner product space with the inherited compatibility and partial inner product.*

3.4.2 Orthocomplemented Subspaces

In our quest for a generalization of the Hilbert space correspondence between projection operators and orthocomplemented subspaces, the first result yields the 'algebraic' part of the property sought for, namely,

Lemma 3.4.7. *A vector subspace W of the nondegenerate PIP-space V is orthocomplemented if and only if it is the range of an orthogonal projection.*

Proof. Let W be orthocomplemented, $V = W \oplus Z$. For every $f \in V$, let $f = f_W + f_Z$ be its unique decomposition. Write $f_W = Pf$. Then P is an orthogonal projection. Indeed, P is idempotent and improves behavior by assumption. In order to verify that P is symmetric, let $g \# h$, $g = g_W + g_Z$, $h = h_W + h_g$. Then

$$\langle Pg | h \rangle = \langle g_W | h_W + h_z \rangle = \langle g_W | h_W \rangle = \langle g_W | Ph \rangle = \langle g | Ph \rangle.$$

All the inner products exist since $\#$ is reduced.

Conversely, let $W = PV$ be the range of an orthogonal projection. Then W is clearly orthocomplemented with $Z = (1 - P)V$. ∎

What remains to prove is that an orthocomplemented subspace is a PIP-space in its own right, i.e., a PIP-subspace, exactly as an orthocomplemented subspace in a Hilbert space is closed, hence a Hilbert space. To that effect, we need two new topological notions, that of a topologically regular subspace and of a subspace self-sufficient in linear forms.

Definition 3.4.8. (1) A vector subspace W of a nondegenerate PIP-space V is said to be *topologically regular* if it satisfies the following two conditions:

(i) for every assaying subset $V_r \subseteq V$, the intersections $W_r = W \cap V_r$ and $W_{\bar{r}} = W \cap V_{\bar{r}}$ are a dual pair in V;

(ii) the intrinsic Mackey topology $\tau(W_r, W_{\bar{r}})$ coincides with the Mackey topology $\tau(V_r, V_{\bar{r}})|_{W_r}$ induced by V_r.

(2) A vector subspace W of a nondegenerate PIP-space V is said to be *self-sufficient in linear forms* if the following holds: For every $f \in V$, there exists a $f_W \in W$ such that $\{f_W\}^\# \supseteq \{f\}^\#$ and

$$\langle f_W | _{W \cap \{f\}^\#} = \langle f | _{W \cap \{f\}^\#}.$$

Note that, in the statement (ii) of (1), Mackey topologies can be replaced by weak topologies. As for (2), the condition means that the restriction of all linear forms $\langle f |$ with $f \in W$ can be "internally generated."

Now we can state our main result.

Theorem 3.4.9. *For a vector subspace W of the nondegenerate PIP-space V, the following conditions are equivalent:*

(i) W is orthocomplemented;
(ii) W is the range of an orthogonal projection;
(iii) W is topologically regular.
(iv) W is self-sufficient in linear forms.

We shall divide the proof into several lemmas.

Lemma 3.4.10. *If W is the range of an orthogonal projection, then W is topologically regular.*

Proof. Let $W = PV$. According to (i) and (iii) of Proposition 3.4.4, V_r and $V_{\bar{r}}$ decompose into topological direct sums

$$V_r = PV_r \oplus QV_r \quad \text{and} \quad V_{\bar{r}} = PV_{\bar{r}} \oplus QV_{\bar{r}},$$

with $W_r = PV_r, W_{\bar{r}} = PV_{\bar{r}}, Q = 1 - P$, and $\langle W_r, W_{\bar{r}} \rangle$ is a dual pair. Hence W_r, with the topology induced by $\tau(V_r, V_{\bar{r}})$, is topologically isomorphic to the quotient V_r/QV_r, the latter being equipped with its quotient topology. By (iv) of the same Proposition 3.4.4,

$$QV_r = (PV_{\bar{r}})^\perp \cap V_r = (W_{\bar{r}})^\perp \cap V_r.$$

Since V_r carries its Mackey topology $\tau(V_r, V_{\bar{r}})$ and Mackey topologies are inherited by quotients, the quotient topology on V_r/QV_r is $\tau(V_r/(W_{\bar{r}})^\perp \cap V_r, W_{\bar{r}}) = \tau(W_r, W_{\bar{r}})$ (see Section B.7). ∎

In order to complete the proof of Theorem 3.4.9, we have to show that every topologically regular subspace is the range of an orthogonal projection. To that effect, we need the notion of a subspace self-sufficient in linear forms, given in Definition 3.4.8 (2). First we give two simple properties of such linear forms.

Lemma 3.4.11. *Let W be self-sufficient in linear forms. Then*

$$(W \cap V^{\#})^{\perp} \cap W = \{0\}.$$

Proof. Let $f \in V^{\#}$, then $\{f\}^{\#} = V, W \cap \{f\}^{\#} = W$. By self-sufficiency, there exists at least one $f_w \in W \cap V^{\#}$ such that $\langle f_w |_W = \langle f|_W$. Let $g \in W$ be orthogonal to $W \cap V^{\#}$. Then

$$0 = \langle f_w | g \rangle = \langle f | g \rangle,$$

that is, g is orthogonal to every $f \in V^{\#}$. Hence $g = 0$. ∎

Lemma 3.4.12. *Let W be self-sufficient in linear forms. Given $f \in V$, there exists only one $f_w \in W$ such that $\{f_w\}^{\#} \supseteq \{f\}^{\#}$ and that $\langle f_w | = \langle f |$ on $W \cap \{f\}^{\#}$.*

Proof. Let f_w^1 and f_w^2 be two such vectors and $g = f_w^1 - f_w^2$. Then $g \in W$ and $\langle g | h \rangle = 0$ for every $h \in W \cap \{f\}^{\#}$. A fortiori, g is orthogonal to every $h \in W \cap V^{\#}$, hence $g = 0$ by the preceding lemma. ∎

Then we proceed with the proof of Theorem 3.4.9.

Proposition 3.4.13. *If W is topologically regular, then it is self-sufficient in linear forms.*

Proof. Let $f \in V$. We want to find a $f_w \in W$ such that $\langle f_w |_{W \cap \{f\}^{\#}} = \langle f |_{W \cap \{f\}^{\#}}$. Define r by $V_r = \{f\}^{\#}, V_{\bar{r}} = \{f\}^{\#\#}$, and consider the intersections $W_r = W \cap V_r, W_{\bar{r}} = W \cap V_{\bar{r}}$. By the assumption of regularity, W_r and $W_{\bar{r}}$ are a dual pair, and $\tau(W_r, W_{\bar{r}})$ coincides with the topology inherited from $\tau(V_r, V_{\bar{r}})$. Consequently, the restriction of $\langle f |$ to $W_r = W \cap \{f\}^{\#}$ is continuous for $\tau(W_r, W_{\bar{r}})$. So there exists a unique $f_w \in W_{\bar{r}}$ such that $\langle f_w |_{W_r} = \langle f |_{W_r}$. Since $W_{\bar{r}} \subseteq V_{\bar{r}} = \{f\}^{\#\#}$, one has $\{f_w\}^{\#} \supseteq \{f\}^{\#}$, which proves the statement. ∎

Proposition 3.4.14. *If W is self-sufficient in linear forms, then it is the range of a (unique) orthogonal projection.*

Proof. The correspondence $f \mapsto f_w$ is well defined by virtue of Lemma 3.4.12. Write $f_w = Pf$. Then we claim that P is an orthogonal projection. Since f_w is at least as good as f, P improves behavior. For any $f \in W, f = Pf$; thus W is the range of P, and P is idempotent, since $Pf \in W$, for any $f \in V$, and $P^2 f = Pf$. Finally, let $f \# g$. Since P improves behavior, $f \# Pg, Pf \# g$ and $Pf \# Pg$. By the definition of P, $\langle Pf | h \rangle = \langle f | h \rangle$ for every $f \in W$. In particular, we have, for any $f \# g$,

$$\langle f | Pg \rangle = \langle Pf | Pg \rangle = \langle Pf | g \rangle.$$ ∎

Proposition 3.4.14 completes the proof of Theorem 3.4.9. The latter gives a complete characterization of all orthocomplemented subspaces. Exactly the same result holds for the simple case of inner product spaces [Bog74, Thm.III.7.2].

Examples 3.4.15. (1) Let $V = L^1_{\text{loc}}(X, d\mu)$ as usual. Take any partition of X into two measurable subsets, of nonzero measure, $X = \Omega \cup \Omega'$, as in Example 3.3.10(2). Then V is decomposed in two orthocomplemented subspaces:

$$V = L^1_{\text{loc}}(\Omega, d\mu_\Omega) \oplus L^1_{\text{loc}}(\Omega', d\mu_{\Omega'}),$$

where $\mu_\Omega, \mu_{\Omega'}$ are the restrictions of μ to Ω, resp. Ω'. The orthogonal projection P_Ω is the operator of multiplication by the characteristic function χ_Ω of Ω. Similarly, $P_{\Omega'} = 1 - P_\Omega$ is the operator of multiplication by $\chi_{\Omega'}$. They are both absolute. Similar considerations hold true for $V = \omega$, as in Example 3.3.10(1).

(2) An example of nonabsolute projection may be obtained as follows. Consider again $V = L^1_{\text{loc}}(X, d\mu)$ with the following compatibility relation: $f \# g$ if at least one of them belongs to $L^\infty_c(X, d\mu)$. Then $V^\# = L^\infty_c(X, d\mu)$ and the only assaying subspaces are V and $V^\#$. Let again Ω be a measurable subset of X and P be the multiplication by χ_Ω, the characteristic function of Ω. This P is obviously an orthogonal projection. Take now an element $f \in V$ with support in Ω and another one, g say, with support in the complement $\Omega' = X \setminus \Omega$, but none of them in $V^\#$. Thus f and g are not compatible. By construction, we have $Pf = f$ and $(1 - P)g = g$. Thus Pf and $(1 - P)g$ are not compatible, i.e., P is not absolute.

Remark 3.4.16. (1) *Quotients*: Let W be an orthocomplemented subspace of V. Let P be the orthogonal projection corresponding to $W : W = PV$ and Q the projection $1 - P$.

Consider the vector space V/W. Notice that $g \in W$ is equivalent to $Qg = 0$, so that the vector Qf is independent of the choice of f in a W-coset. If f and g are two elements of V/W, they will be called compatible (in V/W) if and only if Qf and Qg are compatible (in V). The scalar product of f and g is defined as $\langle Qf|Qg\rangle$. In this way V/W becomes a partial inner product space, which clearly may be identified with $W^\perp$. So, as for inner product spaces (in particular, Hilbert spaces), the notions of orthocomplemented subspaces and quotient spaces coincide.

(2) *Orthogonal sums*: Let V_1 and V_2 be nondegenerate partial inner product spaces, with compatibilities $\#_1$ and $\#_2$. Let $\#_{12}$ be a compatibility between V_1 and V_2, such that every vector in $V_1^{\#_1}$ is $\#_{12}$-compatible to all of V_2, and every vector in $V_2^{\#_2}$ is $\#_{12}$-compatible to all of V_1. Consider the direct sum $V = V_1 \oplus V_2$, and define in it a compatibility following the procedure of Example 1.2.2. Then, for all compatible pairs, define the following inner product, which is obviously nondegenerate:

$$\langle\{v_1, v_2\}|\{v_1', v_2'\}\rangle = \langle v_1|v_1'\rangle + \langle v_2|v_2'\rangle.$$

In order to extend this definition to infinitely many summands $\{V_j\}_{j\in J}$, one has to further restrict the compatibility in $V = \Pi_{j\in J}V_j$ by the requirement

$$\sum_{j\in J}|\langle v_j|w_j\rangle| < \infty.$$

3.4.3 The Order Structure of Proj(V)

Since an orthocomplemented subspace W defines uniquely the corresponding projection P_W, we may define a partial order in $\mathrm{Proj}(V)$ by $P_W \leqslant P_Y$ if and only if $W \subseteq Y$. It is easy to see that this is equivalent to

$$P_W P_Y = P_Y P_W = P_W.$$

There are many questions about $\mathrm{Proj}(V)$ which we are unable to answer in general. For example: When is $\mathrm{Proj}(V)$ a (complete) lattice?

We are able, however, to compare $\mathrm{Proj}(V)$ and the partially ordered set $\mathrm{Proj}(V^\#)$ of all the orthogonal projections in the *inner* product space $V^\#$.

Theorem 3.4.17. *Let V be an nondegenerate partial inner product space and W an orthocomplemented subspace of V. Then $W \cap V^\#$ is an orthocomplemented subspace of the* inner *product space $V^\#$. The correspondence $W \mapsto W\cap V^\#$ defines an injective order-preserving map α from $\mathrm{Proj}(V)$ into $\mathrm{Proj}(V^\#)$. An orthogonal projection $P \in \mathrm{Proj}(V^\#)$ lies in the range of α if and only if it is continuous in every one of the topologies $\tau(V^\#, V_r)$.*

Proof. Let W be an orthocomplemented subspace of V and P_W the corresponding projection. One verifies immediately that the restriction of P_W to $V^\#$ is an orthogonal projection in $V^\#$; as such, it is continuous for $\tau(V^\#, V^\#)$, which is the topology induced on $V^\#$ by $\tau(V, V^\#)$. Consequently, $P_W V^\#$ is an orthocomplemented subspace of $V^\#$. Furthermore, $P_W V^\# = W \cap V^\#$, which proves the first assertion.

Notice that, for each $r \in F$, the restriction of P_W to $V^\#$ is continuous in the topology induced on $V^\#$ by V_r, namely $\tau(V^\#, V_{\bar{r}})$.

Conversely, let $P \in \mathrm{Proj}(V^\#)$ be continuous for every one of the topologies $\tau(V^\#, V_q)$. Then P can be extended to a unique $P \in \mathrm{Proj}(V)$. Indeed P can be considered as the representative (from $V^\#$ to $V^\#$) of a unique element P_W of $\mathrm{Op}(V)$; our assumptions guarantee that this element is an orthogonal projection. ∎

3.4.4 Finite-Dimensional PIP-*Subspaces*

In the case of a finite-dimensional subspace, the situation simplifies. Indeed:

Proposition 3.4.18. *Let V be a nondegenerate partial inner product space, W a finite-dimensional subspace of V. Then W is orthocomplemented if and only if it is contained in $V^{\#}$ and nondegenerate, i.e., $W \cap W^{\perp} = \{0\}$.*

Proof. Let W be finite-dimensional and orthocomplemented. By Proposition 3.4.4 (v), $W \cap V^{\#}$ is dense in W, which means $W \cap V^{\#} = W$, thus $W \subseteq V^{\#}$. By (iii) of the same proposition, $PV = W$ and $PV^{\#} = W$ are a dual pair, i.e., $W \cap W^{\perp} = \{0\}$.

Conversely, if $W \subseteq V^{\#}$, we have $W \cap V^{\#} = W$ and $W \cap V_r = W$ for every $r \in F$. Since $W \cap W^{\perp} = \{0\}$, $\tau(W, W)$ is well-defined and coincides with the topology induced by $\tau(V_r, V_{\overline{r}})$, for any r, since all separated topologies coincide on a finite-dimensional subspace. Thus W is topologically regular, and, therefore, orthocomplemented. ■

Corollary 3.4.19. *An orthogonal projection of finite-dimensional range is absolute.*

Notice that every vector of $W \cap W^{\perp}$ is necessarily orthogonal to itself. So, if the partial inner product is definite (i.e., $f \# f$ and $\langle f | f \rangle = 0$ imply $f = 0$), the condition of nondegeneracy is superfluous. Anyway, it is remarkable that not even all finite-dimensional subspaces of V are orthocomplemented. For instance, the one-dimensional subspace generated by δ in the space $\mathcal{S}^{\times}$ of tempered distributions is of this kind.

For the sake of completeness, it is useful to compare the general case with that of an *inner* product space, not necessarily positive definite. For an arbitrary (nondegenerate) inner product space, one knows that a subspace $W \subseteq V$ is orthocomplemented if and only if the intrinsic weak topology $\sigma(W, W)$ on W coincides with the topology $\sigma(V, V) \restriction W$ induced by V; this happens precisely when the intrinsic and induced Mackey topologies coincide: $\tau(W, W) = \tau(V, V) \restriction W$. Thus, in the case of a general PIP-space, the situation for orthocomplemented subspaces is essentially the same whether the inner product is defined everywhere or not. In addition, we see here that, for finite-dimensional subspaces, there is no difference at all.

Proposition 3.4.18 has important consequences for the structure of bases in PIP-spaces of type (B). First we recall that a sequence $\{e_n, n = 1, 2, \dots\}$ of vectors in a separable Banach space E is a *Schauder basis* if, for every $f \in E$, there exists a unique sequence of scalar coefficients $\{c_k, k = 1, 2, \dots\}$ such that $\lim_{m \to \infty} \| f - \sum_{k=1}^{m} c_k e_k \| = 0$. Then one may write

$$f = \sum_{k=1}^{\infty} c_k e_k. \tag{3.8}$$

The basis is *unconditional* if the series (3.8) converges unconditionally in E (i.e., it keeps converging after an arbitrary permutation of its terms).

A standard problem is to find, for instance, a sequence of functions that is an unconditional basis for *all* the spaces $L^p(\mathbb{R})$, $1 < p < \infty$. In the PIP-space language, this statement means that the basis vectors must belong to $V^\# = \cap_{1<p<\infty} L^p(\mathbb{R})$. Also, since (3.8) means that f may be approximated by finite sums, Proposition 3.4.18 implies that all the basis vectors must belong to $V^\#$. Some examples of unconditional wavelet bases will be given in Section 8.4.

Actually, in the context of signal processing, orthogonal (in the Hilbert sense) bases are not enough, one needs also biorthogonal bases and, more generally, *frames*. We recall that a countable family of vectors $\{g_n\}$ in a Hilbert space $\mathcal{H}$ is called a *frame* if there are two positive constants m,M, with $0 < m \leqslant M < \infty$, such that

$$\mathsf{m}\,\|f\|^2 \;\leqslant\; \sum_{n=1}^{\infty} |\langle g_n|f\rangle|^2 \;\leqslant\; \mathsf{M}\,\|f\|^2, \;\forall\, f \in \mathcal{H}. \tag{3.9}$$

The two constants m,M are called *frame bounds*. If m $=$ M, the frame is said to be *tight*. Consider the analysis operator $C : \mathcal{H} \to \ell^2$ defined by $C : f \mapsto \{\langle g_n|f\rangle\}$ and the frame operator $S = C^*C$. Then the vectors $\widetilde{g}_n = S^{-1}g_n$ also constitute a frame, called the *dual frame*, and one has the (strongly converging) expansions

$$f = \sum_{n=1}^{\infty} \langle g_n|f\rangle \widetilde{g}_n = \sum_{n=1}^{\infty} \langle \widetilde{g}_n|f\rangle\, g_n.$$

Then the considerations made above for bases should apply to frame vectors as well, i.e., the vectors $g_n, \widetilde{g}_n$ should also belong to $V^\#$.

3.4.5 Orthocomplemented Subspaces of Pre-Hilbert Spaces

The aim of this section is to illustrate the above results by specializing them to the familiar case of a pre-Hilbert space E. Theorem 3.4.9 applies immediately: A subspace is orthocomplemented if and only if it is the range of an orthogonal projection, P, i.e., an idempotent ($P^2 = P$), symmetric ($P^* = P$), $\tau(E, E)$-continuous map. But now, E carries also the norm topology ($\|f\|^2 = \langle f|f\rangle$) which is in general finer than the Mackey topology $\tau(E, E)$.

If E is norm-complete (that is, a Hilbert space), both topologies coincide, and the orthocomplemented subspaces are exactly the norm-closed ones. However, if E is not norm-complete, this is no longer true. Every idempotent, symmetric operator P from E into E is automatically *norm*-continuous and its range PE is an orthocomplemented subspace of E, the complement being, of course $(1 - P)E$. This implies that an orthocomplemented subspace is not only $\tau(E, E)$-closed, but also norm-closed (this can also be seen just by comparison of the two topologies). Yet orthocomplemented subspaces may fail to be norm-complete: think of E itself. However

Proposition 3.4.20. *Every norm-complete subspace M of a pre-Hilbert space E is orthocomplemented. In particular, every finite dimensional subspace is orthocomplemented.*

Proof. Let $x \in E$. Put $d = \inf_{y \in M} \|x - y\|$. Then there exists a sequence $\{y_n\} \subset M$ such that $\|x - y_n\| \to d$. Making use of the parallelogram law, we get

$$
\begin{aligned}
\|y_n - y_m\|^2 &= \|(y_n - x) - (y_m - x)\|^2 \\
&= 2\|y_n - x\|^2 + 2\|y_m - x\|^2 - \|-2x + y_n + y_m\|^2 \\
&= 2\|y_n - x\|^2 + 2\|y_m - x\|^2 - 4\|x - 1/2(y_n + y_m)\|^2 \\
&\leqslant 2\|y_n - x\|^2 + 2\|y_m - x\|^2 - 4d^2.
\end{aligned}
$$

The right-hand side tends to zero as $n, m \to \infty$. Hence $\{y_n\}$ is a Cauchy sequence; thus, it converges to an element $z \in M$, since M is complete. It is easily seen that $\|x - z\| = d$ and that z is the unique element in M with this property. Moreover, $x - z \in M^\perp$. Hence $E = M \oplus M^\perp$. ∎

Remark 3.4.21. The orthogonal complement $M^\perp$ of a norm-complete subspace M is clearly orthocomplemented, but it is necessarily incomplete with respect to the norm unless E is a Hilbert space.

We have seen in Proposition 3.4.4 that every orthocomplemented subspace is $\tau(E, E)$-closed. But the converse is not true in general: Not every $\tau(E, E)$-closed subspace is orthocomplemented.

In order to give a counterexample, we remind the reader that a vector subspace $W \subseteq E$ is $\tau(E, E)$-closed (or, equivalently, $\sigma(E, E)$-closed) if and only if $W^{\perp\perp} = W$. Consider now $E = \mathcal{S}(\mathbb{R})$, the Schwartz space of test functions. Let $W = \{\varphi \in \mathcal{S} : \varphi(x) = 0 \text{ for } x \leqslant 0\}$. Then $W^\perp = \{\psi \in \mathcal{S} : \psi(x) = 0 \text{ for } x \geqslant 0\}$. So $W^{\perp\perp} = W$. However, W is not orthocomplemented, since every $\chi \in W + W^\perp$ satisfies $\chi(0) = 0$; consequently, $W + W^\perp \neq \mathcal{S}$.

If we require that every $\tau(E, E)$-closed subspace of a pre-Hilbert space be orthocomplemented, we force E to be a Hilbert space (see Proposition 3.4.26 below). However we can prove the same result using only maximal subspaces.

We remind that a subspace M of E is called *maximal* if it is not contained in any other proper subspace of E. In this case we have

Proposition 3.4.22. *Let M be a proper maximal subspace of the pre-Hilbert space E. Then either M is orthocomplemented or $M^\perp = \{0\}$.*

Proof. If $M^\perp \neq \{0\}$ then $M \oplus M^\perp$ is (isomorphic to) a subspace of E which contains M properly. Hence $M \oplus M^\perp = E$. On the other hand, it is clear that if M is a proper orthocomplemented subspace of E, then $M^\perp \neq \{0\}$. ∎

Proposition 3.4.23. *Let M be a $\tau(E, E)$-closed proper maximal subspace of the pre-Hilbert spaceE. Then M is orthocomplemented.*

Proof. Were it not so, by Proposition 3.4.22, we would have $M^\perp = \{0\}$ and then $M^{\perp\perp} = E$. Since $M = M^{\perp\perp}$, we get $M = E$, a contradiction.

As it is known, maximal subspaces are exactly the kernels of linear functionals. Orthocomplemented maximal subspaces correspond to Mackey continuous linear functionals. Indeed, one has

Proposition 3.4.24. *Let f be a linear functional on the pre-Hilbert space E. Then f is $\tau(E, E)$-continuous if, and only if, its kernel $\mathrm{Ker} f$ is orthocomplemented.*

Proof. If $f = 0$, the statement is obvious. If $f \neq 0$, f is $\tau(E, E)$-continuous if and only if its kernel, which is a proper maximal subspace of E, is $\tau(E, E)$-closed and has a nontrivial (one-dimensional) orthogonal complement. So the statement follows from Proposition 3.4.22. On the other hand, f is not $\tau(E, E)$-continuous if and only if its kernel is $\tau(E, E)$-dense. This is equivalent to saying that $(\mathrm{Ker} f)^\perp = \{0\}$. The statement then follows once more from Proposition 3.4.22. ∎

Proposition 3.4.25. *If every norm-closed maximal subspace of the pre-Hilbert space E is orthocomplemented, then E is a Hilbert space.*

Proof. Let f be a nonzero bounded linear functional on E. Then its kernel $\mathrm{Ker} f$ is norm-closed, hence orthocomplemented. This implies that $\mathrm{Ker} f$ is also $\tau(E, E)$-continuous or, equivalently, $\sigma(E, E)$-continuous. In conclusion, every bounded functional in E is weakly continuous. In other words, the weak dual of E, which is E itself, coincides with the strong dual, which is the Hilbert space completion of E. Hence E is complete. ∎

Since the norm-closed maximal subspaces of E coincide with the $\tau(E, E)$-closed maximal subspaces of E, Proposition 3.4.25 then implies the following standard result, derived in the context of quantum logic.

Proposition 3.4.26. *Let E be a pre-Hilbert space. Assume that every $\sigma(E, E)$-closed vector subspace (or, equivalently, every $\tau(E, E)$-closed vector subspace) is orthocomplemented. Then E is a Hilbert space.*

3.4.6 PIP-*Spaces with many Projections*

Proposition 3.4.26 extends to PIP-spaces in the following sense. Let V be nondegenerate, positive definite, PIP-space (thus $V^\#$ is a pre-Hilbert space). If every $\tau(V, V)$-closed subspace is orthocomplemented, then V is a Hilbert space, i.e., all elements of V are compatible and V is complete in the norm topology.

We shall prove a stronger statement in which the assumption is only that certain $\tau(V, V)$-closed subspaces are orthocomplemented.

Theorem 3.4.27. *Let V be a nondegenerate* PIP-*space such that $\langle f|f \rangle > 0$ for every nonzero $f \in V^\#$. Assume that the following condition holds:*

- *If a vector subspace $W \subseteq V$ is the $\tau(V, V^\#)$-closure of its "infinitely good core" $W \cap V^\#$, then W is orthocomplemented.*

Then V is a Hilbert space.

Proof. Let S be any $\tau(V^\#, V^\#)$-closed subspace of $V^\#$. The $\tau(V, V^\#)$-closure of S is then, according to Proposition 3.4.4, $W = (S^\perp \cap V^\#)^\perp$ and therefore,

$$W \cap V^\# = (S^\perp \cap V^\#)^\perp \cap V^\#$$
$$= \tau(V^\#, V^\#)\text{-closure of } S = S.$$

The first equality follows from the density of $V^\#$ in V (see Remark 3.3.3).

By assumption, W is orthocomplemented in V. By Theorem 3.4.17, $W \cap V^\# = S$ is orthocomplemented in $V^\#$. Since S was arbitrary, it follows from Proposition 3.4.26, that $V^\#$ is a Hilbert space.

Now, since $V^\#$ is dense in V, the topology induced on $V^\#$ by $\tau(V, V^\#)$ is $\tau(V^\#, V^\#)$. But $\tau(V^\#, V^\#)$ is the norm topology on $V^\#$, since $V^\#$ is a Hilbert space. Thus $V^\#$ is complete in the topology induced by V; hence it is closed in V. Since it is also dense, we have necessarily $V^\# = V$.
∎

Our theorem says, in particular, that the space of tempered distributions $\mathcal{S}^\times(\mathbb{R})$ (which is certainly not a Hilbert space) contains subspaces that are the closure of their infinitely good core and yet are not orthocomplemented. An example is provided by the set of distributions with support in a closed interval; the nonexistence of a complementary (orthogonal) subspace is easily proved with the help of Theorem 3.4.17 and the example given in Section 3.4.5.

The preceding result applies, in particular, to a Banach space. Thus we recover a well-known result.

Proposition 3.4.28. *Assume that every (norm-)closed subspace of a Banach space E is orthocomplemented. Then E is a Hilbert space.*

In turn, Proposition 3.4.28 implies a similar result for Fréchet spaces.

Proposition 3.4.29. *A separable infinite-dimensional Fréchet spaces in which every closed subspace is orthocomplemented is isomorphic to* ℓ^2, ω *or* $\omega \oplus \ell^2$.

The general conclusion is that very few spaces do have "many projections".

Notes for Chapter 3

Section 3.1. This section is largely based on Antoine-Grossmann [18], where operators on PIP-spaces were defined.

- It is worth remarking that the argument leading to the definition of the adjoint operator is exactly the same as in the case of a bounded operator on a Hilbert space.
- The machinery of representatives was developed by Grossmann [114] in the context of nested Hilbert spaces (compare also Svetlichny [181]). In particular, regular linear functionals were introduced in that paper.
- The use of the set $\mathsf{j}(A)$ is in fact much older, especially in the context of L^p spaces; a systematic presentation can be found in the monograph of Krasnoselski *et al.* [KZPS66], where $\mathsf{j}(A)$ is called the *L*-characteristic of A.

Section 3.2. This section is largely based on Antoine [13], where operators on indexed PIP-spaces were introduced.

Section 3.3. The convention made in this section, allowing a unified treatment of both PIP-spaces and indexed PIP-spaces, was introduced in Antoine-Mathot [27].

- Regular operators were introduced and studied in Antoine [11] and Antoine-Mathot [27]. Examples of spaces V which are not Mackey-quasi-complete have been discussed by Friedrich–Lassner [100, 101].
- The notion of homomorphism defined here differs from that introduced by Grossmann [116] for a NHS, in the sense that the latter imposes only the first condition in (3.3), with the result that A may be a homomorphism without $A^\times$ being one.
- For a technical discussion of the theory of categories, we refer to Mitchell [Mit65].
- The three operator algebras $\mathfrak{A}(V_I), \mathfrak{B}(V_I), \mathfrak{C}(V_I)$ were introduced in Mathot [70].
- The matrix characterization of bounded operators on ℓ^2 is due to Crone [62].
- For the Riesz-Thorin interpolation theorem and the normal interpolation property, see Bergh and Löfström [BL76] or Krein-Petunin [132].

- The version of the KLMN theorem given in Theorem 3.3.25 is that of Simon [Sim71]. Theorem 3.3.34, in terms of a scale of spaces, is due to Nelson [154] and it is discussed in Reed-Simon II [RS75, Sec. X.2].
- Theorem 3.3.28 and Proposition 3.3.32 are due to Mathot (unpublished).
- The commutator theorem was given in Nelson [155], using already a scale of Hilbert spaces. For a proof and a detailed discussion, we refer to Reed-Simon II [RS75, Sec. X.5]. There also one may find counterexamples to the result of the theorem when one of the assumptions is not satisfied, including a case where $D(\mathsf{A}) = \{0\}$.

Section 3.4. This section is largely based on Antoine-Grossmann [19].

- The argument given in Remark 3.3.3 is due to Köthe [Köt69, §22.2].
- In the special case where V is an inner product space, Proposition 3.4.4 reduces to a result stated by Bognar [Bog74, Sec. III. 6].
- The fact, stated in the proof of Lemma 3.4.10, that W_r, with the topology induced by V_r, is topologically isomorphic to the quotient V_r/QV_r, the latter being equipped with its quotient topology, is a standard result, given in Schaefer [Sch71, Chap.1, §2.4].
- For orthocomplemented subspaces of *inner* product spaces, we refer to Bognar [Bog74]. In particular, the equivalent of Proposition 3.4.14 is given in Theorem III.7.2.
- For Schauder and other bases in Banach spaces, and for frames, we refer to Christensen [Chr03, Chap.3].
- The consideration of maximal subspaces of pre-Hilbert spaces and, in particular, Proposition 3.4.25 are due to Epifanio-Trapani [76].
- Proposition 3.4.26 is due to Piron [166] and Amemiya-Araki [7], who proved it in the context of quantum logic.
- Proposition 3.4.28, which remained conjectured for a long time, was proved by Lindenstrauss-Tzafriri [143]. According to Grothendieck [Gro66, Chap.II, p.73], Proposition 3.4.29 is a consequence of Proposition 3.4.28.

Chapter 4
Examples of Indexed PIP-Spaces

This chapter is devoted to a detailed analysis of various concrete examples of PIP-spaces. We will explore sequence spaces, spaces of measurable
functions, and spaces of analytic functions. Some cases have already been
presented in Chapters 1 and 2. We will of course not repeat these discussions,
except very briefly. In addition, various functional spaces are of great interest in signal processing (amalgam spaces, modulation spaces, Besov spaces,
coorbit spaces). These will be studied systematically in a separate chapter
(Chapter 8).

4.1 Lebesgue Spaces of Measurable Functions

4.1.1 L^p Spaces on a Finite Interval

Our first example is the family of Lebesgue spaces over the interval $[0, 1]$
with their usual norm topology:

$$\mathcal{I} = \{L^p := L^p([0,1]; dx),\ 1 < p < \infty\}.$$

These spaces form a chain: $p > q$ implies $L^p \subset L^q$, the embedding is continuous and has dense image. With the involution

$$L^p \leftrightarrow (L^p)^\times = L^{\overline{p}},\ p^{-1} + \overline{p}^{-1} = 1,$$

$\mathcal{I}$ is an involutive covering of the space $V = \bigcup_{1<p<\infty} L^p = \sum_{1<p<\infty} L^p$.[1]
Given the corresponding compatibility #, i.e., $(L^p)^\# = L^{\overline{p}}$, we can compute
explicitly the complete involutive lattice $\mathcal{F}(V, \#)$.

---

[1] Technically, the inductive limit $\varinjlim_{1<p<\infty} L_p$. See Appendix B.

J.-P. Antoine and C. Trapani, *Partial Inner Product Spaces:*
Theory and Applications, Lecture Notes in Mathematics 1986,
DOI 10.1007/978-3-642-05136-4_4, © Springer-Verlag Berlin Heidelberg 2009

Remark: Notice that we do not include the space L^1. By symmetry this would demand inclusion of L^∞ as well, which would invalidate some of the statements made below about duality properties.

First we evaluate "elements of the first generation" of $\mathcal{F}$. Given an arbitrary subset S of $(1, \infty)$, define the spaces

$$L^P(S) := \bigcap_{p \in S} L^p, \qquad L^I(S) := \bigcup_{p \in S} L^p = \sum_{p \in S} L^p,$$

Introducing $r = \inf S, t = \sup S$, and defining $\widehat{S} = (1, t) \bigcup S$, $\widetilde{S} = (r, \infty) \bigcup S$, one shows that:

$$L^P(S) = L^P(\widehat{S}), \qquad L^I(S) = L^I(\widetilde{S}). \tag{4.1}$$

This leaves us with four possible cases:

(i) $t \in S \Rightarrow \widehat{S} = (1, t]$ and $L^P(S) = \bigcap_{1 < q \leqslant t} L^q = L^t$;

(ii) $t \notin S \Rightarrow \widehat{S} = (1, t)$ and $L^P(S) = \bigcap_{1 < q < t} L^q := L^{t-}$;

(iii) $r \in S \Rightarrow \widetilde{S} = [r, \infty)$ and $L^I(S) = \bigcup_{r \leqslant q < \infty} L^q = L^r$;

(iv) $r \notin S \Rightarrow \widetilde{S} = (r, \infty)$ and $L^I(S) = \bigcup_{r < q < \infty} L^q := L^{r+}$.

Thus we get two new types of spaces,[2] namely the spaces $L^{p\pm}$. Their topological properties are based on the observation that, in the definition of L^{p-}, it is enough [by Eq.(4.1)] to consider a cofinal countable subset of the V_q's. Therefore, we get

(i) For $1 < p \leqslant \infty$, $L^{p-} := \bigcap_{1 < q < p} L^q$, with the projective topology, is a non-normable, reflexive Fréchet space, hence barreled and complete, with conjugate dual $(L^{p-})^\times = (L^{p-})^\# = L^{\bar{p}+}$. In particular, $L^{\infty-}$ coincides with the space L^ω of Arens (also called the Arens algebra).

(ii) For $1 \leqslant p < \infty$, $L^{p+} := \bigcup_{p < q < \infty} L^q$, with the inductive topology, is a nonmetrizable (Mackey) complete, barreled topological vector space, with conjugate dual $(L^{p+})^\times = (L^{p+})^\# = L^{\bar{p}-}$.

(iii) Furthermore, the following inclusions are proper: $L^{p+} \subset L^p \subset L^{p-}$ ($1 < p < \infty$), the embeddings are continuous and have dense range.

Proposition 4.1.1. *Let $\mathcal{I}$ be the chain $\{L^p, 1 < p < \infty\}$. Then the complete lattice $\mathcal{F}(V, \#)$ generated by $\mathcal{I}$ is also a chain, obtained by replacing each L^p in $\mathcal{I}$ by the triplet $L^{p+} \subset L^p \subset L^{p-}$, and adding the smallest element $L^{\infty-}$ and the largest element L^{1+}.*

Proof. First we evaluate elements of the form $\{f\}^{\#\#}$, then $\{f\}^\#$, and finally arbitrary elements of $\mathcal{F}(V, \#)$ through the usual relation:

$$\mathcal{F}(V, \#) \ni V_r = \bigcap_{f \in V_{\bar{r}}} \{f\}^\#.$$

---

[2] Nonstandard analysis could also be used here: V_{p+} is really $V_{p+\epsilon}$, ϵ infinitesimal.

Let $f \in V$. Then, by Eq.(1.9),

$$\{f\}^{\#\#} = \bigcap_{L^p \ni f} L^p = L^{\bar{q}} \text{ or } L^{\bar{q}-}, \text{ for some } \bar{q} \in (1, \infty).$$

Therefore $\{f\}^{\#} = L^q$ or L^{q+} for some $q \in (1, \infty)$. Finally

$$V_r = \bigcap_{f \in V_r} \{f\}^{\#} = \left(\bigcap_{q \in S} L^q \right) \cap \left(\bigcap_{r \in T} L^{r+} \right),$$

where S, T are some subsets of $(1, \infty)$, to be replaced by $\widehat{S}, \widehat{T}$ respectively.
For the first term on the right-hand side, we get

$$\bigcap_{q \in S} L^q = \bigcap_{q \in \widehat{S}} L^q = L^s \text{ or } L^{s-}, \text{ with } s = \sup S.$$

As for the second term, we observe that $r_1 < r_2$ implies $L^{r_2+} \subset L^{r_1+}$ with continuous embedding; the set inclusion is obvious, and the embedding $L^{r_2+} \hookrightarrow L^{r_1+}$ can be factorized continuously through L^r, where r is any real number such that $r_1 < r < r_2$. Therefore, all the spaces $L^{r+}, 1 \leqslant r < \infty$, form a chain with continuous embeddings. Thus we get

$$\bigcap_{r \in T} L^{r+} = L^{t+}, \text{ where } t = \sup T.$$

Finally, V_r is either of the form $L^s \cap L^{t+}$, or of the form $L^{s-} \cap L^{t+}$. For $s \leqslant t$, we have $L^{s-} \supset L^s \supset L^{t+}$ and $V_r = L^{t+}$. For $s > t$, we have $L^{t+} \supset L^{s-} \supset L^s$, so that $V_r = L^s$ or $V_r = L^{s-}$. This concludes the proof. ∎

Notice that, in addition to its PIP-space structure, the family $\mathcal{I}$ generates a partial *-algebra under pointwise multiplication.

4.1.2 The Spaces $L^p(\mathbb{R}, dx)$

We turn now to the L^p spaces on $\mathbb{R}$. If we consider the family $\{L^p(\mathbb{R}) \cap L^1(\mathbb{R}), 1 \leqslant p \leqslant \infty\}$, we obtain a scale similar to the previous one (except that the individual spaces are not complete), which may be used to endow $L^1(\mathbb{R})$ with a PIP-space structure.

However, the spaces $L^p(\mathbb{R})$ themselves no longer form a chain, no two of them being comparable. We have only

$$L^p \cap L^q \subset L^s, \text{ for all } s \text{ such that } p < s < q.$$

Hence we have to take the lattice generated by $\mathcal{I} = \{L^p(\mathbb{R}, dx), 1 \leqslant p \leqslant \infty\}$, that we call $\mathcal{J}$. The extreme spaces of the lattice are, respectively:

$$V_\mathsf{J}^\# = \bigcap_{1 \leqslant q \leqslant \infty} L^q, \quad \text{and} \quad V_\mathsf{J} = \bigcup_{1 \leqslant q \leqslant \infty} L^q = \sum_{1 \leqslant q \leqslant \infty} L^q. \qquad (4.2)$$

Here too, the lattice structure allows to give to V_J a structure of a PIP-space and of a locally convex partial *-algebra.

The lattice operations on $\mathcal{J}$ are easily described:

- $L^p \wedge L^q = L^p \cap L^q$ is a Banach space for the projective norm $\|f\|_{p \wedge q} = \|f\|_p + \|f\|_q$.
- $L^p \vee L^q = L^p + L^q$ is a Banach space for the inductive norm

$$\|f\|_{p \vee q} = \inf_{f = g + h} \{\|g\|_p + \|h\|_q; g \in L^p, h \in L^q\}.$$

- For $1 < p, q < \infty$, both spaces $L^p \wedge L^q$ and $L^p \vee L^q$ are reflexive and $(L^p \wedge L^q)^\times = L^{\bar{p}} \vee L^{\bar{q}}$.

At this stage, it is convenient to introduce a unified notation:

$$L^{(p,q)} = \begin{cases} L^p \wedge L^q, & \text{if } p \geqslant q, \\ L^p \vee L^q, & \text{if } p \leqslant q. \end{cases}$$

Thus, for $1 < p, q < \infty$, each space $L^{(p,q)}$ is a reflexive Banach space, with conjugate dual $L^{(\bar{p}, \bar{q})}$. The modifications when p, q equal 1 or ∞ are obvious.

Next, if we represent (p, q) by the point of coordinates $(1/p, 1/q)$, we may associate all the spaces $L^{(p,q)}$ $(1 \leqslant p, q \leqslant \infty)$ in a one-to-one fashion with the points of a unit square $\mathrm{J} = [0, 1] \times [0, 1]$ (see Fig. 4.1). Thus, in this picture, the spaces L^p are on the main diagonal, intersections $L^p \cap L^q$ above it and sums $L^p + L^q$ below. The space $L^{(p,q)}$ is contained in $L^{(p',q')}$ if (p, q) is on the left and/or above (p', q'). Thus the smallest space is

$$V_\mathsf{J}^\# = L^{(\infty,1)} = L^\infty \cap L^1$$

and it corresponds to the upper left corner, the largest one is

$$V_\mathsf{J} = L^{(1,\infty)} = L^1 + L^\infty,$$

corresponding to the lower right corner. Inside the square, duality corresponds to (geometrical) symmetry with respect to the center $(1/2, 1/2)$ of the square, which represents the space L^2. The ordering of the spaces corresponds to the following rule:

$$L^{(p,q)} \subset L^{(p',q')} \quad \Longleftrightarrow \quad (p, q) \leqslant (p', q') \quad \Longleftrightarrow \quad p \geqslant p' \text{ and } q \leqslant q'. \quad (4.3)$$

By the way, this rule shows that the spaces L^p on the main diagonal are not comparable, as we know.

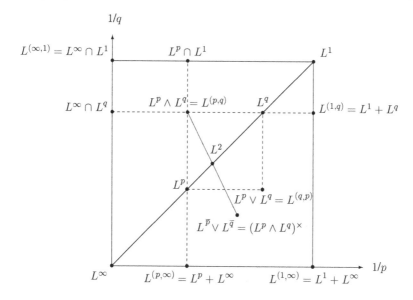

Fig. 4.1 The unit square describing the lattice J

For $\infty \geqslant q_o \geqslant 1$, consider now the horizontal row $q = q_o$, $\{L^{(p,q_o)} : \infty \geqslant p \geqslant 1\}$. It corresponds to the chain:

$$L^{\infty} \cap L^{q_o} \subset \ldots \subset L^r \cap L^{q_o} \subset \ldots \subset L^{q_o} \subset \ldots \subset L^s + L^{q_o} \subset \ldots \subset L^1 + L^{q_o}. \quad (4.4)$$

$$(\infty > r > q_o > s > 1)$$

The point is that all the embeddings in the chain (4.4) are continuous and have dense range. The same holds true for a vertical row $p = p_o$, $\{L^{(p_o,q)} : 1 \leqslant q \leqslant \infty\}$:

$$L^{p_o} \cap L^s \subset \ldots \subset L^{p_o} \cap L^s \subset \ldots \subset L^{p_o} \subset \ldots \subset L^{p_o} + L^r \subset \ldots \subset L^{p_o} + L^{\infty}. \quad (4.5)$$

$$(1 < s < p_o < r < \infty)$$

Combining these two facts, we see that the partial order extends to the spaces $L^{(p,q)}$ ($1 \leqslant p, q \leqslant \infty$), inclusion meaning now continuous embedding with dense range.

Now the set of points contained in the square J may be considered as an involutive lattice with respect to the partial order (4.3), with operations:

$$(p,q) \wedge (p',q') = (p \vee p', q \wedge q')$$
$$(p,q) \vee (p',q') = (p \wedge p', q \vee q')$$
$$\overline{(p,q)} = (\bar{p}, \bar{q}),$$

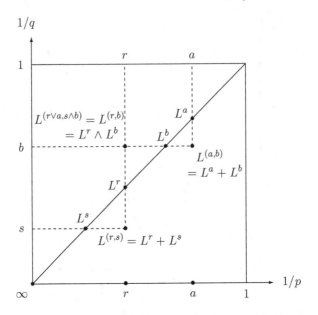

Fig. 4.2 The intersection of two spaces from J

where, as usual, $p \wedge p' = \min\{p, p'\}$, $p \vee p' = \max\{p, p'\}$.

The considerations made above imply that the lattice $\mathcal{J}$ generated by $\mathcal{I} = \{L^p\}$ is already obtained at the first generation. For example, $L^{(r,s)} \wedge L^{(a,b)} = L^{(r \vee a, s \wedge b)}$ (see Fig. 4.2), and the latter may be either above, on or below the diagonal, depending on the values of the indices. For instance, if $p < q < s$, then $L^{(p,q)} \wedge L^{(q,s)} = L^q$, both as sets and as topological vector spaces.

The conclusion is that, using this language, the only difference between the two cases $\{L^p([0,1])\}$ and $\{L^p(\mathbb{R})\}$ lies in the type of order obtained: a chain I (total order) or a partially ordered lattice J. From this remark, the lattice completion of $\mathcal{J}$ can be obtained exactly as before. This introduces again Fréchet and DF-spaces, all reflexive if we start from $1 < p < \infty$, and in natural duality as in the previous case. In particular, for the spaces of the first generation, it suffices to consider intervals $S \subset [1, \infty]$ and define the spaces

$$L^{\mathrm{proj}}(S) = \bigcap_{q \in S} L^q, \quad L^{\mathrm{ind}}(S) = \bigcup_{q \in S} L^q.$$

Then:

- If S is a closed interval $S = [p, q]$, with $p < q$, then $L^{\mathrm{proj}}(S) = L^p \wedge L^q = L^{(q,p)}$ and $L^{\mathrm{ind}}(S) = L^p \vee L^q = L^{(p,q)}$ are Banach spaces. If S is a semi-open or open interval, $L^{\mathrm{proj}}(S)$ is a non-normable Fréchet space and $L^{\mathrm{ind}}(S)$ a DF-space.
- Let $S \subset (1, \infty)$ and define $\overline{S} = \{\overline{q} : q \in S\}$. Then $(L^{\mathrm{proj}}(S))^{\times} = L^{\mathrm{ind}}(\overline{S})$ and $(L^{\mathrm{ind}}(S))^{\times} = L^{\mathrm{proj}}(\overline{S})$.

A special rôle will be played in the sequel by the spaces L^{ind} corresponding to semi-infinite intervals, namely:

$$L^{(p,\infty)} = L^{\mathrm{ind}}([p,\infty]) = \bigcup_{p\leqslant s\leqslant\infty} L^s = L^p + L^\infty, \text{ a nonreflexive Banach space,}$$

$$L^{(p,\omega)} = L^{\mathrm{ind}}([p,\infty)) = \bigcup_{p\leqslant s<\infty} L^s, \text{ a reflexive DF-space.}$$

As for the lattice completion $\mathcal{F}_{\mathrm{J}}$, one can build an 'enriched' or 'nonstandard' square F, exactly as in the previous section. Take first $1 < q < \infty$, that is, the interior J_{o} of the square J. The extreme spaces of the corresponding complete lattice $\mathcal{F}_{\mathrm{o}}$ are:

$$V_{\mathrm{J}_{\mathrm{o}}}^{\#} = \bigcap_{1<q<\infty} L^q, \quad \text{and} \quad V_{\mathrm{J}_{\mathrm{o}}} = \bigcup_{1<q<\infty} L^q = \sum_{1<q<\infty} L^q,$$

with their projective and inductive topologies, respectively. All embeddings are continuous and have dense range.

Similar results are valid when one includes L^1 and L^∞, except for the obvious modifications concerning duality. The extreme spaces of the full lattice $\mathcal{F}_{\mathrm{J}}$ then are $V_{\mathrm{J}} = L_G := L^1 + L^\infty$ and $V_{\mathrm{J}}{}^{\#} = L_G^{\#} = L^1 \cap L^\infty$, with their inductive and projective norms, respectively, which make them into nonreflexive Banach spaces (none of them is the dual of the other). Notice that the space L_G, known as the space of Gould, contains strictly all the L^p, $1 \leqslant p \leqslant \infty$.

Proposition 4.1.2. *The space $L_G := L^1(\mathbb{R}, dx) + L^\infty(\mathbb{R}, dx)$ is a nonreflexive LBS, generated by the family $\mathcal{I} = \{L^p(\mathbb{R}, dx), 1 \leqslant p \leqslant \infty\}$ and the corresponding compatibility $(L^p)^{\#} = L^{\overline{p}}$.*

Exactly as in the case of a finite interval, we may restrict the generating spaces to $\{L^s, \ p \leqslant s \leqslant q)$, which amounts to take a subsquare $\mathrm{J}^{(p,q)}$ of J. The rest is obvious.

4.1.3 Reflexive Chains of Banach Spaces

The chain $\mathcal{I} = \{L^p := L^p([0,1]; dx), 1 < p < \infty\}$ of Lebesgue spaces over the interval $[0,1]$ is the prototype of a whole class of chains of reflexive Banach spaces, with exactly the same properties.

Let $I_o = (a,b)$ be an open interval of $\mathbb{R}$, and for each $p \in I_o$ let there be a reflexive Banach space V_p. We say that the family $\mathcal{I} = \{V_p\}_{p\in I_o}$ is *a (continuous) reflexive chain of Banach spaces* if the two following conditions hold:

(i) $p < q \ \Rightarrow \ V_p \subsetneqq V_q$, the inclusion map is injective and continuous;[3]

---

[3] *Warning:* The order here is direct: $p \mapsto V_p$ is monotone *increasing*, whereas it was *decreasing* in the Lebesgue case (inverse order).

(ii) I_o carries an involution $p \leftrightarrow \overline{p}$ such that $V_{\overline{p}}$ is the conjugate dual of V_p, that is, the norm $\| \cdot \|_{\overline{p}}$ on $V_{\overline{p}}$ is the conjugate of the norm $\| \cdot \|_p$ on V_p.

In other words, a reflexive chain of Banach spaces is a totally ordered LBS. The following properties follow easily from the definition:

(a) $\mathcal{I}$ is an involutive covering of $V := \bigcup_{p \in I_o} V_p = \sum_{p \in I_o} V_p$, corresponding to the compatibility $(V_p)^\# = V_{\overline{p}}$.
(b) Whenever $p < q$, the inclusion map $V_p \to V_q$ has dense range and each V_p is dense in V (considered as the inductive limit $\overrightarrow{\lim}_{p \in I_o} V_p$).
(c) $V^\# := \bigcap_{p \in I_o} V_p$ is a dense subspace of every V_p, and of V as well.

Let now $I = [a, b]$ be a closed interval. The analogous definition is obvious, the only difference being that now $V^\# = V_a, V = V_b$. Within this context, a proposition similar to Proposition 4.1.1. can be formulated. As before we define

$$V_{p-} := \bigcup_{q<p} V_q, \quad V_{p0} := V_p, \quad V_{p+} := \bigcap_{p<q} V_q.$$

Let F_0 be the totally ordered set $F_0 = (a, b) \times \{-, 0, +\}$ with its lexicographic order $(- < 0 < +)$. Then we have:

Proposition 4.1.3. Let $\mathcal{I} = \{V_p\}_{p \in I}$ be a reflexive chain of Banach spaces. Then the complete involutive lattice $\mathcal{F} = \mathcal{F}(V, \#)$ is a chain given explicitly as follows:

(i) If $I = (a, b)$ is an open interval, $\mathcal{F}$ consists of $V^\# := V_{a+}$, $\mathcal{F}_0$, $V := V_{b-}$, where $\mathcal{F}_0$ is a chain indexed by F_0.
(ii) If $I = [a, b]$ is a closed interval, $\mathcal{F}$ is the chain $V^\# := V_a, V_{a+}$, $\mathcal{F}_0$, V_{b-} $V := V_b$.

Proof. The argument is exactly the same as the one given in Proposition 4.1.1 for the case of Lebesgue spaces. First one shows (using the involution $p \leftrightarrow \overline{p}$) that the inclusions $V_{p-} \subset V_p \subset V_{p+}$ are proper. Then, one proves that V_{p+}, with the projective topology, is a reflexive Fréchet space, with dual $V_{\overline{p}-}$, and V_{p-}, with the inductive topology, is a reflexive nonmetrizable, barreled topological vector space, with dual V_{p+}. The rest of the argument is then identical and gives (i); as for (ii), it is obvious. ∎

The situation described above is in fact extremely frequent in applications. The following examples of reflexive chains of Banach spaces are all well-known (the first two have the direct order, the other two the inverse order):

(1) The chain of sequence spaces $\{\ell^p, \ 1<p<\infty\}$ (Example (ii) in Section 1.1);
(2) The chain of ideals $\{\mathcal{C}^p(\mathcal{H}), \ 1 < p < \infty\}$ of compact operators in a Hilbert space $\mathcal{H}$, which is isomorphic to (1) (Example (v) in Section 1.1);

(3) The chain of (generalized) Sobolev spaces (also called Bessel potential spaces) H^s $(-\infty < s < \infty)$ defined as follows : A tempered distribution $f \in \mathcal{S}^\times(\mathbb{R}^n)$ belongs to H^s if its Fourier transform $\hat{f}$ verifies

$$\int_{\mathbb{R}^n} |\hat{f}(\xi)|^2 (1 + |\xi|^2)^s \, d\xi < \infty;$$

the involution here is $(H^s)^\# = (H^s)^\times = H^{-s}$.

(4) The chain of Banach spaces that generalizes the Lebesgue spaces L^p to operator algebras, namely, the noncommutative L^p spaces associated to a Hilbert algebra (with unit), introduced by Segal.

Among reflexive chains of Banach spaces, we find, of course, chains of hilbertian or Hilbert spaces. We have already encountered several examples, such as the canonical scale built on the powers of a self-adjoint operator in a Hilbert space (Section I.1) or the chain of Hilbert spaces $\mathcal{F}^\rho$ $(-\infty < \rho < \infty)$ in Bargmann's space $\mathcal{E}^\times$ (Example (v) in Section 1.1). The latter will be studied in detail in Section 4.6.2 below. As for the former, we will revisit and refine it in Section 5.2.1.

4.2 Locally Integrable Functions

4.2.1 A Generating Subset of Locally Integrable Functions

In this section, we will study in detail the PIP-space $L^1_{\mathrm{loc}}(X, d\mu)$ of locally integrable functions on a measure space (X, μ). First we take $V = L^1_{\mathrm{loc}}(\mathbb{R}^n, dx)$, the space of all functions on $\mathbb{R}^n$ that are locally integrable with respect to the Lebesgue measure. With the compatibility.

$$f \# g \iff \int_{\mathbb{R}^n} |f(x)\, g(x)| \, dx < \infty,$$

one has, as usual, $V^\# = L^\infty_c(\mathbb{R}^n, dx)$, the essentially bounded functions of compact support (Section I.2.3 (ii)).

Let $r : \mathbb{R}^n \to \mathbb{R}^+$ be a measurable, a.e. positive function, such that both r and $\bar{r} := r^{-1}$ are locally square integrable. Denote by $L^2(r)$ the Hilbert space of measurable functions $f : \mathbb{R}^n \to \mathbb{C}$ such that fr^{-1} is square integrable. Then we claim:

Proposition 4.2.1. *(i)* $L_c^\infty \subset L^2(r) \subset L_{\mathrm{loc}}^1$.
(ii) The family $\mathcal{I} := \{L^2(r)\}$ of all such subspaces is a generating sublattice of $\mathcal{F}(V, \#)$.

Proof. Part (i) is a straightforward verification. As for (ii), let $\{K_j, j = 1, 2 \ldots\}$ be a covering of $\mathbb{R}^n$ by an increasing sequence of compact subsets (e.g. closed balls of radius j), and write $\Omega_j = K_j \setminus K_{j-1}$, so that each Ω_j is relatively compact, $\Omega_j \cap \Omega_k = 0$ if $j \neq k$, $\mathbb{R}^n = \bigcup_j \Omega_j$. Let $f, g \in V, f \# g$, and $\{\alpha_j\}, \{\beta_j\}$ two arbitrary sequences of positive numbers such that:

$$\sum_{j=1}^\infty \alpha_j \int_{\Omega_j} |f(x)| \, dx < \infty, \quad \sum_{j=1}^\infty \beta_j \int_{\Omega_j} |g(x)| \, dx < \infty.$$

Define

$$r(x) = \sum_{j=1}^\infty \chi_j(x) r_j(x),$$

where

$$r_j(x)^2 = \max(\beta_j, |f(x)|)/\max(\alpha_j, |g(x)|)$$

and χ_j is the characteristic function of Ω_j. Thus $r^{-2}(x) = \sum_{j=1}^\infty \chi_j(x) r_j^{-2}(x)$, and both r and r^{-1} are locally square integrable.[4] Furthermore, it is easily shown that $f \in L^2(r)$, $g \in L^2(\bar{r})$. For instance:

$$\int_{\mathbb{R}^n} |f|^2 r^{-2} \, dx = \sum_{j=1}^\infty \int_{\Omega_j} |f|^2 \, \frac{\max(\alpha_j, |g|)}{\max(\beta_j, |f|)} \, dx$$

$$\leqslant \sum_{j=1}^\infty \int_{\Omega_j} |f| \max(\alpha_j, |g|) \, dx$$

$$= \sum_{j=1}^\infty \left(\int_{\Omega_j \cap \{x \,|\, |g(x)| \geqslant \alpha_j\}} |fg| \, dx + \int_{\Omega_j \cap \{x \,|\, |g(x)| < \alpha_j\}} \alpha_j |f| \, dx \right)$$

$$\leqslant \sum_{j=1}^\infty \left(\int_{\Omega_j} |fg| \, dx + \alpha_j \int_{\Omega_j} |f| \, dx \right)$$

$$= \int_{\mathbb{R}^n} |fg| \, dx + \sum_{j=1}^\infty \alpha_j \int_{\Omega_j} |f| \, dx < \infty.$$

---

[4] *Remark*: Elements of V are in fact classes of equivalent functions; if in the definition of r_j, f or g is replaced by a equivalent function, so is r_j, and nothing changes in the argument.

Thus the family $\mathcal{I} = \{L^2(r)\}$ is generating in $L^1_{\text{loc}}(\mathbb{R}^n, dx)$. Moreover it is an involutive sublattice of $\mathcal{F}(V, \#)$ with the following lattice operations:

- $L^2(r) \subseteq L^2(s) \iff r(x) \leqslant s(x)$ a.e.,
- $L^2(r) \wedge L^2(s) = L^2(p)$, with $p(x) = \min\{r(x), s(x)\}$,
- $L^2(r) \vee L^2(s) = L^2(q)$, with $q(x) = \max\{r(x), s(x)\}$,
- $[L^2(r)]^\# = L^2(\bar{r})$, with $\bar{r}(x) = r^{-1}(x)$. ▪

Remark 4.2.2. These results generalize to the space $L^1_{\text{loc}}(X, d\mu)$, where X is a locally compact and σ-compact space (i.e., $X = \bigcup_n K_n, K_n \subset K_{n+1}, K_j$ relatively compact) and μ is a non-negative Radon measure on X. Let

$$L^2(r) := \{f : f \text{ is } \mu\text{-measurable}, \int_X |f|^2 r^{-2} d\mu < \infty, \text{ with } r \text{ measurable}, r > 0 \text{ a.e..}\}$$

Then one has, with continuous injections :

$$\begin{aligned} &\text{(i) } L^\infty_c(X, d\mu) \subset L^2(r) \text{ if and only if } r^{-1} \in L^2_{\text{loc}}(X, d\mu), \\ &\text{(ii) } L^2(r) \subset L^1_{\text{loc}}(X, d\mu) \text{ if and only if } r \in L^2_{\text{loc}}(X, d\mu). \end{aligned} \tag{4.6}$$

We denote by I_1 the set of weight functions r such that $r^{\pm 1} \in L^2_{\text{loc}}(X, d\mu)$. Then we have a LHS realization of $L^1_{\text{loc}}(X, d\mu)$ (the proof given in Proposition 4.2.1 extends easily to the general case):

$$L^1_{\text{loc}}(X, d\mu) = \bigcup_{r \in I_1} L^2(r), \quad L^\infty_c(X, d\mu) = \bigcap_{r \in I_1} L^2(r). \tag{4.7}$$

On $L^1_{\text{loc}}(X, d\mu)$, the inductive limit of the norm topologies of $L^2(r)$ coincides with the Mackey topology $\tau(L^1_{\text{loc}}, L^\infty_c)$. However, on $L^\infty_c(X, d\mu)$, the projective limit t_{proj} of the norm topologies is coarser than the Mackey topology $\tau(L^\infty_c, L^1_{\text{loc}})$. Of course, the continuity of the embeddings in (4.6) refers to the Mackey topologies.

Similar results hold with $V = L^2_{\text{loc}}(X, d\mu), V^\# = L^2_c(X, d\mu)$ and $r^{\pm 1} \in L^\infty_{\text{loc}}(X, d\mu)$. Also if X is compact, i.e., $V = L^1(X, d\mu), V^\# = L^\infty(X, d\mu)$, the statement is simply that the family $\{L^2(r), r^{\pm 2} \in L^1(X, d\mu)\}$ is generating.

Let us come back to the simpler case $(X, d\mu) = (\mathbb{R}^n, dx)$. Now, combining (4.7) with the decomposition $\mathbb{R}^n = \bigcup_j \Omega_j, \Omega_j = K_j \setminus K_{j-1}$, we obtain, with obvious identifications, three different realizations:

$$L^1_{\text{loc}}(\mathbb{R}^n, dx) = \bigcup_{r \in I_1} L^2(r) = \bigcap_{j=1}^\infty L^1(K_j) = \bigcap_{j=1}^\infty L^1(\Omega_j), \tag{4.8}$$

$$L^\infty_c(\mathbb{R}^n, dx) = \bigcap_{r \in I_1} L^2(r) = \bigcup_{j=1}^\infty L^\infty(K_j) = \bigcup_{j=1}^\infty L^\infty(\Omega_j). \tag{4.9}$$

In these relations, each intersection is meant to carry the projective limit topology and each union the inductive limit topology. Thus, L^1_{loc}, as countable projective limit of the spaces $L^1(K_j)$, becomes a nonreflexive Fréchet space, with dual L^∞_{c}. The latter, with its strong topology $\beta(L^\infty_{\mathrm{c}}, L^1_{\mathrm{loc}})$, is the inductive limit of the spaces $L^\infty(K_j)$, with their norm topologies. Hence it is an (LF)- and a (DF)-space (see Section B.6).

The Mackey topology $\tau(L^\infty_{\mathrm{c}}, L^1_{\mathrm{loc}})$ coincides with the inductive limit of the Mackey topologies $\tau(L^\infty(K_j), L^1(K_j))$, but it is strictly coarser than the strong topology $\beta(L^\infty_{\mathrm{c}}, L^1_{\mathrm{loc}})$, since the space is nonreflexive.

In addition, one may equip L^∞_{c} with the topology γ of compact (or precompact since L^1_{loc} is Fréchet) convergence, i.e., the uniform convergence on the compact subsets of L^1_{loc}. Then the various topologies on L^∞_{c} are related to one another as follows (σ denotes the weak topology).

Proposition 4.2.3. *Let $L^\infty_{\mathrm{c}}(\mathbb{R}^n)$ denote the space of measurable, essentially bounded functions with compact support on $\mathbb{R}^n$. Then, with the notations above, one has the following relationships among the various topologies on L^∞_{c}:*

$$\beta \succ \tau \succcurlyeq \mathrm{t}_{\mathrm{proj}} \succ \gamma \succ \sigma. \tag{4.10}$$

The space $L^\infty_{\mathrm{c}}(\mathbb{R}^n)$ is complete for all topologies, except the weak one σ.

Proof. We know already that $\beta \succ \tau \succcurlyeq \mathrm{t}_{\mathrm{proj}} \succ \sigma$ and it is a standard result that $\tau \succcurlyeq \gamma \succ \sigma$. Also $\tau \succ \gamma$, since L^1_{loc} contains convex balanced subsets that are weakly compact but not compact. So it remains to prove that $\mathrm{t}_{\mathrm{proj}} \succ \gamma$. By Eq.(4.8) and Tychonoff's theorem, a basis of compact subsets of L^1_{loc} is given by sets of the form $M = \prod_{j=1}^\infty M_j$, where M_j is a compact subset of $L^1(\Omega_j)$. Each M_j is bounded and therefore contained in a ball $B_j := \{f \in L^1(\Omega_j) : \int_{\Omega_j} |f| dx < c_j, c_j > 0\}$. Define

$$r(x) = \sum_{j=1}^\infty \chi_j(x) r_j(x),$$

where

$$r_j(x)^2 = 2^j c_j \max(1, |f(x)|)$$

and χ_j is the characteristic function of Ω_j. Then $r^{\pm 1} \in L^1_{\mathrm{loc}}$, and we have, for every $f \in B := \prod_{j=1}^\infty B_j$:

$$\int |f|^2 r^{-2} dx = \sum_{j=1}^\infty \int_{\Omega_j} |f|^2 r_j^{-2} dx$$

$$\leqslant \sum_{j=1}^\infty 2^{-j} (c_j)^{-1} \int_{\Omega_j} |f| dx \leqslant \sum_{j=1}^\infty 2^{-j} = 1.$$

Thus every compact set $M = \prod_j M_j$ is contained in the unit ball of $L^2(r)$ for some r. Taking polars, we see that every neighborhood of zero in $L_c^\infty[\gamma]$ contains the unit ball of some $L^2(\bar{r})$, which is a neighborhood of zero in $L_c^\infty[\mathsf{t}_{\mathrm{proj}}]$. Thus $\mathsf{t}_{\mathrm{proj}} \succcurlyeq \gamma$. Conversely, since the unit ball of $L^2(r)$ contains $B = \prod_{j=1}^\infty B_j$ which is not weakly compact, *a fortiori* not compact, in L_{loc}^1, it cannot be contained in a compact set, and therefore there are neighborhoods of 0 in $L_c^\infty[\mathsf{t}_{\mathrm{proj}}]$ which contain no neighborhoods of 0 of $L_c^\infty[\gamma]$, i.e., $\mathsf{t}_{\mathrm{proj}} \succ \gamma$.

Concerning completeness, since L_{loc}^1 is Fréchet, its dual L_c^∞ is complete in the topology γ, and also for the topologies β and τ (see [Köt69] §21.6 (4)). It is also complete for the projective topology $\mathsf{t}_{\mathrm{proj}}$, as projective limit of the Hilbert spaces $L^2(r)$.   ∎

As for the last inequality, $\tau \succcurlyeq \mathsf{t}_{\mathrm{proj}}$, it is not known whether it is proper or not. Anyway, it follows from Proposition 4.2.3 that L_c^∞, with any of the topologies $\tau, \mathsf{t}_{\mathrm{proj}}, \gamma, \sigma$ is semi-reflexive, but nonreflexive, hence not (quasi)-barreled.

We turn now to the pair $\langle L_c^2, L_{\mathrm{loc}}^2 \rangle$. Denote by I_2 the set of weight functions r such that $r^{\pm 1} \in L_{\mathrm{loc}}^\infty$. Then, exactly as before, we have both a LHS realization and a natural one:

$$L_{\mathrm{loc}}^2(\mathbb{R}^n, dx) = \bigcup_{r \in I_2} L^2(r) = \bigcap_{j=1}^\infty L^2(K_j) = \bigcap_{j=1}^\infty L^2(\Omega_j),$$

$$L_c^2(\mathbb{R}^n, dx) = \bigcap_{r \in I_2} L^2(r) = \bigcup_{j=1}^\infty L^2(K_j) = \bigcup_{j=1}^\infty L^2(\Omega_j).$$

As before L_{loc}^2 is a Fréchet space, but this one is reflexive, as the inductive limit of a sequence of reflexive (DF)-spaces. Also its strong dual is $L_c^2[\mathsf{t}_{\mathrm{proj}}]$, i.e., the three topologies β, τ and $\mathsf{t}_{\mathrm{proj}}$ coincide on L_c^2. This can also be seen directly in two ways. First, the identity map $i : L_c^2[\mathsf{t}_{\mathrm{proj}}] \to L_c^2[\tau] = $ ind. $\lim L^2(K_j)$ factorizes continuously through any $L^2(r)$ with $r \in L_{\mathrm{loc}}^\infty$ and some $L^2(K_j)$ (there are plenty of such functions r in I_2). Second, one may repeat the argument of Proposition 4.2.3, replacing γ by τ or β, i.e., starting from sets $B = \prod_{j=1}^\infty B_j$, with B_j a closed ball in $L^2(\Omega_j)$, thus bounded and weakly compact. It follows that $\beta \preccurlyeq \mathsf{t}_{\mathrm{proj}}$, and so they must coincide. So we have:

Proposition 4.2.4. *On $L_c^2(\mathbb{R}^n, dx)$, the space of square integrable functions with compact support, the following relationship holds for the various topologies:*

$$\beta = \tau = \mathsf{t}_{\mathrm{proj}} \succ \gamma \succ \sigma.$$

Again the space $L_c^2(\mathbb{R}^n, dx)$ is complete for the first four topologies.

4.2.2 Functions or Sequences of Prescribed Growth

The results of Section 4.2.1 can be improved if V is restricted to those locally integrable functions which satisfy a growth condition at infinity (such as functions, or sequences, of polynomial or exponential growth), in the sense that the weight functions $r(x)$ can now be assumed to have the same type of growth.

More precisely, for X and μ as above, let A be a partially ordered set and let $\{F^{(\alpha)}, \alpha \in A\}$ be a family of positive μ-locally integrable functions, indexed by A and monotonically increasing in α:

$$\alpha \leqslant \beta \Longleftrightarrow F^{(\alpha)}(x) \leqslant F^{(\beta)}(x) \quad (\mu\text{--a.e.}).$$

Assume furthermore that, given $\alpha, \beta \in A$, there exists a positive square integrable function $s_{\alpha\beta}$ which verifies the following inequality for some positive constant $C_{\alpha\beta}$:

$$0 < s_{\alpha\beta}^2(x) \leqslant C_{\alpha\beta} F^{(\alpha)}(x) F^{(\beta)}(x) \quad (\mu\text{--a.e.}). \tag{4.11}$$

Define V as the vector space of those functions $f \in L^1_{\mathrm{loc}}(X, d\mu)$ which grow no faster than the functions $F^{(\alpha)}$, i.e., $f \in V$ if there exists $\alpha \in A$ and a constant $c > 0$ such that:

$$|f(x)| \leqslant c F^{(\alpha)}(x) \quad (\mu\text{--a.e.}).$$

Equip V with the compatibility # inherited from L^1_{loc}. Then:

Proposition 4.2.5. *Let V as above. Consider the family $\mathcal{I}_A := \{L^2(r)\}$, where each r is a weight function that verifies the following inequalities for some $\alpha, \beta \in A$ and positive constants c', c'':*

$$c'[s_{\alpha\beta}(x)/F^{\beta}(x)] \leqslant r(x) \leqslant c''[F^{(\alpha)}(x)/s_{\alpha\beta}(x)] \tag{4.12}$$

Then the family $\mathcal{I}_A$ is a generating subset of $\mathcal{F}(V, \#)$.

Proof. Let $|f(x)| \leqslant c F^{(\alpha)}(x), |g(x)| \leqslant c' F^{(\beta)}(x)$, and $\int_x |f(x)g(x)| d\mu < \infty$. We choose a function $s_{\alpha\beta} \in L^2(X, d\mu)$ that verifies Eq. (4.11). Then we proceed as in Section 1.4.1, dividing X into four disjoint subsets $X_j, j = 1, ..., 4$, depending on whether

$$|f(x)| - [s_{\alpha\beta}^2(x)/F^{(\beta)}(x)] \geqslant 0 \ (X_1 \text{ and } X_3) \text{ or } \leqslant 0 \ (X_2 \text{ and } X_4),$$
$$|g(x)| - [s_{\alpha\beta}^2(x)/F^{(\alpha)}(x)] \geqslant 0 \ (X_1 \text{ and } X_2) \text{ or } \leqslant 0 \ (X_3 \text{ and } X_4).$$

We define a weight function $r_{\alpha\beta}$ as follows:

$$r_{\alpha\beta}(x) = \begin{cases} |f(x)|^{1/2}\, |g(x)|^{-1/2}, & \text{for } x \in X_1, \\ s_{\alpha\beta}(x)|\, |g(x)|^{-1}, & \text{for } x \in X_2, \\ |f(x)|\, s_{\alpha\beta}^{-1}(x), & \text{for } x \in X_3, \\ \text{arbitrary for } x \in X_4, & \text{provided Eq.(4.12) is verified.} \end{cases}$$

It is then straightforward to verify that $f \in L^2(r_{\alpha\beta})$, $g \in L^2(\overline{r_{\alpha\beta}})$, and also that $r_{\alpha\beta}$ verifies Eq. (4.12) on all of X. ∎

This proposition covers many cases of interest. Let us give a few examples.

(i) *Functions of polynomial growth in* $L^1_{\mathrm{loc}}(\mathbb{R}^n, dx)$

$$F^{(\alpha)}(x) = (1 + |x|^2)^{\alpha/2}, \quad \alpha \in \mathbb{Z} \text{ or } \mathbb{R},$$

$$s_{\alpha\beta}(x) = (1 + |x|^2)^{-\gamma/2} \text{ with } \gamma > n/2 \text{ and } \gamma \geqslant -\frac{1}{2}(\alpha + \beta).$$

The assaying subsets $L^2(r_{\alpha\beta})$ obtained in this example actually form an involutive sublattice of $\mathcal{F}(V, \#)$.

(ii) *Slowly increasing sequences (in* ω*)*

By the same reasoning, we find that the family $\{\ell^2(r)\}$, where $r = (r_n)$ is a sequence of tempered weights, i.e.,

$$c'(1 + n)^{-\beta} \leqslant r_n \leqslant c''(1 + n)^{\alpha}, \ c', c'' > 0,$$

is generating in the space $s^\times$ of slowly increasing sequences, equipped with the standard compatibility from ω.

(iii) *Functions of exponential growth in* $L^1_{\mathrm{loc}}(\mathbb{R}^n, dx)$

$$F^{(\alpha)}(x) = e^{\alpha|x|}, \ \alpha \in \mathbb{R},$$

$$s_{\alpha\beta}(x) = e^{\gamma|x|} \text{ with } \gamma \in \mathbb{R}, \ \gamma > \alpha + \beta.$$

(iv) *Entire functions of order 2 in Bargmann's space* $\mathcal{E}^\times$

As in Example (v) in Section 1.1, take $X = \mathbb{C}^n$, with Gaussian measure $d\mu(z) = e^{-|z|^2}\, dz$. Then $\mathcal{E}^\times$ consists also of those Borel functions with growth indexed by the following family:

$$F^{(\alpha)}(z) = e^{1/2|z|^2}(1 + |z|^2)^{-\alpha/2}, \ \alpha \in \mathbb{R} \text{ or } \mathbb{Z},$$

$$s_{\alpha\beta}(z) = 1, \ \forall\, \alpha, \beta.$$

4.2.3 Operators on Spaces of Locally Integrable Functions

Finally we discuss operators on $L^1_{\mathrm{loc}}(\mathbb{R}^n, dx)$ (the same analysis applies to $L^2_{\mathrm{loc}}(\mathbb{R}^n, dx)$). This is a case where the various classes of operators can be characterized reasonably well, hence it is worth treating it in detail. More precisely, we would like to give conditions under which an operator on the PIP-space L^1_{loc} is regular or belongs to one of the three algebras $\mathfrak{A}, \mathfrak{B}$ or $\mathfrak{C}$.

The crucial remark is that each of $L^1_{\mathrm{loc}}, L^\infty_{\mathrm{c}}[\tau], L^\infty_{\mathrm{c}}[\gamma]$ is a normal space of distributions,[5] i.e., a subspace E of $\mathcal{D}^\times(\mathbb{R}^n)$ such that $\mathcal{D}(\mathbb{R}^n) \subset E \subset \mathcal{D}^\times(\mathbb{R}^n)$, where both inclusions are continuous and the first one has dense range. This fact allows us to use Schwartz's theory of kernels. Indeed every continuous linear map $A : L^\infty_{\mathrm{c}}[\tau] \to L^1_{\mathrm{loc}}[\tau]$ is then the extension of a continuous map $A : \mathcal{D}(\mathbb{R}^n) \to \mathcal{D}^\times(\mathbb{R}^n)$ which is uniquely determined by a kernel $A(x, y) \in \mathcal{D}^\times(\mathbb{R}^{2n})$ i.e., a distribution in two variables. Consequently, every operator on the PIP-space $L^1_{\mathrm{loc}}(\mathbb{R}^n)$ is given by a unique kernel $A(x, y)$. The problem is to identify the corresponding class of kernels, and here we may exploit the results of Schwartz. But we will get in general sufficient conditions only since Schwartz uses the topology γ on L_{c}. A continuous map $A : L^\infty_{\mathrm{c}}[\gamma] \to L^1_{\mathrm{loc}}$ is a fortiori continuous on $L^\infty_{\mathrm{c}}[\tau]$. Thus it defines an PIP-space operator, but the converse need not be true. For regular operators, however, the two classes of maps coincide: a map $B : L^\infty_{\mathrm{c}} \to L^\infty_{\mathrm{c}}$ is continuous for τ if and only if it is for γ (since $B = B^{\times\times}$).

We may distinguish several classes of operators:

(1) *General operators,* given by kernels $A(x, y) \in \mathcal{L}(L^\infty_{\mathrm{c}}, L^1_{\mathrm{loc}})$, the space of all linear maps from L^∞_{c} into L^1_{loc}.

   In fact, we will consider only *integral* operators given by a kernel $A(x, y) \in L^1_{\mathrm{loc}}(\mathbb{R}^{2n})$. This is *a priori* a subset of $\mathcal{L}(L^\infty_{\mathrm{c}}, L^1_{\mathrm{loc}}) \simeq \mathrm{Op}(L^1_{\mathrm{loc}})$, since neither $L^\infty_{\mathrm{c}}[\tau]$, nor L^1_{loc} is a *nuclear* space. Note that the kernel of $A^\times$ is $A^\times(x, y) := \overline{A(y, x)}$.

(2) *Convolution operators,* given by kernels $A(x, y) = C(x - y)$.

(3) *Multiplication operators* given by diagonal kernels

$$A(x, y) = \delta(x - y)f(y), \ f \in L^1_{\mathrm{loc}}.$$

(4) *Finite rank operators,* given by separable kernels

$$A(x, y) = \sum_{j=1}^{n} \overline{f_j(y)}g_j(x), \ f_j, g_j \in L^1_{\mathrm{loc}}.$$

---

[5] This notion will be generalized in Section 5.4.1 under the name of *interspaces.*

Table 4.1 Various classes of operators on $L^1_{\text{loc}}(\mathbb{R}^n)$. The symbol $\star$ in a box means that the condition stated is necessary and sufficient for the operator to belong in the corresponding class

kernel	Op	Reg	$\mathfrak{A}$	$\mathfrak{B}$	$\mathfrak{C}$
integral operator $A(x,y)$	$L^1_{\text{loc}} \otimes L^1_{\text{loc}}$	$L^\infty_c(L^\infty_c)$	$\Rightarrow L^\infty_c(L^\infty_c)$	?	$\times$
convolution operator $C(x-y)$	L^1_{loc}	L^1_c	$\Rightarrow L^1_c$	?	$\times$
multiplication operator $\delta(x-y)\,f(y)$	$\star\, L^1_{\text{loc}}$	$\star\, L^\infty_{\text{loc}}$	$\star\, L^\infty$	$\Rightarrow \star\, L^\infty$	$\Rightarrow \star\, L^\infty$
rank 1 operator $\bar{f} \otimes g$	$\star\, L^1_{\text{loc}} \otimes L^1_{\text{loc}}$	$\star\, L^\infty_c \otimes L^\infty_c$	$\Rightarrow \star\, L^\infty_c \otimes L^\infty_c$	(?)	$\times$

For each class we will state sufficient conditions on the kernel for the operator to be in Op, Reg, $\mathfrak{A}, \mathfrak{B}$ or $\mathfrak{C}$. The results are summarized in Table 4.1 below that we now comment (the algebra $\mathfrak{C}$ will be discussed separately).

(1) *Integral operators:* every function $A(x,y)$, locally integrable in both variables, defines an operator on L^1_{loc}, by [Sch57, Prop.33]. By the remark following the same proposition, such an operator is regular when the following two conditions are satisfied:

   (i) the kernel $A(x,y)$ is compact, i.e., for every compact subset K of $\mathbb{R}^n$, $\text{supp}\, A \cap (\mathbb{R}^n \times K)$ and $\text{supp}\, A \cap (K \times \mathbb{R}^n)$ are compact (note that $\text{supp}\, A$ itself need not be compact, as the example of convolution operators shows), which implies that both A and $A^\times$ map L^∞_c into L^1_c.
   (ii) For every compact $K \subset \mathbb{R}^n$, $\int_K |A(x,y)|\,dx$ is bounded in y and $\int_K |A(x,y)|\,dy$ is bounded in x (both functions are in L^1_c by (i)).

We denote by $L^\infty_c(L^\infty_c)$ the class of kernels that verify those two conditions. In fact, the resulting regular operators are automatically in $\mathfrak{A}$ (symbol $\Rightarrow$) and they are even Hilbert-Schmidt operators in every $L^2(r)$. For $\mathfrak{B}$, however, we have no result.

(2) *Convolution operators:* the results follow immediately from the two well-known facts:

   (a) $\text{supp}(f * g) \subset \text{supp}\, f + \text{supp}\, g$;
   (b) $L^1 * L^p \subset L^p, 1 \leqslant p \leqslant \infty$.

For $\mathfrak{B}$, we have again no result.

(3) *Multiplication operators:* all results are obvious, and all conditions are actually necessary *and* sufficient (symbol $\star$)

(4) *Finite rank operators*: obvious again. As for $\mathfrak{B}$, there are finite rank operators that do not belong to it. Let $\overline{f} \otimes g \equiv |g\rangle\langle f|$ be of rank one, with $f, g \in L_c^\infty$ and normalized in L^2. Its norm in $L^2(r)$ is:

$$\|\overline{f} \otimes g\|_{rr} = \sup_{h \in L^2(r)} \frac{|\langle f|h\rangle|\|g\|_r}{\|h\|_r} = \|g\|_r\|f\|_{\overline{r}}.$$

Then $\overline{f} \otimes g \in \mathfrak{B}$ if and only if $\sup_r \|g\|_r \|f\|_{\overline{r}} < \infty$. Consider, for instance, the family of weights $b(x) = \exp bx$, $b \in \mathbb{R}$ (we work in one dimension for simplicity). So:

$$\|g\|_b = \int_{\mathbb{R}} |g(x)|^2 e^{bx} dx \geqslant \exp\left(b \int_{\mathbb{R}} |g(x)|^2 x\, dx\right)$$

by Jensen's inequality. Thus we have:

$$\sup_{r \in I_1} \|g\|_r \|f\|_{\overline{r}} \geqslant \sup_{b \in \mathbb{R}} \|g\|_b \|f\|_{\overline{b}}$$

$$\geqslant \sup_{b \in \mathbb{R}} \exp\left(b \int_{\mathbb{R}} \left(|g(x)|^2 - |f(x)|^2\right) x\, dx\right)$$

and the last supremum is infinite, unless the integral vanishes. Therefore $\overline{f} \otimes g$ cannot belong to $\mathfrak{B}$, except perhaps for special choices of f and g. We may observe also that $\|g\|_b$ is a convex function of b, and often $\|g\|_b = \|g\|_{\overline{b}}$ (for instance when g is even, or odd). In such a case, again $\overline{g} \otimes g \notin \mathfrak{B}$. We do not know, however, if $\mathfrak{B}$ contains any finite rank operator at all.

We finally come to the algebra $\mathfrak{C}$, and here of course we need the Riesz operators $U_{(r)} : L^2(r) \to L^2(\overline{r})$. Clearly $U(r)$ is the multiplication by $r^{-1}(x)$, and it is regular if and only if $r^{\pm 1} \in L_{\text{loc}}^\infty$. Thus some of the Riesz operators are regular, the others are not. Note also that these spaces $L^2(r)$ with $r^{\pm 1} \in L_{\text{loc}}^\infty$ are precisely those for which $r^{\pm n} \in L_{\text{loc}}^\infty$ for all n, so that the whole scale $V(r)$ generated by each such multiplication operator is contained in $\mathcal{I}$. In other words, since all those operators $U(r), r^{\pm 1} \in L_{\text{loc}}$ commute, they generate an abelian *-subalgebra of $\text{Reg}(L_{\text{loc}}^1)$. On the other hand, the family $\mathcal{R}$ of all Riesz operators is not an algebra.

Now a regular operator belongs to $\mathfrak{C}$ if and only if it commutes with every element of $\mathcal{R}$. But it is easy to check that no integral operator can commute with all multiplication operators by $r \in L_{\text{loc}}^\infty$ unless it is itself a multiplication operator. This explains the last column of Table 1. Of course we have not proved that $\mathfrak{B} = \mathfrak{C}$, but all indications point into that direction.

The analysis of the various classes of operators on L_{loc}^2 is entirely parallel to the one above, so we will not do it explicitly.

4.3 Köthe Sequence Spaces

As in Section 1.1, consider $V = \omega$, the space of all complex sequences, with the compatibility

$$(x_n) \# (y_n) \iff \sum_n |x_n y_n| < \infty.$$

This is called α-*duality* and the corresponding assaying subsets are called Köthe's *perfect* sequence spaces. This example shows how big and unpractical the complete lattice $\mathcal{F}(V, \#)$ can be. Indeed, this set $\mathcal{F}(\omega, \#)$ contains almost all possible types of topological vector spaces, many of them with rather awkward properties (it is Köthe's main source of counterexamples!). Thus it is imperative to restrict ourselves to suitable subsets of $\mathcal{F}(\omega, \#)$.

4.3.1 Weighted ℓ^2 Spaces

Our familiar example is the family $\mathcal{I}$ of all assaying subsets of the form

$$\ell^2(r) = \{(x_n) \in \omega : (x_n r_n^{-1}) \in \ell^2\},$$

where $r = (r_n)$ is an arbitrary sequence of positive numbers. The set $\mathcal{I}$ is an involutive covering of ω. Indeed, given $(x_n) \in \omega$ there is a weight sequence $r = (r_n)$ such that $(x_n) \in \ell^2(r)$. Take, for instance, the following weights:

- $r_n = n |x_n|$, whenever $x_n \neq 0$,
- r_n arbitrary, whenever $x_n = 0$.

Then we have shown in Chapter 1 that $\mathcal{F}(\omega, \#)$ is an involutive lattice and that $\mathcal{I}$ is a generating subset of it (see Section 1.4.1), and in fact an involutive sublattice.

Indeed, $\mathcal{I}$ is a lattice for the following operations:

$$\ell^2(r) \wedge \ell^2(s) = \ell^2(u), \quad \text{where } u_n = \min\{r_n, s_n\},$$
$$\ell^2(r) \vee \ell^2(s) = \ell^2(v), \quad \text{where } v_n = \max\{r_n, s_n\}.$$

Let us show that the norms of $\ell^2(u)$ and $\ell^2(v)$ are equivalent, respectively, to the projective and inductive norms defined in (I.4), (I.5).

First, on $\ell^2(u) = \ell^2(r) \cap \ell^2(s)$, the norm $\| \cdot \|_u$ defines the same topology as the projective topology given by $\|x\|_r^2 + \|x\|_s^2$. One has indeed:

$$
\begin{aligned}
\|x\|_u^2 &= \sum_n \frac{|x_n|^2}{\min\{r_n^2, s_n^2\}} = \sum_n |x_n|^2 \max\left\{\frac{1}{r_n^2}, \frac{1}{s_n^2}\right\} \\
&\leq \sum_n |x_n|^2 \left(\frac{1}{r_n^2} + \frac{1}{s_n^2}\right) = \|x\|_{r \wedge s}^2 \\
&\leq \sum_n |x_n|^2 \, 2 \, \max\left\{\frac{1}{r_n^2}, \frac{1}{s_n^2}\right\} = 2\|x\|_u^2.
\end{aligned}
$$

Then, on $\ell^2(v) = \ell^2(r) + \ell^2(s)$, the norm $\|\cdot\|_v$ defines a topology equivalent to the inductive topology given by $\|x\|_{r\vee s}^2 = \inf_{x=y+z}(\|y\|_r^2 + \|z\|_s^2)$. One has indeed:

$$\|x\|_{r\vee s}^2 = \inf_{x=y+z} \left(\sum_n \frac{|y_n|^2}{r_n^2} + \sum_n \frac{|z_n|^2}{s_n^2} \right)$$

$$= \sum_n \inf_{x_n=y_n+z_n} \left(\frac{|y_n|^2}{r_n^2} + \frac{|z_n|^2}{s_n^2} \right) = \sum_n \inf_{y_n} \left(\frac{|y_n|^2}{r_n^2} + \frac{|x_n-y_n|^2}{s_n^2} \right).$$

Since all terms are non-negative, we may compute the minimum of each term separately, by equating to zero the partial derivatives with respect to y_n, z_n, where $x_n = y_n + z_n$. An easy computation shows that this infimum is reached for

$$y_n = x_n \frac{r_n^2}{r_n^2 + s_n^2}, \quad z_n = x_n \frac{s_n^2}{r_n^2 + s_n^2}.$$

From this, we get:

$$\|x\|_{r\vee s}^2 = \inf_{x=y+z}(\|y\|_r^2 + \|z\|_s^2) = 2 \sum_n \frac{|x_n|^2}{r_n^2 + s_n^2}$$

and

$$\frac{1}{2}\|x\|_v^2 = \sum_n \frac{|x_n|^2}{2 \max\{r_n^2, s_n^2\}} \leqslant \sum_n \frac{|x_n|^2}{r_n^2 + s_n^2} = \frac{1}{2}\|x\|_{r\vee s}^2 \leqslant \sum_n \frac{|x_n|^2}{\max\{r_n^2, s_n^2\}} = \|x\|_v^2 .$$

Thus finally

$$\|x\|_u \asymp \|x\|_{r\wedge s}, \qquad \|x\|_v \asymp \|x\|_{r\vee s},$$

where the symbol $\asymp$ denotes equivalence of norms.

4.3.2 Norming Functions and the ℓ_ϕ Spaces

Besides the ℓ^p and the $\ell^2(r)$ spaces, there are many other types of perfect spaces. In the sequel of this section, we will describe an interesting class of them, which constitutes a PIP-space of type (B) (and in fact a LBS). Throughout the following sections, sequences are added and multiplied componentwise: for $x = (x_n), y = (y_n) \in \omega$, we write $x + y = (x_n + y_n)$, $\alpha x = (\alpha x_n)$ $(\alpha \in \mathbb{C})$ and $x \cdot y = (x_n y_n)$. Also, we quote the results without proofs, which may be found in the references given in the Notes.

Definition 4.3.1. A real-valued function ϕ defined on the space φ of finite sequences is said to be a *norming function* if

    (n₁) $\phi(x) > 0$ for $x \neq 0$;
    (n₂) $\phi(\alpha x) = |\alpha|\phi(x)$, $\forall \alpha \in \mathbb{C}$;

(n_3) $\phi(x+y) \leqslant \phi(x) + \phi(y)$;
(n_4) $\phi(1,0,0,0,\dots) = 1$.

A norming function ϕ is *symmetric* if

(n_5) $\phi(x_1, x_2, \dots, x_n, 0, 0, \dots) = \phi(|x_{j_1}|, |x_{j_2}|, \dots, |x_{j_n}|, 0, 0, \dots)$,

where $j_1, j_2, \dots, j_n$ is an arbitrary permutation of $1, 2, \dots, n$.

From property (n_5), it is clear that a symmetric norming function ϕ is entirely determined by its values on the set $[\varphi]$ of finite, positive, nonincreasing sequences. Hence, from conditions (n_2) and (n_4), we deduce that

$$\phi_\infty(x) \leqslant \phi(x) \leqslant \phi_1(x), \ \forall x \in \varphi,$$

where $\phi_\infty(x) = \max_{i=1,\dots,n} |x_i|$ and $\phi_1(x) = \sum_{i=1}^{n} |x_i|$.

Lemma 4.3.2. *The set of all symmetric norming functions possesses a maximal element, ϕ_1, and a minimal one, ϕ_∞.*

To every symmetric norming function ϕ, one can associate a Banach space ℓ_ϕ as follows. Given a sequence $x \in \omega$, define its nth section as $x^{(n)} = (x_1, x_2, \dots, x_n, 0, 0, \dots)$. Then the sequence $(\phi(x^{(n)}))$ is nondecreasing, so that one can define

$$\ell_\phi = \{x \in \omega : \sup_n \phi(x^{(n)}) < \infty\}$$

and then extend the norming function ϕ to the whole of ℓ_ϕ by putting $\phi(x) = \lim_n \phi(x^{(n)})$. This relation defines a norm ϕ on ℓ_ϕ, for which it is complete, hence a Banach space. In other words, we can also say that $\ell_\phi = \{x \in \omega : \phi(x) < \infty\}$ is the natural domain of definition of the extended norming function ϕ.

Clearly, one has $\ell_{\phi_\infty} = \ell^\infty$ and $\ell_{\phi_1} = \ell^1$. Similarly, $\ell^p = \ell_{\phi_p}$, where $\phi_p(x) = \left(\sum_n |x_n|^p\right)^{1/p}$. Thus every space ℓ_ϕ contains ℓ^1 and is contained in ℓ^∞. Moreover, every space ℓ_ϕ is a two-sided ideal in ℓ^∞. Indeed, one has, for every $x \in \ell_\phi$ and $y \in \ell^\infty$,

$$\begin{aligned} x \cdot y \in \ell_\phi \ \text{and} \quad & \phi(x \cdot y) \leqslant \phi(x) \|y\|_\infty , \\ y \cdot x \in \ell_\phi \ \text{and} \quad & \phi(y \cdot x) \leqslant \phi(x) \|y\|_\infty . \end{aligned} \tag{4.13}$$

In addition, the set of Banach spaces ℓ_ϕ constitutes a lattice. Given two symmetric norming functions ϕ and ψ, one defines their infimum and supremum, exactly as for the general case:

- $\phi \wedge \psi := \max\{\phi, \psi\}$, which defines on the space $\ell_{\phi \wedge \psi} := \ell_\phi \cap \ell_\psi$ a norm equivalent to $\phi(x) + \psi(x)$;
- $\phi \vee \psi := \min\{\phi, \psi\}$, which defines on the space $\ell_{\phi \vee \psi} := \ell_\phi + \ell_\psi$ a norm equivalent to $\inf_{x=y+z}\{\phi(y) + \psi(z)\}, x \in \ell_\phi + \ell_\psi, y \in \ell_\phi, z \in \ell_\psi$.

It remains to analyze the relationship of the spaces ℓ_ϕ and the PIP-space structure of ω. Given a symmetric norming function ϕ and a finite sequence y, consider the function

$$F_y(x) := \frac{\langle x|y \rangle}{\phi(x)} = \frac{\sum_n \overline{x_n} y_n}{\phi(x)} \qquad \text{(for } x = 0, \text{ one puts } F_y(0) = 0\text{)}.$$

This function is bounded on $[\varphi]$ and reaches is maximum, since

$$F_y(x) \leqslant F_y((x_1, x_2, \ldots, x_n, 0, 0, \ldots)),$$

where n is the largest index such that $y_n \neq 0$. Thus we may define, for $y \in [\varphi]$:

$$\overline{\phi}(y) := \max_{x \in [\varphi]} \frac{\sum_n \overline{x_n} y_n}{\phi(x)} = \max_{x \in [\varphi]} \frac{\langle x|y \rangle}{\phi(x)}.$$

The function $\overline{\phi}$ thus defined is a symmetric norming function, hence it can be extended to the corresponding Banach space $\ell_{\overline{\phi}}$. The function $\overline{\phi}$ is said to be *conjugate* to ϕ and the space $\ell_{\overline{\phi}}$ is the dual of ℓ_ϕ with respect to the partial inner product, i.e., $\ell_{\overline{\phi}} = (\ell_\phi)^\#$. Clearly one has $\overline{\overline{\phi}} = \phi$, hence $\ell_{\overline{\overline{\phi}}} = (\ell_\phi)^{\#\#} = \ell_\phi$.

In addition, it is easy to show that $\ell_{\overline{\phi \wedge \psi}} = \ell_{\overline{\phi} \vee \overline{\psi}}$ and $\ell_{\overline{\phi \vee \psi}} = \ell_{\overline{\phi} \wedge \overline{\psi}}$. In other words,

Proposition 4.3.3. *The family of Banach spaces ℓ_ϕ, where ϕ is a symmetric norming function, is an involutive sublattice of the lattice $\mathcal{F}(\omega, \#)$ and a LBS.*

Actually, since every ϕ satisfies the inclusions $\ell^1 \subset \ell_\phi \subset \ell^\infty$, the family $\{\ell_\phi\}$ is also an involutive sublattice of the lattice $\mathcal{F}(\ell^\infty, \#)$ obtained by restricting to ℓ^∞ the PIP-space structure of ω.

The interest of the spaces ℓ_ϕ is their close connection with ideals of compact operators. Let indeed A be a compact operator in a Hilbert space $\mathcal{H}$. The *singular values* of A are the (positive) eigenvalues $s_j, j = 1, 2, \ldots$ of the positive self-adjoint operator $|A| := (A^* A)^{1/2}$. Then the set of all compact operators A for which the sequence (s_j) belongs to ℓ_ϕ constitutes a two-sided ideal $\mathcal{C}_\phi$ of the space $\mathcal{C}^\infty$ of all compact operators. In particular, ℓ^1 corresponds to the ideal $\mathcal{C}^1$ of nuclear or trace-class operators, ℓ^2 to the ideal $\mathcal{C}^2$ of Hilbert-Schmidt operators, and ℓ^∞ to $\mathcal{C}^\infty$ itself.

As a final remark, we may also notice that the family of Banach spaces $\ell_\phi \subset \omega$ has a counterpart in the space $L^1_{\text{loc}}(\mathbb{R}^n, dx)$ of locally integrable functions, namely the so-called Köthe function spaces, that we shall discuss in detail in Section 4.4 below.

More general sequence spaces may be defined, in terms of normed ideals. These are defined as follows.

Definition 4.3.4. A *normed ideal* is a couple (l, ν) where
(ni$_1$) l is a subspace of ω such that $\varphi \subseteq l \subseteq \ell^\infty$;
(ni$_2$) ν is a norm on l and one has $\nu(1, 0, \ldots) = 1$;
(ni$_3$) for every $x \in l$ and $y \in \ell^\infty$, one has

$$x \cdot y \in l \text{ and } \nu(x \cdot y) \leqslant \nu(x) \|y\|_\infty ,$$
$$y \cdot x \in l \text{ and } \nu(y \cdot x) \leqslant \nu(x) \|y\|_\infty ;$$

(ni$_4$) if $x \in l$, then $(x_{j_1}, x_{j_2}, \ldots, x_{j_n}, \ldots) \in l$ for any permutation of the indices and $\nu(x) = \nu(x_{j_1}, x_{j_2}, \ldots, x_{j_n}, \ldots)$.

If, in addition, l is complete for the norm ν, it is called a *Banach ideal*.

An order relation may be defined on the set of all Banach ideals as follows. We say that $(l, \nu) \prec (m, \mu)$ if $l \subset m$ and the induced norm $\mu \restriction l$ is equivalent to ν.

Definition 4.3.5. A Banach ideal (l, ν) is said to be *maximal* if there exists no Banach ideal (m, μ) such that $(l, \nu) \prec (m, \mu)$.

To make contact with the previous case, take an arbitrary normed ideal (l, ν) and consider the restriction of the norm ν to the space φ, which is obviously a symmetric norming function. Define, as above, a function $\underline{\nu}$ on ℓ^∞ by $\underline{\nu}(x) = \lim_n \nu(x^{(n)})$. This limit exists, but is not necessarily finite. In addition, one has $\underline{\nu}(x) \leqslant \nu(x)$, $\forall x \in l$.

Definition 4.3.6. ν is said to be a *strong norm* if $\underline{\nu}(x) = \nu(x)$, $\forall x \in l$. Then the ideal (l, ν) is called a strong normed ideal, and a strong Banach ideal if it is complete.

As before, we may extend $\underline{\nu}$ to its natural domain $\ell_\nu := \ell_{\underline{\nu}} = \{x \in \ell^\infty : \underline{\nu}(x) < \infty\}$. Then ℓ_ν is a maximal strong Banach ideal. The interest of this notion in the present context is the next result.

Proposition 4.3.7. *For every normed ideal (l, ν), the following three properties are equivalent and each of them implies that (l, ν) is a strong ideal:*

(i) (l, ν) *is maximal;*
(ii) $(l, \nu) = \ell_\nu$;
(iii) (l, ν) *is perfect, i.e., $(l, \nu)^{\#\#} = (l, \nu)$, i.e., (l, ν) is an assaying subspace in $\mathcal{F}(\ell^\infty, \#)$.*

For a counterexample, take the space c_o of sequences converging to 0. The space c_o is a Banach ideal for the norm $\phi_\infty(x) = \sup_n |x_n|$, but it is not maximal, since ℓ^∞ is a (maximal) ideal containing c_o. It is not perfect either, thus not an assaying subspace of ℓ^∞, since one has $(c_o)^\# = \ell^1$, $(\ell^1)^\# = \ell^\infty$, $(\ell^\infty)^\# = \ell^1$, hence $c_o \subsetneqq (c_o)^{\#\#} = \ell^\infty$.

4.3.3 Other Types of Sequence Ideals

Besides the ideals ℓ_ϕ, there exist many other types of sequence ideals, perfect or not. We mention three different types, namely, quotient ideals, root ideals and weighted function spaces.

(i) Quotient ideals

Definition 4.3.8. Given two normed ideals (l, ν) and (m, μ), consider the pair $(l : m, \nu : \mu)$, where $l : m$ is the set:

$$l : m := \{x \in \ell^\infty : x \cdot m \subseteq l \text{ and } x : (m, \mu) \to (l, \nu) \text{ is a bounded map}\}$$

and $\nu : \mu$ is the norm

$$(\nu : \mu)(x) := \sup_{y \in m, \mu(y) \leqslant 1} \nu(x \cdot y).$$

Then $(l : m, \nu : \mu)$ is a normed ideal called the *quotient ideal*.

This quotient ideal has some interesting properties.

  (i) If l is complete for ν, then $l : m$ is complete for $\nu : \mu$.
 (ii) If ν is a strong norm on l, $\nu : \mu$ is a strong norm on $l : m$.
(iii) If (l, ν) is a maximal strong Banach ideal, so is $(l : m, \nu : \mu)$.
 (iv) If (l, ν) is perfect (= assaying), so is $(l : m, \nu : \mu)$, for any normed ideal (m, μ), and one has $l : m = (l^\# \cdot m)^\#$.
  (v) If $l = \ell^1$, then $\ell^1 : m = m^\#$ and $\nu_1 : \mu = \overline{\mu}$, the norming function conjugate to ν.

Thus this notion of quotient allows one to construct many assaying subspaces in $(\ell^\infty, \#)$. Indeed, if l is assaying, $l^{\#\#} = l$, then the quotient $l : m$ is assaying for any normed ideal m, i.e., $(l : m)^{\#\#} = l : m$.

(ii) Root ideals

Definition 4.3.9. Given a normed ideal (l, ν), its pth *root* $(l^p, \nu_{(p)})$ $(1 < p < \infty)$ is the normed ideal defined as follows:

$$l^p = \{x \in \ell^\infty : |x|^p \in l\},$$
$$\nu_{(p)}(x) = \left(\nu(|x|^p)\right)^{1/p}.$$

Then one shows that $(l^p, \nu_{(p)})$ is perfect if and only if (l, ν) is perfect.
    As examples, we may mention $\ell^p = (\ell^1)^p$ and $(\ell_\phi)^p = \{x \in \ell^\infty : |x|^p \in \ell_\phi\}$. We note the relations

$$\ell_\phi : (\ell_\phi)^p = (\ell_\phi)^q, \text{ where } \tfrac{1}{p} + \tfrac{1}{q} = 1,$$

$$(\ell_\phi)^p : (\ell_\phi)^q = (\ell_\phi)^r, \text{ where } \tfrac{1}{p} + \tfrac{1}{q} = \tfrac{1}{r},$$

$$(\ell_\phi)^{p\#} = (\ell_{\overline{\phi}})^q.$$

(iii) Weighted sequence spaces

Following the familiar example of the weighted Hilbert spaces $\ell^2(r)$, we may introduce weighted ℓ_ϕ spaces as follows. Given an arbitrary sequence $r = (r_n)$ of positive numbers, we define

$$\ell_\phi(r) := \{x \in \omega : \phi(xr^{-1}) < \infty\}, \text{ where } xr^{-1} = (x_n r_n^{-1}).$$

These spaces generalize the ℓ_ϕ spaces, but they need not be contained in ℓ^∞. Yet they are assaying subsets of $(\omega, \#)$, since one has

$$\ell_\phi(r)^\# = \ell_{\overline{\phi}}(r^{-1}),$$

$$\ell_\phi(r)^{\#\#} = \ell_\phi(r).$$

The most common example, which will exploited systematically in Chapter 8, is that of the weighted ℓ^p spaces, called $\ell_m^p \equiv \ell^p(m^{-1})$, for a given sequence $m := (m_k)$, with norm

$$\|x\|_{\ell_m^p} = \left(\sum_k |x_k|^p m_k^p \right)^{1/p}.$$

This notion allows one to compute explicitly the assaying subspaces generated by a single element $x \in \omega$. Let first $x_n \neq 0$ for every n. Then one has

$$\{x\}^\# = \{y \in \omega : \sum_n |x_n|\,|y_n| < \infty\} = \ell^1(|x|^{-1}),$$

$$\{x\}^{\#\#} = \{z \in \omega : |z_n| \leqslant c|x_n|\} = \ell^\infty(|x|),$$

which is the smallest assaying subspace containing x. On the other hand, if x is a finite sequence, then $\{x\}^\# = \omega$ and $\{x\}^{\#\#} = \varphi$. Thus, in general, $\{x\}^\#$ is the direct sum of a weighted ℓ^1 space and ω, whereas $\{x\}^{\#\#}$ is the direct sum of a weighted ℓ^∞ space and φ.

An interesting example of that construction is given by the so-called echelon spaces. Let $a_{(1)}, a_{(2)}, \ldots \in \omega$ be an increasing sequence of positive sequences, called *steps*. Then one calls *echelon space* the set

$$l := \bigcap_k \{a_{(k)}\}^\# = \bigcap_k \ell^1(a_{(k)}^{-1})$$

and *co-echelon space* the #-dual

$$l^{\#} := \bigcup_k \{a_{(k)}\}^{\#\#} = \bigcup_k \ell^{\infty}(a_{(k)}),$$

both of which are assaying subspaces of ω. An example is given by the spaces on entire analytic functions Exp and 3 described in Section 4.6.1.

4.4 Köthe Function Spaces

A nontrivial (i.e., not a chain) example of indexed PIP-space of type (B), actually a LBS, is the family of the so-called Köthe function spaces. Since it is highly instructive, we feel it worthwhile to discuss it extensively. We will begin by repeating the basic definitions.

Let (X, μ) be a σ-finite measure space, M^+ the set of all measurable, nonnegative functions on X, where two functions are identified if they differ at most on a μ-null set. A *function norm* is a mapping $\rho : M^+ \to \overline{\mathbb{R}}$ such that:

(i) $0 \leqslant \rho(f) \leqslant \infty$, $\forall f \in M^+$ and $\rho(f) = 0$ if and only if $f = 0$;
(ii) $\rho(f_1 + f_2) \leqslant \rho(f_1) + \rho(f_2)$, $\forall f_1, f_2 \in M^+$;
(iii) $\rho(af) = a\rho(f)$, $\forall f \in M^+$, $\forall a \geqslant 0$;
(iv) $f_1 \leqslant f_2 \Rightarrow \rho(f_1) \leqslant \rho(f_2)$, $\forall f_1, f_2 \in M^+$.

A function norm ρ is said to have the *Fatou property* if and only if $0 \leqslant f_1 \leqslant f_2 \leqslant \ldots, f_n \in M^+$ and $f_n \to f$ pointwise, implies $\rho(f_n) \to \rho(f)$.

Given a function norm ρ, it can be extended to all complex measurable functions on X by defining $\rho(f) = \rho(|f|)$ (for simplicity, we keep the same notation). Denote by L_ρ the set of all measurable f such that $\rho(f) < \infty$. With the norm $\|f\| = \rho(f)$, L_ρ is a normed space and a subspace of the vector space V of all measurable, μ-a.e. finite, functions on X. In addition, the space L_ρ is *solid*, that is, if $f \in V$, $g \in L_\rho$ and $|f(x)| \leqslant |g(x)|$ a.e., then $f \in L_\rho$ and $\|f\| \leqslant \|g\|$. Furthermore, if ρ has the Fatou property, L_ρ is complete, i.e., a Banach space. This is a generalization of the spaces $L^p(X, \mu)$, which correspond to $\rho(f) = (\int_X |f|^p d\mu)^{1/p}$ for $1 \leqslant p < \infty$ and $\rho(f) = \sup |f|$ for $p = \infty$.

A function norm ρ is said to be *saturated* if, for any measurable set $E \subset X$ of positive measure, there exists a measurable subset $F \subset E$ such that $\mu(F) > 0$ and $\rho(\chi_F) < \infty$ (χ_F is the characteristic function of F).

Let ρ be a saturated function norm with the Fatou property. Define:

$$\rho'(f) = \sup \left\{ \int_X |f\,g|\,d\mu : \rho(g) \leqslant 1 \right\} \tag{4.14}$$

Then ρ' is a saturated function norm with the Fatou property and $\rho'' \equiv (\rho')' = \rho$. Hence, $L_{\rho'}$ is a Banach space. Moreover, one has also:

$$\rho'(f) = \sup\left\{\left|\int_X f\,g\,d\mu\right| : \rho(g) \leqslant 1\right\} \tag{4.15}$$

In our language these results can be restated as follows. The vector space V of all measurable, a.e.-finite functions on X carries a natural PIP-space structure, with compatibility

$$f \# g \iff \int_X |fg|\,d\mu < \infty \tag{4.16}$$

and partial inner product

$$\langle f|g\rangle = \int_X \overline{f}\,g\,d\mu. \tag{4.17}$$

V is clearly the largest space on which the partial inner product (4.17) may be defined, but it is too large. Indeed, $V^\# = \{0\}$ and the partial inner product is degenerate. However, there are plenty of subspaces of V which are nondegenerate, such as $L^1_{\mathrm{loc}}, L^2_{\mathrm{loc}}$ or the space L_{ρ_0} to be defined below. Furthermore, for each ρ as above, L_ρ is a Banach space and $L_{\rho'} = (L_\rho)^\#$, i.e., each L_ρ is assaying. The pair $\langle L_\rho, L_{\rho'}\rangle$ is actually a dual pair, although $\langle V^\#, V\rangle$ is not. The space $L_{\rho'}$ is called the *Köthe dual* or *α-dual* of L_ρ and denoted by $(L_\rho)^\alpha$. However, $L_{\rho'}$ is in general only a closed subspace of the Banach conjugate dual $(L_\rho)^\times$, thus the Mackey topology $\tau(L_\rho, L_{\rho'})$ is coarser than the ρ-norm topology, which is $\tau(L_\rho, (L_\rho)^\times)$. This defect can be remedied by further restricting ρ. A function norm ρ is called *absolutely continuous* if $\rho(f_n) \searrow 0$ for every sequence $f_n \in L_\rho$ such that $f_1 \geqslant f_2 \geqslant \ldots \searrow 0$ pointwise a.e. on X. For instance, the Lebesgue L^p-norm is absolutely continuous for $1 \leqslant p < \infty$ but the L^∞-norm is *not*! Also, even if ρ is absolutely continuous, ρ' need not be. Yet, this is the appropriate concept, in view of the following results:

(i) $L_{\rho'} = (L_\rho)^\alpha = (L_\rho)^\times$ if and only if ρ is absolutely continuous;
(ii) L_ρ is reflexive if and only if ρ *and* ρ' are absolutely continuous and ρ has the Fatou property.

Let us denote by J the set of saturated, absolutely continuous function norms ρ on X, with the Fatou property and such that ρ' is also absolutely continuous. Then, for every $\rho \in J, \langle L_\rho, L_{\rho'}\rangle$ is a reflexive dual pair of Banach assaying subspaces of $(V, \#, \langle\cdot|\cdot\rangle)$. All that remains to do in order to get an indexed PIP-space of type (B) is to restrict the partial inner product to a nondegenerate subspace and perform the lattice construction of Section 2.2. Now the last point is in fact already done:

Lemma 4.4.1. *The set J is an involutive lattice with respect to the following partial order: $\rho_1 \leqslant \rho_2$ if and only if $\rho_1(f) \leqslant \rho_2(f), \forall f \in V$. The lattice operations are the following:*

- $(\rho_1 \vee \rho_2)(f) = \max \{\rho_1(f), \rho_2(f)\}$,
- $(\rho_1 \wedge \rho_2)(f) = \inf \{\rho_1(f_1) + \rho_2(f_2); f_1, f_2 \in M^+, f_1 + f_2 = |f|\}$,
- *involution* : $\rho \leftrightarrow \rho'$.

Proof. Let $\rho_1, \rho_2 \in J$; so are ρ_1', ρ_2'. First we show that $\rho_1 \vee \rho_2$, which is obviously a norm, is saturated. Suppose it is not saturated, i.e., there exists a measurable set E of positive measure, such that $(\rho_1 \vee \rho_2)(\chi_F) = \infty$ for every measurable subset $F \subset E$ of positive measure. Thus, for every such $F \subset E$, $\rho_1(\chi_F) = \infty$ or $\rho_2(\chi_F) = \infty$. Since ρ_1 is saturated, there is a set $G \subset E$ such that $\rho_1(\chi_G) < \infty$ and for every $G_1 \subset G$, $\rho_1(\chi_{G_1}) < \infty$. This implies that $\rho_2(\chi_{G_1}) = \infty$ for every such G_1 and this is impossible for a saturated ρ_2.

Next it is always true [Zaa61, Problem 71.2] that $(\rho_1 \wedge \rho_2)' = \rho_1' \vee \rho_2'$, although $\rho_1 \wedge \rho_2$ as defined could be only a function *seminorm* [i.e., $p(f) = 0 \not\Rightarrow f = 0$]. However, since $\rho_1' \vee \rho_2'$ is a saturated norm, it follows from this equality that $\rho_1 \wedge \rho_2$ is also a saturated norm [Zaa61, Theorem 71.4]. Since ρ_1 and ρ_2 are absolutely continuous, so are all the others. Thus $L_{(\rho_1 \wedge \rho_2)'} = (L_{\rho_1 \wedge \rho_2})^\times$ is reflexive, and, therefore, $L_{\rho_1 \wedge \rho_2}$, is also reflexive, which implies that $\rho_1 \wedge \rho_2$ has the Fatou property. Since $\rho_1 \vee \rho_2$ also has the Fatou property, like any supremum [Zaa61, Theorem 65.4)], the proof is complete.   ∎

It is clear from the construction that we have recovered the general situation, for we have the relations

$$L_{\rho_1 \vee \rho_2} = (L_{\rho_1} \cap L_{\rho_2})_{\mathrm{proj}}, \quad L_{\rho_1 \wedge \rho_2} = (L_{\rho_1} + L_{\rho_2})_{\mathrm{ind}}.$$

It is interesting to notice that for any $\rho_1, \rho_2 \in J$, the ρ_1 norm and the ρ_2 norm are always consistent (see Proposition 2.2.1) on $L_{\rho_1} \cap L_{\rho_2}$ since $\rho_1 \wedge \rho_2$ is a norm.

Finally, for any sublattice I of J, define the space $V_I \equiv \sum_{\rho \in I} L_\rho$; thus $V_I^\# = \bigcap_{\rho \in I} L_\rho$. Then we have:

Proposition 4.4.2. *Let V be the vector space of all measurable, a.e. finite functions on the σ-finite measure space (X, μ). With the compatibility (4.16) and partial inner product (4.17), V becomes a degenerate PIP-space. Denote by J the involutive lattice of all saturated, absolutely continuous function norms ρ on X, which have the Fatou property and are such that ρ' is also absolutely continuous. Let I be any involutive sublattice of J such that:*

(i) $V_I \equiv \sum_{\rho \in I} L_\rho$ is an assaying subset of V;
(ii) $(V_I^\#)^\perp = \{0\}$.

Then V_I with the PIP-space structure induced by V is a nondegenerate indexed PIP-space of type (B) and, actually, a LBS.

Example 4.4.3. The space of locally integrable functions

An interesting example is our familiar space $V_I = L^1_{\mathrm{loc}}(\mathbb{R}^n, dx)$. Then $V_I^\# = L^\infty_c(\mathbb{R}^n, dx)$ and the two conditions above are verified. The corresponding set I is easily characterized: $\rho \in J$ belongs to I if and only if

$L_\rho \subset L^1_{loc}(\mathbb{R}^n, dx)$ with continuous injection. If we write $X = \bigcup_j \Omega_j$, Ω_j compact, as in the proof of Proposition 4.2.1, we have the representations (4.8) and (4.9). Then $L_\rho \subset L^1_{loc}(\mathbb{R}^n, dx)$ with continuous injection, if and only if ρ satisfies the following set of conditions:

$$\int_{\Omega_j} |f| dx \leqslant c_j \rho(f) \quad \text{for each } j = 1, 2, \ldots.$$

For instance, a weighted L^2-space $L^2(r)$ satisfies this condition if $r \in L^2_{loc}(\mathbb{R}^n, dx)$ with $c_j = \left[\int_{\Omega_j} r^2\, dx\right]^{1/2}$. It is thus clear that the family $\{L_\rho, \rho \in I\}$ is generating in $L^1_{loc}(\mathbb{R}^n, dx)$, since it contains the generating subfamily $\{L^2(r), r \text{ and } r^{-1} \in L^2_{loc}(\mathbb{R}^n, dx)\}$.

Example 4.4.4. L^p-spaces

Another example of the construction given in Proposition 4.4.2 is the lattice generated by the spaces $L^p(X, \mu)$, $1 < p < \infty$, where (X, μ) be a σ-finite measure space. Indeed, each L^p norm is saturated and absolutely continuous and has the Fatou property. Thus the family $\{L^p(X, \mu), 1 < p < \infty\}$ generates a LBS with the compatibility (4.16) and the L^2 inner product.

Remember that if $\mu(X) = \infty$ and μ has no atoms, no two L^p spaces are contained in each other, but $L^p \cap L^r \subset L^q$ for all q such that $p \leqslant q \leqslant r$. Hence, we get a genuine lattice in that case, as discussed in Section 4.1.2.

What about the spaces L^1 and L^∞? They are not reflexive, since the L^∞-norm is not absolutely continuous; hence they do not belong to any V_I. But if they are added by hand, one gets an interesting result. Consider the following norm:

$$\rho_0(f) = \sup_E \left\{ \int_E |f| d\mu : \mu(E) = 1 \right\}.$$

Then:

(i) ρ_0 has the Fatou property (since it is a supremum of function norms that have it), hence L_{ρ_0} is a Banach space and, in fact, $L_{\rho_0} = L_G$, the Gould space described in Section 4.1.2;

(ii) $L_{\rho_0} = (L^1 + L^\infty)_{ind}$, $(L_{\rho_0'} = L^1 \cap L^\infty)_{proj}$;

(iii) $\bigcup_{1 \leqslant p \leqslant \infty} L^p$ is properly included in L_{ρ_0}.

So, exactly as for the chain $\{L^p[0, 1], 1 \leqslant p \leqslant \infty\}$ discussed in Section 4.1, the family $\{L^p(X, \mu), 1 \leqslant p \leqslant \infty\}$ generates a lattice with extreme elements L_{ρ_0} and $L_{\rho_0'}$. The completion of that lattice can easily be described along the same lines.

A particular case is that of *weighted L^p spaces*, denoted $L_m^p(\mathbb{R}^d)$, corresponding to $X = \mathbb{R}^d$ and $d\mu(x) = m(x)\,dx$, for some positive, measurable, locally integrable weight function m. The corresponding norm reads

$$\|f\|_{L_m^p} := \left(\int_{\mathbb{R}^d} |f(x)|^p\, m(x)^p\, dx\right)^{1/p}, \ f \in L_m^p(\mathbb{R}^d). \qquad (4.18)$$

These spaces will play a central role in Chapter 8, in the applications to signal processing.

Example 4.4.5. Mixed-norm spaces

Among the Köthe function spaces, an interesting class consists in the so-called L^P spaces with mixed norm. Let (X,μ) and (Y,ν) be two σ-finite measure spaces and $1 \leqslant p,q \leqslant \infty$ (in the general case, one considers n such spaces and n-tuples $P := (p_1, p_2, \ldots, p_n)$). Then, a function $f(x,y)$ measurable on the product space $X \times Y$ is said to belong to $L^{(p,q)}(X \times Y)$ if the number obtained by taking successively the p-norm in x and the q-norm in y, in that order, is finite (exchanging the order of the two norms leads in general to a different space). If $p,q < \infty$, the norm reads

$$\|f\|_{(p,q)} = \left(\int_Y \left(\int_X |f(x,y)|^p\, d\mu(x)\right)^{q/p} d\nu(y)\right)^{1/q}. \qquad (4.19)$$

The analogous norm for p or $q = \infty$ is obvious. For $p = q$, one gets the usual space $L^p(X \times Y)$.

These spaces enjoy a number of properties similar to those of the L^p spaces.

(i) *Completeness*: each space $L^{(p,q)}$ is a Banach space and it is reflexive if and only if $1 < p,q < \infty$.

(ii) *Duality*: the conjugate dual of $L^{(p,q)}$ is $L^{(\overline{p},\overline{q})}$, where, as usual, $p^{-1} + \overline{p}^{-1} = 1$, $q^{-1} + \overline{q}^{-1} = 1$. Thus the topological conjugate dual coincides with the Köthe dual, as for all Köthe function spaces.

(iii) *Generalized Hölder inequality*:

$$\left|\iint_{X \times Y} f_1 f_2 \ldots f_m\, d\mu\, d\nu\right| \leqslant \|f_1\|_{(p_1,q_1)} \|f_2\|_{(p_2,q_2)} \cdots \|f_m\|_{(p_m,q_m)},$$

whenever $\sum_{i=1}^m \frac{1}{p_i} = 1, \sum_{i=1}^m \frac{1}{q_i} = 1$.

(iv) *Interpolation*: if $f \in L^{(p_1,p_2)}$ and $f \in L^{(q_1,q_2)}$, then $f \in L^{(r_1,r_2)}$, where $\frac{1}{r_i} = \frac{t}{p_i} + \frac{1-t}{q_i}$, $i = 1,2$, $0 \leqslant t \leqslant 1$, and

$$\|f\|_{(r_1,r_2)} \leqslant \left(\|f\|_{(p_1,p_2)}\right)^t \left(\|f\|_{(q_1,q_2)}\right)^{1-t}.$$

Notice that there is in general no inclusion relation between two different spaces $L^{(p,q)}$.

When $X = Y = \mathbb{R}$ with Lebesgue measure, additional theorems hold true.

(v) *Young's theorem:* Let $\frac{1}{p_i} + \frac{1}{q_i} = 1 + \frac{1}{r_i}$, $i = 1, 2$. If $f \in L^{(p_1, p_2)}$, $g \in L^{(q_1, q_2)}$, then $f * g \in L^{(r_1, r_2)}$ and

$$\|f * g\|_{(r_1, r_2)} \leqslant \|f\|_{(p_1, p_2)} \|g\|_{(q_1, q_2)}.$$

(vi) *Hausdorff-Young theorem:* Let $f \in L^{(p,q)}$, with $1 \leqslant q \leqslant p \leqslant 2$, and let $\mathcal{F}f(\xi, \eta)$ be the Fourier transform of $f(x, y)$. Then $\mathcal{F}f \in L^{(\bar{p}, \bar{q})}$, where $p^{-1} + \bar{p}^{-1} = 1$, $q^{-1} + \bar{q}^{-1} = 1$, and

$$\|\mathcal{F}f\|_{(\bar{p}, \bar{q})} \leqslant \|f\|_{(p,q)}.$$

On the other hand, the assertion is false if $p < q$.

We have given here the general definition, but in fact only some particular subclasses of these mixed-norm spaces have found applications in signal processing. These spaces will thus be described in Chapter 8 which is entirely devoted to that topic.

4.5 Analyticity/Trajectory Spaces

A different kind of PIP-space is given by the so-called analyticity/trajectory spaces, which were conceived as a substitute to distribution spaces. These spaces are of two types, which are in duality. We treat them briefly in succession.

Let $\mathcal{H}$ be a separable Hilbert space and let A be a non-negative self-adjoint operator in $\mathcal{H}$. The *analyticity space* associated to A, denoted $\mathcal{S}_{\mathcal{H},A}$, consists of all vectors $v \in \mathcal{D}^\infty(A) = \bigcap_{n=1}^\infty D(A^n)$ for which there exists constants a, b such that $\|A^n v\| \leqslant n! a^n b$, $\forall n \in \mathbb{N}$. Such vectors are called *analytic vectors* for A, hence the name of the space (a different, equivalent definition of analytic vectors is given in Section 7.1.2). Let $\{e^{-tA}, t > 0\}$ be the semigroup generated by $-A$. Then the vector v is analytic for A if and only if there exists a $t > 0$ such that $v \in e^{-tA}(\mathcal{H})$. Thus we may write

$$\mathcal{S}_{\mathcal{H},A} = \bigcup_{t>0} e^{-tA}(\mathcal{H}). \tag{4.20}$$

For each $t > 0$, the space $e^{-tA}(\mathcal{H})$ is a Hilbert space for the inner product

$$\langle v | w \rangle_t := \langle e^{tA} v | e^{tA} w \rangle_{\mathcal{H}}, \quad v, w \in e^{-tA}(\mathcal{H}).$$

Since $e^{-tA}(\mathcal{H}) \subset e^{-sA}(\mathcal{H})$ for $0 < s < t$, with continuous embedding, we obtain a continuous scale of Hilbert spaces.

In view of the definition (4.20), one naturally puts on $\mathcal{S}_{\mathcal{H},A}$ the inductive limit topology (but the inductive limit is not strict, see Appendix B). However, one can put on $\mathcal{S}_{\mathcal{H},A}$ an (uncountable) family of seminorms that define the same inductive limit topology, and this fact ensures that the space has nice properties. In particular, $\mathcal{S}_{\mathcal{H},A}$ is complete, bornological and barreled. In addition, $\mathcal{S}_{\mathcal{H},A}$ is nuclear if and only if every operator $e^{-tA}, t > 0$, is Hilbert-Schmidt on $\mathcal{H}$.

Next we introduce the second type of spaces. A *trajectory* is a map $F : (0, \infty) \to \mathcal{H}$ such that

$$F(t + s) = e^{-sA}F(t), \quad \text{for all } t > 0, s \geqslant 0. \tag{4.21}$$

We denote by $\mathcal{T}_{\mathcal{H},A}$ the space of all trajectories. This is in fact the space of all solutions of the evolution equation

$$\frac{dF}{dt} = -AF, \ t > 0,$$

such that $F(t) \in \mathcal{H}$ for all $t > 0$ (but not necessarily $F(0)$). The trajectory space can be given a natural topology with the seminorms $q_t(F) := \|F(t)\|_{\mathcal{H}}$, and for this topology $\mathcal{T}_{\mathcal{H},A}$ is a Fréchet space. As a consequence, $\mathcal{T}_{\mathcal{H},A}$ is complete, bornological and barreled. In addition, $\mathcal{T}_{\mathcal{H},A}$ is nuclear if and only if every operator $e^{-tA}, t > 0$, is Hilbert-Schmidt on $\mathcal{H}$.

The Hilbert space $\mathcal{H}$ can be embedded into $\mathcal{T}_{\mathcal{H},A}$ by the map $\iota : \mathcal{H} \to \mathcal{T}_{\mathcal{H},A}$ defined as $\iota(x) : t \mapsto e^{-tA}x, \ t > 0, x \in \mathcal{H}$. Thus, putting all together, we obtain the following triplet, with continuous embeddings and dense images,

$$\mathcal{S}_{\mathcal{H},A} \subset \mathcal{H} \subset \mathcal{T}_{\mathcal{H},A} . \tag{4.22}$$

Finally one turns $\langle \mathcal{S}_{\mathcal{H},A}, \mathcal{T}_{\mathcal{H},A} \rangle$ into a dual pair, with the sesquilinear form

$$\langle w, F \rangle := \langle e^{tA}w | F(t) \rangle_{\mathcal{H}}, \ w \in \mathcal{S}_{\mathcal{H},A}, \ F \in \mathcal{T}_{\mathcal{H},A}. \tag{4.23}$$

Here t has to be chosen so small that $w \in e^{-tA}(\mathcal{H})$, and then the definition of $\langle w, F \rangle$ is independent of the choice of $t > 0$.

It is easy to see that the sequilinear form $\langle \cdot, \cdot \rangle$ defined in (4.23) is nondegenerate, so that $\langle \mathcal{S}_{\mathcal{H},A}, \mathcal{T}_{\mathcal{H},A} \rangle$ is indeed a dual pair. In addition, both spaces being barreled, the topologies described above coincide with the respective Mackey topologies $\tau(\mathcal{S}_{\mathcal{H},A}, \mathcal{T}_{\mathcal{H},A})$ and $\tau(\mathcal{T}_{\mathcal{H},A}, \mathcal{S}_{\mathcal{H},A})$, respectively, and the same holds for the corresponding strong topologies.

In view of the last result, it is clear that the triplet (4.22) is a RHS of the standard form, including nuclearity in some cases. Thus it may serve as a substitute to the standard RHSs of distributions of Schwartz and Gel'fand (another triplet, based on the so-called Feichtinger algebra, will be described in Section 8.3.2). As such, it is also a PIP-space. But one can also define, in several ways, a compatibility relation directly on $\mathcal{T}_{\mathcal{H},A}$. For instance, one can declare $F, G \in \mathcal{T}_{\mathcal{H},A}$ compatible whenever $\lim_{t\to 0}\langle F(t)|G(t)\rangle_{\mathcal{H}}$ exists. It would be interesting to explore the resulting PIP-space structure.

The formalism of analyticity/trajectory spaces covers a large number of interesting situations. We refer to the literature quoted in the Notes for more information. Let us just give a few examples.

(i) Spherical harmonics on the unit sphere

Let $\mathcal{H} = L^2(S^{n-1})$, where S^{n-1} is the unit sphere in $\mathbb{R}^n$, and write the Laplacian operator in spherical coordinates (r, ϖ):

$$\Delta = -\frac{\partial^2}{\partial r^2} - \frac{n-1}{r}\frac{\partial}{\partial r} + \frac{1}{r^2}\Delta_{\mathrm{LB}},$$

where Δ_{LB} is the Laplace-Beltrami operator on S^{n-1}. Then the spherical harmonics are eigenfunctions of Δ_{LB} and the operators $e^{-t\Delta_{\mathrm{LB}}^{1/2}}$ are Hilbert-Schmidt for all $t > 0$. Hence the space $\mathcal{S}_{L^2(S^{n-1}),\Delta_{\mathrm{LB}}^{1/2}}$ is nuclear.

(ii) Gel'fand-Shilov spaces

For $\mathcal{H} = L^2(\mathbb{R}, dx)$, certain Gel'fand-Shilov spaces S_α^β (see Section 5.4.3) are analyticity spaces, in particular the spaces S_α^α for all $\alpha \geqslant \frac{1}{2}$ (see next item).

(iii) Analyticity spaces based on classical polynomials

A general technique for generating analyticity spaces is to start from an orthonormal basis $\{v_n,\ n \in \mathbb{N}\}$ in $\mathcal{H}$ and the corresponding unitary (Fourier) map $f \mapsto (\langle v_n|f\rangle)$ from $\mathcal{H}$ onto ℓ^2. Let $0 \leqslant \lambda_1 \leqslant \lambda_2 \leqslant \dots$ be a sequence of real numbers such that $\sum_{n=1}^\infty e^{-\lambda_n t}$ converges for all $t > 0$. Consider the space $\mathcal{T}$ of formal series $\sum_{n=1}^\infty a_n v_n$ where the coefficients a_n satisfy the following condition:

$$\sup_{n\in\mathbb{N}} |a_n| e^{-\lambda_n t} < \infty, \ \forall t > 0.$$

Then the (Köthe) α-dual of $\mathcal{T}$, denoted $\mathcal{S}$, can be represented by elements in $\mathcal{H}$ of the form $\sum_{n=1}^\infty b_n v_n$, where the coefficients b_n satisfy the condition

(s)   There exists $s > 0$ such that $\sup_{n\in\mathbb{N}} |b_n| e^{\lambda_n s} < \infty, \ \forall t > 0$.

Consider now the operator $A = \sum_{n=1}^\infty \lambda_n \langle v_n|f\rangle) v_n$ on the domain $D(A) = \{f \in \mathcal{H} : \sum_{n=1}^\infty \lambda_n^2 |\langle v_n|f\rangle) v_n|^2\}$. Clearly A is a non-negative self-adjoint operator in $\mathcal{H}$, with eigenvalues λ_n and eigenvectors v_n. Then one has $\mathcal{S} = \mathcal{S}_{\mathcal{H},A}$ and $\mathcal{T} = \mathcal{T}_{\mathcal{H},A}$.

If we take $\mathcal{H} = L^2(\mathbb{R}, dx)$ and particularize the operator A to $H_{\text{osc}} = \frac{1}{2}\left(-\frac{d^2}{dx^2} + x^2 + 1\right)$ (the Hamiltonian of the quantum harmonic oscillator), so that v_n is the nth Hermite function, one obtains an analyticity space based on Hermite polynomials. In particular, one can show that $\mathcal{S}_{L^2(\mathbb{R}), H_{\text{osc}}^{1/2\alpha}}$ coincides with the Gel'fand-Shilov space S_α^α for all $\alpha \geqslant \frac{1}{2}$. Similar constructions lead to analyticity spaces based on other orthogonal families of classical polynomials, such as the Laguerre or the Jacobi polynomials.

The analyticity/trajectory spaces have been designed as a mathematical tool for obtaining a rigorous formulation of Dirac's formalism of quantum mechanics. We will describe this application in Section 7.1.1 (ii).

4.6 PIP-Spaces of Analytic Functions

Our last class of examples is of a totally different nature. Namely, the spaces to be considered consist of analytic functions, of three different types. In the first case, the space V_I consists of entire functions, the central Hilbert space is the Fock-Bargmann space $\mathfrak{F} \equiv \mathfrak{F}^0$ introduced in Section 1.1.3, Example (v), and the indexing parameter is the rate of growth at infinity. The second class of PIP-spaces is made of functions analytic in a sector and there the indexing parameter is the opening angle of that sector. The interesting feature of these last examples is that the "large" space V is itself a Hilbert space. Finally we turn to spaces of functions analytic in the unit disk. In the sequel we will state the results in substantial details, but omitting all proofs, for which we refer to the original references quoted in the Notes.

4.6.1 A RHS of Entire Functions

There are several situations where one considers Hilbert spaces consisting of analytic functions of a complex variable z. The prime example is the Fock-Bargmann or phase space representation of quantum mechanics, in which $z = q + ip$ (q denotes position and p momentum).

For convenience, we repeat the definition, specializing to one dimension. The Fock-Bargmann Hilbert space is

$$\mathfrak{F} = \{f(z) \text{ entire} : \int_{\mathbb{C}} |f(z)|^2 \, d\mu(z) < \infty\}, \tag{4.24}$$

with inner product

$$\langle f|g \rangle = \int_{\mathbb{C}} \overline{f(z)} g(z) \, d\mu(z). \tag{4.25}$$

Here $d\mu(z) = \pi^{-1} e^{-|z|^2} d\nu(z)$, where $d\nu(z) = \frac{i}{2} dz \wedge d\bar{z} = dx dy$ is the Lebesgue measure on $\mathbb{C}$. The Hilbert space $\mathfrak{F}$ possesses several interesting properties.

- Orthonormal basis: $u_n(z) = (n!)^{-1/2} z^n$, $n = 0, 1, 2, \ldots$. In this basis, the inner product (4.25) reads

$$\langle f | g \rangle = \sum_n n! \, \overline{f_n} \, g_n, \text{ for } f(z) = \sum_n f_n \, z^n, \; g(z) = \sum_n g_n \, z^n. \qquad (4.26)$$

- Principal vectors (coherent states): $e_w(z) = e^{\bar{w}z}$, $w, z \in \mathbb{C}$, so that $f(z) = \langle e_z | f \rangle$, that is, e_z is an evaluation functional (the equivalent of a delta function, but here e_z is a bona fide vector of the Hilbert space).
- Reproducing kernel: $K(w, z) = e^{w\bar{z}} = \langle e_w | e_z \rangle$, which means that

$$f(w) = \int_{\mathbb{C}} K(w, z) f(z) \, d\mu(z), \; \forall f \in \mathfrak{F}. \qquad (4.27)$$

The question that arises now is, how to build a RHS or a LHS around the Fock-Bargmann space $\mathfrak{F}$. Since $\mathfrak{F}$ consists of entire functions, one may identify immediately possible extreme spaces:

- The maximal space 3, consisting of all entire functions. With uniform convergence on compact sets, the space 3 is a nuclear Fréchet space. Its conjugate dual $3^\times$, the space of antianalytic functionals, is thus a nuclear, complete DF-space, exactly what is needed for applying the nuclear spectral theorem.
- The minimal space Exp consisting of entire functions of exponential type:

$$\text{Exp} = \{ f \in 3 : \exists\, a, c > 0 \text{ such that } |f(z)| \leqslant c\, e^{a|z|}, \; \forall z \in \mathbb{C} \}. \qquad (4.28)$$

It is a standard result that $3^\times$ is isomorphic to Exp. The correspondence is

$$\mu \in 3^\times \mapsto \hat{\mu} \in \text{Exp},$$

where $\hat{\mu}(w) = \langle \mu, e_w \rangle$, the Fourier-Borel transform. Thus we get a natural RHS around $\mathfrak{F}$:

$$3^\times \simeq \text{Exp} \subset \mathfrak{F} \subset 3. \qquad (4.29)$$

The duality between Exp and 3 is given indeed by a natural extension of the inner product of $\mathfrak{F}$:

$$\langle \overline{f}, g \rangle = \langle f | g \rangle = \sum_n n! \, \overline{f_n} \, g_n, f \in \text{Exp}, g \in 3. \qquad (4.30)$$

This answers the first question. But how to enrich the RHS (4.29) into a LHS?

4.6.2 A LHS of Entire Functions Around $\mathfrak{F}$

On the Hilbert space $\mathfrak{F}$, the two equivalent forms (4.25) and (4.26) of the inner product define two natural linear compatibilities (see Example 1.5.5):

$$f \,\#_1\, g \iff \int_{\mathbb{C}} |f(z)\, g(z)|\, d\mu(z) < \infty, \tag{4.31}$$

$$f \,\#_2\, g \iff \sum_n n! |f_n\, g_n| < \infty. \tag{4.32}$$

Of course, $\#_1$ and $\#_2$ coincide on $\mathfrak{F}$, but they are not comparable on 3 ! In fact, $\#_1$ is too general, and somewhat pathological, since one has $3^{\#_1} = \{0\}$, whereas $\#_2$ is more regular, and indeed $\mathrm{Exp} \xleftrightarrow{\#_2} 3$, but it applies only to sequences, not analytic functions. Actually, 3 may be considered as an echelon space (see Section 4.3.3) and Exp as the corresponding dual co-echelon space. Indeed, define the steps $a_{(k)}$ by $(a_{(k)})_n = k^n$, $k = 1, 2, \ldots$. Then one has

$$3 = \bigcap_{k=1}^{\infty} \{a_{(k)}\}^{\#_2}, \quad \mathrm{Exp} = \bigcup_{k=1}^{\infty} \{a_{(k)}\}^{\#_2 \#_2}.$$

The solution is to restrict the large space 3 to some smaller, more manageable space. This may be done in two ways.

(i) A chain of Hilbert spaces

For every $\rho \in \mathbb{R}$, consider again the Hilbert space $\mathfrak{F}^{\rho}$ defined in Section 1.1.3, Example (iv):

$$\mathfrak{F}^{\rho} = \{f \in 3 : \|f\|_{\rho}^2 := \int_{\mathbb{C}} |f(z)|^2 \, (1 + |z|^2)^{\rho} \, d\mu(z) < \infty\}. \tag{4.33}$$

The vectors $u_m^{\rho}(z) := (\eta_m^{\rho})^{-1/2} u_m(z)$, $m = 0, 1, 2, \ldots$ form an orthonormal basis of $\mathfrak{F}^{\rho}$, where $\eta_m^{\rho} = \|u_m\|_{\rho}^2$. Thus each space $\mathfrak{F}^{\rho}$ may be realized as a sequence space, namely, a weighted ℓ^2 space.

The family $\{\mathfrak{F}^{\rho}, \rho \in \mathbb{R}\}$ is a chain of Hilbert spaces, corresponding to:

$$\mathfrak{F}^{\rho} \subset \mathfrak{F}^{\sigma} \iff \rho > \sigma, \tag{4.34}$$

and one has $\mathfrak{F}^0 = \mathfrak{F}$, $(\mathfrak{F}^{\rho})^{\times} = \mathfrak{F}^{-\rho} = (\mathfrak{F}^{\rho})^{\#_1} = (\mathfrak{F}^{\rho})^{\#_2}$ (as vector spaces) and $e_w \in \mathfrak{F}^{\rho}$, $\forall \rho \in \mathbb{R}, \forall w \in \mathbb{C}$. Defining the extreme spaces

$$\mathfrak{E} = \bigcap_{\rho \in \mathbb{R}} \mathfrak{F}^{\rho}, \quad \mathfrak{E}^{\times} = \bigcup_{\rho \in \mathbb{R}} \mathfrak{F}^{\rho}. \tag{4.35}$$

one recovers Bargmann's space $\mathfrak{E}^\times$. Then one gets a Hilbert chain, with the structure $(\rho > 0)$:

$$\text{Exp} \subset \mathfrak{E} \subset \dots \mathfrak{F}^\rho \dots \subset \mathfrak{F} \subset \dots \mathfrak{F}^{-\rho} \dots \subset \mathfrak{E}^\times \subset 3, \qquad (4.36)$$

and, by restriction, a RHS isomorphic to the Schwartz triplets $\mathcal{S}(\mathbb{R}) \subset L^2(\mathbb{R}) \subset \mathcal{S}(\mathbb{R})^\times$ and $s \subset \ell^2 \subset s^\times$, namely,

$$\mathfrak{E} \subset \mathfrak{F} \subset \mathfrak{E}^\times. \qquad (4.37)$$

A word of caution is in order here concerning duality. Indeed, in the chain (4.36) above, the space $\mathfrak{F}^\rho$ carries the norm $\| \cdot \|_\rho$ and $\mathfrak{F}^{-\rho}$ the norm $\| \cdot \|_{-\rho}$. The two spaces are in duality, in virtue of the Hölder inequality

$$|\langle g | f \rangle| \leqslant \|g\|_{-\rho} \|f\|_\rho, \quad g \in \mathfrak{F}^{-\rho}, f \in \mathfrak{F}^\rho,$$

where

$$\|g\|_{-\rho}^2 = \sum_m \eta_m^{-\rho} |\langle u_m | g \rangle|^2, \qquad \|f\|_\rho^2 = \sum_m \eta_m^\rho |\langle u_m | f \rangle|^2.$$

Thus, if we define the weights $r_m(\rho) := \left(\eta_m^\rho\right)^{1/2}$, we may identify

$$\mathfrak{F}^\rho \sim \ell^2[r(\rho)], \quad \mathfrak{F}^{-\rho} \sim \ell^2[r(-\rho)].$$

Notice that the "natural" bijection $f(z) \mapsto (1 + |z|^2)^\rho f(z)$ does *not* map $\mathfrak{F}^\rho$ onto $\mathfrak{F}^{-\rho}$, because of the requirement of analyticity. Hence we have to make the detour via sequence spaces for identifying the conjugate dual $(\mathfrak{F}^\rho)^\times$ of $\mathfrak{F}^\rho$. The latter is the set of all continuous, conjugate linear, functionals on $\mathfrak{F}^\rho$. For such a functional $L \in (\mathfrak{F}^\rho)^\times$, define the (dual) norm

$$\|L\|_\rho^\times := \sup_{\|f\|_\rho \leqslant 1} |L(f)|, \ f \in \mathfrak{F}^\rho. \qquad (4.38)$$

By the Riesz Lemma, $L(f)$ is of the form $L(f) = \langle f | h \rangle_\rho$, for a unique element $h \in \mathfrak{F}^\rho$ and

$$\|L\|_\rho^\times = \|h\|_\rho = \left(\sum_m (\eta_m^\rho)^{-1} |L(u_m)|^2\right)^{1/2}. \qquad (4.39)$$

Thus $(\mathfrak{F}^\rho)^\times$ is a Hilbert space for the norm $\| \cdot \|_\rho^\times$ and one may identify it with $\ell^2[r(\rho)^{-1}]$. $(\mathfrak{F}^\rho)^\times$ coincides with $\mathfrak{F}^{-\rho}$ as a vector space, but *not* as Hilbert space. Indeed, every functional $L \in (\mathfrak{F}^\rho)^\times$ is of the form $L(f) = \langle f | g \rangle$, for a unique element $g \in \mathfrak{F}^{-\rho}$. Then the identity

$$(\eta_m^\rho)^{-1} \leqslant \eta_m^{-\rho} \leqslant c_\rho \eta_m^\rho \quad (c_\rho > 0)$$

implies

$$\|L\|_\rho^\times \leq \|g\|_{-\rho} \leq c_\rho \|L\|_\rho^\times,$$

so that the dual norm (4.38) is equivalent to, but different from the norm $\|\cdot\|_{-\rho}$. Since (4.34) implies also $\rho > \sigma \Leftrightarrow (\mathfrak{F}^\rho)^\times \subset (\mathfrak{F}^\sigma)^\times$, we obtain the following results.

Proposition 4.6.1. (i) *The family $\{\mathfrak{F}^\rho, \rho \in \mathbb{R}\}$, is a chain of hilbertian spaces, but not a LHS.*

(ii) *The family $\{\mathfrak{F}^\rho, (\mathfrak{F}^\rho)^\times, \rho \geq 0\}$, is a LHS.*

(iii) *The same holds true for the family $\{(\mathfrak{F}^\rho)^\times, \mathfrak{F}^\rho, \rho \leq 0\}$.*

The obvious conclusion of this proposition is that the notion of LHS is too restrictive in the case of spaces of analytic functions, the structure of a chain of hilbertian spaces is more natural.

(ii) A genuine lattice of hilbertian spaces

One may go one step further and construct a lattice of hilbertian spaces by considering more general weights in (4.33):

$$\mathfrak{F}(\rho) = \{f \in \mathfrak{Z} : \|f\|_{(\rho)}^2 := \int_{\mathbb{C}} |f(z)|^2\, e^{-\rho(z)}\, d\mu(z) < \infty\}, \qquad (4.40)$$

where $\rho : \mathbb{C} \to \mathbb{R}$ is a measurable function. Clearly the space $\mathfrak{F}(\rho)$ reduces to $\mathfrak{F}^\rho$ if $\rho(z) = \ln(1 + |z|^2)^{-\rho}$, $\rho \in \mathbb{R}$. In order to obtain a lattice of hilbertian spaces around $\mathfrak{F}$, we must require that the space $\mathfrak{F}(\rho)$ satisfies several conditions, each of which imposes some restrictions to the weight function ρ (all of them are satisfied by every $\mathfrak{F}^\rho$, $\rho \in \mathbb{R}$):

(i) $\mathfrak{F}(\rho)$ *is a Hilbert space*, i.e., it is complete, which holds true if ρ is locally bounded, that is, bounded on compact sets.

(ii) *The set of polynomials is dense in $\mathfrak{F}(\rho)$.* This is a restriction on the growth of ρ: if ρ grows too fast at infinity, $\mathfrak{F}(\rho)$ may become trivial and $\mathfrak{F}(-\rho)$ too large. When (ii) holds, the monomials $\{u_m(z) = (m!)^{-1/2} z^m, m = 1, 2, \ldots\}$ form a basis of $\mathfrak{F}(\rho)$. If ρ is *radial*, i.e. $\rho(z) = \rho(|z|)$, then the functions

$$u_m^{(\rho)}(z) = (\eta_m^{(\rho)})^{-1/2}\, u_m(z), \ m = 1, 2, \ldots$$

form an orthonormal basis of $\mathfrak{F}(\rho)$. The normalization coefficients, defined as

$$\eta_m^{(\rho)} = \int_{\mathbb{C}} |u_m(z)|^2\, e^{-\rho(z)}\, d\mu(z) = \frac{1}{m!} \int_0^\infty t^m\, e^{-\rho(t)-t}\, dt, \qquad (4.41)$$

are clearly related to a moment problem. In that case, all spaces $\mathfrak{F}(\rho)$ may be realized as weighted ℓ^2 sequence spaces and the compatibility #$_2$ is easy to handle.

(iii) *Duality:* $\mathfrak{F}(\rho)^\times$ *is isomorphic to* $\mathfrak{F}(-\rho)$ *(as locally convex spaces).* This is the crucial condition, and it is difficult to verify, unless ρ is radial. There are several partial results:

(a) If ρ is locally bounded, then $\mathfrak{F}(-\rho) = \mathfrak{F}(\rho)^{\#_1} \subseteq \mathfrak{F}(\rho)^\times$;
(b) If ρ is locally bounded and radial, then one has, in addition, $\mathfrak{F}(\rho)^\times = \mathfrak{F}(\rho)^{\#_2}$;
(c) If ρ is locally bounded and radial, then $\mathfrak{F}(-\rho) = \mathfrak{F}(\rho)^\times$ if and only if there exists a positive constant $c(\rho)$ such that

$$1 \leqslant \eta_m^{(\rho)} \eta_m^{(-\rho)} \leqslant c(\rho), \quad \text{for all } m = 0, 1, 2, \dots.$$

Thus we conclude that the duality condition is equivalent to the condition $\mathfrak{F}(\rho)^{\#_1} = \mathfrak{F}(\rho)^{\#_2}$.

(iv) *Principal vectors:* $e_w \in \mathfrak{F}(\rho), \forall\, w \in \mathbb{C}$, which implies that the principal vectors e_w generate a dense subspace of $\mathfrak{F}(\rho)$, and therefore every operator A in the resulting PIP-space is an integral operator, with kernel $A(w, z) = \langle e_w | A e_z \rangle$.

Now we have to examine the relationship between weights ρ and subspaces $\mathfrak{F}(\rho)$. We say that two weights ρ_1, ρ_2 are *equivalent* ($\rho_1 \approx \rho_2$) if $\mathfrak{F}(\rho_1) = \mathfrak{F}(\rho_2)$ as vector spaces. This happens if and only if there exists a constant C such that $|\rho_1(z) - \rho_2(z)| \leqslant C$ almost everywhere or, equivalently, there exist two constants A, B such that

$$A \leqslant e^{\rho_1(z) - \rho_2(z)} \leqslant B, \quad \text{a.e.} \tag{4.42}$$

In that case, the identity map $\mathfrak{F}(\rho_1) \to \mathfrak{F}(\rho_2)$ is bicontinuous, i.e., the two norms are equivalent.

Thus the family of Hilbert spaces $\{\mathfrak{F}(\rho)\}$ is indexed by $L^\infty_{\mathrm{loc}}(\mathbb{C})/\approx$, which is an involutive lattice for the pointwise partial order ($\rho_1 \leqslant \rho_2$ if and only if $\rho_1(z) \leqslant \rho_2(z)$ a.e.) modulo $\approx$. We notice, in particular, the obvious equivalences

$$\rho_1 \wedge \rho_2 := \min(\rho_1, \rho_2) \approx -\ln(e^{-\rho_1} + e^{-\rho_2}),$$
$$\rho_1 \vee \rho_2 := \max(\rho_1, \rho_2) \approx \ln(e^{\rho_1} + e^{\rho_2}).$$

Although some results may be obtained for general weights ρ, a complete answer may be formulated for radial weights only.

Theorem 4.6.2. *Let I be the set of weight functions $\rho(z)$ that satisfy the following three conditions:*

(i) ρ is locally bounded and radial, $\rho(z) = \rho(|z|)$.
(ii) e_w belongs to $\mathfrak{F}(\rho) \cap \mathfrak{F}(-\rho)$, for any $w \in \mathbb{C}$.
(iii) There are positive constants A, B such that

$$A \leqslant \eta_m^{(\rho)} \, \eta_m^{(-\rho)} \leqslant B, \quad \text{for all } m = 0, 1, 2, \ldots, \qquad (4.43)$$

where $\eta_m^{(\rho)}$ is given by (4.41).

Then the family $\{\mathfrak{F}(\rho), \rho \in I\}$ is a lattice of hilbertian spaces with central Hilbert space $\mathfrak{F}(0) = \mathfrak{F}$ and lattice operations

$$\mathfrak{F}(\rho_1 \wedge \rho_2) = \mathfrak{F}(\rho_1) \cap \mathfrak{F}(\rho_2),$$
$$\mathfrak{F}(\rho_1 \vee \rho_2) = \mathfrak{F}(\rho_1) + \mathfrak{F}(\rho_2),$$
$$\mathfrak{F}(\rho)^\times = \mathfrak{F}(-\rho).$$

The lattice is indexed by the set $I/\approx$, where the equivalence relation $\approx$ is defined by (4.42). The compatibility $\#$ coincides with both $\#_1$ and $\#_2$, and the partial inner product with that induced by $\mathfrak{F}$ and ℓ^2, respectively.

For the proof, it suffices to notice that condition (4.43) is necessary and sufficient for the isomorphism $\mathfrak{F}(\rho)^\times \simeq \mathfrak{F}(-\rho)$ as locally convex spaces.

In this discussion, one takes each Hilbert space $\mathfrak{F}(\rho)$ with its norm $\| \cdot \|_{(\rho)}$, but the dual norm is only equivalent to the norm $\| \cdot \|_{(-\rho)}$, as a consequence of the inequalities (4.43). This may be seen exactly as in the case of the spaces $\mathfrak{F}^\rho$ discussed above. If we denote by $\mathfrak{F}(\widehat{\rho})$ the dual Hilbert space $\mathfrak{F}(\rho)^\times$, equipped with the dual norm, then we can identify the three Hilbert spaces $\mathfrak{F}(\pm\rho), \mathfrak{F}(\widehat{\rho})$ with weighted ℓ^2 spaces and the conclusion follows.

In particular, the problem with this construction is that the space $\mathfrak{F}(\widehat{\rho})$ is a space of functions analytic in $\mathbb{C}$, but need not be associated to a weight. The existence of the latter is a moment problem, for which some results are known, but we shall not pursue the point.

In other words, the structure described in Theorem 4.6.2 is that of a lattice of hilbertian spaces and cannot lead to a LHS.

4.6.3 Functions Analytic in a Sector

In this section, we shall describe another LHS of analytic functions, in which the order parameter is the opening angle of a sector, instead of the rate of growth at infinity. This LHS simplifies considerably the formulation of scattering theory, as we shall see in Section 7.2.

Define $G(a, b)$ $(-\pi < a < b < \pi)$ as the space of all functions $f(z)$, $z = re^{i\varphi}$, which are analytic in the open sector $S_{a,b} := \{z = re^{i\varphi}, a < \varphi < b\}$, and such that the integral $\int_0^\infty |f(re^{i\varphi})|^2 \, dr < \infty$ is uniformly bounded in $\varphi \in (a, b)$. Then it is known that every function $f(re^{i\varphi}) \in G(a, b)$ possesses well-defined limiting values (in the sense of L^2 limits) $f(re^{ia}), f(re^{ib})$ on the boundaries of $S_{a,b}$, and furthermore that $G(a, b)$ is complete, thus a Hilbert space, for the inner product

$$\langle f|g\rangle_{ab} = \int_0^\infty \overline{f(re^{ia})}\, g(re^{ia})\, dr + \int_0^\infty \overline{f(re^{ib})}\, g(re^{ib})\, dr. \tag{4.44}$$

Among the bounded linear operators on $G(a,b)$, a distinguished role is played by the van Winter class $\mathfrak{W}$ of those operators A for which the quantity

$$\mathfrak{a}(A,\varphi) = \sup_{f\in G(a,b)} \left[\int_0^\infty |(Af)(re^{i\varphi})|^2\, dr\right]\left[\int_0^\infty |f(re^{i\varphi})|^2\, dr\right]^{-1/2} \tag{4.45}$$

is uniformly bounded in $\varphi \in (a,b)$. It turns out that the class $\mathfrak{W}$ is a Banach *-algebra for the norm $\alpha(A) = \sup_{a<\varphi<b} \mathfrak{a}(A,\varphi)$. Furthermore, $\mathfrak{W}$ contains a two-sided ideal $\mathfrak{K}$ of integral operators which are all Hilbert-Schmidt. Next, we define the Hilbert space

$$\widetilde{G}(a,b) := \{\widetilde{f}(t,\varphi) = e^{\varphi t - i\varphi/2}\widetilde{f}(t),\ t \in \mathbb{R},\ a \leqslant \varphi \leqslant b, \tag{4.46}$$

$$\text{with } \int_{-\infty}^{+\infty} (e^{2at} + e^{2bt})\,|\widetilde{f}(t)|^2\, dt < \infty\}, \tag{4.47}$$

with inner product

$$\langle f|g\rangle_{\widetilde{ab}} = \int_{-\infty}^\infty (e^{2at} + e^{2bt})\, \overline{\widetilde{f}(t)}\widetilde{g}(t)\, dt.$$

Introduce the Mellin transform of f:

$$\widetilde{f}(t) = (2\pi)^{-1/2} \int_0^{+\infty} f(x)\, x^{-it-\frac{1}{2}}\, dx$$

and its inverse

$$f(re^{i\varphi}) = (2\pi)^{-1/2} \int_{-\infty}^{+\infty} \widetilde{f}(t)\, (re^{i\varphi})^{it-\frac{1}{2}}\, dt.$$

Then it turns out that $f(re^{i\varphi}) \in G(a,b)$ if and only if $\widetilde{f}(t,\varphi) \in \widetilde{G}(a,b)$. Furthermore the Mellin transform $f \mapsto \widetilde{f}$ is a unitary map from $G(a,b)$ onto $\widetilde{G}(a,b)$. One has indeed, by a straightforward calculation,

$$\langle f|g\rangle_{ab} = \int_0^\infty \overline{f(re^{ia})}\, g(re^{ia})\, dr + \int_0^\infty \overline{f(re^{ib})}\, g(re^{ib})\, dr$$

$$= \int_{-\infty}^\infty (e^{2at} + e^{2bt})\, \overline{\widetilde{f}(t)}\widetilde{g}(t)\, dt = \langle \widetilde{f}|\widetilde{g}\rangle_{\widetilde{ab}}.$$

Fig. 4.3 The functions
$r_{a \wedge b}$ and $r_{a \vee b}$ for $a < 0 < b$

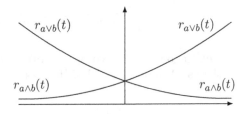

In addition, the Fourier transform is unitary from $\widetilde{G}(a, b)$ onto $\widetilde{G}(-b, -a)$.

We claim that the family $\{G(a, b), -\frac{\pi}{2} \leqslant a < b \leqslant \frac{\pi}{2}\}$ may be identified with a part of a LHS of weighted L^2 spaces. Indeed, for $-\frac{\pi}{2} \leqslant a \leqslant \frac{\pi}{2}$, define the Hilbert space

$$L^2(a) := \left\{ \widetilde{f} : \int_{-\infty}^{+\infty} e^{2at} |\widetilde{f}(t)|^2 \, dt < \infty \right\} = L^2(r_a), \quad \text{with } r_a(t) = e^{-2at}. \tag{4.48}$$

Then consider the lattice generated by the family $\{L^2(a), -\frac{\pi}{2} \leqslant a \leqslant \frac{\pi}{2}\}$. The infimum is $L^2(a) \wedge L^2(b) = L^2(a) \cap L^2(b) = L^2(a \wedge b)$, with $r_{a \wedge b}(t) = \min(r_a(t), r_b(t))$, and the supremum $L^2(a) \vee L^2(b) = L^2(a) + L^2(b) = L^2(a \vee b)$, $r_{a \vee b}(t) = \max(r_a(t), r_b(t))$ (see Fig. 4.3 for the case $a < 0 < b$). As usual, these norms are equivalent to the projective, resp. inductive, norms. Duality is given by $L^2(a \wedge b) \Longleftrightarrow L^2(-a \vee -b)$.

Thus we obtain a LHS, with extreme spaces $V^{\#} = L^2(-\frac{\pi}{2}) \cap L^2(\frac{\pi}{2})$, $V = L^2(-\frac{\pi}{2}) + L^2(\frac{\pi}{2})$, which are themselves Hilbert spaces. In addition, all spaces are obtained at the first generation, i.e., they are all of the form $L^2(c \wedge d)$ or $L^2(c \vee d)$. One has, for instance, with $|c| = \min(|a|, |b|)$:

$$L^2(a \wedge b) \wedge L^2(-b \vee -a) = \begin{cases} L^2(-|c| \vee |c|), & \text{if } a, b \text{ have the same sign,} \\ L^2(-|c| \wedge |c|), & \text{if } a, b \text{ have opposite signs.} \end{cases}$$

Therefore, all spaces may be obtained from $L^2(-\frac{\pi}{2})$ and $L^2(\frac{\pi}{2})$ by interpolation (see Section 5.1.2).

In the case $0 < a < b$, one gets the picture shown in Fig. 4.4. Duality corresponds to symmetry with respect to the center (i.e., L^2): $a \wedge b \Longleftrightarrow -b \vee -a$. Notice that one can work with fixed b also, using only the sublattice generated (in the case depicted in Fig. 4.4) by $L^2(b), L^2(-b)$.

As for operators on this LHS, one may show, using again interpolation theory, that the two operator algebras $\mathfrak{A}$ and $\mathfrak{B}$ defined in Section 3.3.3 coincide, i.e., there is a unique Banach *-algebra of bounded operators:

$$\mathfrak{A} = \mathfrak{B} = \{A \text{ linear } : A \text{ maps } L^2(-\frac{\pi}{2}) \text{ and } L^2(\frac{\pi}{2}) \text{ into themselves continuously}\}. \tag{4.49}$$

Furthermore, this algebra coincides with the Mellin transform of the van Winter algebra $\mathfrak{W}$, which contains $\mathfrak{K}$ as an ideal.

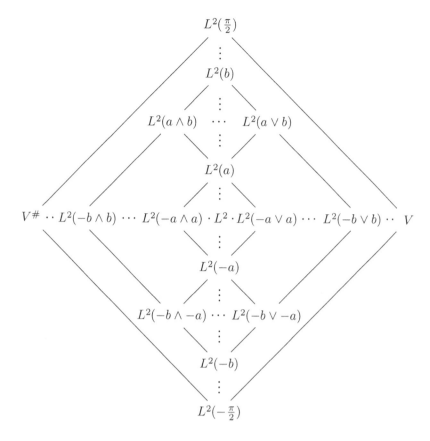

Fig. 4.4 The van Winter LHS

In terms of these spaces, the statement above reads $f(re^{i\varphi}) \in G(a, b)$ if and only if $\widetilde{f}(t, \varphi) = e^{\varphi t - i\varphi/2} \widetilde{f}(t)$, with $\widetilde{f}(t) \in L^2(a) \cap L^2(b) = L^2(a \wedge b)$. As a consequence, although the inverse Mellin transform is unitary from $G(a, b)$ onto $\widetilde{G}(a, b)$, it does not preserve the lattice structure of the left-hand side of the LHS, because of the prefactor in the definition of $\widetilde{f}(t, \varphi)$.

It is true that, given $\varphi \in [a, b]$, every function $f(re^{i\varphi}) \in L^2([0, \infty), dr)$ may be extended to a function $f(re^{i\varphi}) \in G(a, b)$, that is, analytic in the sector $a < \varphi < b$, provided its Mellin transform is $e^{at - ia/2} \widetilde{f}(t)$, where $\widetilde{f}(t) \in L^2(a \wedge b)$. Thus, if the sectors $[a, b]$ and $[c, d]$ have a nontrivial overlap, then $L^2(a \wedge b) \cap L^2(c \wedge d) = L^2(a \wedge d)$ and $G(a, b) \cap G(c, d) = G(a, d)$, by the uniqueness of analytic continuation. But if we take two disjoint sectors $[a, b]$ and $[c, d]$, corresponding to $-\frac{\pi}{2} < a < b < c < d < \frac{\pi}{2}$, there is *a priori* no way of extending analytically a function from one sector to the other, so that we get only $G(a, d) \subseteq G(a, b) \cap G(c, d)$. Nevertheless one still has $L^2(a \wedge d) = L^2(a \wedge b) \cap L^2(c \wedge d)$.

There is a partial solution to this difficulty, however. Given $a < b$, consider the sector $T_{a,b} := \{z : -b < \arg z < \pi - a\}$. Notice that

$$T_{a,b} = \bigcup_{a < \varphi < b} T_{\varphi,\varphi} = \bigcup_{a < \varphi < b} e^{-i\varphi}\mathbb{C}^+, \quad \text{where } \mathbb{C}^+ = \{z \in \mathbb{C} : \operatorname{Im} z > 0\}.$$

Thus, since the opening angle is larger than π, all the sectors $T_{a,b}$ overlap. Then, for a fixed number $\alpha > -1$, define $D(a,b)$ as the space of all functions $f(z)$ analytic in $T_{a,b}$ and such that

$$\int_{\mathbb{C}^+} |f(ze^{-i\varphi})|^2 \, (\operatorname{Im} z)^\alpha \, d\nu(z) \tag{4.50}$$

is uniformly bounded in $\varphi \in (a,b)$. The integral in (4.50) makes sense, since $z \in \mathbb{C}^+$ and $a < \varphi < b$ imply $ze^{-i\varphi} \in T_{a,b}$. It turns out that $D(a,b)$ is a Hilbert space, with inner product

$$\langle f|g\rangle_{D(a,b)} = \int_{\mathbb{C}^+} \overline{f(ze^{-ia})}g(ze^{-ia})(\operatorname{Im} z)^\alpha d\nu(z) + \int_{\mathbb{C}^+} \overline{f(ze^{-ib})}g(ze^{-ib})(\operatorname{Im} z)^\alpha d\nu(z). \tag{4.51}$$

For $\psi \in G(a/2, b/2)$ and $a/2 < \varphi < b/2$, consider the function $V_\alpha\psi$ given by

$$(V_\alpha\psi)(z,\varphi) = \left(2\pi\,\Gamma(\alpha+1)\right)^{-1/2} \int_0^\infty (re^{i\varphi})^{\alpha+3/2} \, e^{iz(re^{i\varphi})^2/2} \, \psi(re^{i\varphi}) \, e^{i\varphi} \, dr. \tag{4.52}$$

This function is analytic in z in the half-plane $R_\varphi := T_{2\varphi,2\varphi} = e^{-2i\varphi}\mathbb{C}^+$. Next one defines a function $f(z) := (\widehat{V}_\alpha\psi)(z)$ as the function analytic in $T_{a,b}$, whose restriction to R_φ is exactly $(V_\alpha\psi)(z,\varphi)$. Then one shows that the function $\widehat{V}_\alpha\psi$ belongs to $D(a,b)$, and moreover, that the map $\widehat{V}_\alpha$ is unitary from $G(a/2, b/2)$ onto $D(a,b)$. The inverse map $\widehat{W}_\alpha : D(a,b) \to G(a/2, b/2)$ reads

$$(\widehat{W}_\alpha f)(re^{i\varphi}) = (2\pi\Gamma(\alpha+1))^{-1/2} \int_{\mathbb{C}^+} (e^{-2i\varphi})^{\alpha/2+5/4} r^{\alpha+3/2} e^{-i\bar{z}r^2/2} f(e^{-2i\varphi}z)(\operatorname{Im} z)^\alpha d\nu(z). \tag{4.53}$$

Thus, for fixed a, b, we get three unitary equivalent Hilbert spaces, where the unitary map T may be computed explicitly from the other two, as shown in Fig. 4.5.

What about the LHS structure of these three families? We have seen above that $\{\widetilde{G}(a,b)\}$ may be identified with the van Winter lattice generated

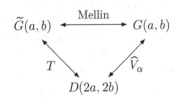

Fig. 4.5 Three unitary
equivalent Hilbert spaces

by the spaces $L^2(c)$, shown in Fig. 4.4, whereas this is impossible for the family $\{G(a, b)\}$. As for the third family, one obtains only the left-hand part of a LHS. Indeed, for any $a \in (-\frac{\pi}{4}, \frac{\pi}{4})$, define the space $D(2a) = D(2a, 2a)$, namely the space of complex-valued functions, analytic in the half-plane $e^{-2ia}\mathbb{C}^+ = T_{2a,2a}$ and such that the following integral converges:

$$\int_{\mathbb{C}+} |f(ze^{-2ia})|^2 (\operatorname{Im} z)^\alpha \, d\nu(z) < \infty.$$

This space is a Hilbert space for the inner product

$$\int_{\mathbb{C}+} \overline{f(ze^{-2ia})} \, g(ze^{-2ia})(\operatorname{Im} z)^\alpha \, d\nu(z).$$

Thus we may conclude that $D(2a, 2b) = D(2a) \cap D(2b)$, with the projective norm. Thus the left-hand side of the van Winter lattice may be transported by the map $T^{-1} : \widetilde{G}(a, b) \to D(2a, 2b)$ and one gets the half-lattice generated, by intersection, by the family $\{D(c), -\frac{\pi}{2} \leqslant c \leqslant \frac{\pi}{2}\}$. However, the identification of the right-hand side of this lattice is problematic, since it involves the duals of the spaces $D(a, b)$, which are not easy to characterize explicitly (of course, since $D(a, b)$ is a Hilbert space, it is anti-isomorphic to its conjugate dual, but we don't know how to characterize the elements of that dual as functions or functionals).

4.6.4 A Link with Bergman Spaces of Analytic Functions in the Unit Disk

However, there is another direction to be explored. Indeed, in the whole analysis so far, the parameter α was fixed. What happens if we let it vary? The starting point is the Bergman space $\mathcal{H}_\alpha := D(0) \equiv A^{2,\alpha+2}$, where the space $A^{p,\beta}$ is defined in (8.23), that is, the space of complex-valued functions f analytic in $\mathbb{C}^+$ and such that

$$\|f\|_{\mathcal{H}_\alpha}^2 := \int_{\mathbb{C}+} |f(z)|^2 (\operatorname{Im} z)^\alpha \, d\nu(z) < \infty. \tag{4.54}$$

With the inner product

$$\langle f|g\rangle_{\mathcal{H}_\alpha} := \int_{\mathbb{C}+} \overline{f(z)} \, g(z)(\operatorname{Im} z)^\alpha \, d\nu(z),$$

$\mathcal{H}_\alpha$ is a Hilbert space, with reproducing kernel

$$\rho_z^\alpha(w) = \frac{\alpha+1}{4\pi}\left(\frac{w-\bar{z}}{2i}\right)^{-\alpha-2}, \quad z,w \in \mathbb{C}^+.$$

The following functions constitute an orthonormal basis of $\mathcal{H}_\alpha$:

$$v_n^\alpha(z) := \frac{2^{\alpha+1}}{\pi^{1/2}}\left(\frac{\Gamma(n+\alpha+2)}{\Gamma(\alpha+1)\Gamma(n+1)}\right)^{1/2}\frac{(z-i)^n}{(z+i)^{n+\alpha+2}}, \quad n \in \mathbb{N}.$$

The interesting point is that each space $\mathcal{H}_\alpha$ is unitarily equivalent to a space of analytic functions in the unit disk $\mathbb{D} := \{w \in \mathbb{C} : |w| < 1\}$, namely, the Bergman space $\mathcal{K}_\alpha$ of all functions g analytic in $\mathbb{D}$ such that the following integral converges:

$$\|g\|_{\mathcal{K}_\alpha}^2 := \int_{\mathbb{D}}|g(w)|^2\left(\frac{1-|w|^2}{2}\right)^\alpha d\nu(w) < \infty. \tag{4.55}$$

With the norm $\|\cdot\|_{\mathcal{K}_\alpha}$ and the corresponding inner product, $\mathcal{K}_\alpha$ is a Hilbert space with reproducing kernel

$$\hat{\rho}_z^\alpha(w) = \frac{\alpha+1}{4\pi}\left(\frac{1-\bar{z}w}{2}\right)^{-\alpha-2}, \quad z,w \in \mathbb{D}.$$

Here again there is an orthonormal basis, consisting of the functions:

$$u_n^\alpha(w) := \frac{2^{\alpha/2}}{\pi^{1/2}}\left(\frac{\Gamma(n+\alpha+2)}{\Gamma(\alpha+1)\Gamma(n+1)}\right)^{1/2}w^n, \quad n \in \mathbb{N}.$$

The unitary maps connecting the two spaces, namely, $B_\alpha : \mathcal{H}_\alpha \to \mathcal{K}_\alpha$ and $B_\alpha^{-1} : \mathcal{K}_\alpha \to \mathcal{H}_\alpha$, read, respectively:

$$B_\alpha f(w) = 2^{\alpha+1/2}\left(\frac{1-w}{i}\right)^{-\alpha-2}f\left(i\frac{1+w}{1-w}\right), \quad f \in \mathcal{H}_\alpha, \tag{4.56}$$

$$B_\alpha^{-1}g(z) = 2^{\alpha+1/2}(z+i)^{-\alpha-2}g\left(i\frac{z-i}{z+i}\right), \quad g \in \mathcal{K}_\alpha. \tag{4.57}$$

The sequel of the analysis follows closely that of Section 4.6.2, the spaces $\mathcal{K}_\alpha$ playing the role of the $\mathfrak{F}^\rho$. In particular, on the space $\mathcal{K}(\mathbb{D})$ of all functions which are analytic in $\mathbb{D}$, as for entire functions of Bargmann type, one has two natural linear compatibilities:

$$f \#_1 g \iff \int_{\mathbb{D}}|f(z)\,g(z)|\,d\nu(z) < \infty, \tag{4.58}$$

$$f \#_2 g \iff \sum_n \frac{\pi}{n+1}|f_n\,g_n| < \infty. \tag{4.59}$$

Then, given a Lebesgue-measurable function p on $\mathbb{D}$, one defines the space $\mathcal{K}(p)$ as the space of functions f analytic in $\mathbb{D}$ and such that the following integral converges:

$$\|f\|^2_{\mathcal{K}(p)} := \int_{\mathbb{D}} |f(z)|^2 e^{-p(z)} \, d\nu(z) < \infty \qquad (4.60)$$

With the associated inner product, the space $\mathcal{K}(p)$ is a pre-Hilbert space, which obviously reduces to $\mathcal{K}_\alpha$ for the weight $p_\alpha(z) = -\alpha \ln \left(\frac{1-|z|^2}{2} \right)$. In order to proceed, we have to impose several conditions on the spaces $\mathcal{K}(p)$, which are translated into restrictions on the weights p:

(i) *Completeness:* $\mathcal{K}(p)$ is complete, thus a Hilbert space, if p is locally bounded (that is, bounded on compact sets), i.e., $p \in L^\infty_{\mathrm{loc}}(\mathbb{D})$.

(ii) *Density of the set of polynomials in* $\mathcal{K}(p)$: Consider the monomials $\{u_n(z) = (\frac{n+1}{\pi})^{1/2} z^n, \, n = 1, 2, \ldots\}$. Assume that p is radial, $p(z) = p(|z|)$, and that

$$\eta_n^{(p)} := \int_{\mathbb{D}} |u_n(z)|^2 \, e^{-p(z)} \, d\nu(z) < \infty, \; n = 1, 2, \ldots. \qquad (4.61)$$

Then the functions

$$u_n^{(p)}(z) = (\eta_n^{(p)})^{-1/2} \, u_n(z), \; n = 1, 2, \ldots$$

form an orthonormal basis of $\mathcal{K}(p)$. In that case, all spaces $\mathcal{K}(p)$ may be realized as weighted ℓ^2 sequence spaces. Thus, for the spaces $\mathcal{K}(p)$ the compatibility $\#_2$ is easy to handle.

(iii) *Duality:* $\mathcal{K}(p)^\times$ is isomorphic to $\mathcal{K}(-p)$. Here too there are several results:

(a) if p is locally bounded, then $\mathcal{K}(-p) = \mathcal{K}(p)^{\#_1} \subseteq \mathcal{K}(p)^\times$;

(b) if p is locally bounded and radial, and $\eta_n^{(p)} < \infty, \forall n \in \mathbb{N}$, then $\mathcal{K}(p)^\times = \mathcal{K}(p)^{\#_2}$;

(c) if p is locally bounded and radial, and $\eta_n^{(\pm p)} < \infty, \forall n \in \mathbb{N}$, then $\mathcal{K}(p)^\times = \mathcal{K}(-p)$ if and only if there exists a positive constant $c(p)$ such that

$$1 \leqslant \eta_n^{(p)} \eta_n^{(-p)} \leqslant c(p), \text{ for all } n = 0, 1, 2, \ldots. \qquad (4.62)$$

Considering now the family of Hilbert spaces $\mathcal{K}(p)$, we may state a theorem exactly parallel to Theorem 4.6.2

Theorem 4.6.3. *Let I be the set of weight functions $p(z)$ that satisfy the following three conditions:*

(i) p is locally bounded and radial, $p(z) = p(|z|)$.
(ii) $u_n^{(p)}(z)$ belongs to $\mathcal{K}(p)$ and $\mathcal{K}(-p)$, for every $n \in \mathbb{N}$.
(iii) There are positive constants A, B such that

$$A \leqslant \eta_n^{(p)} \, \eta_n^{(-p)} \leqslant B, \quad \text{for all } n = 0, 1, 2, \ldots, \qquad (4.63)$$

where $\eta_n^{(p)}$ is given by (4.61).

Then the family $\{\mathcal{K}(p), \ p \in I\}$ is a chain of hilbertian spaces with central Hilbert space $\mathcal{K}(0)$. The compatibility $\#$ coincides with both $\#_1$ and $\#_2$, and the partial inner product with that induced by $\mathcal{K}(\mathbb{D})$ and ℓ^2, respectively.

The order structure and the lattice operations on this chain of hilbertian spaces are exactly the same as in the previous case, replacing weights ρ by weights p. In particular, the same difficulty arises with duality. Let us indeed identify $\mathcal{K}(p)$ with a sequence space:

$$f \in \mathcal{K}(p) \Longleftrightarrow f = \sum_n a_n u_n^{(p)}, \text{ with } \|f\|_{\mathcal{K}(p)}^2 := \sum_n \eta_n^{(p)-1} |a_n|^2 < \infty,$$

where the series representing f converges uniformly on compact subsets of $\mathbb{D}$. Then the dual space, denoted $\mathcal{K}(\widehat{p})$, corresponds to the dual sequence space

$$f \in \mathcal{K}(\widehat{p}) \Longleftrightarrow f = \sum_n b_n u_n^{(p)}, \text{ with } \|f\|_{\mathcal{K}(\widehat{p})}^2 := \sum_n \eta_n^{(p)} |b_n|^2 < \infty.$$

The point is that, if the condition (4.63) is satisfied, then the norm $\| \cdot \|_{\mathcal{K}(\widehat{p})}^2$ is equivalent to, but different from the norm $\| \cdot \|_{\mathcal{K}(-p)}^2$. Hence we have again to make a distinction between Hilbert spaces and hilbertian spaces, thus between a chain of hilbertian spaces and a LHS.

Once again, the problem is that the space $\mathcal{K}(\widehat{p})$ is a space of functions analytic in $\mathbb{D}$, but need not be associated to a weight. The existence of the latter is a moment problem, as in the previous case. This is why we prefer to stay with the spaces $\mathcal{K}(p)$ and the chain of hilbertian spaces they constitute.

If we particularize these results to the spaces $\mathcal{K}_\alpha$, $-1 < \alpha < 1$, we obtain the following results. First we notice the equivalences

$$\alpha_2 > \alpha_1 \Longleftrightarrow \mathcal{K}_{\alpha_1} \subset \mathcal{K}_{\alpha_2} \Longleftrightarrow \mathcal{K}_{\widehat{\alpha}_2} \subset \mathcal{K}_{\widehat{\alpha}_1},$$

where $\mathcal{K}_{\widehat{\alpha}}$ denotes the dual of $\mathcal{K}_\alpha$ as above. Then we have

Theorem 4.6.4. *(i) The family $\{\mathcal{K}_\alpha, \ -1 < \alpha < 1\}$, is a chain of hilbertian spaces.*

(ii) The family $\{\mathcal{K}_\alpha, \mathcal{K}_{\hat\alpha}, 0 \leqslant \alpha < 1\}$, is a LHS:

$$\ldots \mathcal{K}_{\hat\alpha_2} \subset \mathcal{K}_{\hat\alpha_1} \subset \mathcal{K}_{\hat{0}} \simeq \mathcal{K}_0 \subset \mathcal{K}_{\alpha_1} \subset \mathcal{K}_{\alpha_2} \ldots \qquad (\alpha_2 > \alpha_1).$$

Finally we come back to the original spaces $\mathcal{H}_\alpha$. Since the unitary map $B_\alpha : \mathcal{H}_\alpha \to \mathcal{K}_\alpha$ linking the two spaces depends on α, the following result does not contradict Theorem 4.6.4.

Proposition 4.6.5. *Let $\alpha_2 > \alpha_1 > -1$. Then the two spaces $\mathcal{H}_{\alpha_1}$ and $\mathcal{H}_{\alpha_2}$ are not comparable.*

However, there *is* a continuous embedding $U_{\alpha_2\alpha_1} : \mathcal{H}_{\alpha_1} \to \mathcal{H}_{\alpha_2}$. Consider the maps $V_\alpha : L^2(\mathbb{R}^+) \to \mathcal{H}_\alpha$ and $W_\alpha : \mathcal{H}_\alpha \to L^2(\mathbb{R}^+)$ defined from (4.52) and (4.53), respectively, for $\varphi = 0$. Then one defines

$$U_{\alpha_2\alpha_1} = V_{\alpha_2} W_{\alpha_1} \quad\text{and}\quad U_{\alpha_2\alpha_1}^{-1} = V_{\alpha_1} W_{\alpha_2} = U_{\alpha_1\alpha_2}.$$

The map $U_{\alpha_2\alpha_1}$ may be computed explicitly. Given $f_1 \in \mathcal{H}_{\alpha_1}$, one has

$$f_2(z) := [U_{\alpha_2\alpha_1} f_1](z)$$

$$= \frac{1}{4\pi} \left(\frac{\Gamma(\frac{\alpha_1+\alpha_2}{2}+2)}{\Gamma(\alpha_1+1)\Gamma(\alpha_2+1)} \right)^{1/2} \int_{\mathbb{C}^+} \left(\frac{z-\overline{w}}{2i} \right)^{-(\frac{\alpha_1+\alpha_2}{2}+2)} f_1(w) \, (\operatorname{Im} w)^{\alpha_1} \, d\nu(w).$$

Equivalently, if $f_1 = V_{\alpha_1} \psi$, with $\psi \in L^2(\mathbb{R}^+)$, that is,

$$f_1(z) = \left(2\pi\,\Gamma(\alpha_1+1) \right)^{-1/2} \int_0^\infty r^{\alpha_1+3/2} \, e^{izr^2/2} \, \psi(r) \, dr,$$

then

$$f_2(z) := [U_{\alpha_2\alpha_1} f_1](z) = \left(2\pi\,\Gamma(\alpha_1+1) \right)^{-1/2} \int_0^\infty r^{\alpha_1+3/2} \, e^{izr^2/2} \, r^{(\alpha_2-\alpha_1)} \psi(r) \, dr.$$

Combining now the spaces $\mathcal{H}_{\alpha_j}$ and $\mathcal{K}_{\alpha_j}$, we get the following commutative diagram:

$$\alpha_2 \geqslant \alpha_1 > -1: \qquad \begin{array}{ccc} \mathcal{K}_{\alpha_1} & \xrightarrow{\ i_{\alpha_2\alpha_1}\ } & \mathcal{K}_{\alpha_2} \\[2mm] {\scriptstyle B_{\alpha_1}}\big\uparrow & & \big\downarrow{\scriptstyle B_{\alpha_2}^{-1}} \\[2mm] \mathcal{H}_{\alpha_1} & \xrightarrow{\ \widehat{i}_{\alpha_2\alpha_1}\ } & \mathcal{H}_{\alpha_2} \end{array}$$

where $\widehat{i}_{\alpha_2\alpha_1} = B_{\alpha_2}^{-1} \cdot i_{\alpha_2\alpha_1} \cdot B_{\alpha_1}$ and the maps B_α, B_α^{-1} are defined in (4.56),(4.57). Since $i_{\alpha_2\alpha_1}$, i.e., the natural embedding, is continuous with dense range, so is $\widehat{i}_{\alpha_2\alpha_1}$. In addition, these maps satisfy the composition law:

For $\alpha_3 \geqslant \alpha_2 \geqslant \alpha_1 > -1$,   one has $\widehat{i}_{\alpha_3\alpha_1} = \widehat{i}_{\alpha_3\alpha_2}\widehat{i}_{\alpha_2\alpha_1}$.

Actually, the maps $\widehat{\imath}_{\alpha_2\alpha_1}$ can be calculated explicitly.

Proposition 4.6.6. *Let $\alpha_2 \geqslant \alpha_1 > -1$. Then the injection $\widehat{\imath}_{\alpha_2\alpha_1} : \mathcal{H}_{\alpha_1} \to \mathcal{H}_{\alpha_2}$ is the operator of multiplication by $2^{(\alpha_2-\alpha_1)/2} (z+i)^{(\alpha_1-\alpha_2)}$. This operator is injective and continuous, and has dense range.*

In order to understand the structure of the family $\{\mathcal{H}_\alpha\}$, we must introduce a new notion. Given a directed set A, an *inductive spectrum* is a family $\{E_\alpha, \pi\}_{\alpha\in A}$, where each E_α is a locally convex space and π is a collection of continuous linear maps $\pi_{\beta\alpha} : E_\alpha \to E_\beta$, $\alpha \leqslant \beta$, satisfying the composition rule $\pi_{\gamma\beta} \circ \pi_{\beta\alpha} = \pi_{\gamma\alpha}$, $\alpha \leqslant \beta \leqslant \gamma$ (thus $\pi_{\alpha\alpha}$ is the identity on E_α). Given the inductive spectrum $\{E_\alpha, \pi\}_{\alpha\in A}$, one may construct the algebraic inductive limit of the family $\{E_\alpha\}_{\alpha\in A}$ as explained for nestings in Section 2.4.1 (iv). From this, we see that a nesting is an inductive spectrum with an involution $\alpha \leftrightarrow \overline{\alpha}$.

Coming back to the spaces $\{\mathcal{H}_\alpha\}$, we may state

Proposition 4.6.7. *The family $\{\mathcal{H}_\alpha, -1 < \alpha < 1,\}$ is an inductive spectrum of Hilbert spaces with respect to the maps $\widehat{\imath}_{\alpha_2\alpha_1}$.*

In order to get a genuine NHS, we need that $\mathcal{H}_{-\alpha}$ be the conjugate dual of $\mathcal{H}_\alpha$, but that is not obvious. In any case, we have here another example of nestings which are not natural embeddings.

4.6.5 Hardy Spaces of Analytic Functions in the Unit Disk

There is another classical family of spaces consisting of functions analytic in the unit disk, the Hardy spaces H^p, $0 < p \leqslant \infty$. For reasons of simplicity, we shall restrict ourselves to the range $1 \leqslant p < \infty$.

Given a function f analytic in the unit disk $\mathbb{D}$, define the quantity

$$M_p(r, f) := \left(\frac{1}{2\pi} \int_0^{2\pi} |f(re^{i\theta})|^p d\theta \right)^{1/p}, \quad 0 < r < 1, 1 \leqslant p < \infty, \quad (4.64)$$

which is a nondecreasing function of r. Then the function f is said to be of class H^p if $M_p(r, f)$ remains bounded as $r \to 1$.

For any function $f \in H^p$, the limit $f(e^{i\theta}) = \lim_{r\to 1} f(re^{i\theta})$ exists almost everywhere and belongs to the space $L^p(S^1, d\theta)$, where S^1 is the unit circle. Then H^p is a Banach space for the norm

$$\|f\|_p := \lim_{r\to 1} M_p(r, f) = \left(\frac{1}{2\pi} \int_0^{2\pi} |f(e^{i\theta})|^p d\theta \right)^{1/p}, \quad 1 \leqslant p < \infty.$$

This allows to identify H^p with a closed subspace of $L^p(S^1)$. This subspace consists of all functions $f \in L^p$ such that

$$\int_0^{2\pi} e^{in\theta} f(e^{i\theta}) \, d\theta = 0, \ n = 1, 2, 3, \ldots$$

In particular, the polynomials are dense in H^p. Clearly, $H^{p_1} \subset H^{p_2}$ if $p_1 > p_2$, the embedding is continuous and has dense range.

Let us look now at the dual spaces. We know that the conjugate dual $(L^p)^\times$ of L^p is $L^{\bar p}$, with $1/p + 1/\bar p = 1$. Thus every bounded linear functional ϕ on L^p $(1 \leqslant p < \infty)$ has a unique representation

$$\phi(f) = \frac{1}{2\pi} \int_0^{2\pi} f(e^{i\theta}) \, g(e^{i\theta}) d\theta, \text{ with } g \in L^{\bar p}.$$

Since H^p is subspace of L^p, the conjugate dual $(H^p)^\times$ is the quotient $L^{\bar p}/H_0^{\bar p}$, where $H_0^{\bar p}$ is the annihilator of H^p. An element $g(e^{i\theta})$ of that subspace is the boundary function of some $g(z) \in H^{\bar p}$ such that $g(0) = 0$. Actually one can replace $H_0^{\bar p}$ by $H^{\bar p}$ in the quotient, so that finally $(H^p)^\times \simeq L^{\bar p}/H^{\bar p}$ (isometric isomorphism). In addition, for $1 < p < \infty$, each $\phi \in (H^p)^\times$ is representable by a *unique* function $g \in H^{\bar p}$ for which

$$\phi(f) = \frac{1}{2\pi} \int_0^{2\pi} f(e^{i\theta}) \overline{g(e^{i\theta})} d\theta, \ \forall \, f \in H^p.$$

Since the conjugate dual of $L^{\bar p}/H_0^{\bar p}$ is H^p, the latter is reflexive. The conclusion is that the family $\{H^p, \ 1 < p < \infty\}$, is an inductive spectrum and a chain of reflexive Banach spaces, but not a LBS.

There are two variants of the Hardy spaces that ought to be mentioned. First one can consider functions analytic in the upper half-plane $\mathbb{C}^+$, as for the Bergman spaces. Thus one defines the space $\mathfrak{H}_+^p$ $(1 \leqslant p < \infty)$ of functions analytic in $\mathbb{C}^+$, such that $|f(x + iy)|^p$ is integrable for each $y > 0$ and

$$\mathfrak{M}_p^{(+)}(y, f) := \left(\int_{-\infty}^{\infty} |f(x + iy)|^p \, dx \right)^{1/p} \tag{4.65}$$

is bounded for $0 < y < \infty$. The space $\mathfrak{H}_+^p$ is again a reflexive Banach space, which has the same properties as H^p. In particular, the limit property takes the following form. If $f \in \mathfrak{H}^p$, then the boundary function $f(x) = \lim_{y \to 0} f(x + iy)$ exists almost everywhere and belongs to $L^p(\mathbb{R})$. For duality, it turns out that the dual of $\mathfrak{H}_+^p$ is isometrically isomorphic to $L^{\bar p}/\mathfrak{H}_+^{\bar p}$ and the same conclusion follows.

Similar considerations apply to the Hardy spaces $\mathfrak{H}_-^p$ $(1 \leqslant p < \infty)$ of functions analytic in the lower half-plane $\mathbb{C}^-$.

Another case of interest is the space $L_a^p(\mathbb{D})$ of L^p functions analytic in the unit disk $\mathbb{D}$. This is a closed subspace of $L^p(\mathbb{D})$, hence a Banach space. Its dual is simply $L_a^{\overline{p}}(\mathbb{D})$, so that here we get a genuine reflexive lattice of Banach spaces.

As a final remark, we may add that most of the results of this section are valid for Hardy spaces of harmonic functions.

Notes for Chapter 4

Section 4.1. The results of this section are based on the extensive study by Davis *et al.* [68, 69]. Notice that we differ slightly from these authors, in that we do not include the space L^1. However, their results apply to our case also, because the first Eq. (4.1) (shown in [68, 69]) holds true also with the definition $\widehat{S} = (1, t) \bigcup S$.

- In the terminology of Floret–Wloka [FW68, §9], L^{p-} is a strict FG-space.
- The space L^ω has been introduced by Arens [33]. It is usually called the Arens algebra in the literature.
- A thorough analysis of partial *algebras may be found in the monograph of Antoine–Inoue–Trapani [AIT02]. The case of the L^p spaces is discussed in Section 6.3.1, that we follow closely. Note that there are some slight errors in the computation of the so-called multiplier spaces. They have been corrected in Antoine [16].
- Further information about Sobolev spaces may be found in the books of Hörmander [Hör63], Trêves [Tre67], or Adams [Ada75].
- The noncommutative L^p spaces associated to a von Neumann algebra were introduced by Segal [176], and they were later studied by Inoue in the context of algebras of unbounded operators [127, 128].
- Chains of hilbertian spaces were defined and studied by Palais [162] and Krein–Petunin [132]. Concerning the distinction between chains and scales, we follow the terminology of Palais [162]. On the contrary, Krein–Petunin [132] use the term 'scales of Banach spaces' for the general case, Hilbert scales being distinguished by their interpolation properties.
- Bargmann's space $\mathcal{E}^\times$ was introduced in [42, 43] for the purpose of the formulation of quantum mechanics in the phase space representation and the consequences of the latter for the theory of distributions (see the note under Section 4.6).

Section 4.2.

- For the Schwartz theory of kernels, see [175, Sch57]. Further information on nuclear spaces may also be found in Trêves [Tre67].
- For the assaying subsets $L^2(r_{\alpha\beta})$ of $L^1_{\mathrm{loc}}(\mathbb{R}^n, dx)$, see Hörmander [Hör63, Theorem 2.1.1].
- For the topology γ of compact convergence and for standard results on the various topologies on L^1_{loc} and L_c^∞, see for instance the textbooks of Köthe

[Köt69], Schwartz [Sch57] or Trèves [Tre67]. In particular, LF-spaces are discussed at length in Trèves [Tre67, Chap. 13].

- Related results on the representation of operators on L^2 spaces have been obtained by Ascoli *et al.* [36].

Section 4.3.

- Köthe has considered also another duality on the space ω, corresponding to the so-called β-compatibility, namely

$$(x_n)\#_\beta(y_n) \iff \sum_n \overline{x_n}y_n \text{ converges.}$$

This correspondence associates to each subspace l its β-*dual* [Köt69, §30.10]. The two compatibilities will be studied in more detail in Section 5.5.7.

- In fact, Proposition 4.3.7 allows us to dispense of the notions of strong norm and strong Banach ideals, the notion of perfect spaces or assaying subspaces of ω is equivalent. For further details on normed ideals and their relationship with spaces of compact operators, we refer to the classical texts of Schatten [Sch70], Simon [Sim79], Pietsch [Pie80] or Gohberg–Krein [GK69]. See also the work of Cigler [58] and Oostenbrink [Oos73].
- Echelon and co-echelon spaces have been introduced by Köthe, see [Köt69, §30.8].

Section 4.4. Köthe function spaces have been introduced (and given that name) by Dieudonné [73]. The corresponding lattice has been introduced and studied by Luxemburg–Zaanen (see [Zaa61] and references therein), that we follow here. These spaces are also called normed Köthe spaces or Banach function spaces.

- The space L_G was introduced by Gould [111], who calls it Ω; see also Zaanen [Zaa61, §30, Exercises], who denotes it by L_ρ.

Section 4.5. Analyticity and trajectory spaces were introduced and studied systematically by van Eijndhoven–de Graaf in a series of papers, later synthesized in van Eijndhoven's PhD thesis [Eij83] and two books [EG85,EG86]. The avowed goal of this work was to produce a mathematical framework capable of producing a rigorous version of Dirac's bra-and-ket formalism of quantum mechanics. More about this will be found in Section 7.1.1 (ii). These spaces also play a role in the representation theory of Lie groups, see Section 7.4.

Section 4.6.

- The analysis of this section is largely due to Antoine–Vause [31].
- Besides the familiar position (q) and momentum (p) representations of quantum mechanics, the Fock-Bargmann representation offers an attractive alternative [42, 43]. It is based on the canonical (or oscillator) coherent states and it is characterized by the fact that its wave functions

are entire analytic functions of $z = q + ip$. It is therefore a *phase space* representation. As such it is useful for studying the quantum-to-classical transition, quantum optics, path integrals, geometric quantization, etc. The properties of the Hilbert space $\mathfrak{F}$ are in fact characteristic of all phase space representations. For further details on the latter and, in particular, the Fock-Bargmann representation, see for instance Ali *et al.* [6] or the monograph of Ali–Antoine–Gazeau [AAG00].

- The duality between the spaces $\mathfrak{Z}$ and Exp is discussed in detail in the textbook of Trêves [Tre67, Chap. 22], in particular Exercise 22.5.
- The lattice of hilbertian spaces constituted from the spaces $\mathfrak{F}(\rho)$, which is *not* a scale, has been studied in detail by Antoine–Vause [31]. This lattice of hilbertian spaces has found interesting applications in the study of Weyl quantization by Daubechies [65–67], that is, the setting of a correspondence between functions on phase space $f(q + ip) \equiv \tilde{f}(q, p)$ and operators on $L^2(\mathbb{R}, dx)$.
- The spaces $G(a, b)$ and $\widetilde{G}(a, b)$, for fixed a, b, were introduced and studied systematically by van Winter in a series of papers devoted to a reformulation of quantum scattering theory [188, 189]. The Mellin transform and its inverse are usually defined as follows:

$$
M(s) = \int_0^\infty f(x)\, x^{s-1}\, dx, \quad f(x) = \frac{1}{2\pi i} \int_{c-i\infty}^{c+i\infty} M(s)\, x^{-s}\, ds,
$$

but the change of variables $s = -it + \frac{1}{2}$ leads to the version given in the text.

- The van Winter LHS $\{L^2(a)\}$ and its connection with the spaces $G(a, b)$ and $\widetilde{G}(a, b)$ was first described in the work of Gollier [Gol82]. The space $D(a, b)$ and its connection with the three unitary Hilbert spaces were introduced by Klein [Kle87].
- Part of the van Winter LHS $\{L^2(a)\}$ has been considered, for similar purposes in quantum scattering theory, by Horwitz–Katznelson [125] and by Skibsted [180]. Note, however, that the analysis of Horwitz–Katznelson is not entirely correct.
- The spaces $\mathcal{H}_\alpha$ and $\mathcal{K}_\alpha$, for integer values of α, including their mutual relationship, are standard in the representation theory of the groups SL(2,$\mathbb{R}$) and SU(1,1). The corresponding spaces for arbitrary $\alpha \in (-1, 1)$ have been introduced by Paul [163]. As for the spaces $\mathcal{K}(p)$, they have been introduced and studied in detail by Klein [Kle87]. There one may find several examples and counterexamples of weights.
- The notion of inductive spectrum is studied in detail in the monograph of Floret–Wloka [FW68, §23].
- Hardy spaces are essential tools in harmonic analysis. Detailed treatments may be found in the monographs of Duren [Dur70, Chaps. 3,7 and 11] and Koosis [Koo80, Chaps. IV and VII]. The space L_a^p and its generalizations are described by Luecking [144].

Chapter 5
Refinements of PIP-Spaces

5.1 Construction of PIP-Spaces

5.1.1 The Refinement Problem

We have seen in Section 1.5, that the compatibility relation underlying a PIP-space may always be coarsened, but not refined in general. There is an exception, however, namely the case of a scale of Hilbert spaces and analogous structures. We shall describe it in this section.

We begin with the simplest example of LHS, namely, the discrete scale of Hilbert spaces $H_I = \{\mathcal{H}_\alpha, \alpha \in \mathbb{Z}\}$ generated by a self-adjoint operator A in a Hilbert space $\mathcal{H}$, described in Section 5.2.1. Using the spectral theorem for self-adjoint operators, we get a Hilbert space $\mathcal{H}_\alpha$ for every $\alpha \in \mathbb{R}$, and thus a *continuous scale of Hilbert spaces* $H_{\overline{I}} = \{\mathcal{H}_\alpha, \alpha \in \mathbb{R}\}$. This continuous scale $H_{\overline{I}}$ contains the discrete scale $H_I = \{\mathcal{H}_\alpha, \alpha \in \mathbb{Z}\}$ and one or the other is used in applications, depending on the problem at hand. In the terminology of Chapter 1, $H_{\overline{I}}$ is a (proper) refinement of H_I, in the sense that the complete involutive lattice generated by H_I is an involutive sublattice of the one generated by $H_{\overline{I}}$.

Alternatively, given the discrete scale, the continuous scale $H_{\overline{I}}$ can be reconstructed uniquely ("functorially") by the method of *quadratic interpolation*, as we shall see below. Thus quadratic interpolation may be viewed as a particular method for refining the PIP-space H_I. Our aim in this section is to extend this construction to the more general case of an involutive lattice of Hilbert spaces. We know that the problem of refining an arbitrary PIP-space has in general no solution. But we will show here that such a solution (and in fact infinitely many solutions) always exists for the case of a countable LHS. The method of the proof is a straightforward combination of quadratic interpolation and the spectral theorem for self-adjoint operators. It is worth remarking that the same procedure can be applied to any of the Hilbert scales discussed in Chapter 4.

Now the LHS H_I or $H_{\overline{I}}$ lives between the extreme spaces of a rigged Hilbert space (RHS) which can be constructed in a natural way starting from A.

J.-P. Antoine and C. Trapani, *Partial Inner Product Spaces:*
Theory and Applications, Lecture Notes in Mathematics 1986,
DOI 10.1007/978-3-642-05136-4_5, © Springer-Verlag Berlin Heidelberg 2009

This fact motivates the possibility of considering a more general set-up and posing the question as to whether any RHS can be refined to a LHS or to a general PIP-space. The notion of interspaces will pave the way towards this construction. Since any family $\mathcal{O}$ of closable operators on a dense domain $\mathcal{D}$ of a Hilbert space leads to a RHS on one hand and to a LHS on the other, we will also consider cases (e.g., when $\mathcal{O}$ is an O*-algebra) where the two structures fit together in a reasonable way.

We will proceed in several steps. We treat first a scale of Hilbert spaces, then a general chain (in both cases, we denote assaying subspaces by $\mathcal{H}_n$) and finally a general lattice, as described above. In addition, we will describe another method of refinement, totally independent of the previous ones. In a certain case, this method leads to the so-called Bessel potential spaces (also called fractional Sobolev spaces).

5.1.2 Some Results from Interpolation Theory

For the convenience of the reader, we collect first some basic facts about interpolation between Hilbert spaces.

Let X_1, X_0 be two Hilbert spaces, with inner products $\langle \cdot | \cdot \rangle_1$ and $\langle \cdot | \cdot \rangle_0$, respectively, such that $X_1 \hookrightarrow X_0$, the injection $X_1 \to X_0$ is continuous with norm not greater than 1 and has dense range. Then there exists a *unique* self-adjoint operator A in X_0, with $A \geqslant 1$, such that $X_1 = Q(A) := D(A^{1/2})$, the form domain of A, and

$$\langle f | f \rangle_1 = \langle f | Af \rangle_0$$

$$= \langle A^{1/2} f | A^{1/2} f \rangle_0, \ f \in X_1 = Q(A).$$

Then, for $0 \leqslant t \leqslant 1$, one defines the space $X_t := Q_t(X_1, X_0)$ as the form domain of the self-adjoint operator A^t equipped with the inner product

$$\langle f | f \rangle_t = \langle f | A^t f \rangle_0 = \langle A^{t/2} f | A^{t/2} f \rangle_0, \ f \in Q(A^t) = D(A^{t/2}).$$

The spaces X_t $(0 \leqslant t \leqslant 1)$ have the following properties:

(i) X_t is a Hilbert space, and for $0 \leqslant t \leqslant s \leqslant 1$, one has $X_1 \subseteq X_s \subseteq X_t \subseteq X_0$, where all injections are continuous and have dense range.

(ii) The definition of X_t *is intrinsic* in the following sense: Replacing the inner products on X_1 and X_0 by equivalent ones results in a different operator A, but the *same* vector space with an equivalent topology.

(iii) *Interpolation property*: Let $X_1 \subseteq X_0, Y_1 \subseteq Y_0$ two such pairs, $\{X_t\}, \{Y_t\}$ being the corresponding interpolating scales. Let T be a linear bounded operator from X_0 into Y_0 and also from X_1 into Y_1. Then T is linear and bounded from X_t into Y_t for every t.

(iv) *Norm convexity property*: Given the operator T of (iii), its norm satisfies the following convexity property:

$$\|T\|_t \leqslant \|T\|_0^{1-t}\|T\|_1^t, \ 0 \leqslant t \leqslant 1, \ \text{where } \|\cdot\|_t \text{ denotes the norm of } T{:}X_t \to Y_t.$$

(v) *Duality property* : Let $X_1^\times, X_0^\times$ be the (conjugate) duals of X_1, X_0. Then $[Q_t(X_1, X_0)]^\times = Q_{1-t}(X_0^\times, X_1^\times)$.

As an example we may take $X_1 = L^2(\mathbb{R}, (1+x^2)dx)$ and $X_0 = L^2(\mathbb{R}, dx)$. Then we have $X_t = L^2(\mathbb{R}, (1+x^2)^t dx)$. More generally, with $X_n = L^2(\mathbb{R}, (1+x^2)^n dx), n \in \mathbb{N}$, we have $Q_t(X_n, X_{-n}) = X_{n(2t-1)}$. In particular, $Q_{1/2}(X_n, X_{-n}) = X_0$ for every $n \in \mathbb{N}$.

This method of constructing the continuous scale $\{X_t\}_{0 \leqslant t \leqslant 1}$ from the pair $\{X_1, X_0\}$ is called *quadratic interpolation*.

5.2 Refining Scales of Hilbert Spaces

5.2.1 The Canonical Scale of Hilbert Spaces Generated by a Self-Adjoint Operator

We begin with the simplest example of LHS, namely, the canonical scale of Hilbert spaces generated by a self-adjoint operator, already introduced in Section I.1

Let A be a self-adjoint operator in a Hilbert space $\mathcal{H}$, with dense domain $D(A)$. By the spectral theorem, the operator A^n is well-defined and self-adjoint for each $n \in \mathbb{N}$. As usual, we put

$$\mathcal{D}^\infty(A) = \bigcap_{n=1}^\infty D(A^n),$$

which is a dense subspace of $\mathcal{H}$. Endowed with the topology t_A generated by the set of seminorms $f \mapsto \|A^n f\|$, $n \in \mathbb{N}$, $\mathcal{D}^\infty(A)$ is a reflexive Fréchet space. We denote by $\mathcal{D}_{\overline{\infty}}(A)$ its conjugate dual with respect to the inner product of $\mathcal{H}$ and endow it with the strong dual topology $\mathsf{t}_A^\times$.

Let now

$$\mathcal{H}_n = \bigcap_{k=0}^n D(A^k) \quad \text{and} \quad \langle f|g \rangle_n = \sum_{k=0}^n \langle A^k f|A^k g\rangle, \ f,g \in \mathcal{H}_n.$$

Then $(\mathcal{H}_n, \langle \cdot|\cdot \rangle_n)$ is a Hilbert space.

Since $D((1 + A^2)^{n/2}) = \bigcap_{k=0}^{n} D(A^k) = D(A^n)$ and $\langle f|g \rangle_n = \langle (1 + A^2)^{n/2} f$

$(1 + A^2)^{n/2} g \rangle$, then

$$\mathcal{D}^\infty(A) = \bigcap_{n=1}^{\infty} D((1 + A^2)^{n/2})$$

and the topology of $\mathcal{D}^\infty(A)$ can equivalently be described by the norms:

$$f \mapsto \|(1 + A^2)^{n/2} f\|, \quad n \in \mathbb{N}.$$

In other words, we could have required $A \geqslant 1$ from the beginning without loss of generality.

Now since $(1 + A^2)^{1/2} \geqslant 1$, it makes sense to consider, again with the help of the spectral theorem, all its positive and negative real powers. We then define $\mathcal{H}_\alpha$ to be $D((1 + A^2)^{\alpha/2}) = Q((1 + A^2)^\alpha)$, endowed with the inner product

$$\langle f|g \rangle_\alpha = \langle (1 + A^2)^{\alpha/2} f | (1 + A^2)^{\alpha/2} g \rangle = \langle f | (1 + A^2)^\alpha g \rangle, \ \alpha \geqslant 0.$$

For every $\alpha \in \mathbb{R}^+ \cup \{0\}$, we denote by $\mathcal{H}_{\overline{\alpha}} := \mathcal{H}_{-\alpha}$ the conjugate dual of $\mathcal{H}_\alpha$. It is easily shown that $\mathcal{H}_{\overline{\alpha}}$ is isometrically isomorphic to the completion of $\mathcal{H}$ with respect to the inner product $\langle f|g \rangle_{\overline{\alpha}} = \langle (1 + A^2)^{-\alpha/2} f | (1 + A^2)^{-\alpha/2} g \rangle$.

Then we have, for $\alpha < \beta$,

$$\mathcal{D}^\infty(A) \subset \ldots \hookrightarrow \mathcal{H}_\beta \hookrightarrow \mathcal{H}_\alpha \hookrightarrow \ldots \hookrightarrow \mathcal{H} := \mathcal{H}_0 \hookrightarrow \ldots \hookrightarrow \mathcal{H}_{\overline{\alpha}} \hookrightarrow \mathcal{H}_{\overline{\beta}} \hookrightarrow \ldots \subset \mathcal{D}_{\overline{\infty}}(A).$$
$$(5.1)$$

Clearly, the family of spaces $\{\mathcal{H}_\alpha\}_{\alpha \in \mathbb{R}}$ forms a chain of Hilbert spaces interpolating between $\mathcal{D}^\infty(A)$ and $\mathcal{D}_{\overline{\infty}}(A)$, in which all the embeddings are continuous and have dense range (such spaces will be called *interspaces* in Section 5.4.1).

The operator A leaves $\mathcal{D}^\infty(A)$ invariant, and so, being symmetric, $A \upharpoonright \mathcal{D}^\infty(A)$ has a unique extension $\hat{A} : \mathcal{D}_{\overline{\infty}}(A) \to \mathcal{D}_{\overline{\infty}}(A)$ continuous in the strong dual topology of $\mathcal{D}_{\overline{\infty}}(A)$, which coincides with its Mackey topology. The same holds true for the positive integer powers of A and the real powers of $(1 + A^2)^{1/2}$.

In fact, we have here not only a chain, but a scale of Hilbert spaces. To make things precise, let us give a formal definition.

Definition 5.2.1. A continuous chain of Hilbert spaces $\{\mathcal{H}_\alpha\}_{\alpha \in \mathbb{R}}$ is called a *scale* if there exists a self-adjoint operator $B \geqslant 1$ such that $\mathcal{H}_\alpha = Q(B^\alpha)$, $\forall \alpha \in \mathbb{R}$, with the form norm $\|f\|_\alpha^2 = \langle f | B^\alpha f \rangle$. A similar definition holds for a discrete chain $\{\mathcal{H}_n\}_{n \in \mathbb{Z}}$, which is a particular case of a *countably Hilbert space*.

To give a counterexample, take the chain $\{\mathfrak{F}^\rho,\ \rho \in \mathbb{R}\}$ of Hilbert spaces of entire functions defined in Section 4.6.2. This chain is *not* a scale, in the sense we have defined. Indeed it would correspond to the operator of multiplication by $(1 + |z|^2)^{1/2}$, but the latter has domain $\{0\}$ in $\mathfrak{F}$, since $\mathfrak{F}$ contains only entire analytic functions.

The scale of Hilbert spaces (5.1) possesses interesting properties, quite similar to those of the chain of the Lebesgue spaces described in Section 4.1.

Let us start with $\alpha = 1$, that is, the scale of Hilbert spaces (as usual, we write $\overline{1} = -1$)

$$\mathcal{H}_1 \hookrightarrow \mathcal{H} \hookrightarrow \mathcal{H}_{\overline{1}}.$$

In what follows it is essential that A be unbounded. Without loss of generality, we assume $A \geqslant 1$.

For every $\alpha > 0$, A^α is still a self-adjoint positive operator and $A^\alpha \geqslant 1$. Thus, if $\alpha, \beta \in (0,1)$, with $\beta > \alpha$, we get

$$\mathcal{H}_1 \hookrightarrow \mathcal{H}_\beta \hookrightarrow \mathcal{H}_\alpha \hookrightarrow \mathcal{H} \hookrightarrow \mathcal{H}_{\overline{\alpha}} \hookrightarrow \mathcal{H}_{\overline{\beta}} \hookrightarrow \mathcal{H}_{\overline{1}}. \tag{5.2}$$

As for the norms, we notice that, if $\alpha, \beta \in (0,1)$ and $\alpha < \beta$, then, for every $f \in \mathcal{H}_\beta$, one has

$$\|f\|_\alpha \leqslant \|f\|_\beta. \tag{5.3}$$

For $\beta \in [0,1]$, consider the space

$$\mathcal{H}_{\beta-} := \bigcap_{\alpha \in (0,\beta)} \mathcal{H}_\alpha.$$

This is, in general, no longer a Hilbert space, but only a reflexive Fréchet space, whose dual we denote as

$$\mathcal{H}_{\overline{\beta}+} := \bigcup_{\alpha \in (0,\beta)} \mathcal{H}_{\overline{\alpha}}.$$

Then $\mathcal{H}_{\overline{\beta}+}$ is a reflexive DF-space.

For obvious reasons, we will refer to $\mathcal{H}_{\beta-}$ and $\mathcal{H}_{\overline{\beta}+}$ as *nonstandard* spaces. Actually the standard spaces may be recovered from the nonstandard ones.

Proposition 5.2.2. *The following statement holds true:*

$$\mathcal{H}_\beta = \left\{ f \in \mathcal{H}_{\beta-} : \sup_{\alpha \in (0,\beta)} \|f\|_\alpha < +\infty \right\} \quad (0 \leqslant \beta \leqslant 1).$$

Furthermore, if $f \in \mathcal{H}_\beta$, one has

$$\|f\|_\beta = \sup_{\alpha \in (0,\beta)} \|f\|_\alpha.$$

Proof. If $f \in \mathcal{H}_\beta$, it follows from (5.3) that

$$\sup_{\alpha \in (0,\beta)} \|f\|_\alpha \leqslant \|f\|_\beta.$$

To prove the converse inequality, we consider the spectral resolution of the operator A:

$$A = \int_1^{+\infty} \lambda dE(\lambda),$$

to be understood, as usual, in the sense of the strong convergence. By the Lebesgue monotone convergence theorem, we get

$$\lim_{\alpha \to \beta} \|f\|_\alpha^2 = \lim_{\alpha \to \beta} \int_1^{+\infty} \lambda^{2\alpha} d\langle E(\lambda)f|f\rangle = \int_1^{+\infty} \lambda^{2\beta} d\langle E(\lambda)f|f\rangle.$$

But

$$\lim_{\alpha \to \beta} \int_1^{+\infty} \lambda^{2\alpha} d\langle E(\lambda)f|f\rangle = \sup_{\alpha \in (0,\beta)} \|f\|_\alpha^2.$$

Therefore

$$\int_1^{+\infty} \lambda^{2\beta} d\langle E(\lambda)f|f\rangle = \|f\|_\beta^2.$$

∎

Moreover, if A is unbounded, then $\mathcal{H}_\alpha \neq \mathcal{H}_\beta$ if $\alpha \neq \beta$, and the nonstandard space $\mathcal{H}_{\beta-}$ is necessarily different from $\mathcal{H}_\beta$. Indeed, we have

Proposition 5.2.3. *(i) If $\alpha < \beta$ and $\mathcal{H}_\alpha = \mathcal{H}_\beta$, then A is bounded. (ii) If $\mathcal{H}_{\beta-} = \mathcal{H}_\beta$, then A is bounded.*

Proof. (i) The assumption $\alpha < \beta$ implies the inequality

$$\|A^\alpha f\| \leqslant \|A^\beta f\|, \ f \in \mathcal{H}_\beta,$$

and, therefore, since the two spaces coincide, the identity map $i : f \in \mathcal{H}_\beta \mapsto f \in \mathcal{H}_\alpha$ is continuous and onto. By the open mapping theorem, it follows that i^{-1} is continuous. Thus there exists $c > 0$ such that

$$\|A^\beta f\| \leqslant c\|A^\alpha f\|, \quad \forall f \in \mathcal{H}_\beta.$$

This may be rewritten as

$$\|A^{\beta-\alpha} f\| \leqslant c\|f\|, \quad \forall f \in \mathcal{H}_\beta.$$

Therefore $A^{\beta-\alpha}$ is bounded, and so A is bounded as well.

(ii) The argument is the same, using the identity map $i : f \in \mathcal{H}_\beta \mapsto f \in \mathcal{H}_{\beta-}$, and the definition, namely, $\mathcal{H}_{\beta-} = \bigcap_{\alpha \in (0,\beta)} \mathcal{H}_\alpha$ with the projective topology. ∎

Let $\mathcal{B}(\mathcal{H}_1, \mathcal{H}_{\overline{1}})$ be the Banach space of bounded operators from $\mathcal{H}_1$ into $\mathcal{H}_{\overline{1}}$ with its natural norm $\| \cdot \|_{1,\overline{1}}$. In $\mathcal{B}(\mathcal{H}_1, \mathcal{H}_{\overline{1}})$ define an involution $A \mapsto A^\times$ by

$$< A^\times f, g > = \overline{< Ag, f >}, \quad \forall f, g \in \mathcal{H}_1,$$

where $< \cdot, \cdot >$ is the form that puts $\mathcal{H}_1$ and $\mathcal{H}_{\overline{1}}$ in conjugate duality. If $\alpha, \beta \in (-1, 1)$, we can also consider the Banach space $\mathcal{B}(\mathcal{H}_\alpha, \mathcal{H}_\beta)$ of bounded operators from $\mathcal{H}_\alpha$ into $\mathcal{H}_\beta$ with its natural norm $\| \cdot \|_{\alpha,\beta}$.

Because of (5.2), the restriction to $\mathcal{H}_1$ of an operator of $\mathcal{B}(\mathcal{H}_\alpha, \mathcal{H}_\beta)$ may be identified with an operator in $\mathcal{B}(\mathcal{H}_1, \mathcal{H}_{\overline{1}})$. We consider this as an inclusion and, therefore,

$$\mathcal{B}(\mathcal{H}_\alpha, \mathcal{H}_\beta) \subset \mathcal{B}(\mathcal{H}_1, \mathcal{H}_{\overline{1}}), \quad \forall \alpha, \beta \in [-1, 1].$$

Moreover, $\mathcal{B}(\mathcal{H}_\alpha, \mathcal{H}_\beta)^\times = \mathcal{B}(\mathcal{H}_{\overline{\beta}}, \mathcal{H}_{\overline{\alpha}})$ for every $\alpha, \beta \in [-1, 1]$.

Let now $\mathcal{B}(\mathcal{H}_{\beta-}, \mathcal{H}_{\overline{\beta}+})$ be the space of all continuous linear maps from $\mathcal{H}_{\beta-}$ into $\mathcal{H}_{\overline{\beta}+}$. Then, it is easily seen that, if $X \in \mathcal{B}(\mathcal{H}_{\beta-}, \mathcal{H}_{\overline{\beta}+})$, then $X \upharpoonright \mathcal{H}_\beta$ is an element of $\mathcal{B}(\mathcal{H}_\beta, \mathcal{H}_{\overline{\beta}})$. The converse, however, is not true in general.

To make notations lighter, we will use the same notation for an operator X and for its restriction or extension to another space, if no confusion arises.

Proposition 5.2.4. $X \in \mathcal{B}(\mathcal{H}_{\beta-}, \mathcal{H}_{\overline{\beta}+})$ *if and only if there exists* $\alpha \in (0, \beta)$ *such that* $X \in \mathcal{B}(\mathcal{H}_\alpha, \mathcal{H}_{\overline{\alpha}})$.

Proof. If $X \in \mathcal{B}(\mathcal{H}_\alpha, \mathcal{H}_{\overline{\alpha}})$, with $\alpha \in (0, \beta)$, it is obvious that $X \in \mathcal{B}(\mathcal{H}_{\beta-}, \mathcal{H}_{\overline{\beta}+})$.

Conversely, let $X \in \mathcal{B}(\mathcal{H}_{\beta-}, \mathcal{H}_{\overline{\beta}+})$, then, by the continuity of X, for every $\alpha \in (0, 1)$ there exists $\gamma \in (0, 1)$ and $c > 0$ such that

$$\|Xf\|_{\overline{\alpha}} \leqslant c\|f\|_\gamma, \quad \forall f \in \mathcal{H}_{\beta-}.$$

Let now $\delta = \max\{\alpha, \gamma\}$ then we have:

$$\|Xf\|_{\overline{\delta}} \leqslant \|Xf\|_{\overline{\alpha}} \leqslant c\|f\|_\gamma \leqslant \|f\|_\delta, \quad \forall f \in \mathcal{H}_{\beta-}.$$

Therefore $X \in \mathcal{B}(\mathcal{H}_\delta, \mathcal{H}_{\overline{\delta}})$. ∎

Of course, the statement of the previous proposition can be rewritten as

$$\mathcal{B}(\mathcal{H}_{\beta-}, \mathcal{H}_{\overline{\beta}+}) = \bigcup_{\alpha \in (0,\beta)} \mathcal{B}(\mathcal{H}_\alpha, \mathcal{H}_{\overline{\alpha}}).$$

Proposition 5.2.5. *If* $\mathcal{B}(\mathcal{H}_\beta, \mathcal{H}_{\overline{\beta}}) = \mathcal{B}(\mathcal{H}_{\beta-}, \mathcal{H}_{\overline{\beta}+})$, *then* A *is bounded.*

Proof. Let U_β denote the unitary operator from $\mathcal{H}_\beta$ into $\mathcal{H}_{\overline{\beta}}$ given by the Riesz lemma. Clearly, $U_\beta \in \mathcal{B}(\mathcal{H}_\beta, \mathcal{H}_{\overline{\beta}})$. By the assumption U_β has a continuous extension $\widetilde{U}_\beta$ from $\mathcal{H}_{\beta-}$ onto $\mathcal{H}_{\overline{\beta}+}$. If $F \in \mathcal{H}_{\overline{\beta}+}$, then there exists $f \in \beta^-$ such that $\widetilde{U}_\beta f = F$. But $F_o := F{\upharpoonright}\mathcal{H}_\beta$ is an element of $\mathcal{H}_{\overline{\beta}}$, therefore $F_o = U_\beta f'$ for some $f' \in \mathcal{H}_\beta$. This implies that $f = f'$ and so $\mathcal{H}_\beta = \mathcal{H}_{\beta-}$. The statement then follows from Proposition 5.2.3. ∎

Next, we can repeat the same procedure for every $n \in \mathbb{N}$, getting the scale of Hilbert spaces $\mathcal{H}_n \subset \mathcal{H} \subset \mathcal{H}_{\overline{n}}$, and finally the full scale (5.1). In conclusion, all the spaces in that scale are different and, if one includes the nonstandard spaces, one obtains the complete lattice $\mathcal{F}(\mathcal{D}_{\overline{\infty}}(A))$, which is still totally ordered, i.e., it is a chain.

Such scales generated by a self-adjoint operator are frequent in applications. To mention a few examples :

(i) The Nelson scale associated to a unitary representation U of a Lie group G, built on $A = 1 + \overline{U(\Delta)}$, where Δ is the Nelson operator. This scale is discussed in detail in Section 7.4.

(ii) The hamiltonian scale used in Quantum Field Theory, that is, the scale built on the powers of $1 + H$, where H is the Hamiltonian, which is a positive self-adjoint operator as a consequence of the spectrum condition (Section 7.3). This scale is the natural environment for Euclidean field theory (Section 7.3.3) and the proper definition of unsmeared fields (Section 7.3.4). Also the operator $(-\Delta + m^2)^{1/2}$ and its powers are ubiquitous in scattering theory.

(iii) An interesting scale of Hilbert spaces was used by Ginibre and Velo in their work on the classical field limit of a nonrelativistic many-boson system [105]. Instead of considering the standard scale built on the powers of the number operator N in the (usual) Fock space $\mathcal{H}_F$, they use the scale $\{\mathcal{H}_\delta, \delta \in \mathbb{R}\}$ defined as follows. For every $\delta > 0$, they define the operator $f(\delta, N) = \{\Gamma(N+\delta+1)/\Gamma(N+1)\}^{1/2}$, which is self-adjoint and larger than 1 on the domain $D(N^\delta)$ (here Γ is the usual Gamma function). Then, for $\delta \geqslant 0$, $\mathcal{H}_\delta$ is $D(N_\delta)$ with the norm $\|\psi\|_\delta = \|f(\delta, N)\psi\|$ and its dual is $\mathcal{H}_{\overline{\delta}}$, i.e., the completion of $\mathcal{H}_0 = \mathcal{H}_F$ in the norm $\mathcal{H}_{\overline{\delta}} = \|f(\delta, N)^{-1}\psi\|$. The norm $\|\cdot\|_\delta$ is equivalent to the graph norm of N^δ, but more convenient for the purposes of the authors, in particular for solving perturbatively a Dyson integral equation (the corresponding evolution operators actually belong the algebra $\mathfrak{A}$ of the scale).

5.2.2 Refinement of a Scale of Hilbert Spaces

In the following sections, we will present several techniques for refining the PIP-space associated to a scale of Hilbert spaces, such as the

canonical scale (5.1). Two ingredients are available to that effect, interpolation theory and the spectral theorem for self-adjoint operators. Let us
begin with the former.

We start with the simplest situation, namely the familiar triplet of Hilbert
spaces

$$\mathcal{H}_1 \hookrightarrow \mathcal{H}_0 \hookrightarrow \mathcal{H}_{\bar{1}} , \tag{5.4}$$

where $\mathcal{H}_{\bar{1}} = \mathcal{H}_1^{\times}$ is the conjugate dual of $\mathcal{H}_1$. This scale defines a PIP-space
H_I on $\mathcal{H}_{\bar{1}}$, with index set $I = \{1, 0, \bar{1}\}$, complete lattice $\mathcal{F}$ identical to (5.4)
and the natural partial inner product inherited from $\mathcal{H}_0$. By quadratic inter-
polation, we get a continuous scale of Hilbert spaces between $\mathcal{H}_1$ and $\mathcal{H}_{\bar{1}}$:

$$\mathcal{H}_1 \hookrightarrow \ldots \hookrightarrow \mathcal{H}_\alpha \hookrightarrow \ldots \hookrightarrow \mathcal{H}_0 \hookrightarrow \ldots \hookrightarrow \mathcal{H}_{\bar{\alpha}} \hookrightarrow \ldots \hookrightarrow \mathcal{H}_{\bar{1}} \tag{5.5}$$

where $\mathcal{H}_\alpha = Q_\alpha(\mathcal{H}_1, \mathcal{H}_0)$, $0 \leqslant \alpha \leqslant 1$, and $\mathcal{H}_{\bar{\alpha}} = Q_{1-\alpha}(\mathcal{H}_0, \mathcal{H}_{\bar{1}}) = (\mathcal{H}_\alpha)^{\times}$.
The new scale (5.5) defines another PIP-space $H_{\tilde{I}}$ on $\mathcal{H}_{\bar{1}}$, with index set
$\tilde{I} = [1, -1]$ and the same partial inner product as H_I. Obviously, $H_{\tilde{I}}$ is a
proper refinement of the original PIP-space H_I. The point is that the new
PIP-space $H_{\tilde{I}}$ coincides with the continuous scale constructed above with the
sole spectral theorem. Of course, this is to be expected, since after all the
quadratic interpolation method itself is based on the spectral theorem for
self-adjoint operators.

Remark 5.2.6. (a) The same construction works if $\mathcal{H}_1$, $\mathcal{H}_{\bar{1}}$ are reflexive
Banach spaces, dual of each other, but not for more general spaces. For
instance, Girardeau [106] has shown that no normed space exists between
Schwartz spaces $\mathcal{S}$ and $\mathcal{S}^{\times}$, such that the interpolation property (iii) of
Section 5.1.2 holds for every operator T.

(b) On the example discussed so far, the intrinsic character of the spaces
$\mathcal{H}_\alpha$ stated in item (ii) of Section 5.1.2 can be seen very explicitly. For
instance, the multiplication operators by $(1 + x^2)$ and $(1 + |x| + x^2)$
have the same form domain $\mathcal{H}_1$ and their form graph norms $\langle f | A f \rangle$ are
different, but equivalent. Thus they both define the same intermediate
spaces $\mathcal{H}_\alpha$, with different but equivalent norms.

Now we can go further, with help of the spectral theorem for self-adjoint
operators.

Lemma 5.2.7. *The* PIP*-space* $H_{\tilde{I}}$ *admits infinitely many proper refinements
of type* (H).

Proof. The statement follows from a straightforward application of the spectral theorem to the self-adjoint operator A which interpolates between $\mathcal{H}_1$
and $\mathcal{H}_0$. One has indeed, for each $0 \leqslant \alpha \leqslant 1$,

$$\mathcal{H}_\alpha := Q(A^\alpha) = \left\{ f \in \mathcal{H}_0 : \int_1^\infty s^\alpha \, d\langle f | E(s) f \rangle < \infty \right\}$$

with inner product

$$\langle f|g \rangle_\alpha = \langle f|A^\alpha g \rangle, f, g \in \mathcal{H}_\alpha.$$

Now let φ be any continuous, positive function on $[1, \infty)$ such that $\varphi(t)$ is unbounded for $t \to \infty$, but increases slower than any power $t^\alpha (0 < \alpha \leqslant 1)$. An example is $\varphi(t) = \log t$ $(t \geqslant 1)$. Then $\varphi(A)$ is a well-defined self-adjoint operator, with form domain

$$Q(\varphi(A)) = \left\{ f \in \mathcal{H}_0 : \int_1^\infty (1 + \varphi(s)) \, d\langle f|E(s)f \rangle < \infty \right\}.$$

With the corresponding inner product

$$\langle f|g \rangle_\varphi = \langle f|g \rangle + \langle f|\varphi(A)g \rangle,$$

$Q(\varphi(A))$ becomes a Hilbert space $\mathcal{H}_\varphi$. For every α, $0 < \alpha \leqslant 1$, one has, with proper inclusions and continuous embeddings,

$$\mathcal{H}_\alpha \hookrightarrow \mathcal{H}_\varphi \hookrightarrow \mathcal{H}_0. \tag{5.6}$$

Taking duals, one gets thus a new chain indexed by $\widetilde{\widetilde{I}} := \widetilde{I} \cup \{\varphi, \overline{\varphi}\}$ where $\mathcal{H}_{\overline{\varphi}}$ is the conjugate dual of $\mathcal{H}_\varphi$. One has, in fact, with proper inclusions,

$$\mathcal{H}_\alpha \hookrightarrow \mathcal{H}_{0+} := \bigcup_{\alpha > 0} \mathcal{H}_\alpha \hookrightarrow \mathcal{H}_\varphi \hookrightarrow \mathcal{H}_0 \hookrightarrow \mathcal{H}_{\overline{\varphi}} \hookrightarrow \mathcal{H}_{0-} := \bigcap_{\alpha < 0} \mathcal{H}_\alpha \hookrightarrow \mathcal{H}_{\overline{\alpha}}. \tag{5.7}$$

Indeed, $\mathcal{H}_\varphi, \mathcal{H}_{\overline{\varphi}}$ being Hilbert spaces, they cannot coincide with $\mathcal{H}_{0\pm}$ which are not. Indeed $\mathcal{H}_{0-}$ is a nonnormable Fréchet space, whereas $\mathcal{H}_{0+}$ is a complete DF space. Hence $H_{\widetilde{\widetilde{I}}}$ is a proper refinement of $H_{\widetilde{I}}$. One can go further by interpolating between $\mathcal{H}_\varphi$ and $\mathcal{H}_0$ and then iterating the construction. In the same fashion, additional spaces can be inserted between any fixed $\mathcal{H}_{\alpha_o}$ and all the $\mathcal{H}_\beta$, $\beta > \alpha_o$. Clearly this process can be repeated indefinitely with different functions φ, which proves the assertion. ∎

Let us turn now to an infinite discrete chain H_I

$$\ldots \hookrightarrow \mathcal{H}_2 \hookrightarrow \mathcal{H}_1 \hookrightarrow \mathcal{H}_0 \hookrightarrow \mathcal{H}_{\overline{1}} \hookrightarrow \mathcal{H}_{\overline{2}} \hookrightarrow \ldots \tag{5.8}$$

As before, there exists for each $j = 1, 2, \ldots$ a self-adjoint operator $A_j \geqslant 1$ on $\mathcal{H}_0$ such that $\mathcal{H}_j = Q(A_j)$ and $\|f\|_j^2 = \langle f|A_j f \rangle$. Thus the chain (5.8) is a *scale* if $A_j = (A_1^j), j = 1, 2, \ldots$.

Remark: If we write $A_1 = 1 + A_1'$, with $A_1' \geqslant 0$, then standard estimates show that $(A_1)^j = (1 + A_1')^j$ and $1 + (A_1')^j$ define equivalent norms on $\mathcal{H}_j$, i.e., $\|f\|_j^2 = \langle f|A_j f \rangle$ is equivalent to the form graph norm $\|f\|_j^2 + \langle f|(A_1')^j f \rangle$ of $(A_1')^j$.

We assume now H_I to be a discrete scale ($I \subseteq \mathbb{Z}$). Then quadratic interpolation (or the spectral theorem) yields a unique continuous scale of Hilbert spaces $H_{\tilde{I}}$, with $\tilde{I} = \mathbb{R}$, which is obviously a proper refinement of H_I. The corresponding index set $\tilde{F}$ is the familiar "nonstandard" real line.

Then we proceed exactly as before. For a suitable function φ, we consider the self-adjoint operator $\varphi(A_1)$ and the scale

$$\mathcal{H}_{\varphi,k} := Q\big((\varphi(A_1))^k\big), \quad k \in \mathbb{J}^+ \quad (\mathbb{J} = \mathbb{R} \text{ or } \mathbb{Z}),$$

with norms

$$\langle f | f \rangle_{\varphi,k} = \int \big(1 + \varphi(s)\big)^k \, d\langle f | E(s) f \rangle.$$

Here k is positive and may take either integer values ($\mathbb{J} = \mathbb{Z}$) or arbitrary positive values ($\mathbb{J} = \mathbb{R}$). This yields a new chain, indexed by $\tilde{\tilde{I}} := I \cup \mathbb{J}^+$, with proper inclusions, valid for every $0 < \alpha < 1$ and every $k > 0$:

$$\dots \hookrightarrow \mathcal{H}_1 \hookrightarrow \mathcal{H}_\alpha \hookrightarrow \mathcal{H}_{\varphi,k} \hookrightarrow \mathcal{H}_0 \hookrightarrow \mathcal{H}_{\overline{\varphi,k}} \hookrightarrow \mathcal{H}_{\overline{\alpha}} \hookrightarrow \mathcal{H}_{\overline{1}} \hookrightarrow \dots$$

Then the same procedure may be repeated for each interval $[n-1, n]$ or, more generally, $[\alpha, \beta]$ with $\alpha < \beta$. It can also be iterated or applied with a different function φ. Thus we have proved the following general result:

Proposition 5.2.8. *Every scale of Hilbert spaces possesses infinitely many proper refinements which are themselves chains of Hilbert spaces.*

The standard example is, of course, again obtained with $A_1 = $ multiplication by $(1 + x^2)$ on $\mathcal{H}_0 = L^2(\mathbb{R})$, and $\varphi(s) = \log s$ ($s \geqslant 1$).

It follows from this result that no chain of Hilbert spaces can be maximal; it can always be refined into another such chain. Yet maximal compatibility relations do exist, as a consequence of Zorn's lemma (see Section 1.5). What Proposition 5.2.8 tells us, then, is that the corresponding PIP-spaces have by necessity a more complicated structure than a chain of Hilbert spaces. This result falls in line with the general philosophy behind PIP-spaces. The choice of a given (indexed) PIP-space is dictated by the problem at hand, but never uniquely; there always remains the freedom of refining a given structure in order to answer more precise questions.

5.2.3 Refinement of a LHS and Operator Algebras

We consider again a discrete chain of Hilbert spaces $\{\mathcal{H}_n, n \in \mathbb{Z}\}$ or any symmetric subset H_I of that scale ($I \subseteq \mathbb{Z}$), in particular, the pair $\{\mathcal{H}_n, \mathcal{H}_{\overline{n}}\}$. A useful object for the study of operators in H_I is the *-algebra $\mathfrak{A}(H_I)$, introduced in Section 3.3.3, which consists of all the operators that map every $\mathcal{H}_n, n \in I$, continuously into itself. Let $H_{\widetilde{I}}$ be the continuous scale obtained from H_I by quadratic interpolation. Then the interpolation property (iii) of Section 5.1.2 implies that every operator $T \in \mathfrak{A}(H_I)$ maps each $\mathcal{H}_\alpha, \alpha \in \widetilde{I}$, also continuously into itself. In other words, the algebras $\mathfrak{A}(H_I)$ and $\mathfrak{A}(H_{\widetilde{I}})$ coincide. In particular, they have the same idempotent elements, i.e., orthogonal projections: $\mathrm{Proj}(H_I) = \mathrm{Proj}(H_{\widetilde{I}})$. This in turn implies that the orthocomplemented subspaces of the chains H_I and $H_{\widetilde{I}}$ are in one-to-one correspondence with each other. This was known so far only for the finite dimensional subspaces, as a result of Proposition 3.4.18, since a finite dimensional subspace is orthocomplemented if and only if it is contained in $V^\#$.

Remark 5.2.9. Let P be a projection in H_I. This means, in particular, that $P\mathcal{H}_1 \subset \mathcal{H}_1$ or $A^{1/2}PA^{-1/2}$ is bounded in $\mathcal{H}_0$. This is certainly true if P commutes with A (then P is absolute), but that condition is not necessary. Take, for instance, a rank-1 projection on $\mathcal{H}_1, P = |\psi\rangle\langle\psi|$ with $\psi \in \mathcal{H}_1$, i.e., $\psi = A^{-1/2}\varphi$ for a unique $\varphi \in \mathcal{H}_0$. Then $A^{1/2}PA^{-1/2} = |\varphi\rangle\langle\varphi|A^{-1}$, which is obviously bounded in $\mathcal{H}_0$. On the other hand, P does not necessarily commute with A. Indeed, $AP = |A\psi\rangle\langle\psi|, PA = |\psi\rangle\langle A\psi|$. So if χ is such that $\langle\psi|\chi\rangle = 0$, but $\langle A\psi|\chi\rangle \neq 0$ (this is always possible unless ψ is an eigenvector of A, in which case $AP = PA$), then $AP = 0$, but $PA \neq 0$. In other words, H_I contains nondiagonal nonabsolute projections.

5.2.4 The LHS Generated by a Self-Adjoint Operator

In fact, one can go much further with help of the spectral theorem. Indeed, a positive self-adjoint operator $A \geqslant 1$ in a Hilbert space $\mathcal{H}$ does not only generate scales or chains of Hilbert spaces, but a full lattice, that is, a LHS $H_I(A)$, as we show now.

5.2.4.1 Constructing the LHS $H_I(A)$

Thus, let $A \geqslant 1$ be a positive self-adjoint operator in $\mathcal{H}$, with dense domain $D(A)$, and spectrum $\sigma(A)$. We denote by Σ the set of all real valued functions ϕ defined on the spectrum $\sigma(A)$, which are measurable with respect to the spectral measure $E(\cdot)$ of A, and such that ϕ and $\check{\phi} := 1/\phi$ are bounded on every bounded set. For every $\phi \in \Sigma$, we denote by $\mathcal{H}_\phi$ the Hilbert space obtained as completion of $D(\phi(A))$ with respect to the inner product:

$$\langle f|g\rangle_\phi := \langle\phi(A)f|\phi(A)g\rangle.$$

For an arbitrary positive measurable function ϕ, the restriction of the inner product of $\mathcal{H}$ to $D(\phi(A)) \times D(\check{\phi}(A))$ extends uniquely to a continuous sesquilinear form on $\mathcal{H}_\phi \times \mathcal{H}_{\check\phi}$, which will also be denoted by $\langle \cdot | \cdot \rangle$. Moreover, using this form, $\mathcal{H}_{\check\phi}$ can be identified with the conjugate dual space of $\mathcal{H}_\phi$, and one has:

$$\mathcal{H}_\phi \hookrightarrow \mathcal{H} \hookrightarrow \mathcal{H}_{\check\phi},$$

where $\hookrightarrow$ denotes, as usual, a continuous and dense embedding. We notice that $\check\phi(A) = \phi(A)^{-1}$, thus $D(\check\phi(A)) = \operatorname{Ran} \phi(A)$.

If $\phi, \chi \in \Sigma$, then the corresponding Hilbert spaces $\mathcal{H}_\phi$, $\mathcal{H}_\chi$ need not be comparable. However, the family $\{\mathcal{H}_\phi\}_{\phi \in \Sigma}$ is a lattice of Hilbert spaces. Indeed, let us define in Σ the following partial order:

$$\phi \preccurlyeq \chi \iff \exists \gamma > 0 : \phi \leqslant \gamma \chi.$$

If $\phi \preccurlyeq \chi$, then $\mathcal{H}_\chi \hookrightarrow \mathcal{H}_\phi$ and the continuous dense embedding is given by the identity; one also has $\check\chi \preccurlyeq \check\phi$, and so $\phi \mapsto \check\phi$ is an order-reversing involution in Σ. The set Σ is an involutive lattice, with lattice operations $\vee$ and $\wedge$ defined as

$$(\phi \vee \chi)(x) := \max\{\phi(x), \chi(x)\} \quad \text{and} \quad (\phi \wedge \chi)(x) := \min\{\phi(x), \chi(x)\}.$$

The lattice $\{\mathcal{H}_\phi\}_{\phi \in \Sigma}$ is a LHS. The PIP-space is given by the inductive limit $H_I(A)$ of the family $\{\mathcal{H}_\phi\}_{\phi \in \Sigma}$, the compatibility relation is defined by

$$f, g \in \mathcal{H}_I, \ f \# g \iff \text{there exists } \phi \in \Sigma \text{ such that } f \in \mathcal{H}_\phi \text{ and } g \in \mathcal{H}_{\check\phi},$$

and the partial inner product $\langle f | g \rangle$ is defined on compatible pairs.
Furthermore, the set

$$\mathcal{E}_A = \{E(\Delta)f : f \in \mathcal{H}, \ \Delta \text{ a Borel set of } \mathbb{R}\}$$

is a dense subset of $D(\phi(A))$ for all $\phi \in \Sigma$. Indeed, by the spectral theorem,

$$D(\phi(A)) = \left\{ f \in \mathcal{H} : \int_{\sigma(A)} \phi(\lambda)^2 \, d\langle E(\lambda)f | f \rangle < \infty \right\};$$

putting $g = E(\Delta)f$, one has

$$\int_{\sigma(A)} \phi(\lambda)^2 \, d\langle E(\lambda)E(\Delta)f | E(\Delta)f \rangle = \int_{\Delta \cap \sigma(A)} \phi(\lambda)^2 \, d\langle E(\lambda)f | f \rangle < \infty.$$

It follows that $\mathcal{E}_A$ is dense in the projective limit $\mathcal{H}_I^\# = \bigcap_{\phi \in \Sigma} \mathcal{H}_\phi$ and in the space $\mathcal{H}_I$, that is, the PIP-space $\mathcal{H}_I$ is nondegenerate. Finally, the family $\{\mathcal{H}_\phi\}_{\phi \in \Sigma}$ is stable under intersection. Indeed, one has $D(\phi(A) + \chi(A)) = D(\phi(A)) \cap D(\chi(A))$. Since the topology generated by the norm $\| \cdot \|_{\phi + \chi}$ is equivalent to the projective topology on $D(\phi(A)) \cap D(\chi(A))$ defined by the norms $\|\cdot\|_\phi$ and $\|\cdot\|_\chi$, their completions $\mathcal{H}_{\phi+\chi}$ and $\mathcal{H}_\phi \cap \mathcal{H}_\chi$ can be identified, that is, $\mathcal{H}_\phi \cap \mathcal{H}_\chi \simeq \mathcal{H}_{\phi+\chi}$.

5.2.4.2 Norms of Operators in $H_I(A)$

We will now consider operators on the LHS $H_I(A)$ and derive several convexity properties for the norms of their representatives. Let $X \in \mathrm{Op}(H_I(A))$ be an operator on $H_I(A)$, with a representative $X_{\chi,\phi} \in \mathcal{B}(\mathcal{H}_\phi, \mathcal{H}_\chi)$. The operator $\chi(A) X_{\chi,\phi} \breve{\phi}(A)$ is easily shown to be bounded in $\mathcal{H}$ and, by definition, $\|X_{\chi,\phi}\|_{\phi,\chi} = \|\chi(A) X_{\chi,\phi} \breve{\phi}(A)\|$.

Given $\psi \in \Sigma$, we can define the multiplication map $L_\psi : \Sigma \to \Sigma$ as

$$L_\psi : \phi \mapsto \psi\phi.$$

This map defines a correspondence $\widetilde{L}_\psi$ in the lattice $\{\mathcal{H}_\phi\}_{\phi \in \Sigma}$:

$$\mathcal{H}_\phi \mapsto \mathcal{H}_{L_\psi(\phi)} = \mathcal{H}_{\psi\phi}.$$

In general, the Hilbert spaces $\mathcal{H}_\phi$ and $\mathcal{H}_{\psi\phi}$ are not comparable (since ϕ and $\psi\phi$ are not, in general), unless ψ is constant, $\psi(x) = \gamma > 0$. However, the following result can be proved.

Proposition 5.2.10. *Let $X \in \mathrm{Op}(H_I(A))$. Let us consider $\phi, \chi, \psi \in \Sigma$ with $\psi > 0$ such that $(\phi, \chi) \in j(X)$ and $(\psi\phi, \psi\chi) \in j(X)$. Then for every $t \in [0, 1]$, one has $(\psi^t\phi, \psi^t\chi) \in j(X)$. Furthermore,*

$$\|X_{\psi^t\chi, \psi^t\phi}\| \leqslant \max\{\|X_{\chi,\phi}\|, \|X_{\psi\chi, \psi\phi}\|\}. \tag{5.9}$$

The proof is rather technical, so we refer the reader to the original paper [185].

Proposition 5.2.11. *Let $X \in \mathrm{Op}(H_I(A))$. Let us consider $\phi, \chi, \psi \in \Sigma$ with $\psi > 0$ such that $(\phi, \chi) \in j(X)$ and $(\psi\phi, \psi\chi) \in j(X)$. Then, the following inequality holds for every $t \in [0, 1]$:*

$$\|X_{\psi^t\chi, \psi^t\phi}\| \leqslant \|X_{\chi,\phi}\|^{1-t} \|X_{\psi\chi, \psi\phi}\|^t. \tag{5.10}$$

Proof. By Proposition 5.2.10, one has $(\psi^t\phi, \psi^t\chi) \in j(X)$ and it follows from the proof that the function $z \mapsto \phi_z(B) = \psi^z(A) B \psi^{-z}(A)$ is bounded and continuous on the strip $S_1 := \{z \in \mathbb{C} : 0 \leqslant \mathrm{Re}\, z \leqslant 1\}$ and analytic on the interior $\overset{\circ}{S}_1$. Furthermore, for every $z \in \mathbb{C}$ such that $\mathrm{Re}\, z = 0$, one has:

$$\|\phi_z(B)\| = \|\psi^{iy}(A) \chi(A) X_{\chi,\phi} \breve{\phi}(A) \psi^{-iy}(A)\|$$
$$\leqslant \|\exp i(y \log \psi(A))\| \cdot \|\chi(A) X_{\chi,\phi} \breve{\phi}(A)\| \cdot \|\exp -i(y \log \psi(A))\|$$
$$= \|\chi(A) X_{\chi,\phi} \breve{\phi}(A)\| = \|X_{\chi,\phi}\|.$$

On the other hand, for every $z \in \mathbb{C}$ such that $\operatorname{Re} z = 1$, one has:

$$\|\phi_z(B)\| = \|\psi(A)\,\psi^{iy}(A)\,\chi(A)X_{\psi\chi,h\phi}\check{\phi}(A)\,\psi^{-1}(A)\,\psi^{-iy}(A)\|$$
$$\leqslant \|\exp(iy\log\psi(A))\| \cdot \|\psi(A)\,\chi(A)\,X_{\psi\chi,\psi\phi}\check{\phi}(A)\,\psi^{-1}(A)\| \cdot \|\exp(-iy\log\psi(A))\|$$
$$= \|(\psi\chi)(A)\,X_{\psi\chi,\psi\phi}(\psi\check{\phi})(A)\| = \|X_{\psi\chi,\psi\phi}\|.$$

Then, by Hadamard's three lines theorem, one has

$$\|\psi^z(A)\,\chi(A)X_{\chi,\phi}\check{\phi}(A)\,\psi^{-z}(A)\| \leqslant \|X_{\chi,\phi}\|^{1-\operatorname{Re} z} \cdot \|X_{\psi\chi,\psi\phi}\|^{\operatorname{Re} z}.$$

The statement follows by putting $z = t \in [0,1]$. ∎

Corollary 5.2.12. *Under the same assumptions as in Proposition 5.2.11, the following inequality holds:*

$$\|X_{\psi^t\chi,\psi^t\phi}\| \leqslant (1-t)\|X_{\chi,\phi}\| + t\|X_{\psi\chi,\psi\phi}\|, \text{ for every } t \in [0,1]. \tag{5.11}$$

Proof. The statement is an easy consequence of (5.10) and of Young's inequality: for every $x, y \in \mathbb{R}^+$, and for every $r, s \in \mathbb{R}$ such that $\frac{1}{r} + \frac{1}{s} = 1$, one has:

$$x \cdot y \leqslant \frac{x^r}{r} + \frac{x^s}{s}.$$

With the substitution $s = \frac{1}{t}$, $r = \frac{1}{1-t}$, it follows that:

$$\|X_{\psi^t\chi,\psi^t\phi}\| \leqslant \|X_{\chi,\phi}\|^{1-t}\|X_{\psi\chi,\psi\phi}\|^t \leqslant (1-t)\|X_{\chi,\phi}\| + t\|X_{\psi\chi,\psi\phi}\|. \quad ∎$$

5.2.4.3 Back to the Canonical Scale of Hilbert Spaces

We can recover the continuous scale of Hilbert spaces (5.1) by considering the functions $\phi_\alpha = x^\alpha$, $\alpha \in \mathbb{R}$, that clearly belong to Σ. In this case, considering the projective limit $\mathcal{D}^\infty(A) = \bigcap_{\alpha\in\mathbb{R}} D(A^\alpha)$, its dual space $\mathcal{D}_{\overline{\infty}}(A)$ plays the role of the LHS denoted by $\mathcal{H}_I$ up to now.

In this case, if $X : \mathcal{D}_{\overline{\infty}}(A) \to \mathcal{D}_{\overline{\infty}}(A)$, we denote by $X_{\beta,\alpha} : \mathcal{H}_\alpha \to \mathcal{H}_\beta$ its representatives, and write $\|X_{\beta,\alpha}\| := \|X_{\beta,\alpha}\|_{\beta,\alpha} := \|A^\beta X_{\beta,\alpha}A^{-\alpha}\|$. Let us define the set

$$\mathsf{z}(X) := \{(\alpha,\beta) : (x^\alpha, x^\beta) \in \mathsf{j}(X)\} \subseteq \mathbb{R}^2.$$

Then an immediate consequence of Proposition 5.2.11 is the following:

Corollary 5.2.13. *Let $X : \mathcal{D}_{\overline{\infty}}(A) \to \mathcal{D}_{\overline{\infty}}(A)$ and $\alpha,\beta,\gamma \in \mathbb{R}$ such that $\alpha \leqslant \beta$, $(\alpha,\beta) \in \mathsf{z}(X)$, and $(\alpha+\gamma,\beta+\gamma) \in \mathsf{z}(X)$. Then, for every $t \in [0,1]$, one has $(\alpha+\gamma t,\beta+\gamma t) \in \mathsf{z}(X)$. Furthermore,*

$$\|X_{\beta+\gamma t,\alpha+\gamma t}\| \leqslant \|X_{\beta,\alpha}\|^{1-t}\|X_{\beta+\gamma,\alpha+\gamma}\|^t. \tag{5.12}$$

Proof. The proof follows from Proposition 5.2.11, putting $\phi = x^\alpha$, $\psi = x^\beta$, $\chi = x^\gamma$. ∎

In a similar way, one has the following convexity result:

Corollary 5.2.14. *The following inequality holds:*

$$\|X_{\beta+\gamma t, \alpha+\gamma t}\| \leqslant (1-t)\|X_{\beta,\alpha}\| + t\|X_{\beta+\gamma,\alpha+\gamma}\|, \ \forall t \in [0,1]. \tag{5.13}$$

Corollary 5.2.14 has some interesting consequences in the case of operators acting on the discrete scale of Hilbert spaces

$$\mathcal{D}^\infty(A) \subset \ldots \hookrightarrow \mathcal{H}_2 \hookrightarrow \mathcal{H}_1 \hookrightarrow \mathcal{H}_0 = \mathcal{H} \hookrightarrow \mathcal{H}_{\overline{1}} \hookrightarrow \mathcal{H}_{\overline{2}} \hookrightarrow \ldots \mathcal{D}_{\overline{\infty}}(A)$$

For example, by (5.13), we can compare a norm $\|X_{n,n}\|$ with the next one, putting $\alpha = \beta = 1$, $\gamma = n$ and $t = \frac{n-1}{n}$:

$$\|X_{n,n}\| \leqslant \frac{1}{n}\|X_{1,1}\| + \frac{n-1}{n}\|X_{n+1,n+1}\|$$

(this result can be easily proved by an induction argument). Similarly, putting $\alpha = \beta = 1$, $\gamma = n+p-1$, with $p \in \mathbb{N}$ and $t = \frac{n-1}{n+p-1}$ one gets:

$$\|X_{n,n}\| \leqslant \frac{1}{n+p-1}\Big[p\|X_{1,1}\| + (n-1)\|X_{n+p,n+p}\|\Big].$$

For $n = 2$, in particular, this gives:

$$\|X_{2,2}\| \leqslant \frac{1}{p+1}\Big[p\|X_{1,1}\| + \|X_{2+p,2+p}\|\Big].$$

From this estimate we can derive an interesting result for the algebra $\mathfrak{A} := \mathfrak{A}(\mathcal{D}_{\overline{\infty}}(A))$ of totally regular operators, described in Section 3.3.3.

Corollary 5.2.15. *If $X \in \mathfrak{A}$ and $\|X_{2,2}\| > \|X_{1,1}\|$, one has*

$$\lim_{n\to\infty} \|X_{n,n}\| = \infty.$$

Remark 5.2.16. In the case of the discrete scale of Hilbert spaces $\{\mathcal{H}_n\}_{n\in\mathbb{N}}$, we have, by Corollary 5.2.15, that $\|X_{2,2}\| > \|X_{1,1}\|$ implies that $X \in \mathfrak{A} \setminus \mathfrak{B}$.

5.3 Refinement of General Chains and Lattices of Hilbert Spaces

If we turn to cases more general than a scale, the situation gets more involved. Let us start again with a finite chain:

$$\mathcal{H}_2 \hookrightarrow \mathcal{H}_1 \hookrightarrow \mathcal{H}_0 \hookrightarrow \mathcal{H}_{\overline{1}} \hookrightarrow \mathcal{H}_{\overline{2}}. \tag{5.14}$$

For $j = 1, 2$ there exists a unique self-adjoint operator $A_j \geqslant 1$ in $\mathcal{H}_0$ such that $\mathcal{H}_j = Q(A_j)$, but we assume $A_2 \neq (A_1)^2$, i.e., the chain (5.14) is *not* a scale. First we can easily characterize such chains.

Lemma 5.3.1. $\mathcal{H}_2 \subset \mathcal{H}_1$ *if and only if* $A_1^{1/2} A_2^{-1/2}$ *is bounded.*

Proof. (a) Let $A_1^{1/2} A_2^{-1/2}$ be bounded, Since $A_2 \geqslant 1$, $A_2^{-1/2}$ is bounded and $\mathcal{H}_2 = D(A_2^{1/2}) = A_2^{-1/2} \mathcal{H}_0$. Since $A_1^{1/2} A_2^{-1/2}$ is well defined for every $f \in \mathcal{H}_0$, it follows that $D(A_2^{1/2}) \subset D(A_1^{1/2}) = \mathcal{H}_1$.
(b) As for the converse, notice that $A_1^{1/2}$ and $A_2^{-1/2}$ are closed operators, $A_1^{-1/2}$ is bounded, and hence $A_1^{1/2} A_2^{-1/2}$ is closed. Now, if $\mathcal{H}_2 \subset \mathcal{H}_1$, $A_1^{1/2} A_2^{-1/2}$ is defined on all of $\mathcal{H}_0$ and therefore it is a bounded operator. ∎

Notice that the embedding $\mathcal{H}_2 \to \mathcal{H}_1$ is continuous, under the same conditions, if $\mathcal{H}_j, j = 1, 2$, is equipped, as usual, with the norm $\langle f|f \rangle_j = \langle f|A_j f \rangle$.

Let now B be the interpolating operator between $\mathcal{H}_2$ and $\mathcal{H}_1$. B has the form domain $\mathcal{H}_2$, it is self-adjoint in $\mathcal{H}_1$ (but not necessarily in $\mathcal{H}_0$), and it is larger than 1. lt verifies the relations

$$\langle f|f \rangle_2 = \langle f|Bf \rangle_1 = \langle B^{1/2} f|B^{1/2} f \rangle_1 \quad (f \in \mathcal{H}_1).$$

By the uniqueness of the interpolation operator, one has $A_2 = (B^{1/2})^* A_1 B^{1/2}$, where $(B^{1/2})^*$ is the adjoint of $B^{1/2}$ in $\mathcal{H}_0$. Indeed, one gets

$$\langle f|f \rangle_2 = \langle f|Af \rangle_0$$
$$= \langle B^{1/2} f|B^{1/2} f \rangle_1 = \langle B^{1/2} f|AB^{1/2} f \rangle_0 = \langle f|(B^{1/2})^* AB^{1/2} f \rangle_0.$$

Thus we can interpolate between $\mathcal{H}_2$ and $\mathcal{H}_1$ with B, between $\mathcal{H}_1$ and $\mathcal{H}_0$ with A_1 and also directly between $\mathcal{H}_2$ and $\mathcal{H}_0$ with A_2. Defining the interpolating spaces as follows,

$$\left.\begin{array}{l} \mathcal{H}_{20}^\alpha = Q(A_2^\alpha) \text{ in } \mathcal{H}_0 \\ \mathcal{H}_{10}^\alpha = Q(A_1^\alpha) \text{ in } \mathcal{H}_0 \\ \mathcal{H}_{21}^\alpha = Q(B^\alpha) \text{ in } \mathcal{H}_1 \end{array}\right\} \quad (0 \leqslant \alpha \leqslant 1).$$

we obtain the situation depicted in Fig. 5.1. So the question arises, how do the two chains $\{\mathcal{H}_{21}^\alpha, \mathcal{H}_{10}^\beta\}$ and $\{\mathcal{H}_{20}^\alpha\}$, both between $\mathcal{H}_2$ and $\mathcal{H}_0$, compare to each other?

Fig. 5.1 Interpolating in
a finite chain of Hilbert
spaces

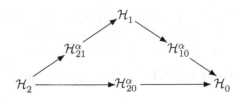

We have to distinguish two cases:

(1) $\mathcal{H}_1 = \mathcal{H}_{20}^\gamma \equiv Q(A_2^\gamma)$ *for some* γ $(0 < \gamma < 1)$. Equality here means equality as vector spaces with equivalent norms (comparable norms would be sufficient, of course). By property (ii) of Section 5.1.2, we can replace the norm of $\mathcal{H}_1$ by that of $Q(A_2^\gamma)$ without changing any of the interpolation spaces $\mathcal{H}_{21}^\alpha, \mathcal{H}_{10}^\beta$. This amounts to writing $A_1 = A_2^\gamma, B = A_2^{1-\gamma}$. In that case, there is only one continuous interpolating scale between $\mathcal{H}_2$ and $\mathcal{H}_0$ as follows from the interpolation property (iii) of Section 5.1.2 applied to the identity operator.

(2) *There is no* γ *such that* $\mathcal{H}_1 = Q(A_2^\gamma)$. In that case, the three scales $\{\mathcal{H}_{21}^\alpha\}, \{\mathcal{H}_{10}^\alpha\}, \{\mathcal{H}_{20}^\gamma\}$ generate, by intersection and vector sum, a genuine lattice, i.e., *not* a chain.

Putting all together, we obtain the following result.

Proposition 5.3.2. *Let H_1 be the following discrete finite chain of Hilbert spaces:*

$$\mathcal{H}_2 \hookrightarrow \mathcal{H}_1 \hookrightarrow \mathcal{H}_0 \hookrightarrow \mathcal{H}_{\bar{1}} \hookrightarrow \mathcal{H}_{\bar{2}}.$$

Let $A_j (j = 1, 2)$ be the interpolating operator between $\mathcal{H}_j$ and $\mathcal{H}_0$. Let $H_{\bar{I}}$ be the lattice generated from H_I by quadratic interpolation. Then:

(i) *If for some γ, $0 < \gamma < 1, Q(A_2^\gamma) = \mathcal{H}_1$, with equivalent norms, then $H_{\bar{I}}$ is a continuous scale of Hilbert spaces.*

(ii) *If no such γ exists, $H_{\bar{I}}$ is a continuous involutive lattice, but not a chain.*

In both cases $H_{\bar{I}}$ is a proper refinement of H_I.

This construction shows clearly that quadratic forms are more natural than operators in a PIP-space context. For instance, one has:

$$\mathcal{H}_{20}^\alpha \cap \mathcal{H}_{10}^\beta = Q(A_2^\alpha) \cap Q(A_1^\beta) = Q(A_2^\alpha \dotplus A_1^\beta),$$

i.e., the form domain of the *form sum* $\dotplus$ of the two operators.

We conclude this discussion with an example. Let $\mathcal{H}_0 = L^2(\mathbb{R}, dx)$, $A_1 = 1 + x^2$, $A_2 = 2 + x^2 + p^2$, with $p = -i\frac{d}{dx}$. Then $B = 1 + (1 + x^2)^{-1}(1 + p^2)$, which is indeed self-adjoint in $\mathcal{H}_1 = L^2(\mathbb{R}, (1 + x^2)dx)$. Thus the two chains generated by (B, A_1) and A_2, respectively, have no element in common, and so generate a genuine lattice, with typical elements of the form

$$\mathcal{H}_{20}^{\alpha} \cap \mathcal{H}_{10}^{\beta} = Q(A_2^{\alpha} \dotplus A_1^{\beta})$$

$$= Q((2 + x^2 + p^2)^{\alpha} \dotplus (1 + x^2)^{\beta}).$$

The same argument applies *a fortiori* to an infinite chain; so we may state in general:

Proposition 5.3.3. *Let H_I be a discrete chain of Hilbert spaces $\{\mathcal{H}_n, n \in I \subseteq \mathbb{Z}\}$. Then quadratic interpolation between $\mathcal{H}_j (j > 0)$ and $\mathcal{H}_0$ generates a* PIP-*space $H_{\tilde{I}}$ of type (H) with the following structure:*

$$V^{\#} = \bigcap_{n \in I} \mathcal{H}_n \subset \{\mathcal{H}_{\alpha}\} \subset \mathcal{H}_0 \subset \{\mathcal{H}_{\overline{\alpha}}\} \subset V = \bigcup_{n \in I} \mathcal{H}_n,$$

where both $\{\mathcal{H}_{\alpha}\}$ and $\{\mathcal{H}_{\overline{\alpha}}\}$ are infinite continuous lattices. These two lattices are chains if, and only if, for every $j = 1, 2, 3, \ldots$, there exists $\gamma_j (0 < \gamma_j < 1)$ such that $Q(A_j^{\gamma_j}) = \mathcal{H}_j$, with equivalent norms, where A_j is the interpolating operator between $\mathcal{H}_j$ and $\mathcal{H}_0$. In any case, $H_{\tilde{I}}$ is a proper refinement of H_I.

Combining this result with the construction of Section 5.2.2, we see that every chain of Hilbert spaces actually possesses arbitrarily many proper refinements by LHSs, which are in general no longer chains. *A fortiori*, the situation gets worse if we start from a genuine lattice of Hilbert spaces, not even a chain. We will not go into the details, but only quote the final result.

Theorem 5.3.4. *Every countable LHS admits infinitely many proper refinements, each of which is a LHS.*

This result underlines once again the versatility of the PIP-space approach. The choice of a given structure is to a large extent a matter of taste and commodity, even for a given problem. For instance, the example discussed above leads to a whole class of LHSs built around the Schwartz triplet $\mathcal{S} \subset L^2 \subset \mathcal{S}^{\times}$; among them one finds scales, discrete or continuous, such as $\{Q((1 + p^2 + x^2)^{\alpha}), \alpha \in \mathbb{Z} \text{ or } \mathbb{R}\}$, as well as various lattices built on powers of $(1 + x^2)$ and $(1 + p^2)$. It should be noticed, however, that for a given Fréchet space Φ, there does not always exist a *scale* interpolating between Φ, and $\Phi^{\times}$. It may be necessary to use the powers of several self-adjoint operators for reproducing the topology of Φ. Thus the general case discussed in this section is necessary, even for countably Hilbert spaces.

5.4 From Rigged Hilbert Spaces to PIP-Spaces

The scale of Hilbert spaces generated by a self-adjoint operator A, discussed in Section 5.2.1, may also be regarded from another point of view. Indeed, the space $\mathcal{D}^{\infty}(A)$, endowed with the topology t_A, is one end of the familiar triplet

$$\mathcal{D}^{\infty}(A) \hookrightarrow \mathcal{H} \hookrightarrow \mathcal{D}_{\overline{\infty}}(A),$$

where, as usual, $\hookrightarrow$ stands for continuous embedding with dense range and $\mathcal{D}_{\overline{\infty}}(A)$ is the conjugate dual of $\mathcal{D}^{\infty}(A)$.

This triplet is certainly one of the best known structures enlarging Hilbert space: the rigged Hilbert Spaces or Gel'fand triplets, familiar from the theory of distributions. Of course, a RHS is also a PIP-space, but its structure is very poor, since it consists of three spaces only. In the simple case where the triplet is obtained from a self-adjoint operator A, as we have seen so far, this structure can be *refined* and the final result is a scale of Hilbert spaces. In a similar way, starting from an arbitrary RHS, arises the desire to enrich it, by adding, in appropriate way, intermediate spaces. The notion of interspace is the answer to that query. We will now describe this approach in a fairly detailed fashion.

5.4.1 Rigged Hilbert Spaces and Interspaces

Let $\mathcal{D}$ be a dense subspace of a Hilbert space $\mathcal{H}$. Let us endow $\mathcal{D}$ with a locally convex topology t, finer than that induced on $\mathcal{D}$ by the Hilbert norm and let $\mathcal{D}^{\times}$ be its topological conjugate dual. Then $\mathcal{D}^{\times}$ contains $\mathcal{H}$, in the sense that $\mathcal{H}$ can be identified with a subspace of $\mathcal{D}^{\times}$. Indeed, for every $h \in \mathcal{H}$ the conjugate linear functional $\widehat{h}(f) = \langle f|h \rangle$, $f \in \mathcal{D}$, is continuous with respect to t. The map $h \in \mathcal{H} \mapsto \widehat{h} \in \mathcal{D}^{\times}$ is linear, injective and continuous. A common simplification, which we will adopt from now on, consists in identifying the element $h \in \mathcal{H}$ with the conjugate linear functional $\widehat{h} \in \mathcal{D}^{\times}$. With this identification, if we denote with $< \cdot, \cdot >$ the sesquilinear form that puts $\mathcal{D}$ and $\mathcal{D}^{\times}$ in conjugate duality (i.e., $< f, F >= F(f)$ for every $f \in \mathcal{D}$ and $F \in \mathcal{D}^{\times}$), it follows that $< f, h >= \langle f|h \rangle$, if $h \in \mathcal{H}$ and $f \in \mathcal{D}$. This means that $< \cdot, \cdot >$ is an extension of the inner product of $\mathcal{D}$; thus, we indicate both of them with the same notation $\langle \cdot|\cdot \rangle$, i.e., we write $F(f) = \langle f|F \rangle$, $f \in \mathcal{D}, F \in \mathcal{D}^{\times}$. The space $\mathcal{D}^{\times}$ is endowed with the strong dual topology $\mathsf{t}^{\times}$ defined by the set of seminorms

$$\Phi \mapsto \|\Phi\|_{\mathcal{M}} := \sup_{\phi \in \mathcal{M}} |\langle \phi|\Phi \rangle|, \tag{5.15}$$

where $\mathcal{M}$ runs over the family of all bounded subsets of $\mathcal{D}[t]$. In this way we get the familiar triplet

$$\mathcal{D} \hookrightarrow \mathcal{H} \hookrightarrow \mathcal{D}^{\times}$$

called a *rigged Hilbert space (RHS)*.

Let now $\mathcal{E}$ be a subspace of $\mathcal{D}^{\times}$, containing $\mathcal{D}$ as subspace. Let us assume that $\mathcal{E}$ is endowed with a locally convex topology $\mathsf{t}_{\mathcal{E}}$ such that

$(t^{\times})_{\restriction \mathcal{D}} \preceq (t_{\mathcal{E}})_{\restriction \mathcal{D}} \preceq t$ and $\mathcal{D}$ is dense in $\mathcal{E}$. Then the identity maps of $\mathcal{D}$ into $\mathcal{E}$ and of $\mathcal{E}$ into $\mathcal{D}^{\times}$ are continuous embeddings with dense range. Hence,

$$\mathcal{D} \hookrightarrow \mathcal{E} \hookrightarrow \mathcal{D}^{\times} \tag{5.16}$$

By duality, the conjugate dual $\mathcal{E}^{\times}$ of $\mathcal{E}[t_{\mathcal{E}}]$ is continuously embedded in $\mathcal{D}^{\times}$ and the embedding has dense range. It is easily seen that $\mathcal{E}^{\times}$ can be identified with the following subspace of $\mathcal{D}^{\times}$

$$\mathcal{E}^{\times} \simeq \{\Phi \in \mathcal{D}^{\times} : \Phi{\restriction}\mathcal{D} \text{ is } t_{\mathcal{E}}\text{-continuous}\}.$$

The space $\mathcal{E}^{\times}$ is endowed with its own strong dual topology $t_{\mathcal{E}^{\times}}$. The space $\mathcal{D}$ is also continuously embedded in $\mathcal{E}^{\times}$ but in this case the image of $\mathcal{D}$ is not necessarily dense in $\mathcal{E}^{\times}$, unless $\mathcal{E}$ is semi-reflexive. Thus $\mathcal{E}^{\times}[t_{\mathcal{E}^{\times}}]$ does not necessarily satisfy (5.16). In order to avoid this difficulty we endow the space $\mathcal{E}$ with the Mackey topology $\tau_{\mathcal{E}} := \tau(\mathcal{E}, \mathcal{E}^{\times})$ and the space $\mathcal{E}^{\times}$ with the Mackey topology $\tau_{\mathcal{E}^{\times}} := \tau(\mathcal{E}, \mathcal{E}^{\times})$. The same can be done, of course, with the spaces $\mathcal{D}$ and $\mathcal{D}^{\times}$ themselves. If (5.16) holds for the initial topologies, then it holds also when each space is endowed with its Mackey topology.

Such subspaces $\mathcal{E}$ will play a crucial role in the construction of PIP-spaces, hence they deserve a special name.

Definition 5.4.1. Let $\mathcal{E}$ be a subspace of $\mathcal{D}^{\times}$ satisfying (5.16) with respect to a given locally convex topology $t_{\mathcal{E}}$ and let $\mathcal{E}^{\times}$ be the conjugate dual of $\mathcal{E}$ with respect to $t_{\mathcal{E}}$. Then the subspaces $\mathcal{E}$ and $\mathcal{E}^{\times}$, each endowed with its own Mackey topology, are called *interspaces*.

Taking into account the previous discussion, the following is obvious.

Lemma 5.4.2. *Let $\mathcal{E}$ be an interspace. The form $b_{\mathcal{E}}(\cdot, \cdot)$ that puts $\mathcal{E}$ and $\mathcal{E}^{\times}$ in conjugate duality is an extension of the inner product of $\mathcal{D}$.*

Let now $\mathsf{F} = \{\mathcal{E}_{\alpha}, \alpha \in I\}$ be a family of interspaces. Assume that the family F is stable under the operation of taking conjugate duals, i.e., if $\mathcal{E}_{\alpha} \in \mathsf{F}$, then $\mathcal{E}_{\alpha}^{\times} \in \mathsf{F}$. For simplifying the notations, we put $\mathcal{E}_{\bar{\alpha}} := \mathcal{E}_{\alpha}^{\times}$. This means, in other words, that we have introduced in I an *involution* $\alpha \mapsto \bar{\alpha}$ in such a way that $\mathcal{E}_{\bar{\alpha}}$ is the conjugate dual of $\mathcal{E}_{\alpha}$. The natural question arises, is it possible to introduce in $\mathcal{D}^{\times}$ a structure of PIP-space in such a way that the members of F are assaying subspaces?

First of all, we need to define a compatibility $\#$ on $\mathcal{D}^{\times}$. The family $\{\mathcal{D}, \{\mathcal{E}_{\alpha}\}, \mathcal{D}^{\times}\}$ does not constitute, in general, an involutive covering of $\mathcal{D}^{\times}$, unless it is stable under finite intersections (Section 1.3). But for this we need that the intersection of two interspaces be still an interspace, which is not always true. This is due to the fact that $\mathcal{D}$ is not necessarily dense in the intersection $\mathcal{E} \cap \mathcal{F}$ of two interspaces $\mathcal{E}, \mathcal{F}$, endowed with the projective topology $\tau_{\mathcal{E}} \wedge \tau_{\mathcal{F}}$. Here is a counterexample.

Example 5.4.3. Let $\mathcal{S}(\mathbb{R})$ denote the Schwartz space of rapidly decreasing C^∞-functions, endowed with its usual topology t and $\mathcal{S}^\times(\mathbb{R})$ its conjugate dual endowed with the strong dual topology $\mathsf{t}^\times$. If (a,b) is an interval of the real line we denote by $W^{k,2}(a,b)$, $k \in \mathbb{Z}$, the usual Sobolev space and by $\|\cdot\|_{k,2}^{(a,b)}$ its norm, defined, for $k \geqslant 0$, as

$$W^{k,2}(a,b) := \{f \in \mathcal{S}^\times(\mathbb{R}) : f^{(j)} \in L^2([a,b], dx),\ j = 0, 1, \ldots, k\}, \qquad (5.17)$$

$$\|f\|_{k,2}^{(a,b)} := \left(\sum_{j=0}^{k} \int_a^b |f^{(j)}(x)|^2 \right)^{1/2}, \qquad (5.18)$$

the space $W^{-k,2}$ being the (conjugate) dual of $W^{k,2}$.

Let $\mathfrak{X}$ and $\mathfrak{X}^\times$ be, respectively, the completions of $\mathcal{S}(\mathbb{R})$ with respect to the norms

$$\|\cdot\|_{\mathfrak{X}} = \left((\|\cdot\|_{1,2}^{(0,+\infty)})^2 + (\|\cdot\|_{-1,2}^{(-\infty,+0)})^2 \right)^{1/2},$$

$$\|\cdot\|_{\mathfrak{X}^\times} = \left((\|\cdot\|_{-1,2}^{(0,+\infty)})^2 + (\|\cdot\|_{1,2}^{(-\infty,+0)})^2 \right)^{1/2}.$$

Then $\mathfrak{X}$ and $\mathfrak{X}^\times$ may be regarded as Hilbert spaces, in conjugate duality to each other. Moreover, the embedding $\mathcal{S}(\mathbb{R}) \hookrightarrow \mathcal{S}^\times(\mathbb{R})$ may be extended to continuous embeddings of $\mathfrak{X} \hookrightarrow \mathcal{S}^\times(\mathbb{R})$ and $\mathfrak{X}^\times \hookrightarrow \mathcal{S}^\times(\mathbb{R})$, so that we may regard $\mathfrak{X}$ and $\mathfrak{X}^\times$ as vector subspaces of $\mathcal{S}^\times(\mathbb{R})$. It is easily seen that $\mathfrak{X} \cap \mathfrak{X}^\times = W^{1,2}(\mathbb{R} \setminus \{0\})$. Since $W^{1,2}(\mathbb{R})$ is a closed subspace of $W^{1,2}(\mathbb{R} \setminus \{0\})$ and $\mathcal{S}(\mathbb{R})$ is dense in $W^{1,2}(\mathbb{R})$, it follows that $\mathfrak{X} \cap \mathfrak{X}^\times$ is not an interspace.

The following statement, based on standard facts of duality theory, characterizes interspaces whose intersection is again an interspace.

Proposition 5.4.4. *Let $\mathcal{E}, \mathcal{F}$ be interspaces. The following statements are equivalent:*

(i) $\mathcal{D}$ is dense in $(\mathcal{E} \cap \mathcal{F})[\tau_\mathcal{E} \wedge \tau_\mathcal{F}]$.
(ii) $\mathcal{D}_{\mathcal{E} \cap \mathcal{F}}^\perp = \{\Phi \in (\mathcal{E} \cap \mathcal{F})^\times : \Phi(f) = 0,\ \forall f \in \mathcal{D}\} = \{0\}$.
(iii) $(\mathcal{E} \cap \mathcal{F})[\tau_{\mathcal{E} \cap \mathcal{F}}]$ is an interspace.
(iv) $(\mathcal{E} \cap \mathcal{F})^\times[\tau_{(\mathcal{E} \cap \mathcal{F})^\times}]$ is an interspace.

Such interspaces are crucial for defining properly the product of two operators in $\mathrm{Op}(V)$. We will study this problem in Section 6.3.1.

Remark 5.4.5. Let us say that a PIP-space structure is compatible with the duality properties of a family of interspaces, whenever the sesquilinear form that puts two interspaces $\mathcal{E}$ and $\mathcal{E}^\times$ in conjugate duality extends the inner product of $\mathcal{D}$. Then, clearly, if we want a family of interspaces to be assaying subspaces of a compatible PIP-space, the intersection of any two of them, endowed with the projective topology, must contain $\mathcal{D}$ as a dense subspace, since this is always true in a PIP-space. Thus if two interspaces do not satisfy this condition, we can conclude that there is no PIP-space structure on $\mathcal{D}^\times$ compatible with the duality of the interspaces.

Definition 5.4.6. A family $\mathfrak{L}$ of interspaces in the rigged Hilbert space $(\mathcal{D}[t], \mathcal{H}, \mathcal{D}^\times[t^\times])$ is called *generating* if

(i) $\mathcal{D} \in \mathfrak{L}$;
(ii) $\forall \mathcal{E} \in \mathfrak{L}$, its conjugate dual $\mathcal{E}^\times$ also belongs to $\mathfrak{L}$;
(iii) $\forall \mathcal{E}, \mathcal{F} \in \mathfrak{L}$, $\mathcal{E} \cap \mathcal{F} \in \mathfrak{L}$.

Theorem 5.4.7. *Assume the rigged Hilbert space $(\mathcal{D}[t], \mathcal{H}, \mathcal{D}^\times[t^\times])$ contains a generating family $\mathfrak{L}$ of interspaces. Then there exist a compatibility $\#$ on $\mathcal{D}^\times$ and a partial inner product $\langle \cdot | \cdot \rangle$ associated to it.*

Proof. The compatibility $\#$ is defined by

$$f \# g \iff (f, g) \in \bigcup_{\mathcal{E} \in \mathfrak{L}} \mathcal{E} \times \mathcal{E}^\times.$$

The partial inner product is defined by

$$\langle f | g \rangle = b_\mathcal{E}(f, g), \ \forall (f, g) \in \mathcal{E} \times \mathcal{E}^\times,$$

where $b_\mathcal{E}$ is the sesquilinear form defined in Lemma 5.4.2. The density of $\mathcal{D}$ in the intersection of any two interspaces also guarantees that this definition does not depend on the choice of $\mathcal{E}$.   ∎

Example 5.4.8. The simplest situations occur, obviously, when the family $\{\mathcal{E}_\alpha\}$ consists of very few elements, as in the following examples.

(i) The RHS as PIP-space Any RHS $(\mathcal{D}, \mathcal{H}, \mathcal{D}^\times)$ may be viewed as a PIP-space, the compatibility being defined by

$$f \# g \iff \{f, g\} \in (\mathcal{D} \times \mathcal{D}^\times) \cup (\mathcal{D}^\times \times \mathcal{D}) \cup (\mathcal{H} \times \mathcal{H}),$$

so that $V = \mathcal{D}^\times$, $V^\# = \mathcal{D}$, and $\mathcal{F}(V) = \{\mathcal{D}, \mathcal{H}, \mathcal{D}^\times\}$.

(ii) Forgetting $\mathcal{H}$...: More generally, if $\mathcal{E}$ is an interspace, such that $\mathcal{D}$ is dense in $\mathcal{E} \cap \mathcal{E}^\times$, one may define a different PIP-space by taking $\mathfrak{L}$ as consisting of the interspaces $\mathcal{D}$, $\mathcal{E}$, $\mathcal{E}^\times$, $\mathcal{E} \cap \mathcal{E}^\times$, $\mathcal{E} + \mathcal{E}^\times$ and $\mathcal{D}^\times$. In this case also $V = \mathcal{D}^\times$, $V^\# = \mathcal{D}$ and $\mathcal{F}(V)$ consists exactly of the same the six elements. It is clear that the Hilbert space $\mathcal{H}$, which is always included in the RHS, is not included in this PIP-space (unless $\mathcal{E} = \mathcal{H}$), although it is necessary for the construction of the whole structure.

5.4.2 Bessel-Like Spaces as Refinement of the LHS Generated by a Single Operator

The canonical scale generated by a single self-adjoint operator, discussed in Section 5.2.1, allows a more general construction coupling it with certain families of Banach spaces. The outcome is an abstract construction that produces the so-called Bessel potential spaces in familiar situations. This is another method for refining the scale, totally independent from the previous ones.

We start by considering the strongly continuous one-parameter unitary group $U(t)$ which has A as infinitesimal generator. This group, as we shall see, extends to the conjugate dual space $\mathcal{D}_{\overline{\infty}}(A)$. Since the sesquilinear form which puts $\mathcal{D}^{\infty}(A)$ and $\mathcal{D}_{\overline{\infty}}(A)$ in conjugate duality is an extension of the inner product of $\mathcal{D}^{\infty}(A)$, we denote both with the same symbol $\langle \cdot | \cdot \rangle$. As it is well-known,

$$D(A) = \left\{ f \in \mathcal{H} : \lim_{t \to 0} \frac{U(t) - 1}{t} f \text{ exists in } \mathcal{H} \right\},$$

$$U(t)f = e^{itA}f, \quad \forall f \in \mathcal{H}.$$

Since $U(t) = e^{itA} : \mathcal{D}^{\infty}(A) \to \mathcal{D}^{\infty}(A)$ continuously, it admits an extension $\widehat{U}(t)$ to the whole $\mathcal{D}_{\overline{\infty}}(A)$, continuous as well. Namely,

$$\langle \widehat{U}(t)F | f \rangle = \langle F | U(-t)f \rangle, \quad F \in \mathcal{D}_{\overline{\infty}}(A), f \in \mathcal{D}^{\infty}(A).$$

For every $\phi \in C_0^{\infty}(\mathbb{R})$ and $f \in \mathcal{H}$, we put

$$T_\phi f = \int_{\mathbb{R}} \phi(t) U(t) f \, dt.$$

By the strong continuity of $U(t)$, the integral on the right-hand side can be taken in Riemann's sense.

Proposition 5.4.9. *For each fixed $\phi \in C_0^{\infty}(\mathbb{R})$, the operator T_ϕ is linear and bounded in $\mathcal{H}$. Moreover, $T_\phi f \in \mathcal{D}^{\infty}(A)$, for every $f \in \mathcal{H}$, and T_ϕ is a continuous operator also from $\mathcal{H}$ into $\mathcal{D}^{\infty}(A)$.*

Proof. Linearity is obvious. $U(t)$ is a strongly continuous group of unitary operators, so the boundedness follows from the estimate

$$\|T_\phi f\| = \left\| \int_{\mathbb{R}} \phi(t) U(t) f \, dt \right\| \leqslant \int_{\mathbb{R}} \|\phi(t) U(t) f\| \, dt$$

$$\leqslant \sup_{\mathbb{R}} |\phi(t)| \int_{\text{supp}\,\phi} \|U(t)f\| \, dt \ \leqslant \ \sup_{\mathbb{R}} |\phi(t)| \, \mu(\text{supp}\,\phi) \, \|f\|,$$

where μ denotes the Lebesgue measure on $\mathbb{R}$. Furthermore, $T_\phi f \in D(A)$ and $AT_\phi f = -iT_{-\phi'}f$, hence $T_\phi f \in D(A^n)$, for every $f \in \mathcal{H}$ and $\forall n \in \mathbb{N}$, so $T_\phi f \in \mathcal{D}^\infty(A)$, $\forall f \in \mathcal{H}$. To prove that $T_\phi : \mathcal{H} \to \mathcal{D}^\infty(A)$ is continuous, we note that

$$A^n T_\phi f = (-i)^n T_{(-1)^n \phi^{(n)}} f = i^n T_{\phi^{(n)}} f.$$

Hence,

$$\|A^n T_\phi f\| = \|T_{\phi^{(n)}}f\| = \left\| \int_\mathbb{R} \phi^{(n)}(t)U(t)f\,dt \right\| \leqslant \int_\mathbb{R} |\phi^{(n)}(t)| \|U(t)f\|\,dt$$
$$= \|f\| \int_\mathbb{R} |\phi^{(n)}(t)|\,dt. \qquad \blacksquare$$

By Proposition 5.4.9, there exists a linear map $T_\phi^\ddagger : \mathcal{D}_{\overline{\infty}}(A) \to \mathcal{H}$ such that

$$\langle T_\phi^\ddagger F | g \rangle = \langle F | T_\phi g \rangle, \ \forall g \in \mathcal{H}, \ \forall F \in \mathcal{D}_{\overline{\infty}}(A).$$

Then $T_\phi^\ddagger$ is an extension of the Hilbert adjoint T_ϕ^* of T_ϕ, that is, $T_\phi^\ddagger \upharpoonright \mathcal{H} = T_\phi^*$.

Now, given $f, g \in \mathcal{D}^\infty(A)$, an explicit expression of T_ϕ^* can be obtained as follows:

$$\langle f | T_\phi g \rangle = \left\langle f \,\Big|\, \int_\mathbb{R} \phi(t)U(t)g\,dt \right\rangle = \int_\mathbb{R} \phi(t)\langle f | U(t)g \rangle\,dt = \int_\mathbb{R} \phi(t)\langle U(-t)f | g \rangle\,dt$$
$$= \left\langle \int_\mathbb{R} \overline{\phi(t)}U(-t)f\,dt \,\Big|\, g \right\rangle = \left\langle \int_\mathbb{R} \overline{\phi(-s)}U(s)f\,ds \,\Big|\, g \right\rangle = \langle T_{\tilde{\phi}} f | g \rangle,$$

where we have put $\widetilde{\phi(t)} = \overline{\phi(-t)}$, $t \in \mathbb{R}$. It follows that $T_\phi^* = T_{\tilde{\phi}}$. Hence we can define an extension $\widehat{T}_\phi$ of T_ϕ to the whole space $\mathcal{D}_{\overline{\infty}}(A)$ by the formula:

$$\langle \widehat{T}_\phi F | h \rangle = \langle F | T_{\tilde{\phi}} h \rangle, \quad F \in \mathcal{D}_{\overline{\infty}}(A), \ h \in \mathcal{H},$$

or, what is the same, $\widehat{T}_\phi := T_\phi^\ddagger$. The map $\widehat{T}_\phi : \mathcal{D}_{\overline{\infty}}(A) \to \mathcal{H}$ is continuous, because it is a continuous extension of T_ϕ.

In a certain sense, $\widehat{T}_\phi$ regularizes the space $\mathcal{D}_{\overline{\infty}}(A)$. Indeed we have

Lemma 5.4.10. *The following statements hold true.*

(i) $\widehat{T}_\phi F \in \mathcal{D}^\infty(A)$, *for any* $F \in \mathcal{D}_{\overline{\infty}}(A)$.
(ii) *The map* $\widehat{T}_\phi : \mathcal{D}_{\overline{\infty}}(A) \to \mathcal{D}^\infty(A)$ *is continuous.*

Proof. (i): Let $h \in D(A)$. Then,

$$\langle \widehat{T}_\phi F | Ah \rangle = \langle F | T_{\tilde{\phi}} Ah \rangle = \langle F | AT_{\tilde{\phi}} h \rangle = -i\langle F | T_{-\tilde{\phi}'} h \rangle = \langle -i\widehat{T}_{\phi'} F | h \rangle.$$

Hence $\widehat{T}_\phi F \in D(A^*) = D(A)$ and $A\widehat{T}_\phi F = -i\widehat{T}_{\phi'}F$. This implies that $\widehat{T}_\phi F \in \mathcal{D}^\infty(A)$, and $A^n \widehat{T}_\phi F = (-i)^n \widehat{T}_{\phi^{(n)}} F$, for any integer n.

(ii) is a consequence of (i) and of the continuity of $\widehat{T}_\phi$ from $\mathcal{D}_{-\infty}(A)$ into $\mathcal{H}$. ∎

Remark 5.4.11. If $\phi \in \mathcal{D}^\infty(A)$ and $F \in \mathcal{D}_{\overline{\infty}}(A)$, then $\widehat{T}_\phi F$ can be viewed as a *generalized convolution* induced by A. Indeed, it coincides with the ordinary convolution when $A = i\frac{d}{dx}$.

Proposition 5.4.12. *Let* $j \in C_0^\infty(\mathbb{R})$ *be a regularizing function and* j_ϵ *the corresponding approximate identity. Then the net* $\widehat{T}_{j_\epsilon}F$ *converges weakly to* F *in* $\mathcal{D}_{\overline{\infty}}(A)$.

Proof. By Lemma 5.4.10, $\widehat{T}_{j_\epsilon}F \in \mathcal{D}^\infty(A)$ for all $F \in \mathcal{D}_{\overline{\infty}}(A)$; hence for the continuity of F, one has, for some $\gamma > 0$,

$$|\langle\phi|\widehat{T}_{j_\epsilon}F - F\rangle| = |\langle\phi|\widehat{T}_{j_\epsilon}F\rangle - \langle\phi|F\rangle| = |\langle T_{\widetilde{j}_\epsilon}\phi|F\rangle - \langle\phi|F\rangle| \leqslant \gamma\|A^n(T_{\widetilde{j}_\epsilon}\phi - \phi)\|,$$

where $\widetilde{j}_\epsilon(t) := \overline{j_\epsilon(-t)}$ is also an approximate identity.

Now we notice that, for fixed $\phi \in C_0^\infty(\mathbb{R})$, the following representation holds:

$$\widehat{T}_\phi F = \int_{\mathbb{R}} \phi(t)\widehat{U}(t)F dt,$$

$\widehat{U}(t)$ being the continuous extension of $U(t)$ to $\mathcal{D}_{\overline{\infty}}(A)$. Indeed, we have

$$\langle F|\widehat{T}_{\widetilde{\phi}}h\rangle = \left\langle F\Big|\int_{\mathbb{R}}\widetilde{\phi}(t)U(t)h dt\right\rangle = \int_{\mathbb{R}}\widetilde{\phi}(t)\langle F|U(t)h\rangle dt = \int_{\mathbb{R}}\widetilde{\phi}(t)\langle\widehat{U}(-t)F|h\rangle dt$$

$$= \left\langle\int_{\mathbb{R}}\phi(-t)\widehat{U}(-t)F dt\Big|h\right\rangle = \left\langle\int_{\mathbb{R}}\phi(t)\widehat{U}(t)F dt\Big|h\right\rangle$$

and the result follows from the equality $\langle\widehat{T}_\phi F|h\rangle = \langle F|\widehat{T}_{\widetilde{\phi}}h\rangle$. ∎

Lemma 5.4.13. *Let* $\phi \in \mathcal{D}^\infty(A)$, *then the net* $\widehat{T}_{j_\epsilon}\phi$ *converges to* ϕ *in* $\mathcal{D}^\infty(A)$ *with respect to the topology* t_A.

Proof. Indeed, taking into account the strong continuity of $U(t)$, we have, for each n,

$$\|A^n(T_{j_\epsilon}\phi - \phi)\| = \left\|\int_{\mathbb{R}} j_\epsilon(t)(U(t)A^n\phi - A^n\phi)dt\right\|$$

$$\leqslant \sup_{t\in[-\epsilon,\epsilon]}\|U(t)A^n\phi - A^n\phi\|\int_{\mathbb{R}} dt\, j_\epsilon(t) \longrightarrow 0,$$

which implies the statement. ∎

Let us consider now a family $\{\mathcal{E}_\alpha\}_{\alpha \in I}$ of interspaces, where each $\mathcal{E}_\alpha$ is a Banach space with norm $\|\cdot\|_\alpha$. If the family is closed under duality, then we can define an involution $\alpha \mapsto \overline{\alpha}$ in the set of indices, such that $\mathcal{E}_{\overline{\alpha}} \simeq (\mathcal{E}_\alpha)^\times$.

We assume that the family $\{\mathcal{E}_\alpha\}_{\alpha \in I}$ is *compatible* with A in the following sense:

(1) $\widehat{U}(t)\mathcal{E}_\alpha = \mathcal{E}_\alpha, \ \forall \, \alpha \in I$;
(2) $\lim_{t \to 0} \|\widehat{U}(t)F - F\|_\alpha = 0, \ \forall \, F \in \mathcal{E}_\alpha, \ \forall \, \alpha \in I$.

Theorem 5.4.14. *Let $\{\mathcal{E}_\alpha\}$ be a family of interspaces compatible with A and let $F \in \mathcal{E}_\alpha$, for some $\alpha \in I$. Then $\widehat{T}_{j_\epsilon} F \to F$ with respect to the norm $\|\cdot\|_\alpha$ and $\mathcal{D}^\infty(A)$ is dense in $\mathcal{E}_\alpha \cap \mathcal{E}_\beta$ with the projective topology.*

Proof. Indeed,

$$\left\| \int_{\mathbb{R}} j_\epsilon(t)\widehat{U}(t)F dt - F \right\|_\alpha = \left\| \int_{\mathbb{R}} j_\epsilon(t)(\widehat{U}(t)F - F)dt \right\|_\alpha \leqslant \sup_{t \in [-\epsilon, \epsilon]} \|\widehat{U}(t)F - F\|_\alpha \int_{\mathbb{R}} j_\epsilon(t)dt \, .$$

For $\epsilon \to 0$, one has $t \to 0$, so the result follows from condition (2).

The density of $\mathcal{D}^\infty(A)$ in $\mathcal{E}_\alpha \cap \mathcal{E}_\beta$ follows from the fact that if $F \in \mathcal{E}_\alpha \cap \mathcal{E}_\beta$, the same net $\{\widehat{T}_{j_\epsilon} F\}$ can be used to approximate F with respect to both norms. $\blacksquare$

Definition 5.4.15. Let $\{\mathcal{E}_\alpha\}$ be a family of interspaces compatible with A. We define, for every $s \in \mathbb{R}$ and $\alpha \in I$, the set $\mathcal{L}_A^{s,\alpha}$:

$$\mathcal{L}_A^{s,\alpha} := \{F \in \mathcal{D}_{\overline{\infty}}(A) : \ (1 + A^2)^{s/2}F \in \mathcal{E}_\alpha\}.$$

Proposition 5.4.16. *The following statements hold true:*

(1) $\mathcal{L}_A^{s,\alpha}[\|\cdot\|_{s,\alpha}]$, endowed with the norm $F \mapsto \|F\|_{s,\alpha} = \|(1 + A^2)^{s/2}F\|_\alpha$, is a Banach space and $(1 + A^2)^{s/2}$ is an isometry of $\mathcal{L}_A^{s,\alpha}$ into $\mathcal{E}_\alpha$.
(2) The map $i : f \in \mathcal{D}^\infty(A) \mapsto f \in \mathcal{L}_A^{s,\alpha}$ is continuous.
(3) $(\mathcal{L}_A^{s,\alpha})^\times \simeq \mathcal{L}_A^{-s,\overline{\alpha}}$.
(4) For any $s \in \mathbb{R}$ and $\alpha \in I$, $\mathcal{D}^\infty(A)$ is dense in $\mathcal{L}_A^{s,\alpha}$.

Proof. (1) The proof is simple.
(2) If $f \in \mathcal{D}^\infty(A)$, then $(1 + A^2)^{s/2}f \in \mathcal{D}^\infty(A) \subset \mathcal{E}_\alpha$; furthermore, there exists $\gamma > 0$, $n \in \mathbb{N}$, such that

$$\|(1 + A^2)^{s/2}f\|_\alpha \leqslant \gamma \|(1 + A^2)^{n/2}(1 + A^2)^{s/2}f\| \leqslant \gamma \|(1 + A^2)^{\frac{1}{2}(n+[s]+1)}f\|,$$

where $[s]$ denotes the integer part of s.
(3) Let $\Phi \in (\mathcal{L}^{s,\alpha})^\times$, then there exists $\gamma > 0$ such that

$$|\langle \Phi | F \rangle| \leqslant \gamma \|F\|_{s,\alpha} = \gamma \|(1 + A^2)^{s/2}F\|_\alpha, \quad \forall \, F \in \mathcal{L}_A^{s,\alpha}.$$

Let $G = (1 + A^2)^{s/2}F$, so that $G \in \mathcal{E}_\alpha$. Let us define $\langle \Phi_s | G \rangle = \langle \Phi | F \rangle$. Then, for some $\gamma > 0$,

$$|\langle \Phi_s | G \rangle| = |\langle \Phi | F \rangle| \leqslant \gamma \|F\|_{s,\alpha} = \gamma \|G\|_\alpha,$$

hence $\Phi_s \in \mathcal{E}_{\overline{\alpha}}$. Since $\Phi_s(G) = \Phi((1+A^2)^{-s/2}G) \equiv ((1+A^2)^{-s/2} \cdot \Phi)(G)$, it follows that $\Phi_s = (1 + A^2)^{-s/2}\Phi \in \mathcal{E}_{\overline{\alpha}}$, so $\Phi \in \mathcal{L}_A^{-s,\overline{\alpha}}$.

(4) Let $\Phi \in \mathcal{L}_A^{s,\alpha}$ with $\Phi(f) = 0$, $\forall f \in \mathcal{D}^\infty(A)$. But $\Phi \in \mathcal{D}_{\overline{\infty}}(A)$; from the density of $\mathcal{D}^\infty(A)$, it follows that $\Phi = 0$. ∎

Proposition 5.4.17. *For every $\alpha, \alpha' \in I$ and every $s, s' \in \mathbb{R}$, $\mathcal{D}^\infty(A)$ is dense in $\mathcal{L}_A^{s,\alpha} \cap \mathcal{L}_A^{s',\alpha'}$ with the projective topology.*

Proof. It suffices to prove that for all $s \in \mathbb{R}$ and $\alpha \in I$:

$$\|(1 + A^2)^{s/2}(\widehat{T}_{j_\epsilon}F) - (1 + A^2)^{s/2}F\|_\alpha \to 0,$$

where $F \in \mathcal{L}_A^{s,\alpha} \cap \mathcal{L}_A^{s',\alpha'}$ and $j_\epsilon \in \mathcal{D}^\infty(A)$ is an approximate identity. First of all, we prove that

$$\widehat{T}_{j_\epsilon}((1 + A^2)^{s/2}F) = (1 + A^2)^{s/2}(\widehat{T}_{j_\epsilon}F).$$

Indeed,

$$\langle (1+A^2)^{s/2}(\widehat{T}_{j_\epsilon}F)|g\rangle = \langle \widehat{T}_{j_\epsilon}F|(1+A^2)^{s/2}g\rangle = \Big\langle \int_{\mathbb{R}} j_\epsilon(t)\widehat{U}(t)F dt \Big| (1+A^2)^{s/2}g \Big\rangle$$

$$= \int_{\mathbb{R}} j_\epsilon(t)\langle \widehat{U}(t)F|(1+A^2)^{s/2}g\rangle dt$$

$$= \int_{\mathbb{R}} j_\epsilon(t)\langle (1+A^2)^{s/2}\widehat{U}(t)F|g\rangle dt$$

$$= \int_{\mathbb{R}} j_\epsilon(t)\langle \widehat{U}(t)(1+A^2)^{s/2}F|g\rangle = \langle \widehat{T}_{j_\epsilon}(1+A^2)^{s/2}F|g\rangle.$$

Now we have

$$\|(1+A^2)^{s/2}(\widehat{T}_{j_\epsilon}F) - (1+A^2)^{s/2}F\|_\alpha = \|\widehat{T}_{j_\epsilon}((1+A^2)^{s/2}F) - (1+A^2)^{s/2}F\|_\alpha.$$

But $(1+A^2)^{s/2}F \in \mathcal{E}_\alpha$, and $\mathcal{D}^\infty(A)$ is dense in every $\mathcal{E}_\alpha$, hence, by Theorem 5.4.14, one has $\|\widehat{T}_{j_\epsilon}((1 + A^2)^{s/2}F) - (1+A^2)^{s/2}F\|_\alpha \to 0$, $\forall \alpha \in I$. ∎

Let us now consider a subspace $\mathcal{D} \subset \mathcal{D}^\infty(A)$ endowed with a locally convex topology t, with the following properties:

(d1) the topology t of $\mathcal{D}$ is finer than the topology induced on $\mathcal{D}$ by $\mathcal{D}^\infty(A)$;

(d2) $(1 + A^2)^{1/2} : \mathcal{D}[t] \to \mathcal{D}[t]$ continuously;

(d3) for all $n \in \mathbb{N}$, $\mathcal{D}$ is a core for $(1 + A^2)^{n/2}$, that is, $\overline{(1 + A^2)^{n/2} \upharpoonright_{\mathcal{D}}} = (1 + A^2)^{n/2}$.

Then, the identity map $i : \mathcal{D} \to \mathcal{D}^\infty(A)$ is continuous and has dense range as a consequence of (d3). If $\mathcal{D}^\times$ denotes the conjugate dual of $\mathcal{D}[t]$, then

$$\mathcal{D} \hookrightarrow \mathcal{D}^\infty(A) \hookrightarrow \mathcal{H} \hookrightarrow \mathcal{D}_{\overline{\infty}}(A) \hookrightarrow \mathcal{D}^\times.$$

If $\mathcal{I} = \{G_\alpha\}$ is a family of interspaces (stable under duality) between $\mathcal{D}^\infty(A)$ and $\mathcal{D}_{\overline{\infty}}(A)$, then $\mathcal{I}$ is a family of interspaces between $\mathcal{D}$ and $\mathcal{D}^\times$ too. Indeed, if G_α is in $\mathcal{I}$, then the embedding of $\mathcal{D}$ in G_α is continuous (as composition of continuous maps). Density follows from the fact that, if $\mathcal{D}$ is not dense in G_α, then there exists $\Phi \in G_{\overline{\alpha}}$, $\Phi \neq 0$, $\Phi(f) = 0$, $\forall f \in \mathcal{D}$, in contradiction with the fact that $G_{\overline{\alpha}} \subset \mathcal{D}_{\overline{\infty}}(A) \subset \mathcal{D}^\times$. Thus, in conclusion, if $\{G_\alpha\}$ is a generating family of interspaces in the rigged Hilbert space $\mathcal{D}^\infty(A) \hookrightarrow \mathcal{H} \hookrightarrow \mathcal{D}_{\overline{\infty}}(A)$, it is also generating in the rigged Hilbert space $\mathcal{D} \hookrightarrow \mathcal{H} \hookrightarrow \mathcal{D}^\times$.

Now, since $(1 + A^2)^{1/2} : \mathcal{D} \to \mathcal{D}$ is continuous, by (d2), it can be extended to $\mathcal{D}^\times$, and the same holds true for its integer or real powers, as done in Section 5.2.1. This allows us to define another class of Banach spaces:

$$L_A^{s,\alpha} = \{F \in \mathcal{D}^\times : (1 + A^2)^{s/2} F \in \mathcal{E}_\alpha\}.$$

Of course,

$$\mathcal{L}_A^{s,\alpha} = L_A^{s,\alpha} \cap \mathcal{D}_{\overline{\infty}}(A).$$

The next result shows indeed that these new spaces can be identified with those introduced earlier.

Theorem 5.4.18. $L_A^{s,\alpha} \simeq \mathcal{L}_A^{s,\alpha}$ for every $s \in \mathbb{R}$, and for every $\alpha \in I$.

Proof. For simplicity, put $B := (1 + A^2)^{1/2}$. Assume that $F \in L_A^{s,\alpha}$. Then $F \in \mathcal{D}^\times$ and $B^s F \in \mathcal{E}_\alpha$. But $\mathcal{E}_\alpha \subset \mathcal{D}_{\overline{\infty}}(A)$ and hence $B^s F \in \mathcal{D}_{\overline{\infty}}(A)$. This implies that $B^{-s}(B^s F) \in \mathcal{D}_{\overline{\infty}}(A)$, so $F \in \mathcal{D}_{\overline{\infty}}(A)$. ∎

Taking into account Proposition 5.4.16, we list some properties of the spaces $L_A^{s,\alpha}$:

Proposition 5.4.19. *The following statements hold:*

(1) $L_A^{s,\alpha}$ is a Banach space with respect to the norm

$$\|F\|_{s,\alpha} = \|(1 + A^2)^{s/2} F\|_\alpha;$$

(2) $\mathcal{D} \hookrightarrow L_A^{s,\alpha}$;

(3) $(L_A^{s,\alpha})^\times \simeq L_A^{-s,\overline{\alpha}}$;

(4) $\mathcal{D}$ is dense in $L_A^{s,\alpha}$.

Proof. It remains only to prove that $\mathcal{D}$ is dense in $L_A^{s,\alpha}$. Were it not so, then there would exist $\Phi \in L_A^{-s,\overline{\alpha}}$ such that $\Phi(f) = 0$, for any $f \in \mathcal{D}$. But $\Phi \in \mathcal{D}^\times$, so necessarily $\Phi = 0$. ∎

As consequence of the previous properties we have

Proposition 5.4.20. *The following statements hold true.*

(i) *For every $\alpha, \alpha' \in I$ and every $s, s' \in \mathbb{R}$, $\mathcal{D}$ is dense in $L_A^{s,\alpha} \cap L_A^{s',\alpha'}$, with the projective topology.*

(ii) *If $t < s$, then*

$$L_A^{s,\alpha} \hookrightarrow L_A^{t,\alpha}.$$

Proof. We only check that $L_A^{s,\alpha} \subset L_A^{t,\alpha}$. For this, put $s = t + \epsilon$ with $\epsilon > 0$; then $(1 + A^2)^{s/2} = (1 + A^2)^{\epsilon/2}(1 + A^2)^{t/2}$. Since $\|\cdot\|_{t,\alpha} \leqslant \|\cdot\|_{s,\alpha}$, the identity map provides a continuous embedding of $L_A^{s,\alpha}$ into $L_A^{t,\alpha}$. ∎

In the construction made so far, we have only assumed that the family $\{\mathcal{E}_\alpha\}_{\alpha \in I}$ of Banach spaces is closed under duality and compatible with the operator A. Of course, it might well happen that this family satisfies additional conditions, making the corresponding structure of the family $\{L_A^{s,\alpha}\}$ richer.

First of all, we notice that the set of indices I that describes the family $\{\mathcal{E}_\alpha\}_{\alpha \in I}$ has a natural partial order defined by

$$\alpha \leqslant \beta \iff \mathcal{E}_\alpha \hookrightarrow \mathcal{E}_\beta.$$

The map $\alpha \mapsto \overline{\alpha}$ becomes then an order-reversing involution in I.

Now, assume that the family $\{\mathcal{E}_\alpha\}_{\alpha \in I}$ has the following property:

(p) $\forall \alpha, \beta \in I$,   there exists $\gamma \in I$ such that $\mathcal{E}_\alpha \cap \mathcal{E}_\beta \hookrightarrow \mathcal{E}_\gamma$.

Then, by Proposition 5.4.20, it follows that

$$L_A^{s,\alpha} \cap L_A^{s,\beta} \hookrightarrow L_A^{t,\gamma}$$

for any $t \leqslant s$ and α, β, γ as in (p). In particular, if $t \leqslant s$ and $\alpha \leqslant \gamma$, we have

$$L_A^{s,\alpha} \hookrightarrow L_A^{t,\gamma}.$$

Example 5.4.21. We will now show a concrete realization of what we have developed so far. In particular, we will show that in the rigged Hilbert space having as extreme elements the Schwartz space $\mathcal{S}$ of rapidly decreasing functions and the space $\mathcal{S}^\times$ of tempered distributions, with a suitable, but natural, choice of the operator A and of the family $\{\mathcal{E}_\alpha\}$, the previous construction leads to the *Bessel potential spaces* $W^{s,p}(\mathbb{R})$ (sometimes denoted $L^{s,p}(\mathbb{R})$). These spaces generalize the Sobolev spaces $W^{k,p}(\mathbb{R})$ to a continuous index (indeed, if $s = k \in \mathbb{N}$, then $L^{k,p}(\mathbb{R}) = W^{k,p}(\mathbb{R})$). Indeed, taking $(a,b) = \mathbb{R}$ and replacing 2 by p in (5.17),(5.18), we obtain

$$W^{k,p}(\mathbb{R}) := \{f \in \mathcal{S}^\times(\mathbb{R}) : f^{(j)} \in L^p(\mathbb{R}, dx), \ j = 0, 1, \ldots, k\}. \tag{5.19}$$

For $k = 2$, the following norm is equivalent to (5.18),

$$\|f\|_{k,2} = \left(\int_{\mathbb{R}} (1 + |\xi|^2)^k |\widehat{f}(\xi)|^2 \, d\xi \right)^{1/2},$$

$$= \|\mathcal{F}^{-1}((1 + |\cdot|^2)^{k/2} \mathcal{F}(f))\|_2 .$$

Thus, we define the space $W^{s,p}(\mathbb{R})$ as

$$W^{s,p}(\mathbb{R}) := \{f \in \mathcal{S}^\times(\mathbb{R}) : \|f\|_{s,p} < \infty\}, \tag{5.20}$$

where

$$\|f\|_{s,p} := \|\mathcal{F}^{-1}((1 + |\cdot|^2)^{s/2} \mathcal{F}(f))\|_p, \quad s \in \mathbb{R}. \tag{5.21}$$

Let us consider the rigged Hilbert space:

$$\mathcal{S}(\mathbb{R}) \hookrightarrow L^2(\mathbb{R}) \hookrightarrow \mathcal{S}^\times(\mathbb{R}).$$

As it is well-known, $\mathcal{S}(\mathbb{R})$ coincides with the space of C^∞-vectors of the operator $H_{\text{osc}} = -\frac{d^2}{dx^2} + x^2$, i.e., $\mathcal{D}^\infty(H_{\text{osc}}) = \mathcal{S}(\mathbb{R})$ and the topology $t_{H_{\text{osc}}}$ (defined as in Section 5.2.1) is equivalent to the usual topology of the Schwartz space $\mathcal{S}(\mathbb{R})$.

To begin our construction, we take as A the operator P defined on the Sobolev space $W^{1,2}(\mathbb{R})$ by

$$(Pf)(x) = -if'(x), \quad f \in W^{1,2}(\mathbb{R}),$$

where f' stands for the weak derivative. The operator P is self-adjoint on $W^{1,2}(\mathbb{R})$ and

$$\mathcal{D}^\infty(P) = \{f \in C^\infty(\mathbb{R}) : f^{(k)} \in L^2(\mathbb{R}), \ \forall k \in \mathbb{N}\}.$$

Clearly, $\mathcal{S}(\mathbb{R}) \subset \mathcal{D}^\infty(P)$. It is easily seen that the usual topology of $\mathcal{S}(\mathbb{R})$ is finer than the one induced on it by t_P. Furthermore, with the help of the Fourier transform, one sees easily that the operator $(1 + P^2)^{1/2}$ leaves $\mathcal{S}(\mathbb{R})$ invariant and it is continuous on it.

In a similar way, taking into account the corresponding properties of the multiplication operator $(1 + x^2)^{1/2}$, one can show that $\mathcal{S}(\mathbb{R})$ is a core for any power of $(1 + P^2)^{1/2}$. Hence the conditions (d1)-(d3) are all satisfied.

Now, we should choose the spaces $\{\mathcal{E}_\alpha\}$. For this purpose, we take the spaces $L^p(\mathbb{R})$ with $1 < p < \infty$. These spaces are compatible with $P = i\frac{d}{dx}$. Indeed, $(U(t)f)(x) = (e^{iPt}f)(x) = f(x - t)$, $f \in L^p(\mathbb{R})$. By [Bre86, Lemma IV.3], one has:

$$\lim_{t \to 0} \|f(x - t) - f(x)\|_{L^p} = 0.$$

By definition,

$$L_P^{s,p} = \{F \in \mathcal{S}^\times : (1 + P^2)^{s/2} F \in L^p(\mathbb{R})\}.$$

This amounts to say that

$$u := \left(1 - \frac{d^2}{dx^2}\right)^{s/2} F \in L^p(\mathbb{R}).$$

Then, taking the Fourier transform $\mathcal{F}$ (in $\mathcal{S}^\times$), we get

$$\mathcal{F}\left(\left(1 - \frac{d^2}{dx^2}\right)^{s/2} F\right) = \mathcal{F}(u),$$

that is,

$$\left(1 + |\xi|^2\right)^{s/2} \mathcal{F}(F) = \mathcal{F}(u).$$

Thus, finally, taking the inverse Fourier transform, we get

$$u = \mathcal{F}^{-1}\left(\left(1 + |\xi|^2\right)^{s/2} \mathcal{F}(F)\right).$$

In conclusion, if F is a tempered distribution,

$$F \in L_P^{s,p} \quad \text{if, and only if,} \quad \mathcal{F}^{-1}\left(\left(1 + |\xi|^2\right)^{s/2} \mathcal{F}(F)\right) \in L^p(\mathbb{R}).$$

The condition on the right-hand side is exactly the one which defines the Bessel potential space $W^{s,p}(\mathbb{R})$ given in (5.20), whose properties (such as inclusions, density of $\mathcal{S}(\mathbb{R})$, etc.) can therefore be derived from the abstract ones discussed above.

Of course, similar results hold, with obvious modifications, if we work in $\mathbb{R}^n$ instead of $\mathbb{R}$.

Example 5.4.22. Another example of RHS is the triplet built on Feichtinger's algebra $\mathcal{S}_0$,

$$\mathcal{S}_0 \hookrightarrow L^2 \hookrightarrow \mathcal{S}_0^\times,$$

that we will study in Section 8.3.2.

5.4.3 PIP-*Spaces of Distributions*

An obvious question is, whether one may equip the space $\mathcal{S}^\times(\mathbb{R}^n)$ of tempered distributions with the structure of a PIP-space, but it is not immediately obvious what the compatibility should be. A "conservative" definition is: $f \# g$ if and only if at least one member of the pair is a test function from $\mathcal{S}$, that is, coming back to Example 5.4.8. It does make $\mathcal{S}^\times$ into a nondegenerate partial inner product space. Apart from that, it is not very useful.

A natural alternative is to realize $\mathcal{S}^\times$ as the space $s^\times$ of sequences of slow increase, that is, the Hermite realization, described in the example (ii,b) of Section 1.1.3. It associates to every $f \in \mathcal{S}^\times$ the sequence of coefficients $\langle h_k | f \rangle$, where the h_k are normalized Hermite functions. Among assaying subspaces one finds here all Sobolev spaces.

Another possibility is to consider the scale generated by the Hamiltonian of the harmonic oscillator, $H_{\text{osc}} = x^2 + p^2$, as we did in Example 5.4.21, or the lattice generated by the mixed powers of the position and momentum (quantum) operators, as we explained in Section I.1 of the Introduction.

The next question is how to extend all this to general, nontempered, distributions (or even ultra-distributions). Thus we consider the quintuplet of spaces (on $\mathbb{R}$ or on $\mathbb{R}^d$)

$$\mathcal{D} \hookrightarrow \mathcal{S} \hookrightarrow L^2 \hookrightarrow \mathcal{S}^\times \hookrightarrow \mathcal{D}^\times, \tag{5.22}$$

where $\mathcal{D}$ is the space of test functions (C^∞ functions of compact support) and $\mathcal{D}^\times$ is the space of distributions. For interpolating between $\mathcal{D}$ and $\mathcal{S}$, one possibility is to use the Gel'fand-Shilov spaces $S_\alpha, \alpha \geq 0$, defined as follows (for simplicity, we restrict ourselves to one dimension):

$$S_\alpha := \{C^\infty \text{ functions } \phi : |x^k \phi^{(q)}(x)| \leq c_q \, a^k k^{k\alpha}, \, k, q = 0, 1, 2, \ldots\}$$
$$= \{C^\infty \text{ functions } \phi : |\phi^{(q)}(x)| \leq c_q \, \exp(-b|x|^{1/\alpha}), \, q = 0, 1, 2, \ldots\},$$

where the constants c_q, a, b depend on ϕ. This space is an inductive limit of countably Hilbert spaces, hence complicated. In addition, it is not invariant under Fourier transform, since one has:

$$S_\alpha \xleftrightarrow{\mathcal{F}} S^\alpha,$$

where the space S^β, $\beta > 0$, is defined as

$$S^\beta := \{C^\infty \text{ functions } \phi : |x^k \phi^{(q)}(x)| \leq c_k \, d^q q^{q\beta}, \, q = 0, 1, 2, \ldots\}.$$

For $\beta \leq 1$, S^β consists of analytic functions, hence it does not contain $\mathcal{D}$. One can also consider the Gel'fand-Shilov space $S_\alpha^\alpha (\alpha \geq 1/2)$, which is invariant under Fourier transform (and time-frequency shifts), but this one is contained in $\mathcal{D}$, so that elements of its dual are ultra-distributions.

Instead, Grossmann has designed a family of Hilbert spaces, called Hilbert spaces of type S, which are closely related to the Gel'fand-Shilov spaces and contain nontempered distributions. The construction, which generalizes the Hermite realization of $\mathcal{S}^\times$ mentioned above, goes as follows (we restrict ourselves again to one dimension). For $n = 0, 1, 2, \ldots$, define recursively the following operators acting on $\mathcal{S}(\mathbb{R})$:

$$M_0 := 1$$
$$M_1 := (x^2 + p^2)$$
$$\vdots$$
$$M_n := (xM_{n-1}x + pM_{n-1}p)$$

The following properties are immediate:

(i) Positivity: $\langle u|M_n u\rangle > 0$, for every $n \geqslant 0$ and every $u \in \mathcal{S}, u \neq 0$.
(ii) Orthogonality: $\langle h_k|M_n h_m\rangle = 0$, if $k \neq m$, where h_k are the normalized Hermite functions.
(iii) Fourier invariance: $\langle u|M_n u\rangle = \langle \mathcal{F}u|M_n \mathcal{F}u\rangle$, for every $n \geqslant 0$ and every $u \in \mathcal{S}$.

Define the numbers $c(n; k) := \langle h_k|M_n h_k\rangle$. Then one has, for every $u \in \mathcal{S}$ and every $n \geqslant 0$,

$$\langle u|M_n u\rangle = \sum_k c(n; k) \, |\langle h_k|u\rangle|^2.$$

Now we can construct the Hilbert spaces we are looking for. Given a sequence $\beta =: (\beta_n)$ of positive numbers, we denote by $S(\beta)$ the set of all $u \in \mathcal{S}$ such that $\sum_{n=0}^{\infty} \beta_n \langle u|M_n u\rangle < \infty$.

Alternatively, define the numbers λ_k $(0 \leqslant \lambda_k < \infty)$ by

$$\lambda_k^{-2} := \sum_{n=0}^{\infty} \beta_n c(n; k). \tag{5.23}$$

Then $S(\beta)$ consists of those $u \in \mathcal{S}$ for which $\sum_k |\langle h_k|u\rangle|^2 \lambda_k^{-2} < \infty$.

The space $S(\beta)$ is obviously a pre-Hilbert space. Then:

Theorem 5.4.23. *(i) Let the sequence* $\beta = (\beta_n)$ *be such that, for every* k, $\lambda_k^{-2} = \sum_{n=0}^{\infty} \beta_n c(n; k) < \infty$ *(i.e.* $\lambda_k \neq 0$*). Then* $S(\beta)$ *is an infinite-dimensional Hilbert space, with inner product*

$$\langle u|v\rangle_\beta = \sum_{n=0}^{\infty} \beta_n \langle u|M_n v\rangle = \sum_k \langle u|h_k\rangle \lambda_k^{-2} \langle h_k|v\rangle. \tag{5.24}$$

The family $\{\lambda_k h_k\}$ *is an orthonormal basis of* $S(\beta)$.

(ii) Conversely, if, for every k, $\sum_{n=0}^{\infty} \beta_n c(n; k) = \infty$, *then* $S(\beta) = \{0\}$.

The family of Hilbert spaces $S(\beta)$, indexed by sequences β of positive numbers, has a partial order.

Proposition 5.4.24. *Let the sequences β and β' be such that $N^2 := \sum_{n=0}^{\infty} \beta'_n/\beta_n < \infty$. Then $S(\beta) \subset S(\beta')$ and the norm of the embedding does not exceed N.*

Next we define the conjugate dual $S(\bar\beta) := S(\beta)^{\times}$, with respect to the L^2 inner product, so that

$$S(\beta) \hookrightarrow L^2 \hookrightarrow S(\bar\beta).$$

In addition, the natural embeddings $E_{o\beta} : S(\beta) \to L^2$ and $E_{\bar\beta o} : L^2 \to S(\bar\beta)$ are trace class operators.

The spaces $S(\bar\beta)$ have the following properties:

(i) $\mathcal{S}^{\times}$ is contained in $S(\bar\beta)$, so that we have

$$S(\beta) \hookrightarrow \mathcal{S} \hookrightarrow L^2 \hookrightarrow \mathcal{S}^{\times} \hookrightarrow S(\bar\beta).$$

(ii) Define the Fourier transform $\mathcal{F}$ on $S(\bar\beta)$, as on $\mathcal{S}^{\times}$, by

$$\langle f|\mathcal{F}F\rangle = \langle \mathcal{F}f|F\rangle, \; \forall\, f \in S(\beta),\, F \in S(\bar\beta).$$

Then the Fourier transform is a unitary operator on $S(\bar\beta)$.

Among the (huge) family of the spaces $S(\beta)$, there is a subclass, corresponding to a particular choice of the sequence β, that constitutes a LHS. Given positive numbers α, a, we denote by $\mathcal{H}(\alpha; a)$ the space $S(\beta)$ associated with the sequence

$$\beta_n = [a^n \Gamma(\alpha n)]^{-2} \quad (\Gamma \text{ is the usual Gamma function}).$$

Thus the numbers λ_k^{-2} of (5.23) become

$$\lambda_{k,\alpha}^{-2} := \sum_{n=0}^{\infty} a^{-2n} \Gamma(\alpha n)^{-2n} c(n;k), \tag{5.25}$$

so that $\mathcal{H}(\alpha; a)$ consists of the functions $u \in \mathcal{S}$ such that

$$\langle u|u\rangle_\alpha = \sum_k \langle u|h_k\rangle \lambda_{k,\alpha}^{-2} \langle h_k|u\rangle < \infty, \tag{5.26}$$

with the corresponding inner product $\langle u|v\rangle_\alpha$.

Proposition 5.4.24 shows that the spaces $\mathcal{H}(\alpha; a)$ are totally ordered:

$$\mathcal{H}(\alpha'; a') \subset \mathcal{H}(\alpha''; a'') \quad \Longleftrightarrow \quad \alpha' \leqslant \alpha'' \text{ or } \alpha' = \alpha'' \text{ and } a' \leqslant a'',$$

in other words, if $\{\alpha'; a'\} \leqslant \{\alpha''; a''\}$, in lexicographic order. Then one deduces from Theorem 5.4.23 that

- for $\{\alpha; a\} \leqslant \{\frac{1}{2}; \frac{1}{\sqrt{2}}\}$, $\mathcal{H}(\alpha; a) = \{0\}$;
- for $\{\alpha; a\} > \{\frac{1}{2}; \sqrt{8e}\}$, $\mathcal{H}(\alpha; a)$ is infinite-dimensional.

The situation is unknown for the spaces $\mathcal{H}(\frac{1}{2}; a)$ with $\frac{1}{\sqrt{2}} < a \leqslant \sqrt{8e}$. Also, Proposition 5.4.24 implies that, if $\alpha < \alpha'$, then the embedding of $\mathcal{H}(\alpha; a)$ into $\mathcal{H}(\alpha'; a')$ is of trace class (for arbitrary a, a').

Now, we denote by $\mathcal{H}(\overline{\alpha}; \overline{a})$ the conjugate dual of $\mathcal{H}(\alpha; a)$. Thus, if $\{\alpha'; a'\} \leqslant \{\alpha''; a''\}$, one has $\mathcal{H}(\overline{\alpha}''; \overline{a}'') \subset \mathcal{H}(\overline{\alpha}'; \overline{a}')$. In conclusion, taking all spaces $\mathcal{H}(\alpha; a), \mathcal{H}(\overline{\alpha}; \overline{a})$, together with L^2, one obtains a LHS. Within the latter, one finds several sublattices which are chains of Hilbert spaces. Such is, for instance, the chain corresponding to $\frac{1}{2} < \alpha < 1$, $a = 1$:

$$\mathcal{H}(\tfrac{1}{2}; 1) \ldots \hookrightarrow \mathcal{H}(\alpha; 1) \hookrightarrow \ldots \mathcal{H}(1; 1) \hookrightarrow L^2 \hookrightarrow \mathcal{H}(\overline{1}; \overline{1}) \hookrightarrow \ldots \mathcal{H}(\overline{\alpha}; \overline{1}) \hookrightarrow \ldots \mathcal{H}(\overline{\tfrac{1}{2}}; \overline{1})$$
$$(5.27)$$

Since one has

$$\mathcal{H}(1; 1) \hookrightarrow \mathcal{S} \hookrightarrow L^2 \hookrightarrow \mathcal{S}^{\times} \hookrightarrow \ldots H(\overline{\alpha}; \overline{1}),$$

it follows that (5.27) is a chain of Hilbert spaces, consisting of Schwartz functions on the left and nontempered distributions on the right. In addition, the extreme spaces are themselves Hilbert spaces.

Another example is the (open ended) chain

$$\{\mathcal{H}(1; a)\}_{0 < a < \infty} \subset L^2 \subset \{\mathcal{H}(\overline{1}; \overline{a})\}_{0 < a < \infty}.$$

The interest of all these spaces is that their elements have entire analytic continuations and often admit entire analytic functions as multipliers from one space into another one of the same family. For further details, we refer to the original paper.

5.5 The PIP-Space Generated by a Family of Unbounded Operators

5.5.1 Basic Idea of the Construction

The lesson of Section 5.3 is that LHSs are intimately connected with positive self-adjoint operators and positive quadratic forms. To make this precise, take a LHS H_I and consider an assaying subspace $\mathcal{H}_r$ not comparable to $\mathcal{H}_0$:

$$\mathcal{H}_{r \wedge 0} := \mathcal{H}_r \cap \mathcal{H}_0 \hookrightarrow \left\{ \begin{matrix} \mathcal{H}_r \\ \mathcal{H}_0 \end{matrix} \right\}$$

Let $A = 1 + R$ be the interpolating operator between $\mathcal{H}_{r \wedge 0} = \mathcal{H}_r \cap \mathcal{H}_0$ and $\mathcal{H}_0$:

$$\langle f | f \rangle_{r \wedge 0} = \langle f | f \rangle + \langle f | f \rangle_r = \langle f | (1 + R) f \rangle. \tag{5.28}$$

By construction, R is a non-negative self-adjoint operator, with form domain $Q(R) = \mathcal{H}_r \cap \mathcal{H}_0$. The space $\mathcal{H}_r$ is the completion of $\mathcal{H}_r \cap \mathcal{H}_0$ in the norm

$$\langle f | f \rangle_r = \langle f | R f \rangle, \quad f \in \mathcal{H}_r \cap \mathcal{H}_0.$$

Thus $R^{1/2}$ is isometric from $\mathcal{H}_r \cap \mathcal{H}_0$ into $\mathcal{H}_0$, so its closure $\overline{R^{1/2}}$ is unitary from $\mathcal{H}_r$ onto $\mathcal{H}_0$. Since $\| \cdot \|$ is a norm, the operator R is invertible, i.e., 0 does not belong to its point spectrum. Indeed, the dual space $\mathcal{H}_{\bar{r}} = (\mathcal{H}_r)^\times$ corresponds precisely to the operator R^{-1}, in the sense that the space $\mathcal{H}_{\bar{r}}$ is the completion of $\mathcal{H}_{\bar{r}} \cap \mathcal{H}_0$ in the norm

$$\langle f | f \rangle_{\bar{r}} = \langle f | R^{-1} f \rangle, \quad f \in \mathcal{H}_{\bar{r}} \cap \mathcal{H}_0.$$

and again $\overline{R^{-1/2}}$ is unitary from $\mathcal{H}_{\bar{r}}$ onto $\mathcal{H}_0$. Two cases may happen:

(i) R and R^{-1} are both unbounded (in $\mathcal{H}_0$); then $\mathcal{H}_r$ and $\mathcal{H}_0, \mathcal{H}_{\bar{r}}$ and $\mathcal{H}_0$ are mutually noncomparable. Take, for example, $\mathcal{H}_0 = L^2(\mathbb{R}, dx)$, $A = 1 + x^2$, then $R = x^2, \mathcal{H}_r = L^2(\mathbb{R}, x^2 dx)$, $\mathcal{H}_r \cap \mathcal{H}_0 = L^2(\mathbb{R}, (1 + x^2) dx), R^{-1} = x^{-2}, \mathcal{H}_{\bar{r}} = L^2(\mathbb{R}, x^{-2} dx)$.
(ii) One of them, say R^{-1}, is bounded. Then one has a triplet $\mathcal{H}_r \hookrightarrow \mathcal{H}_0 \hookrightarrow \mathcal{H}_{\bar{r}}$, as discussed in Section 5.2.2.

Of course, if both R and R^{-1} are bounded, the norms $\| \cdot \|_r$ and $\| \cdot \|$ are equivalent, i.e., $\mathcal{H}_r = \mathcal{H}_0 = \mathcal{H}_{\bar{r}}$. We notice also that the norms $\| \cdot \|_r$ and $\| \cdot \|$ are automatically consistent on $\mathcal{H}_r \cap \mathcal{H}_0$ (see Proposition 2.2.1). Finally, all that has been said can be repeated in terms of the closed, positive, nondegenerate quadratic form $r(f, g) := \langle f | R g \rangle$, with $Q(r) = Q(R)$.

These remarks suggest another method for constructing a LHS, starting from a given Hilbert space and a family of self-adjoint operators (or quadratic forms). Let R_1, R_2 be two non-negative invertible self-adjoint operators on $\mathcal{H}_0$, with form domains $Q(R_1), Q(R_2)$. Let $\mathcal{H}_{R_j} (j = 1, 2)$ be the completion of $Q(R_j)$ with respect to the norm $\| f \|_j = \langle f | R_j f \rangle$. We say that R_1 and R_2 are *consistent* if:

(i) $Q(R_1) \cap Q(R_2)$ is dense in both $\mathcal{H}_{R_1}$ and $\mathcal{H}_{R_2}$.
(ii) the norms $\| \cdot \|_1$, and $\| \cdot \|_2$ are consistent on $Q(R_1) \cap Q(R_2)$.

Remark 5.5.1. There is a major difference between the present situation and that of Proposition 2.2.1. In that case, indeed, one starts from two Banach spaces X_a, X_b and one consider their intersection $X_a \cap X_b$, which is identified with $X_{[a,b]}$, a *closed* subspace of $X_a \oplus X_b$. Here, instead, one starts from two pre-Hilbert spaces $Q(R_1), Q(R_2)$ and their intersection

$Q(R_1) \cap Q(R_2)$ is in general *not* closed in the direct sum of the completions $\mathcal{H}_{R_1} \oplus \mathcal{H}_{R_2}$, so that the argument does not apply. Therefore, condition (ii) has to be imposed explicitly in order to ensure that the natural embeddings E_{rs} in the resulting PIP-space are continuous and have dense range.

Let I be a family of non-negative, invertible, self-adjoint operators on $\mathcal{H}_0$. We say that I is *admissible* if one has:

(a1) Any two operators of I are consistent;
(a2) The inverse of any element of I belongs to I;
(a3) The sum of any two elements of I belongs to I

Given an admissible family I, denote by $\mathcal{I} = \{\mathcal{H}_R\}_{R \in I}$ the corresponding family of Hilbert spaces. We may, of course, always assume that $\mathcal{H}_0 \in \mathcal{I}$, that is $1 \in I$. Let $V = \sum_{R \in I} \mathcal{H}_R$ be the algebraic inductive limit of $\mathcal{I}$. It is clear that $\mathcal{I}$ is an involutive covering of V (in particular, (a3) implies that $\mathcal{I}$ is closed under finite intersections), and this defines a linear compatibility on it, by the relation $f \# g \iff \exists R \in I$ such that $F \in \mathcal{H}_R, g \in \mathcal{H}_{\overline{R}} := \mathcal{H}_{R^{-1}}$. With the partial inner product

$$\langle f | g \rangle := \langle \overline{R^{1/2} f} | \overline{R^{-1/2} g} \rangle,$$

we therefore obtain a PIP-space of type (H) on V, and in fact a LHS. However, we don't know *a priori* whether this LHS is nondegenerate. This would require that $V^{\#} = \cap_{R \in I} \mathcal{H}_R$ be dense in every $\mathcal{H}_R$ and be a core for each $R \in I$, but that might necessitate additional assumptions. We will come back to this construction in Section 5.5.2. The above discussion is summarized in the following:

Proposition 5.5.2. *Any admissible family I of non-negative, invertible, self-adjoint operators on a Hilbert space $\mathcal{H}_0$ defines an LHS on $V = \sum_{R \in I} \mathcal{H}_R$ with central Hilbert space $\mathcal{H}_0$. The nondegeneracy of this LHS is not guaranteed.*

An entirely similar construction is obtained by starting from an admissible family of closed, non-negative, nondegenerate, quadratic forms, the connection between the two approaches being, of course,

$$r(f, g) = \langle f | Rg \rangle, \quad f, g \in Q(r) := Q(R).$$

As a standard example of this construction, we may take our familiar LHS of locally integrable functions $\{L^2(r)\}$, discussed in Section 4.2.1. Let $\mathcal{H}_0 = L^2(\mathbb{R}^n, dx)$ and define R to be the operator of multiplication by the measurable, a.e. positive, function r^{-1}. This gives

$$\mathcal{H}_r = \{f : \mathbb{R}^n \to \mathbb{C}, \text{ measurable}, \int_{\mathbb{R}^n} |f(x)|^2 \, r(x)^{-2} \, dx < \infty\}.$$

If r and r^{-1} are both locally square integrable, $Q(R^{\pm 1})$ contains the space $L_c^\infty(\mathbb{R}^n, dx)$ of essentially bounded functions of compact support, which is dense in $\mathcal{H}_r$ and $\mathcal{H}_{r^{-1}}$. Thus if we take all such functions r, with $r^{\pm 1} \in L_{loc}^2(\mathbb{R}^n, dx)$, we get an admissible family, which yields the usual PIP-space on $V = L_{loc}^1(\mathbb{R}^n, dx)$. Similarly, if one takes instead functions r such that $r^{\pm 1} \in L_{loc}^\infty(\mathbb{R}^n, dx)$, one gets a PIP-space on $V = L_{loc}^2(\mathbb{R}^n, dx)$ with $V^{\#} = L_c^2(\mathbb{R}^n, dx)$.

5.5.2 The Case of a Family of Unbounded Operators

Now we apply the construction of the previous section to the case of a family of unbounded operators, in particular, an operator algebra. The PIP-space constructed in Section 5.2.1 on the powers of the self-adjoint operator A can be thought as generated by the O*-algebra of polynomials in A on the domain $\mathcal{D}^\infty(A)$ or as a refinement of the RHS associated to this O*-algebra (see Section 3.3.1). An O*-algebra, or even a family $\mathcal{O}$ of closable operators defined on a dense domain $\mathcal{D}$ in Hilbert space (this is called an O-family on $\mathcal{D}$), generates also a RHS having $\mathcal{D}$ as smallest space. More precisely, let $\mathcal{O}$ be an O-family on $\mathcal{D}$. For any $A \in \mathcal{O}$, we write $R_A = 1 + A^* \overline{A}$, where $\overline{A}$ is the closure of A. Each R_A is a self-adjoint, invertible operator, with bounded inverse. The graph topology $t_{\mathcal{O}}$ on $\mathcal{D}$ is then defined by the family of norms

$$ f \in \mathcal{D} \mapsto \|(1 + A^*\overline{A})^{1/2} f\| = \|R_A^{1/2} f\|, \quad A \in \mathcal{O}. $$

Let $\mathcal{D}^\times$ be the conjugate dual of $\mathcal{D}[t_{\mathcal{O}}]$, endowed with the strong dual topology $t_{\mathcal{O}}^\times$. The RHS

$$ \mathcal{D}[t_{\mathcal{O}}] \hookrightarrow \mathcal{H} \hookrightarrow \mathcal{D}^\times[t_{\mathcal{O}}^\times] $$

will be called the RHS *associated to* $\mathcal{O}$.

As is customary, the domain $D(\overline{A})$ of the closure of A can be made into a Hilbert space, to be denoted by $\mathcal{H}(R_A)$, when it is endowed with the graph norm $\|f\|_{R_A} := \|R_A^{1/2} f\|$. Then $D(\overline{A}) = D(R_A^{1/2}) = Q(R_A)$.

The conjugate dual of $\mathcal{H}(R_A)$, with respect to the inner product of $\mathcal{H}$, is (isomorphic to) the completion of $\mathcal{H}$ in the norm $\|R_A^{-1/2} \cdot \|$; we denote it by $\mathcal{H}(R_A^{-1})$. Thus we have

$$ \mathcal{H}(R_A) \hookrightarrow \mathcal{H} \hookrightarrow \mathcal{H}(R_A^{-1}). \tag{5.29} $$

The operator $R_A^{1/2}$ is unitary from $\mathcal{H}(R_A)$ onto $\mathcal{H}$, and from $\mathcal{H}$ onto $\mathcal{H}(R_A^{-1})$. Hence R_A is the Riesz unitary operator mapping $\mathcal{H}(R_A)$ onto its conjugate dual $\mathcal{H}(R_A^{-1})$, and similarly R_A^{-1} from $\mathcal{H}(R_A^{-1})$ onto $\mathcal{H}(R_A)$.

From (5.29) it follows that for every $A \in \mathcal{O}$, $\mathcal{H}(R_A)$ and $\mathcal{H}(R_A^{-1})$ are interspaces in the sense of Definition 5.4.1.

First of all, we show that, to any such family $(\mathcal{O}, \mathcal{D})$, there corresponds a canonical LHS. In a standard fashion, the spaces $\mathcal{H}(R_A)$ generate, by set inclusion and vector sum, a lattice of Hilbert spaces, all dense in $\mathcal{H}$. For any $A, B \in \mathcal{O}$, let us define :

$$R_{A \wedge B} := R_A \dotplus R_B,$$
$$R_{A \vee B} := (R_A^{-1} \dotplus R_B^{-1})^{-1}, \tag{5.30}$$

where $\dotplus$ denotes the form sum, so that the operators on the left-hand side are indeed self-adjoint.

Remark 5.5.3. The left-hand side of the preceding equations should be read as symbols for denoting the right-hand side: indeed, this does not mean that there exist operators $A \wedge B, A \vee B$ defined on $\mathcal{D}$ for which the required equalities hold. Indeed, whereas one may write $R_{A \wedge B} = R_C = 1 + |C|^2$, where $|C|^2 = 1 + |A|^2 + |B|^2$ is indeed positive and self-adjoint, this does not define the operator C uniquely (see Remark 5.5.5 below). Moreover, we do not know whether any such C is defined on $\mathcal{D}$ and, when it is, we do not know whether $\mathcal{D}$ is a dense subspace of $\mathcal{D}(C)$, a fortiori, there is no reason why C would belong to $\mathcal{O}$. Things might improve, however, if we assume that $\mathcal{O}$ is an O*-algebra (see Section 5.5.3).

For the corresponding Hilbert spaces, one has:

$$\mathcal{H}(R_{A \wedge B}) = \mathcal{H}(R_A) \cap \mathcal{H}(R_B),$$
$$\mathcal{H}(R_{A \vee B}) = \mathcal{H}(R_A) + \mathcal{H}(R_B), \tag{5.31}$$

where the first space carries the projective norm, the second the inductive norm. In addition, the norms corresponding to R_A and R_B are consistent on $\mathcal{H}(R_{A \wedge B})$, since the operators R_A, R_B are closed.

Doing the same with the dual spaces $\mathcal{H}(R_A^{-1})$, one gets another lattice, dual to the first one. The conjugate duals of the spaces (5.31) are, respectively:

$$\mathcal{H}(R_{A \wedge B}^{-1}) = \mathcal{H}(R_A^{-1}) + \mathcal{H}(R_B^{-1}),$$
$$\mathcal{H}(R_{A \vee B}^{-1}) = \mathcal{H}(R_A^{-1}) \cap \mathcal{H}(R_A^{-1}). \tag{5.32}$$

We will denote by $\mathcal{R}$ the set of all positive self-adjoint operators $R_A^{\pm 1}$:

$$\mathcal{R} = \mathcal{R}(\mathcal{O}) := \{R_A^{\pm 1} : A \in \mathcal{O}\}.$$

Definition 5.5.4. Given the O-family $\mathcal{O}$ on $\mathcal{D}$ and the corresponding set of (Riesz) operators $\mathcal{R} = \mathcal{R}(\mathcal{O})$, we define $\Sigma_{\mathcal{R}}$ as the minimal set of self-adjoint operators containing $\mathcal{R}$ and satisfying the following conditions:

(c1) for every $R \in \Sigma_{\mathcal{R}}$, $R^{-1} \in \Sigma_{\mathcal{R}}$;

(c2) for every $R, S \in \Sigma_{\mathcal{R}}$, $R \dotplus S \in \Sigma_{\mathcal{R}}$.

Then the set $\Sigma_{\mathcal{R}}$ is said to be an *admissible cone of self-adjoint operators* if, in addition,

(c3) $\mathcal{D}$ is dense in every $\mathcal{H}(R)$, $R \in \Sigma_{\mathcal{R}}$.

In particular, all the operators $R_{A \wedge B}^{\pm 1}, R_{A \vee B}^{\pm 1}$ belong to $\Sigma_{\mathcal{R}}$, so that every element $R \in \Sigma_{\mathcal{R}}$ is the Riesz operator of the dual pair of Hilbert spaces $\mathcal{H}(R), \mathcal{H}(R^{-1})$. Note also that the norms corresponding to any $R, S \in \Sigma_{\mathcal{R}}$ are consistent, since all operators in $\Sigma_{\mathcal{R}}$ are closed. Thus the family $\Sigma_{\mathcal{R}}$ is admissible, in the sense of Section 5.5.1.

Then the family $\Sigma_{\mathcal{R}}$ obtained in this way generates an involutive lattice of Hilbert spaces $\mathcal{I}(\Sigma_{\mathcal{R}})$ indexed by self-adjoint operators. We have the following picture:

$$\mathcal{D} \subseteq V^{\#} = \bigcap_{R \in \Sigma_{\mathcal{R}}} \mathcal{H}(R) = \bigcap_{A \in \mathcal{O}} \mathcal{H}(R_A) \subset \Big\langle \mathcal{H}(R_A), A \in \mathcal{O} \Big\rangle \subset \mathcal{H} \subset \ldots$$

$$\ldots \subset \Big\langle \mathcal{H}(R_A^{-1}), A \in \mathcal{O} \Big\rangle \subset V := \sum_{A \in \mathcal{O}} \mathcal{H}(R_A^{-1})$$

$$= \sum_{R \in \Sigma_{\mathcal{R}}} \mathcal{H}(R), \tag{5.33}$$

where $\big\langle \mathcal{H}(R_A), A \in \mathcal{O} \big\rangle$ denotes the lattice generated by the operators R_A according to the rules (5.31), and similarly for the other one. Actually, this lattice is peculiar, in the sense that each space $\mathcal{H}(R_A)$ is contained in $\mathcal{H}$ and each space $\mathcal{H}(R_A^{-1})$ contains $\mathcal{H}$.

Remark 5.5.5. The space $\mathcal{H}(R_A)$ does not determine the operator R_A, or A, uniquely. Indeed one sees easily that $\mathcal{H}(R_A) = \mathcal{H}(R_B)$ wherever $R_A^{1/2} R_B^{-1/2}$ is bounded with bounded inverse. This defines an equivalence relation $\sim$ on the set of all self-adjoint operators on $\mathcal{H}$, and on $\mathcal{R}$ in particular. Each $\mathcal{H}(R)$ is determined by a family of equivalent self-adjoint operators. By the spectral theorem, this family contains a self-adjoint operator $\tilde{R}$ with purely discrete spectrum (see the references in the Notes). This observation reduces the classification of operator domains to that of sequences of positive numbers. However, although we start with $R \in \mathcal{R}, \tilde{R}$ need not belong to $\mathcal{R}$ anymore. In particular, even if R leaves $\mathcal{D}$ invariant, $\tilde{R}$ need not do so. But this is irrelevant for the present purpose: we consider only the lattice $\mathcal{I}(\Sigma_{\mathcal{R}}) = \{\mathcal{H}(R), R \in \Sigma_{\mathcal{R}}\}$ and this one is fully determined by the family $\mathcal{O}$.

Following the definition given in Chapter 2, the lattice $\mathcal{I}(\Sigma_{\mathcal{R}}) = \{\mathcal{H}(R), R \in \Sigma_{\mathcal{R}}\}$ is a LHS with central Hilbert space $\mathcal{H}$ and total space $V = \sum_{A \in \mathcal{O}} \mathcal{H}(R_A^{-1})$. Thus we may state

Theorem 5.5.6. *Let $\mathcal{O}$ be a family of closable linear operators on a Hilbert space $\mathcal{H}$, with common dense domain $\mathcal{D}$. Assume that the corresponding set $\Sigma_{\mathcal{R}}$ is an admissible cone. Let $\mathcal{I}(\Sigma_{\mathcal{R}})$ be the lattice of Hilbert spaces generated by $\mathcal{O}$, as in Eq. (5.33). Then*

(i) *$\mathcal{O}$ generates a PIP-space, with central Hilbert space $\mathcal{H}$ and total space $V = \sum_{A \in \mathcal{O}} \mathcal{H}(R_A^{-1})$, where $\mathcal{H}(R_A^{-1})$ is the completion of $\mathcal{H}$ in the norm $\|(1 + A^*\overline{A})^{-1/2} \cdot \|$. The compatibility is*

$$f \# g \iff \exists R \in \Sigma_{\mathcal{R}} \text{ such that } f \in \mathcal{H}(R), g \in \mathcal{H}(R^{-1}) \qquad (5.34)$$

and the partial inner product is

$$\langle f | g \rangle = \langle R^{1/2} f | R^{-1/2} g \rangle. \qquad (5.35)$$

(ii) *The lattice $\mathcal{I}(\Sigma_{\mathcal{R}})$ itself is a LHS, with central Hilbert space $\mathcal{H}$, with respect to the compatibility (5.34) and the partial inner product (5.35) inherited from the PIP-space of (i).*

(iii) *One has $V^{\#} = \bigcap_{A \in \mathcal{O}} \mathcal{H}(R_A)$, where $\mathcal{H}(R_A)$ is $\mathcal{D}(\overline{A})$ with the graph norm $\|(1 + A^*\overline{A})^{1/2} \cdot \|$, and $\mathcal{D} \subseteq V^{\#}$.*

By this construction, the space $V^{\#}$ acquires a natural topology $t_{\mathcal{R}}$, as the projective limit of all the spaces $\mathcal{H}(R), R \in \mathcal{R}$ (or, equivalently, $R \in \Sigma_{\mathcal{R}}$). With this topology, $V^{\#}$ is complete and semi-reflexive, with dual V. However the Mackey topology $\tau(V^{\#}, V)$ may be strictly finer than the projective topology. On the dual V, on the contrary, the topology of the inductive limit of all the $\mathcal{H}(R), R \in \mathcal{R}$, coincides with both $\tau(V, V^{\#})$ and $\beta(V, V^{\#})$, i.e., V is barreled.

If the family $\mathcal{O} \equiv \{A_i\}$ is finite, then $V^{\#}$ is the Hilbert space $\mathcal{H}(R_C)$ with $R_C = 1 + \sum_i A_i^* \overline{A_i}$. If $\mathcal{O}$ is countable, $V^{\#}$ is a reflexive Fréchet space (and then $\tau(V^{\#}, V)$ coincides with $t_{\mathcal{R}}$). Otherwise $V^{\#}$ is nonmetrizable.

The most interesting case arises when we start with a *-invariant family $\mathcal{O}$ of closable operators, with a common dense *invariant* domain $\mathcal{D}$. For then $\mathcal{O}$ generates a *-algebra $\mathfrak{M}$ of operators on $\mathcal{D}$, i.e., an O*-algebra. We equip $\mathcal{D}$ with the natural projective topology $t_{\mathfrak{M}}$ defined by $\mathfrak{M}$. Then all the operators $A \in \mathfrak{M}$ are continuous from $\mathcal{D}$ into $\mathcal{D}$. The domain $\mathcal{D}$ need not be complete in topology $t_{\mathfrak{M}}$. In any case, its completion $\widetilde{\mathcal{D}}(\mathfrak{M})$ coincides with the full closure $\widehat{\mathcal{D}}(\mathfrak{M}) := \bigcap_{A \in \mathfrak{M}} D(\overline{A})$, and we have $V^{\#} = \widehat{\mathcal{D}}(\mathfrak{M})$. In other words, we can assume from the beginning that the *-algebra $\mathfrak{M}$ is fully closed.[1]

Theorem 5.5.7. *Let $\mathfrak{M}$ be an O*-algebra on the dense domain $\mathcal{D} \subset \mathcal{H}$. Then $\mathfrak{M}$ generates a PIP-space structure and a LHS structure on $V = \sum_{A \in \mathfrak{M}} \mathcal{H}(R_A^{-1})$. Then the subspace $V^{\#} = \bigcap_{A \in \mathfrak{M}} \mathcal{H}(R_A)$ is the completion*

---

[1] We recall that an O*-family $\mathcal{O}$ is *closed* if $\mathcal{D} = \widetilde{\mathcal{D}}(\mathcal{O})$ and *fully closed* if $\mathcal{D} = \widetilde{\mathcal{D}}(\mathcal{O}) = \widehat{\mathcal{D}}(\mathcal{O})$ (the two notions coincide for an O*-algebra). In addition, $\mathcal{O}$ is *self-adjoint* if $\mathcal{D} = \widehat{\mathcal{D}}(\mathcal{O}) = \mathcal{D}^*(\mathcal{O}) := \bigcap_{A \in \mathcal{O}} D(A^*)$. See our monograph [AIT02, Sec. 2.2] for a detailed discussion.

$\widehat{\mathcal{D}}(\mathfrak{M})$ of $\mathcal{D}$ in the $\mathfrak{M}$-topology and $\widehat{\mathfrak{M}} \subseteq \mathrm{Reg}(V)$, where $\widehat{\mathfrak{M}}$ is the full closure of $\mathfrak{M}$ and $\mathrm{Reg}(V)$ denotes the set of regular operators on V.

Examples 5.5.8. (i) Let $\mathcal{O}$ consist of only one closable operator A. If we assume A to be essentially self-adjoint, then $\mathcal{D} := \bigcap_{n \geq 0} D(|A|^n)$, where $|A| = (A^* \overline{A})^{1/2}$, is an invariant domain and the construction gives the standard Hilbert scale generated by A.

(ii) Let $\mathcal{H} = L^2(\mathbb{R})$ and $\mathcal{O}$ consist of $q = x$ and $p = -i\frac{d}{dx}$ on $\mathcal{D} := \mathcal{D}(q) \cap \mathcal{D}(p)$. Then the finite LHS generated according to Theorem 5.5.6 contains nine spaces ($\hookrightarrow$ denotes a continuous embedding):

$$V^{\#}=\mathcal{D}=\mathcal{H}(2+p^2+q^2) \hookrightarrow \left\{ \begin{array}{c} \mathcal{H}(1+p^2) \\ \mathcal{H}(1+p^2) \end{array} \right\} \hookrightarrow \mathcal{H}\left([(1+p^2)^{-1}+(1+q^2)^{-1}]^{-1}\right) \hookrightarrow \dots$$

$$(5.36)$$

$$\dots \hookrightarrow L^2 \hookrightarrow \mathcal{H}((1+p^2)^{-1}+(1+q^2)^{-1}) \hookrightarrow \left\{ \begin{array}{c} \mathcal{H}((1+p^2)^{-1}) \\ \mathcal{H}((1+q^2)^{-1}) \end{array} \right\} \hookrightarrow \mathcal{H}((2+p^2+q^2)^{-1})=V$$

As for Theorem 5.5.7, the invariant domain is the Schwartz space $\mathcal{S}(\mathbb{R}^3) = D^{\infty}(p^2 + q^2)$ and, of course, the LHS is the lattice generated by the algebra of all polynomials in p and q.

(iii) Let $\mathcal{H} = \ell^2$ and $\mathcal{O}$ the algebra of infinite diagonal matrices, $R = \mathrm{diag}(r_n)$, on the invariant domain $\mathcal{D} = \varphi$, the space of finite sequences. Then the LHS generated is the familiar one, $\langle \varphi, \omega \rangle$ with $\mathcal{H}(R) = \ell^2(\overline{r})$. One may obtain an entirely similar situation if one starts from a suitable family of multiplication operators in the space $L^2(X, d\mu)$.

We now come back to the original RHS structure associated to $\mathcal{O}$ and ask whether the LHS constructed here is compatible (in the sense of Remark 5.4.5) with the duality properties of the family of Hilbert spaces which constitutes it.

The answer is in general negative, since even though $V^{\#}$ is dense in every one of the Hilbert spaces $\mathcal{H}(R)$, $R \in \Sigma_{\mathcal{R}}$, as a consequence of the PIP-space structure, this is no longer true for $\mathcal{D}$ itself; $\mathcal{D}$, in fact, may fail to be dense in $V^{\#}$ endowed with $t_{\mathcal{R}}$. In order to avoid this pathology, we are led once more to impose the condition that the intersection of any finite number of interspaces be an interspace. This is guaranteed, of course, if $\mathcal{D}$ is dense in $V^{\#}$ endowed with $t_{\mathcal{R}}$. As it is well-known, this always happens if $\mathcal{O}$ is an O*-algebra.

Proposition 5.5.9. *Let $\mathcal{D} \subset \mathcal{H} \subset \mathcal{D}^{\times}$ be the RHS associated to the family $\mathcal{O}$. If $\mathcal{D}$ is dense in $V^{\#}[t_{\mathcal{R}}]$, then the LHS generated by $\mathcal{O}$ is compatible with the duality of the RHS and it is, therefore, a refinement of the latter. The RHS generated by an O*-algebra $\mathcal{O}$ can always be refined to a LHS compatible with its duality properties.*

5.5.3 Algebras of Bounded Regular Operators

Let V_I be an arbitrary LHS, with central Hilbert space $\mathcal{H}_o$. The *-algebra $\mathrm{Reg}(V_I)$ of all regular operators contains three remarkable *-subalgebras, which have been described in Section 3.3.3:

$$\mathrm{Reg}(V_I) \supset \mathfrak{A}(V_I) \supset \mathfrak{B}(V_I) \supset \mathfrak{C}(V_I).$$

We will reformulate this description in terms of Riesz operators, along the lines of Section 5.5.2.

Let $t \in I$ be such that $\mathcal{H}_t \subset \mathcal{H}_o$ (here $\mathcal{H}_t$ denotes the Hilbert space, with a well-defined norm, not only the underlying vector space !). Then there exists a unique self-adjoint positive operator T such that $\mathcal{H}_t = D(T)$, with $\|f\|_t^2 = \|f\|_o^2 + \|Tf\|_o^2$; so we may write $\mathcal{H}_t = \mathcal{H}(R_T)$, as in Section 5.5.2. The operator T is unbounded if $\mathcal{H}_t \neq \mathcal{H}_o$. Thus the Riesz operator $U_{(t)} : \mathcal{H}_t \to \mathcal{H}_{\bar{t}}$ is simply $R_T = 1 + |T|^2$.

Let now $\mathcal{H}_s$ be noncomparable with $\mathcal{H}_o$. Then $t := o \wedge s \in I$ and $\mathcal{H}_t = \mathcal{H}_{o \wedge s} = \mathcal{H}_o \cap \mathcal{H}_s$, with norm $\|f\|_t^2 = \|f\|_o^2 + \|f\|_s^2$. On the other hand, since $\mathcal{H}_t \subset \mathcal{H}_o$, we may again write $\|f\|_t^2 = \|f\|_o^2 + \|Tf\|_o^2$, that is, $\mathcal{H}_t = \mathcal{H}(R_T)$. Thus $\mathcal{H}_s$ is the completion of $\mathcal{H}_t$ in the norm $\|f\|_s = \|Tf\|_o$, i.e., $\mathcal{H}_s = \mathcal{H}(R_T')$, where the positive self-adjoint operator T is uniquely defined by the interpolation argument and it is invertible, since $\| \cdot \|_s$ is a norm. The corresponding Riesz operator is $U_{(s)} = |T|^2$. Thus we may as well index the LHS by the set of all Riesz operators.

When V_I is generated by some family $\mathcal{O}$ of unbounded operators, in particular a (partial) O*-algebra $\mathfrak{M}$, the same description applies, with the additional restriction that the lattice $\mathcal{I}_{\Sigma_\mathcal{R}}$, described in (5.33), is generated by a family of spaces $\mathcal{H}(R_T^{\pm 1})$ with $T \in \mathcal{O}$, resp. $T \in \mathfrak{M}$. In that case, the set of all Riesz operators coincides with the family $\Sigma_\mathcal{R}$ (elements of the latter will be denoted generically by R, as usual).

Using this language, the three algebras $\mathfrak{A}, \mathfrak{B}, \mathfrak{C}$ are described as follows (for simplicity, we write $\mathcal{H} := \mathcal{H}_o$).

(i) First one defines : $\mathfrak{A}(V_I) = \{X \in \mathcal{B}(\mathcal{H}) : X : \mathcal{H}(R) \to \mathcal{H}(R)$ continuously, $\forall R \in \Sigma_\mathcal{R}\}$. Since $R^{\pm 1/2}$ are unitary between $\mathcal{H}(R)$ and $\mathcal{H}$, we have equivalently :

$$\mathfrak{A}(V_I) = \{X \in \mathcal{B}(\mathcal{H}) : R^{1/2} X R^{-1/2} \in \mathcal{B}(\mathcal{H}), \, \forall R \in \Sigma_\mathcal{R}\}.$$

The *-algebra $\mathfrak{A}(V_I)$ carries a natural topology, defined by the family of norms

$$\|X\|_R := \|R^{1/2} X R^{-1/2}\|, \, R \in \Sigma_\mathcal{R},$$

and it is complete in that topology as the projective limit of the Banach spaces $\mathcal{B}(\mathcal{H}(R))$.

(ii) $\mathfrak{B}(V_I)$ is the *-subalgebra of $\mathfrak{A}(V_I)$ defined as follows:

$$\mathfrak{B}(V_I) = \{X \in \mathfrak{A}(V_I) : \|X\|_{\mathcal{R}} := \sup_{R \in \Sigma_{\mathcal{R}}} \|R^{1/2}XR^{-1/2}\| < \infty\}.$$

It is complete in the norm $\|.\|_{\Sigma_{\mathcal{R}}}$, i.e., it is a Banach algebra.

(iii) Finally $\mathfrak{C}(V_I)$ consists of those elements of $\mathfrak{B}(V_I)$ which commute with all Riesz operators $U_{(r)} : \mathcal{H}(R) \to \mathcal{H}(R^{-1})$, that is, with all $R \in \Sigma_{\mathcal{R}}$. Since every $R \in \Sigma_{\mathcal{R}}$ is self-adjoint, it follows that an element of $\mathfrak{C}(V_I)$ will commute with every $R^{\pm 1/2}$ as well. Thus we have $\mathfrak{C}(V_I) = (\Sigma_{\mathcal{R}})' = \mathcal{R}'$ and therefore $\mathfrak{C}(V_I)$ is a von Neumann algebra.

From Section 3.3.3, we know that the algebra $\mathfrak{A}(V_I)$ is intrinsic, in the sense that it depends only on the topologies of the assaying spaces V_r, not on their norms, whereas $\mathfrak{B}(V_I)$ and $\mathfrak{C}(V_I)$ are not in general. In the present case, however, the relation $\mathfrak{C}(V_I) = \mathcal{R}'$ implies that $\mathfrak{C}(V_I)$ is also intrinsic.

First we notice that, if V_I is nontrivial, at least two of the three algebras $\mathfrak{A}(V_I), \mathfrak{B}(V_I), \mathfrak{C}(V_I)$ must be distinct.

Proposition 5.5.10. *Let V_I be a LHS such that $\mathfrak{A}(V_I) = \mathfrak{B}(V_I) = \mathfrak{C}(V_I)$. Then V_I is a Hilbert space.*

Proof. Let $\mathcal{H}_t$ be any assaying subspace contained in $\mathcal{H}$, the central Hilbert space. Then $\mathcal{H}_t = \mathcal{H}(R_T)$, for some positive self-adjoint operator T, and the corresponding Riesz operator is $R_T = 1 + T^2$. Let now $f \in V^\#$ with $\|f\| = 1$. Then the projection operator $|f\rangle\langle f|$, which is in $\mathfrak{A}(V_I)$ by definition, is also in $\mathfrak{C}(V_I)$ by assumption, i.e., it commutes with R_T, and thus with T. This means that f is an eigenvector of T. The same is true for *any* $f \in V^\#$, which implies that T must be constant on the dense subspace $V^\#$ of $\mathcal{H}$, and therefore on all of $\mathcal{H}$, i.e., $\mathcal{H}_t = \mathcal{H}$. ∎

We assume now that V_I is the LHS generated by an O*-family on the domain $\mathcal{D}$. Let X belong to the strong commutant $\mathcal{O}'_s$ of $\mathcal{O}$, that is, $X \in \mathcal{B}(\mathcal{H}), X\mathcal{D} \subseteq \mathcal{D}$ and $XAf = AXf, \forall A \in \mathcal{O}, \forall f \in \mathcal{D}$. Then we have, for every $f \in \mathcal{D}$,

$$\begin{aligned}
\|Xf\|^2_{R_A} &= \|Xf\|^2 + \|AXf\|^2 \\
&= \|Xf\|^2 + \|XAf\|^2 \\
&\leqslant \|X\|^2 \|f\|^2_{R_A}.
\end{aligned}$$

Thus X may be extended to a bounded operator from every $\mathcal{H}(R_A)$ into itself, i.e., X commutes with every $A \in \mathcal{O}$ as an unbounded operator. By definition of the topologies defined in Eqs.(5.30)-(5.31), the same is true for all spaces in the lattice generated by $\{\mathcal{H}(R_A), A \in \mathcal{O}\}$, which are all contained in $\mathcal{H}$.

However the same reasoning does *not* apply to the larger space $\mathcal{H}(R_A^{-1})$. If we assume $\mathcal{O}$ to be *-invariant, the argument shows that $X \in \mathcal{O}'_s$ commutes with every $\overline{A^\dagger}$, where $A^\dagger \equiv A^* \restriction \mathcal{D}$, and this may be a proper restriction of A^*. Thus X need not commute with A^*, and thus not with $R_A^{\pm 1/2}$. The reason is that, in general, $\mathcal{O}'_s$ need not be $*$-invariant. We have, however:

Proposition 5.5.11. *Assume that V_I is the LHS generated by an O^*-family $\mathcal{O}$ on the domain $\mathcal{D}$. Let $X \in \mathcal{O}'_s \cap (\mathcal{O}'_s)^*$. Then $X \in \mathfrak{C}(V_I)$.*

Proof. The assumption means that $X = Y^*$ for some $Y \in \mathcal{O}'_s$. By the argument above, the operator $Y : \mathcal{H}(R_A) \to \mathcal{H}(R_A)$ is bounded and Y commutes with A, for every $A \in \mathcal{O}$. Since $\langle \mathcal{H}(R_A), \mathcal{H}(R_A^{-1}) \rangle$ is a dual pair for the inner product of $\mathcal{H}$, this means that $Y^* : \mathcal{H}(R_A^{-1}) \to \mathcal{H}(R_A^{-1})$ is also continuous. Since $Y^* = X$, this means $X \in \mathfrak{A}(V_I)$. Then an easy calculation shows that $X^* \in \mathcal{O}'_s$ implies that X commutes with every A^*. Since X already commutes with A, it follows that X commutes with R_A, thus also with $R_A^{-1}, R_A^{\pm 1/2}$, i.e., $X \in \mathcal{R}' = \mathfrak{C}(V_I)$.   ◼

With help of Lemma 5.5.11, we may obtain a precise information on the spectral properties of essentially self-adjoint regular operators.

Corollary 5.5.12. *Let V_I be the LHS defined by a family $\mathcal{O}$ of closable operators. Let $T = T^\times$ be a symmetric regular operator on V_I, essentially self-adjoint on $V^\#, \overline{T} = \int \lambda dE(\lambda)$ the spectral decomposition of its closure. Assume the algebra $\mathcal{T}$ generated by T satisfies the condition $\mathcal{T}'' \subseteq \mathcal{O}'_s$. Then $E(\lambda) \in \mathfrak{C}(V_I)$ for every $\lambda \in \mathbb{R}$.*

Proof. The result is immediate by Lemma 5.5.11, for every $E(\lambda) \in \mathcal{T}''$, and $\mathcal{T}''$ is a $*$-invariant subset of $\mathcal{O}'_s$, i.e., $E(\lambda) \in \mathcal{O}_s \cap (\mathcal{O}'_s)^*$.   ◼

When the family $\mathcal{O}$ is not related to the PIP-space V_I, the only conclusion is that each $E(\lambda)$ is regular. In the present case, V_I itself is generated by $\mathcal{O}$, and we get much more. In particular, $E(\lambda)$ is now, for every λ, an orthogonal projection in the PIP-space sense, and thus every subspace $E(\lambda)V$ is an orthocomplemented subspace of V_I. Also, if V_I is generated by an O^*-algebra $\mathfrak{M}$ (assumed to be closed for simplicity), then the result holds, in particular, for those elements $T \in \mathfrak{M}$ that satisfy the condition $\mathcal{T}'' \subseteq \mathfrak{M}'_s$, since $\mathfrak{M} \subseteq \mathrm{Reg}(V)$.

Another characterization yet of $\mathfrak{C}(V_I)$ can be given when V_I is generated by an O^*-algebra $\mathfrak{M}$, taken as closed for simplicity. First we have then $V^\# = \bigcap_{A \in \mathfrak{M}} \mathcal{H}(R_A)$, according to Theorem 5.5.7. Next define the set $|\mathfrak{M}| := \{|A| : A \in \mathfrak{M}\}$, where, as usual, $|A| \equiv (A^* \overline{A})^{1/2}$ is a nonnegative self-adjoint operator, with domain $D(|A|) = D(\overline{A})$. Thus $|\mathfrak{M}|$ is a $*$-invariant family of self-adjoint, hence closed, operators, with common dense core $D(|\mathfrak{M}|) = \cap_{A \in \mathfrak{M}} D(|A|) = \bigcap_{A \in \mathfrak{M}} \mathcal{H}(R_A) = V^\#$. This subspace is indeed a core for every $|A| \in |\mathfrak{M}|$, since it is dense in every $\mathcal{H}(R_A)$, which is $D(|A|) = D(\overline{A})$ with the graph norm. However, the set $|\mathfrak{M}|$ is in general neither an algebra, nor a subset of $\mathfrak{M}$. Let $|\mathfrak{M}|'$ be its commutant, the set of bounded operators commuting with the elements of $|\mathfrak{M}|$ in the usual sense, that is, the bounded operators commuting with the spectral projections $E(\lambda)$ of all the self-adjoint operators $|A|$, $A \in \mathfrak{M}$. In that case, $|\mathfrak{M}|'$ consists of all bounded operators which commute with all the operators $(1 + |A|)^{-1}$. Since the latter constitute a *-invariant family of bounded operators, its commutant $|\mathfrak{M}|'$ is a von Neumann algebra.

We have used the standard commutant $|\mathfrak{M}|'$, which is natural for a family of self-adjoint operators. However, we can also use a strong commutant. Since this notion applies only to a family of closable operators on the *same* dense domain, we have to consider the set $|\mathfrak{M}| \upharpoonright D(|\mathfrak{M}|)$. Then an element of its strong commutant, $X \in \left(|\mathfrak{M}| \upharpoonright D(|\mathfrak{M}|) \right)'_s$, satisfies $X : D(|\mathfrak{M}|) \to D(|\mathfrak{M}|)$ and $X|A|f = |A|Xf$, $\forall A \in \mathfrak{M}$, $\forall f \in D(|\mathfrak{M}|)$. Since the latter is a core for every $|A|$, $A \in \mathfrak{M}$, we have $\overline{|A| \upharpoonright D(|\mathfrak{M}|)} = |A|$. Given $A \in \mathfrak{M}$ and $f \in D(|A|)$, take a sequence $f_n \in D(|\mathfrak{M}|)$ such that $f = \lim_n f_n$ and $|A|f = \lim_n |A|f_n$. Then we may write $X|A|f = \lim_n XAf_n = \lim_n |A|Xf_n$. Hence, $X : D(|A|) \to D(|A|)$ and $X|A|f = |A|Xf$, $\forall f \in D(|A|)$. This means that X commutes with $|A|$ in the usual sense. In other words, we have $\left(|\mathfrak{M}| \upharpoonright D(|\mathfrak{M}|) \right)'_s = |\mathfrak{M}|'$.

This being said, we have:

Proposition 5.5.13. *Let V_I be the LHS defined by the closed O*-algebra $\mathfrak{M}$. Let $|\mathfrak{M}| = \{|A| : A \in \mathfrak{M}\}$. Then $\mathfrak{C}(V_I) = |\mathfrak{M}|'$.*

Proof. The proof is immediate. We know that $\mathfrak{C}(V_I) = \mathcal{R}'$, where $\mathcal{R}$ is the family of self-adjoint operators generated from $\{R_A : A \in \mathfrak{M}\}$, as discussed above. Then $\mathcal{R}' = |\mathfrak{M}|'$, since $X \in \mathcal{R}'$ if and only if X commutes with every $R_A = 1 + A^*\overline{A} = 1 + |A|^2$, i.e., X commutes with every $|A|$. ∎

If, in addition, $\mathfrak{M}$ is a self-adjoint O*-algebra, then $\mathfrak{M}'_s = \mathfrak{M}'$ is a von Neumann algebra. Hence, by Lemma 5.5.11, $\mathfrak{M}' \subseteq \mathfrak{C}(V_I) = |\mathfrak{M}|'$.

5.5.4 The Case of the Scale Generated by a Positive Self-Adjoint Operator

Now we particularize the previous discussion where the family $\mathcal{O}$ consists of a single positive self-adjoint operator A. Then the PIP-space generated is the canonical scale (5.1), with $V^\# = \mathcal{D}^\infty(A)$, and A is essentially self-adjoint on $V^\#$. First, the O*-algebra $\mathfrak{M}$ generated by A is then self-adjoint, so that $\mathfrak{M}' \subseteq \mathfrak{C} = |\mathfrak{M}|' = \mathcal{R}'$. This implies that $\mathfrak{C}$ contains the spectral projections of every self-adjoint B commuting (strongly) with A, in particular those of A itself. Therefore, every such $E(\lambda)$ is an orthogonal projection in the PIP-space V_I, and even a totally orthogonal one (see Section 3.4.1). Equivalently, for every $\lambda \in \mathbb{R}$, the subspace $E(\lambda)V$ is orthocomplemented in V_I. Thus we have here a PIP-space with a large number of orthocomplemented subspaces (but not "too many", in the sense of Section 3.4.6).

Let us consider the algebra $\mathfrak{A} = \mathfrak{A}(V_I)$. Every element $T \in \mathfrak{A}$ has a bounded representative T_{00}. Let $\mathfrak{A}_0$ denote the algebra of those representatives (of course $\mathfrak{A}_0$ is isomorphic to $\mathfrak{A}$). Then we have:

Proposition 5.5.14. *Let A be a positive self-adjoint operator in $\mathcal{H}$. In the canonical Hilbert scale defined by A, denote by $\mathfrak{A}_0$ the *-algebra of the $(0,0)$-representatives of the elements of $\mathfrak{A}(V_I)$. Then $\mathfrak{A}_0$ is weakly dense in $\mathcal{B}(\mathcal{H})$, i.e., $(\mathfrak{A}_0)'' = \mathfrak{B}(\mathcal{H})$.*

Proof. By definition, $X \in \mathcal{B}(\mathcal{H})$ belongs to $\mathfrak{A}$ if and only if $X \in \mathcal{B}(\mathcal{H}_n)$ for $n = 1, 2, \ldots$, i.e., $\|R_{A^n}^{1/2} X R_{A^n}^{-1/2}\| < \infty$ for all n. Let X be a self-adjoint element of $\mathcal{B}(\mathcal{H})$; we show that $E(\lambda) X E(\lambda) \in \mathfrak{A}$ for every spectral projection $E(\lambda)$ of A. Using the representation:

$$R_{A^n} = \int_{\mathbb{R}} (1 + \lambda^{2n}) \, dE(\lambda),$$

we have for every $n = 1, 2, \ldots$, and any $f \in \mathcal{H}_n$:

$$\|E(\lambda) X E(\lambda) f\|_n = \|R_{A^n}^{1/2} E(\lambda) X E(\lambda) f\|$$
$$\leqslant (1 + \lambda^{2n})^{1/2} \|X\| \, \|f\|$$
$$\leqslant (1 + \lambda^{2n})^{1/2} \|X\| \, \|f\|_n$$

and, therefore, $E(\lambda) X E(\lambda) \in \mathcal{B}(\mathcal{H}_n)$. Since $E(\lambda) X E(\lambda)$ is self-adjoint in $\mathcal{H}$, it extends to a symmetric operator in the scale, and by duality $E(\lambda) X E(\lambda)$ is bounded in each $\mathcal{H}_{\overline{n}}, n = 1, 2 \ldots$, as well. So it belongs to $\mathfrak{A}$. Now, when $\lambda \to \infty, (E(\lambda) X E(\lambda))_{00}$ converges weakly to X_{00}, so that the weak closure $\mathfrak{A}_0''$ of $\mathfrak{A}_0$ in $\mathcal{B}(\mathcal{H})$ contains all bounded self-adjoint elements of $\mathcal{B}(\mathcal{H})$, i.e. $\mathfrak{A}_0'' = \mathcal{B}(\mathcal{H})$. ∎

In many examples of such scales, the generating self-adjoint operator A is positive. If $A \geqslant 1$, we can say a little more. First, obviously $\mathfrak{A}' = |\mathfrak{A}|' = \mathfrak{C}$. What about $\mathfrak{B}$?

Proposition 5.5.15. *Let A be self-adjoint with $A \geqslant 1$. Then, in the canonical scale generated by A, one has $\mathfrak{B} = \mathfrak{C} = \mathfrak{A}' = \mathcal{R}'$.*

Proof. By the remark above, we know already that $\mathfrak{A}' = \mathfrak{C} = \mathcal{R}' \subseteq \mathfrak{B}$. It is enough to show that any $X \in \mathfrak{B}$ commutes with A. Consider $1 < \lambda_1 < \lambda_2$ and the intervals $\Delta_1 = (1, \lambda_1], \Delta_2 = [\lambda_2, \infty)$, with $E(\Delta_1), E(\Delta_2)$ the corresponding spectral projections of $A = \int \lambda \, dE(\lambda)$. Then, for any $f, g \in \mathcal{H}$, we have:

$$|\langle E(\Delta_1) f | X E(\Delta_2) g \rangle| = |\langle R_{A^n}^{1/2} E(\Delta_1) f \mid R_{A^n}^{-1/2} X E(\Delta_2) g \rangle|$$
$$\leqslant (\sup_n \|R_{A^n}^{-1/2} X R_{A^n}^{1/2}\|).\|R_{A^n}^{1/2} E(\Delta_1) f\|.\|R_{A^n}^{-1/2} E(\Delta_2) g\|$$
$$\leqslant C(1 + \lambda_1^{2n})^{1/2}(1 + \lambda_2^{2n})^{-1/2} \|f\| \, \|g\|, \text{ since } X \in \mathfrak{B}.$$

For $n \to \infty$, this tends to zero since $\lambda_1 < \lambda_2$. So we have

$$E(\Delta_1)XE(\Delta_2) = 0.$$

Proceeding in the same way with X^*, we get $E(\Delta_2)XE(\Delta_1) = 0$. Taking in particular $\lambda_1 = \lambda, \lambda_2 = \lambda + \epsilon$ (for any $\lambda > 0$) and letting $\epsilon \to 0$, these two relations become :

$$E(\lambda)X(1 - E(\lambda)) = (1 - E(\lambda))XE(\lambda) = 0,$$

which means that X commutes with every spectral projection $E(\lambda)$ of A, i.e. $X \in \mathfrak{A}'$. Thus $\mathfrak{B} = \mathfrak{A}'$. ∎

So, for the scale generated by a positive self-adjoint operator, we always have, by Lemma 5.5.10, $\mathfrak{A} \supsetneq \mathfrak{B} = \mathfrak{C}$. This implies, of course, that $\mathfrak{B}$ is also intrinsic. The same result holds for a general LHS provided it contains sufficiently many such chains.

Proposition 5.5.16. *Let V_I be a LHS with lattice $\mathcal{I}$. Given any assaying subspace $\mathcal{H}_t \subset \mathcal{H}$, let T be the unique positive self-adjoint operator such that $\mathcal{H}_t = \mathcal{H}(R_T), R_T = 1 + |T|^2$. Assume that for every such $\mathcal{H}_t \subset \mathcal{H}$, the continuous scale $V(T)$ generated by the corresponding T is contained in $\mathcal{I}$. Then $\mathfrak{A} \supset \mathfrak{B} = \mathfrak{C}$.*

Proof. Let $A \in \mathfrak{B}$, i.e., $\sup_{r \in I} \|A\|_{rr} < \infty$. Given $\mathcal{H}_t \equiv \mathcal{H}(R_T) \subset \mathcal{H}$, A belongs *a fortiori* to the algebra $\mathfrak{B}(V(T))$ of the scale $V(T)$ generated by T. By Proposition 5.5.15, A commutes with the Riesz operators of that scale, in particular with $R_T = 1 + |T|^2$. Let now $\mathcal{H}_s \in \mathcal{I}$ be arbitrary. Then, as discussed at the beginning of Section 5.5.3, the Riesz operator of $\mathcal{H}_s$ is $U_{(S)} = |S|^2$, where $\mathcal{H}(R_S) = \mathcal{H}_s \cap \mathcal{H}$, and so, by the preceding argument, A commutes with $U_{(S)}$. Thus A commutes with every Riesz operator of V_I, i.e., $A \in \mathfrak{C}$. ∎

Conversely, we may know beforehand that $\mathfrak{B} = \mathfrak{C}$; then the same relation holds in the finer LHS obtained by adding all such scales.

Corollary 5.5.17. *Let V_I be a LHS for which $\mathfrak{B}(V_I) = \mathfrak{C}(V_I)$. For any $\mathcal{H}_t = \mathcal{H}(R_T) \subset \mathcal{H}$, let $V(T)$ the continuous scale generated by T. Let $V_{\bar{I}}$ be the finer, and possibly larger, LHS obtained by adding to V_I every such scale $V(T)$. Then $\mathfrak{B}(V_{\bar{I}}) = \mathfrak{C}(V_{\bar{I}})$.*

Proof. Let $A \in (V_{\bar{I}})$. In particular $A \in \mathfrak{B}(V_I) = \mathfrak{C}(V_I)$, so that A commutes with the Riesz operator R_T associated to every $\mathcal{H}_t = \mathcal{H}(R_T)$. Thus it also commutes with all powers of T, that is, all Riesz operators in the scale $V(T)$, and, therefore, with all Riesz operators in $V_{\bar{I}}$. So finally $A \in \mathfrak{C}(V_{\bar{I}})$. ∎

The result applies in particular when V_I is a discrete scale. $V_{\bar{I}}$ is then the continuous scale obtained from V_I by interpolation, and we know that

$\mathfrak{A}(V_{\bar{I}}) = \mathfrak{A}(V_I)$. Then Corollary 5.5.17 implies $\mathfrak{B}(V_I) = \mathfrak{C}(V_I) = \mathfrak{B}(V_{\bar{I}}) = \mathfrak{C}(V_{\bar{I}})$. The same result holds for a discrete lattice. But, in fact, we may go much further.

Proposition 5.5.18. *Let V_I be any LHS generated by a closed O^*-algebra $\mathfrak{M}$ on $V^{\#}$. Then $\mathfrak{B}(V_I) = \mathfrak{C}(V_I)$.*

Proof. Let A be any element of $\mathfrak{M}$, $\mathcal{H}(R_A) = \mathcal{H}(R_{|A|})$ the corresponding Hilbert space in V_I with Riesz operator $R_{|A|} = 1 + |A|^2$. The operator $A^*\overline{A} = |A|^2$ is self-adjoint and positive. Let $V(|A|^2)$ be the corresponding scale, which obviously verifies the condition $\mathfrak{B} = \mathfrak{C}$, by Proposition 5.5.15. The restriction of $|A|^2$ to $V^{\#}$ belongs to $\mathfrak{M}$, and similarly for all powers $|A|^{2n}, n = 1, 2, \ldots$. Thus every space in $V(|A|^2)$ belongs to V_I. So every $T \in \mathfrak{B}(V_I)$ belongs also to $\mathfrak{B}(V(|A|^2))$ which coincides with $\mathfrak{C}(V(|A|^2))$. In other words, T commutes with $|A|^2$, and therefore also with $|A|$ and $R_{|A|}$. Since the same is true for every $A \in \mathfrak{M}, T$ commutes with all Riesz operators of V_I, that is, $T \in \mathfrak{C}(V_I)$. ∎

This last proposition applies to many familiar cases. For instance:

(i) $\omega \simeq \{\ell^2(r)\}$, which is generated by the algebra of all infinite diagonal matrices, acting on φ.

(ii) $\mathcal{S}^{\times} \simeq \{\ell^2(r), r \text{ tempered}\}$, corresponding to diagonal tempered matrices acting on $\mathcal{S}$.

(iii) $\mathcal{S}^{\times}$ generated by the algebra of polynomials in p and q acting on $\mathcal{S}$.

As for the LHS $L^1_{\text{loc}} \simeq \{L^2(r)\}$, it is not generated by an algebra, and does not satisfy the condition of Proposition 5.5.16 either ($r^{\pm 1} \in L^2_{\text{loc}}$ obviously does not imply $r^{\pm n} \in L^2_{\text{loc}}$ for all n). In fact it is an open question whether $\mathfrak{B}(L^1_{\text{loc}}) = \mathfrak{C}(L^1_{\text{loc}})$, as we have already seen in Section 4.2.3.

To conclude this section, we consider a totally different situation, namely a *finite* Hilbert scale, discrete or continuous, e.g. obtained by truncation of the scale $\{\mathcal{H}_n\}_{n \in \mathbb{Z}}$:

$$\mathcal{H}_m \hookrightarrow \ldots \hookrightarrow \mathcal{H} \hookrightarrow \ldots \hookrightarrow \mathcal{H}_{\overline{m}}.$$

Every regular operator S maps both $\mathcal{H}_m$ and $\mathcal{H}_{\overline{m}}$ continuously into themselves. Then by interpolation, S maps every $\mathcal{H}_r$ ($m \leqslant r \leqslant \overline{m}$) into itself continuously, and moreover the norm $\|S\|_{rr}$ is a logarithmically convex function of r. Therefore $\text{Reg}(\mathcal{H}_{\overline{m}}) = \mathfrak{A} = \mathfrak{B}$ for such a finite scale and thus $\mathfrak{B} \neq \mathfrak{C}$ by Lemma 5.5.10. In particular, every regular operator S is bounded. Hence $D(\overline{S}) = \mathcal{H}$ and therefore $\text{Reg}(\mathcal{H}_{\overline{m}})$ is *not* a closed algebra: its closure has $\mathcal{H}$ for domain, and is a *-subalgebra of $\mathcal{B}(\mathcal{H})$. Notice that the last result remains true if the family $\{\mathcal{H}_r, m \leqslant r \leqslant \overline{m}\}$ is only assumed to be a chain, instead of a scale. $\text{Reg}(\mathcal{H}_{\overline{m}})$ is again not closed, but it does not necessarily coincide with $\mathfrak{A}$, since no interpolation theorem is available in general.

The same situation prevails for certain chains of Banach spaces. Such is, for instance, the chain of sequence spaces $\{\ell^p, 1 \leqslant p \leqslant \infty\}$. For every

$S \in \text{Reg}(\ell^\infty)$, one has $S : \ell^1 \to \ell^1$ and $S : \ell^\infty \to \ell^\infty$, continuously for the respective weak and Mackey topologies, and also for the strong (i.e. norm) topology on ℓ^∞, since $\ell^\infty[\tau(\ell^\infty, \ell^1)]$ is semi-reflexive. Then again, by the Riesz-Thorin interpolation theorem, $S : \ell^p \to \ell^p$ continuously and $\text{Reg}(\ell^\infty) = \mathfrak{A} = \mathfrak{B}$. The same result holds true for the chain $\{\mathcal{C}^p, 1 \leqslant p \leqslant \infty\}$ of ideals of compact operators on a Hilbert space, as well as for the whole lattice generated by the family $\{L^p(X, \mu), 1 \leqslant p \leqslant \infty\}$, where (X, μ) is any measure space (Example 4.4.4). More generally, for any scale of Banach spaces which has the so-called normal interpolation property.

5.5.5 The PIP-Space Generated by Regular Operators

Let $V_I = \{V_r, r \in I\}$ be an arbitrary PIP-space around $\mathcal{H}$, with involution $\# : V_r \leftrightarrow V_{\overline{r}}$. Then the algebra $\text{Reg}(V)$ of all regular operators on V_I may be identified with an O*-algebra acting on $V^\#$. By the construction described in Section 5.5.2, this algebra generates a new PIP-space structure around $\mathcal{H}$, in fact a LHS, with lattice $\mathcal{I}_{\Sigma_R} = \{\mathcal{H}(R), R \in \Sigma_R\}$ and involution $\#_R : \mathcal{H}(R) \leftrightarrow \mathcal{H}(R^{-1})$, possibly different from $\#$. $\mathcal{I}_{\Sigma_R}$ is generated by the spaces $\mathcal{H}(R_A), A \in \text{Reg}(V)$, and their duals $\mathcal{H}(R_A^{-1})$, where $\mathcal{H}(R_A)$ is the form domain of $R_A = 1 + A^*\overline{A}$ with the norm $\langle \cdot | R_A \cdot \rangle^{1/2}$. How do these two PIP-space structures compare to each other ?

We will first concentrate on the extreme spaces $V^\#$ and V. The original space $V^\#$ carries now three natural topologies:

(i) the Mackey topology $\tau(V^\#, V)$;
(ii) the projective topology t_{proj} from the lattice $\{V_r : r \in I\}$;
(iii) the projective topology t_R defined by the algebra $\text{Reg}(V)$, i.e., by the lattice $\mathcal{I}_R = \{\mathcal{H}(R)\}$.

The topology t_{proj} is coarser than τ, and both give V as dual (see Chapter 2). If V_I is reflexive, i.e., $\langle V_r, V_{\overline{r}} \rangle$ is a reflexive dual pair for every $r \in I$, then $V^\#$ is semi-reflexive for both t_{proj} and τ, and V is barreled, $\tau(V, V^\#) = \beta(V, V^\#)$. If, in addition, every V_r is complete (e.g. for type (B) or (H)), then $V^\#$ is complete for t_{proj}. In general, the dual V need not be quasi-complete for its Mackey topology. However, in the following discussion, we will suppose the quasi-completeness of V. This assumption is quite strong, but it is sufficient to cover a large class of familiar examples. In this case, $\text{Reg}(V)$ can be identified with $L^\dagger(V^\#)$, the maximal O*-algebra on $V^\#$.

The topology t_R is also coarser than τ, since it is the coarsest topology on $V^\#$ for which every regular operator is continuous from $V^\#$ into $\mathcal{H}$. However, t_R and t_{proj} need not be comparable. Furthermore, $V^\#[\mathsf{t}_R]$ need not be complete, unless the algebra $\text{Reg}(V)$ is closed. Its completion is $(V^\#)_R := \overline{V^\#[\mathsf{t}_R]} = \bigcap_{R \in \mathcal{R}} \mathcal{H}(R)$. Its dual is the space $V_R := \sum_{R \in \mathcal{R}} \mathcal{H}(R^{-1})$. The space $(V^\#)_R$ is semi-reflexive, and its dual

V_R is barreled, $\tau(V_R, (V^\#)_R) = \beta(V_R, (V^\#)_R)$. Finally we note that $(V^\#)_R$ and V_R correspond to each other under the involution $\#_R$.

Let us endow each of the spaces V_R and V with its Mackey topology $\tau(., V^\#)$. Then the whole discussion may be summarized by the following diagram (in the case where $V^\#[t_{\text{proj}}]$ is complete):

$$V^\#[\tau] \hookrightarrow \left\{ \begin{array}{c} V^\#[t_R] \hookrightarrow (V^\#)_R \\ \\ V^\#[t_{\text{proj}}] \end{array} \right\} \hookrightarrow \mathcal{H} \hookrightarrow V_R \hookrightarrow V \qquad (5.37)$$

We will first assume that $V^\#[\tau]$ is complete. This happens, for instance, when every V_r is complete, in particular for all PIP-spaces of type (B) or (H). Then we have :

Proposition 5.5.19. *Let V_I be a PIP-space. Let $V^\#[t_R]$ be barreled. Then $V_R = V$. If, in addition, $V^\#[\tau]$ is complete, then the dual pair $\langle V^\#, V \rangle$ is reflexive and the algebra $\text{Reg}(V)$ is closed.*

Proof. Since $V^\#[t_R]$ is barreled, its dual $V_R[\tau(V_R, V^\#)]$ is semi-reflexive and quasi-complete. Since $\tau(V_R, V^\#)$ coincides with the topology induced on V_R by $\tau(V, V^\#)$ and $V^\#$ is dense in $V[\tau]$, we have $V_R = V$. Furthermore the topology t_R coincides with $\beta(V^\#, V)$ and *a fortiori* with $\tau(V^\#, V)$. Thus, $V^\#[\tau] = V^\#[t_R] = (V^\#)_R$. Hence, if $V^\#[\tau]$ is complete, $\text{Reg}(V)$ is closed and $V^\#[\tau]$ is semi-reflexive. Since $V[\tau]$ is also semi-reflexive, the dual pair $\langle V^\#, V \rangle$ is reflexive.   ∎

Before continuing the discussion we give two examples of this situation.

(1) $V^\#[t_R]$ is a Fréchet space (notice that completeness is essential: a non-complete metrizable space need not be barreled !). Such is the case of the scale generated by a self-adjoint operator A. Then $V^\# = \bigcap_{n \geqslant 0} D(A^n)$ and t_R coincides with $t_{\mathfrak{A}}$, where $\mathfrak{A}$ is the abelian algebra generated by A, hence $t_R = t_{\text{proj}} = \tau(V^\#, V)$. More generally, the proposition applies whenever $\text{Reg}(V)$ contains a closed O*-algebra that generates on $V^\#$ the same, metrizable, topology (for instance, when $\text{Reg}(V)$ is countably dominated in the sense of Schmüdgen [Sch90]).

(2) Let $V = \omega, V^\# = \varphi$ (Section 1.1.3, Example (ii)). Then $\text{Reg}(\omega)$ consists of all infinite matrices with a finite number of nonzero entries in each row and in each column. Let $\mathfrak{D}$ be the subalgebra of all diagonal infinite matrices. Then $\mathfrak{D}$ dominates $\text{Reg}(\omega)$; indeed one shows easily that every regular operator is continuous from $\varphi[t_{\mathfrak{D}}] = \text{proj lim}\{\ell^2(r)\}$ into ℓ^2. Since t_R is, by definition, the coarsest topology on φ with that property, it follows that $t_{\mathfrak{D}} = t_R$ on φ. Thus $\varphi[t_R]$ is complete. Furthermore, $t_{\mathfrak{D}}$ also coincides with $\tau(\varphi, \omega) = \beta(\varphi, \omega)$, i.e., the direct sum topology, and so $\varphi[t_R]$ is barreled.

Remark 5.5.20. If we drop completeness of $V^{\#}[\tau]$, the latter need not be quasi-complete. If we assume, in addition, that $V^{\#}[\tau] = V^{\#}[t_R]$ is semi-reflexive, then the dual pair $\langle V^{\#}, V \rangle$ is reflexive and both spaces are τ-quasi-complete, but not necessarily complete, so that the algebra $\mathrm{Reg}(V)$ need not be closed. Let us assume instead that $V^{\#}[\tau]$ is complete, but $V^{\#}[t_R]$ not necessarily barreled. Then the pair $\langle V^{\#}, V \rangle$ need not be reflexive: $V^{\#}[\tau]$ could be complete and semi-reflexive but not reflexive (hence not (quasi)-barreled); this happens if and only if $V[\tau]$ is barreled, but not semi-reflexive, for instance if V is a non-reflexive Fréchet space. An example is the space $V = L^1_{\mathrm{loc}}(X, d\mu)$; then $V^{\#} = L^{\infty}_c(X, d\mu)$, with its Mackey topology, has the required properties (see Section 1.1.3, Example (iv), and Section 4.2.1).

When $V^{\#}[t_R]$ is not barreled, not only reflexivity may fail, but the three other properties stated in Proposition 5.5.19 as well. One may have $V_R \neq V$, V non quasi-complete or $\mathrm{Reg}(V)$ nonclosed. We will describe four examples, all nonreflexive, with $V^{\#}[\tau]$ complete, where at least one of those pathologies occurs. The first three have a closed algebra $L^{\dagger}(V^{\#})$ and yet $V^{\#}[\tau]$ and $V^{\#}[t_+]$ are nonreflexive.

(1) *The space* $\omega_d : V_R \neq V$, V *non quasi-complete*
Let $\omega_d = \mathbb{C}^d$ be the coordinate space of uncountable power d. With the same compatibility as for the usual space ω, one gets the triplet $\varphi_d \subset \ell^2_d \subset \omega_d$. As in the countable case, the dual pair $\langle \varphi_d, \omega_d \rangle$ is reflexive, the bounded sets in φ_d are finite dimensional, so that all three topologies σ, τ and β coincide on ω_d, which is complete for any of them. Also, the algebra $\mathrm{Reg}(\omega_d)$ consists of all $d \times d$ matrices with finite rows and finite columns, and it is a closed O*-algebra on φ_d. Then again $t_+ = t_R = t_{\mathfrak{D}}$, where $\mathfrak{D}$ denotes the algebra of $d \times d$ diagonal matrices, but this topology is strictly coarser that $\tau(\varphi_d, \omega_d) = \beta(\varphi_d, \omega_d)$, which is the locally convex direct sum topology. Hence $\varphi_d[t_R]$ is complete and semi-reflexive, but nonreflexive. Its dual is $\omega_d^{(o)}$, the space of sequences of power d with at most countably many nonzero entries. This space is not τ-quasi-complete, and its quasi-completion is ω_d.

(2) *The space* $\omega_d^{(o)} : V_R = V$, V *non quasi-complete*
If, in the previous example, we restrict the compatibility to $\omega_d^{(o)}$, then we get the triplet $\varphi_d \subset \ell^2_d \subset \omega_d^{(o)}$. The algebra $\mathrm{Reg}(\omega_d^{(o)}) = L^{\dagger}(\varphi_d)$ remains the same and closed. But now $V_R = V$, and it is not quasi-complete. Actually one can make the situation even more pathological : if V is any subspace of ω_d containing $\omega_d^{(o)}$ (e.g. the subspace generated by $\omega_d^{(o)}$ and a single, constant, sequence from ω_d), then one gets: $V_R = \omega_d^{(o)} \neq V \neq \hat{V} = \omega_d$, where $\hat{V}$ denotes the quasi-completion of V.

(3) *Kürsten's domain :* $V_R = V$, V *non quasi-complete*
The two previous examples involve the nonseparable Hilbert space ℓ^2_d. Kürsten [134] has exhibited a similar (but very complicated) example in the separable case, namely with $\mathcal{H} = \ell^2(L)$, where L is the (countable) set of all finite sequences of zeros and ones. Starting from a Cantor set, he

constructs a dense domain $\mathcal{D}$ in $\mathcal{H}$ with the following properties. Let as usual t_+ denote the $L^\dagger(\mathcal{D})$-topology on $\mathcal{D}$, and $\mathcal{D}^\times := \mathcal{D}[t_+]^\times$. Then $\mathcal{D}[t_+]$ is complete (i.e., $L^\dagger(\mathcal{D})$ is closed), nonseparable, semi-reflexive, but nonreflexive; the three topologies $t_+ \prec \tau(\mathcal{D}, \mathcal{D}^\times) \prec \beta(\mathcal{D}, \mathcal{D}^\times)$ are all different on $\mathcal{D}$, so that $\mathcal{D}[\tau]$ is also complete and nonreflexive. Furthermore, all bounded sets in $\mathcal{D}[t_+]$ are finite-dimensional, so that $\mathcal{D}^\times[\beta] = \mathcal{D}^\times[\sigma]$. So $\mathcal{D}^\times[\tau]$ cannot be quasi-complete, since $\mathcal{D}[t_+]$ is not barreled. Thus any PIP-space with $V = \mathcal{D}^\times, V^\# = \mathcal{D}$ has the required properties.

(4) *The interpolation chains :* Reg(V) *nonclosed,* $V_R \neq V$
Take for instance the chain $\{\ell^p, 1 \leqslant p \leqslant \infty\}$ (Example (ii) in Section 1.1.3). Then every regular operator is bounded, so that Reg(ℓ^∞) is not closed : $\mathcal{H}(R) = \ell^2$ for every $R \in \mathbb{R}$, so that $V_R = \ell^2$. Notice also that ℓ^∞ is a nonreflexive Banach space, so that $\ell^\infty[\tau(\ell^\infty, \ell^1)]$ is complete and semi-reflexive, but nonreflexive.

In the analysis so far we have considered only the extreme spaces corresponding to the two compatibilities $\#$ and $\#_R, V^\# \subseteq (V^\#)_R$ and $V_R \subseteq V$, and we have discussed under what conditions these two inclusions are in fact equalities. However, if we replace $\#$ by any coarser compatibility, we still have the same dual pair $\langle V^\#, V \rangle$, hence the same algebra Reg(V) and compatibility $\#_R$ as well. In particular, nothing changes in the discussion above if we replace $\#$ by the trivial compatibility $\#_o$, which has $V^\#$ and V as only assaying subsets. However, a full exploitation of the compatibility $\#$ yields more information and new criteria for the equalities $V = V_R, V^\# = (V^\#)_R$. This we will see in the next section.

5.5.6 Comparison Between a PIP-Space and Its Associated LHS

Let again V_I be a positive definite PIP-space, and $\#_R$ the compatibility defined by the algebra Reg(V) of regular operators on V_I. As discussed in Section 5.5.5, we have the following structure

$$V^\# \subseteq (V^\#)_R \subseteq \mathcal{H} \subseteq V_R \subseteq V,$$

where $(V^\#)_R$ is the completion of $V^\#[t_R]$ and V_R is its dual. By definition $V_R = (V^\#)_R^{\#_R}$ and $(V^\#)_R = (V_R)^{\#_R} = (V^\#)^{\#_R\#_R}$ are the extreme spaces of the LHS generated by $\#_R$. Individual Hilbert spaces of this LHS will be denoted as before by $\mathcal{H}(R)$, and the corresponding lattice by $\mathcal{I}(\Sigma_R) = \{\mathcal{H}(R)\}$. In this section we will compare the two PIP-space structures corresponding to $\#_R$ and $\#$, respectively. Before proceeding, we may point out that the same discussion could be done with Reg(V) replaced by $L^\dagger(V^\#)$ (if they are different) or by any subalgebra $\mathfrak{A}$. For simplicity we will treat only the case of $\#_R$.

The compatibility $\#_R$ is not defined on V, but only on the dense subspace V_R, so that we must restrict ourselves to V_R. Four cases, mutually exclusive, are possible for the two compatibilities $\#_R$ and $\#$ on V_R (examples will be given later).

(1) $\#_R = \#$:

As we will see in Proposition 5.5.21 below, this relation implies $V_R = V$. This means than the PIP-space V_I has an equivalent realization as a LHS, defined by its regular operators. In other words, the families $\{\mathcal{H}(R_A), R_A \in \mathrm{Reg}(V)\}$ and, *a fortiori*, $\mathcal{I}(\Sigma_\mathcal{R})$ are generating families of assaying subspaces of V_I. Clearly this is the best possible case for applications, since a LHS is much more convenient than a general PIP-space.

(2) $\#_R \rhd \#$:

This is the most frequent situation, and here too $V_R = V$ necessarily. So the LHS $\{\mathcal{H}(R)\}$ is a (canonical) refinement of the original PIP-space (as we know, such refinements of PIP-spaces do not always exist) and this might be useful for applications. Typical examples of Case (2) are those where the original PIP-space structure is poor, for instance when V_I is a chain or $\# = \#_o$, the trivial compatibility defined in Section 1.5.1.

(3) $\#_R \lhd \#$:

This relation means that every $\mathcal{H}(R)$ is assaying, but their family is not generating; in other words, the algebra $\mathrm{Reg}(V)$ is too small. An extreme example is that of a PIP-space where $\mathrm{Reg}(V)$ contains only bounded operators, such as the interpolation chains : then $(V^\#)_R = \mathcal{H} = V_R$.

(4) $\#_R$ and $\#$ are not comparable:

This case is also pathological; it may happen e.g. if $(V_R)^\# \neq (V_R)^{\#_R} = (V^\#)_R$, but we have no example.

We will now discuss each case in detail; in particular, we examine under what conditions each of them may arise, and we also give concrete examples.

Cases (1) and (2) are very similar, so we discuss them together. First we notice that they exclude the worst pathologies automatically.

Proposition 5.5.21. *In the notation above, let V_I be such that $\#_R \rhd \#$ on $V_R := (V^\#[t_R])^\times$. Then $V_R = V$ and the algebra $\mathrm{Reg}(V)$ is closed.*

Proof. $\#_R \rhd \#$ implies $S^\# \subseteq S^{\#_R}$ and $S^{\#\#} \supseteq S^{\#_R\#_R}$ for every $S \subseteq V_R$ (Section 1.5). Let $S = V^\#$. Then the first relation gives $V \subseteq S_R^{\#\#} = V_R$, and the second gives $V^\# \supseteq V^{\#\#_R\#_R} = (V_R)^{\#_R} = (V^\#)_R$. Since the reverse inclusions hold, we get equality on both sides, i.e., $V = V_R$. Hence, the corresponding Mackey topologies on $V^\#$ coincide. Thus, $V^\#$ is t_R-complete. ■

So in Case (1) or (2), the algebra $\mathrm{Reg}(V)$ is necessarily closed and generates the whole space V. However the examples discussed in Section 5.5.5 show that one may still have $V^\#[t_R]$ nonreflexive and V not quasi-complete.

As we said above, a typical example of a Case (2) situation is that of a chain. Since $\mathrm{Reg}(V)$ is never abelian, the LHS it generates must be a genuine lattice, and cannot be a chain (Section 5.3). Take for instance the scale $s^\times$ generated by the (number) operator $N = \mathrm{diag}(n)$ in ℓ^2. Then the LHS generated by $\mathrm{Reg}(s^\times)$ contains already all spaces $\ell^2(r)$, with r a tempered weight sequence. A less trivial example is the space ω of all sequences. It carries two natural compatibilities: the familiar one, denoted $\#_\alpha$, defined by the condition $\sum_o^\infty |x_n y_n| < \infty$, and another one $\#_\beta$ where absolute convergence is replaced by ordinary convergence (see the Notes to Section 4.3). We will study this example in Section 5.5.7 below and show that $(\omega, \#_\alpha)$ is a Case (2), whereas $(\omega, \#_\beta)$ is a Case (1).

Now we want to find sufficient conditions for Cases (1) and (2) to happen. Because of Proposition 5.5.21, we may assume from the beginning that $\mathrm{Reg}(V)$ is closed and $V_R = V$. Then we may state:

Proposition 5.5.22. *Let $\mathrm{Reg}(V)$ be closed and $V_R = V$. Assume there exists a generating subset $\{V_r, r \in J\}$ of V_I such that, for every $r \in J$, the dual pair $\langle V_r, V_{\overline{r}} \rangle$ coincides, as vector spaces, with a dual pair $\langle \mathcal{H}(R), \mathcal{H}(R^{-1}) \rangle$ of $\mathcal{I}_R$. Then $\#_R \trianglerighteq \#$.*

Proof. Let $V_r = \mathcal{H}(R), V_{\overline{r}} = \mathcal{H}(R^{-1})$ as vector spaces. The space $V_r[\tau(V_r, V_{\overline{r}})]$ and the Hilbert space $\mathcal{H}(R)$ have the same dual $V_{\overline{r}} = \mathcal{H}(R^{-1})$, hence $\tau(V_r, V_{\overline{r}})$ coincides with $\tau(\mathcal{H}(R), \mathcal{H}(R^{-1}))$, which is the hilbertian topology of $\mathcal{H}(R)$. So every $V_r, r \in J$, is a Hilbert space, and the lattice generated by the family $\{V_r, r \in J\}$ is an involutive sublattice of $\mathcal{I}_R$. The same is true for their respective lattice completions, i.e., $\# \trianglelefteq \#_R$.   ∎

The condition of Proposition 5.5.22 is in general only sufficient, but it becomes necessary in certain cases. Consider the lattice completion $\mathcal{F}_R$ of $\mathcal{I}_R$. If $\# \trianglelefteq \#_R$, every $V_r, r \in I$, belongs to $\mathcal{F}_R$. Let first V_r be an arbitrary intersection of spaces $\mathcal{H}(R)$, that is, $V_r = \bigcap_{R \in J} \mathcal{H}(R)$. If we assume that t_R is metrizable, the index set J may be taken as countable. Then there are two cases:

(i) either V_r is a Hilbert space, if it can be represented as a *finite* intersection of $\mathcal{H}(R)$, thus a $\mathcal{H}(R_0)$ itself, and then $V_{\overline{r}} = \mathcal{H}(R_0^{-1})$;

(ii) or V_r is a Fréchet non-normable space, in which case, $V_{\overline{r}} = \sum_{R \in J} \mathcal{H}(R^{-1})$ is a (DF)-space.

In the second case, V_r is a Fréchet space for two comparable topologies, its own norm topology and the non-normable Fréchet topology as intersection of Hilbert spaces (both being Mackey topologies). This is impossible.

Now, if V_r be an arbitrary sum of spaces $\mathcal{H}(R)$, then $V_{\overline{r}}$ is an intersection of such spaces and, by the preceding argument, $V_{\overline{r}} = \mathcal{H}(R_1)$ for some R_1, thus $V_r = \mathcal{H}(R_1^{-1})$. In conclusion, $\mathcal{F}_R$ contains no Hilbert space besides the original elements $\mathcal{H}(R)$ of $\mathcal{I}_R$. Thus we may conclude:

Corollary 5.5.23. *Let V_I be of type (H). If t_R is metrizable, then $\# \trianglelefteq \#_R$ if and only if every $V_r, r \in I$ with $V_r \subseteq \mathcal{H}$ coincides with some $\mathcal{H}(R_A), A \in \mathrm{Reg}V$. This is possible only if V_I is a LHS.*

When V_I is a LHS (whatever t_R), a sufficient condition for $\# \trianglelefteq \#_R$ may also be formulated in terms of Riesz operators, as in Section 5.5.2. Let $\mathcal{H}_r \subset \mathcal{H}$. Then $\mathcal{H}_r = \mathcal{H}(R_{A_{(r)}})$ for a unique positive self-adjoint operator $A_{(r)}$ and the corresponding Riesz operator is $U_{(r)} = 1 + |A_{(r)}|^2$. Notice that $A_{(r)}$ is unbounded if $\mathcal{H}_r \neq \mathcal{H}$.

On the other hand, in the lattice $\mathcal{I}_R$, the Riesz operators are of the form R_A with $A \in \mathrm{Reg}(V)$. Then we reformulate our criterion as follows (here again, it is enough to take $\mathcal{H}_r \subset \mathcal{H}$) :

Proposition 5.5.24. *Let* $V_I = \{\mathcal{H}_r, r \in I\}$ *be a LHS. Assume that, for every* $\mathcal{H}_r \subseteq \mathcal{H}$ *in a generating family, the corresponding Riesz operator is of the form* $U_{(r)} = 1 + |A_{(r)}|^2$, *with* $A_{(r)}$ *self-adjoint in* $\mathcal{H}$ *and regular. Then* $\# \trianglelefteq \#_R$, *and moreover* $t_{\mathrm{proj}} \precsim t_R$ *on* $V^\#$.

In fact, only the subspaces $\mathcal{H}_r$ matter, not the norms themselves, which may be replaced by equivalent ones. So the assumptions of Proposition 5.5.24 may be weakened with the same conclusions.

Corollary 5.5.25. *Let* $V_I = \{\mathcal{H}_r, r \in I\}$ *be a LHS. Assume the following condition holds : for every* $\mathcal{H}_r \subset \mathcal{H}$ *in a generating family, with Riesz operator* $U_{(r)}$, *there exists a regular operator* $A_{(r)}$ *and a bounded operator* $C_{(r)}$ *with bounded inverse, such that :*

$$U_{(r)}^{1/2} = C_{(r)} R_{A(r)}^{1/2}.$$

Then $\# \trianglelefteq \#_R$.

The assumption of Proposition 5.5.24 holds, for instance, in the case of the sequence LHS, $\omega = \{\ell^2(r)\}$, resp. $s^\times = \{\ell^2(r), r \text{ tempered}\}$, where the Riesz operators are diagonal, resp. tempered diagonal, matrices. It holds also, of course, for a LHS generated by an O*-algebra. More generally, whenever $V^\#[t_{\mathrm{proj}}]$ is barreled, the condition of Proposition 5.5.24 implies that $\tau = t_{\mathrm{proj}} = t_R$ on $V^\#$ (but not that $\# = \#_R$!), thus $\mathrm{Reg}(V)$ is dominated by the Riesz operators. Such is the case, e.g., for $s^\times$ where $\mathrm{Reg}(s^\times)$ is dominated by the powers of $N = \mathrm{diag}(n)$.

There are also cases where the condition is not satisfied. For instance, for a finite scale or an interpolation chain (see Section 1.1.3), every regular operator is bounded, whereas not all Riesz operators can be. Another example is the space $L^1_{\mathrm{loc}} = \{L^2(r)\}$ discussed in Section 4.2 for which some Riesz operators are regular, and some are not.

Now we turn to Case (3), which is the converse of (1) and (2) : $\# \triangleright \#_R$ on V_R (here we may have $V_R \neq V$), i.e., the lattice $\mathcal{I}_R$ is an involutive sublattice of $\mathcal{F}(V, \#)$, but it is not generating. A first criterion, analogous to Proposition 5.5.22, is obvious.

Proposition 5.5.26. *Let* V_I *be an arbitrary* PIP-*space. If for every regular operator* A, *the pair* $\langle \mathcal{H}(R_A), \mathcal{H}(R_A^{-1}) \rangle$ *is a dual pair of assaying subsets,* then $\# \trianglerighteq \#_R$ on V_R and $t_{\mathrm{proj}} \succcurlyeq t_R$ on $V^\#$. If, in addition, $V_R \neq V$ or $\mathrm{Reg}(V)$ is not closed, then $\# \triangleright \#_R$ on V_R.

As before, in Corollary 5.5.23 and the discussion preceding it, let V_I be of type (B) or (H), but assume now that t_{proj} is metrizable. Then $\mathcal{F}(V, \#)$ contains as only normable spaces the elements of the original family $\mathcal{I} = \{V_r, r \in I\}$, so that we have again:

Corollary 5.5.27. *Let V_I by a* PIP-*space of type (B) or (H) with t_{proj} metrizable. Then $\# \trianglerighteq \#_R$ on V_R if and only if every $\mathcal{H}(R_A), A \in \mathrm{Reg}(V)$, coincides with some $V_r, r \in I$.*

Finally, if V_I is a LHS, we have a criterion in terms of Riesz operators, a sort of converse of Propositions 5.5.24 and 4.1.3.

Proposition 5.5.28. *Let $V_I = \{\mathcal{H}_r, r \in I\}$ be a LHS. Assume that, for each regular A, there exists $r \in I$ such that $R_A^{1/2} = DU_{(r)}^{1/2}$ where D is bounded with bounded inverse, and $U_{(r)}$ is the Riesz operator of $\mathcal{H}_r$. Then $\# \trianglerighteq \#_R$ on V_R and $t_{\mathrm{proj}} \trianglerighteq t_R$ on $V^\#$.*

This last situation is verified, e.g., when every regular operator A is bounded, i.e., $\mathcal{H}(R_A) = \mathcal{H}$. Then one may choose, for each A, $\mathcal{H}_r = \mathcal{H}$ and $U_{(r)} = 1$; indeed, R_A is bounded with bounded inverse. Since $\mathrm{Reg}(V)$ is not closed in those cases, this is a genuine example of Case (3), and in fact, we have no other explicit example.

As for Case (4), one can design artificial examples, such as the direct sum of a Case (2) and a Case (3), but this is not very interesting. Nontrivial examples are still missing.

5.5.7 An Example: PIP-*Spaces of Sequences*

The PIP-space (φ, ω) of all sequences, already considered in Sections 1.1.3 and 4.3.1, offers a nice example of the various situations described in Section 5.5.6. On the space ω of all sequences, one may define the following four compatibilities:

(i) *The Köthe compatibility $\#_\alpha$:*

$$x \#_\alpha y \iff \sum_{n=1}^{\infty} |x_n y_n| < \infty$$

According to the results of Section 4.3.1, this coincides with the compatibility defined by the lattice $\{\ell^2(r)\}$, itself generated by the algebra of *diagonal* regular operators, i.e., the diagonal infinite matrices.

(ii) *The compatibility $\#_d$, generated by the family of all self-adjoint regular operators with purely discrete spectrum.*

(iii) *The compatibility* $\#_R$, *generated by the algebra* $\mathrm{Reg}(\omega)$, *with consists of all infinite matrices with a finite number of nonzero entries in each row and in each column.*

(iv) *The generalized Köthe compatibility* $\#_\beta$:

$$x \#_\beta y \iff \sum_{n=1}^{\infty} \overline{x_n}\, y_n \text{ is convergent.}$$

But in fact, only two of these are distinct.

Proposition 5.5.29. *Let* $\#_\alpha, \#_d, \#_R, \#_\beta$ *be the four compatibilities just defined on the space* ω. *Then one has:*

$$\#_\alpha \vartriangleleft \#_d = \#_R = \#_\beta.$$

Proof. The relation $\#_\alpha \vartriangleleft \#_\beta$ is a standard result.

Next, the relation $\#_R \trianglelefteq \#_\beta$ is immediate. Indeed $x \#_R y$ means there exists a self-adjoint regular operator R such that $x \in D(R^{-1/2})$ and $y \in D(R^{1/2})$. Then $\langle x|y \rangle = \langle R^{-1/2} x | R^{1/2} y \rangle_{\ell^2}$ exists, i.e., is given by a convergent series, that is, $x \#_\beta y$. Since the Hilbert spaces $\mathcal{H}(R), \mathcal{H}(R^{-1})$ are obviously β-assaying, the assertion follows. As for $\#_d \trianglelefteq \#_R$, it is obvious.

As stated above, $x \#_\alpha y$ means there exists a weight sequence r such that $x \in \ell^2(r) = D(R^{-1/2}), y \in \ell^2(\bar{r}) = D(R^{1/2})$, where $R = \mathrm{diag}(r_n)$ is a diagonal, positive, infinite matrix, i.e., an invertible, self-adjoint, regular operator on ω. On the other hand, $x \#_d y$ means that $x \in D(S^{-1/2})$ and $y \in D(S^{1/2})$, where S is an invertible, positive, self-adjoint regular operator with purely discrete spectrum. Thus $x \#_\alpha y$ implies $x \#_d y$, that is, $\#_\alpha \trianglelefteq \#_d$.

Now such a matrix S can be diagonalized by a unitary matrix $U : S = U^{-1} S_o U$, where $S_o = \mathrm{diag}(s_n)$ and $s_n > 0, \forall n$. Since S is self-adjoint, the same matrix U diagonalizes $S^{\pm 1/2}$ as well : $S^{\pm 1/2} = U^{-1} S_o^{\pm 1/2} U$. It follows that $x \in D(S^{-1/2})$ if and only if $Ux \in D(S_o^{-1/2}) = \ell^2(s)$, or $D(S^{-1/2}) = U^{-1}\ell^2(s)$. Similarly $D(S^{1/2}) = U^{-1}\ell^2(\bar{s})$. So $x \#_d y$ implies there exists a unitary matrix U such that $Ux \#_\alpha Uy$. However, it is known that a space $U\ell^2(s)$ is α-perfect, i.e., α-assaying, if and only if $U\ell^2(s) = \ell^2(u)$, for some u, that is, if and only if U is diagonal. Hence not all d-assaying spaces $U\ell^2(s)$ are α-assaying, i.e., $\#_\alpha \vartriangleleft \#_d$, so that we get $\#_\alpha \vartriangleleft \#_d \trianglelefteq \#_R \trianglelefteq \#_\beta$.

Thus it remains to prove that $\#_\beta \trianglelefteq \#_d$. Given $x \#_\beta y$, we will construct a regular, block-diagonal, unitary matrix U such that $Ux \#_\alpha Uy$. This will imply there exists a weight sequence r such that $Ux \in \ell^2(r) = D(R_o^{-1/2}), y \in \ell^2(\bar{r}) = D(R_o^{1/2})$, where $R_o = \mathrm{diag}(r_n)$. Thus $x \in D(R^{-1/2}), y \in D(R^{1/2})$, where $R^{\pm 1/2} = U^{-1} R_o^{\pm 1/2} U$. Since U is regular, so is R, that is, $x \#_d y$.

Now we proceed to construct such a unitary matrix U. First assume there is a partition of $\mathbb{N}$ into finite blocks $(j_{k-1} + 1, \ldots j_k)$, $j_o = 0$, $k = 1, 2 \ldots$, such that:

$$\sum_{k=1}^{\infty} |\sigma^{(k)}| < \infty,$$

where

$$\sigma^{(k)} = \sum_{n=j_{k-1}+1}^{j_k} \overline{x}_n y_n \neq 0.$$

Obviously, we may assume $y_{j_k} \neq 0$ for every $k = 1, 2, \ldots$. For each k, let $U^{(k)}$ be the unitary finite dimensional matrix which is uniquely defined by the following action:

$$U^{(k)} : (y_{j_{k-1}+1}, \ldots, y_{j_k}) \mapsto (0, 0, \ldots, 0, N_k \sigma^{(k)})$$

$$U^{(k)} : (x_{j_{k-1}+1}, \ldots, x_{j_k}) \mapsto (z_{j_{k-1}+1}, \ldots, z_{j_k-1}, N_k^{-1})$$

where

$$N_k^2 = \left(\sum_{n=j_{k-1}+1}^{j_k} |y_n|^2 \right) |\sigma^{(k)}|^{-2}$$

and $\{z_{j_{k-1}+1}, \ldots, z_{j_k-1}\}$ are chosen arbitrarily so that $U^{(k)}$ is unitary. We denote by U the infinite unitary matrix made up of the blocks $U^{(1)}, U^{(2)}, \ldots$. Then we have :

$$\langle Ux|Uy \rangle = \sum_{k=1}^{\infty} \sigma^{(k)} = \sum_{n=1}^{\infty} \overline{x}_n y_n ,$$

which converges by assumption. On the other hand :

$$\sum_{n=1}^{\infty} |(Ux)_n (Uy)_n| = \sum_{k=1}^{\infty} |\sigma^{(k)}| < \infty$$

by construction, i.e., $(Ux) \#_\alpha (Uy)$, so that the matrix U just constructed answers the question. It remains to show that a suitable partition of $\mathbb{N}$ always exists, but this follows easily from the convergence of the series $\sum_n \overline{x}_n y_n$. Notice that the length of the successive blocks $(j_{k-1} + 1, \ldots, j_k)$ may increase arbitrarily, but always remains finite. This concludes the proof of the proposition.   ∎

If we compare this result with the discussion of Section 5.5.6, we find that $(\omega, \#_\beta)$ is a Case (1), whereas $(\omega, \#_\alpha)$ is a Case (2), as announced.

Notes for Chapter 5

Section 5.1. Further details on interpolation theory may be found in Krein–Petunin [132], Palais [162], Lions–Magenes [LM68] and Bergh–Löfström [BL76]. In particular, the name 'quadratic interpolation' is due to Palais [162].

Section 5.2.

- For various applications of the powers of $(-\Delta + m^2)^{1/2}$ in scattering theory, see Reed–Simon's treatise [RS75, RS79], for instance, Sections X.13 or XI.10.
- The results of Section 5.2.1 largely come from the work of Trapani [182].
- The construction of Section 5.2.2 is due to Antoine–Karwowski [23]. We may recall that a *countably Hilbert space*, as introduced by Gel'fand–Vilenkin [GV64], is a complete locally convex topological vector space with topology given by a countable family of mutually consistent Hilbert norms.
- In questions related to the limiting absorption principle of scattering theory, Agmon–Hömander [1, 59, 60] have also used (general) interpolation theory for refining the chain of weighted L^2-spaces $\{L_s^2(\mathbb{R}^n)\}$. More precisely they consider a space $\mathfrak{B}$ such that

$$\bigcup_{s>1/2} L_s^2 = L_{1/2+}^2 \subset \mathfrak{B} \subset L_{1/2}^2.$$

$\mathfrak{B}$ is the inverse Fourier transform of the Besov space $B_{2,1}^{1/2}$ (see Section 8.4), which is a nonreflexive Banach space [BL76, Tri78a]. The elementary Hilbert space method just described, of course, cannot yield such a space (although it does yield $B_{2,2}^{1/2} = L_{1/2}^2$), but it allows one to construct many Hilbert spaces sitting between $L_{1/2+}^2$ and $L_{1/2}^2$, which could possibly be used for the same purposes as $\mathfrak{B}$.

- The family $H_{\overline{T}} = \{\mathcal{H}_\alpha\}_{\alpha \in \mathbb{R}}$ is a continuous scale of Hilbert spaces in the sense of Palais [162], Krein–Petunin [132], or Berezanskii [Ber68]. Infinite discrete chains like the one given in (5.8) are also studied by he same authors.
- Hadamard's three lines theorem may be found in Reed–Simon [RS75, App. to Sec. IX.4]).

Section 5.3. For the construction of refinements of an arbitrary LHS, we refer to Antoine–Karwowski [23]. The case of a countably Hilbert space was treated previously in Antoine–Karwowski [22].

- O*-algebras have been introduced by Lassner [138]. For a complete information, we refer the reader to our monograph [AIT02].
- For the notion of form sum or, more generally, the relation between quadratic forms and self-adjoint operators, see Reed–Simon [RS72, Section VIII.6] and [RS75, Section X.3]. The use of form sums in the present context has been advocated systematically by Antoine–Karwowski [23].
- The pathology quoted at the end of Section 5.3 has been discovered by Lassner–Timmermann [141, 142], who have exhibited counterexamples.

Section 5.4. The notion of interspace and their crucial role in the construction of PIP-spaces were emphasized by Epifanio–Trapani in [77]. The discussions in Section 5.4.1 and Section 5.4.2 come partly from Tschinke's

PhD Thesis [Tsc02] and the papers of Trapani–Tschinke [183], [184] and Tschinke [185]. There these ideas were exploited for studying the problem of multiplication of distributions in the framework of the so-called duality method, dating back to L. Schwartz [175]. In particular, the crucial condition that $\mathcal{D}$ be dense in $\mathcal{E} \cap \mathcal{F}$ (Proposition 5.4.4 (i)) is already stated in [175].

- For Bessel potential spaces $L^{s,p}(\mathbb{R})$, see for instance the monographs of Adams [Ada75] or Triebel [Tri78a, Tri78b]. Note that some authors call these spaces fractional Sobolev spaces. The relation $AT_\phi f = -iT_{-\phi'} f$ and its iterations follow from Stone's theorem given, for instance, in Reed–Simon [RS72, Theorem VIII.8].
- The Gel'fand-Shilov spaces of type S, namely, S_α, S^β and S_α^β, are described in Gel'fand–Shilov's treatise [GS64, Vol. II]. The space S_α^α has recently found interesting applications in time-frequency analysis, for the characterization of the so-called localization operators. Further information and numerous references may be found in Cordero-Rodino [61].
- Hilbert spaces of type S, i.e., $S(\beta)$ and $\mathcal{H}(\alpha; a)$, have been introduced by Grossmann [115]. His construction, described here, is totally in the spirit of PIP-spaces, which is not surprising, since this family spaces is precisely the one that led Grossmann to the concept of NHS!

Section 5.5. The idea of constructing a LHS starting from a given Hilbert space and a family of self-adjoint operators (or quadratic forms) is due to A. Grossmann (unpublished manuscript and private communication).

- The graph topology is defined in the textbook of Schmüdgen [Sch90]
- This construction described in Theorem 5.5.6 is a refinement of the one performed in an unpublished report by Antoine–Mathot [28], and due originally to Roberts [171] and Friedrich–Lassner [100]. Part of it was used already by Mathot [71]. It should be noticed also that the construction of a LHS from a family of self-adjoint operators was described in detail in Antoine–Karwowski [23]. Actually, the family $\mathcal{R}$ is a *cone* of positive self-adjoint operators (or of closed quadratic forms). In this version, the construction goes back in part to the classical paper of L. Schwartz on Hilbertian subspaces of topological vector spaces [175].
- As pointed in the discussion, each $\mathcal{H}(R_A)$ is the range of a closable operator, namely $R_A^{-1/2}$. It is a standard result, due to Dixmier and Mackey (see the review article of Fillmore–Williams [94]) that such operator ranges constitute a lattice under intersection and vector sum. Indeed the formulas (5.31), (5.32) are equivalent to those given in [94].
- The reduction of the classification of operator domains to that of sequences of positive numbers, mentioned in Remark 5.5.5, is due to Fillmore–Williams [94] and Lassner–Timmermann [141, 142].
- For a complete information on O*-algebras and partial O*-algebras, including the notions of strong and weak commutants, we refer the reader to Schmüdgen's book [Sch90] and to our monograph [AIT02].

- Theorem 5.5.7 is a refinement of Proposition 2.6 of Antoine–Mathot [27]: the latter asserted only the existence of a PIP-space V_I, whereas the present one gives the detailed structure of V_I. As compared to the situation of Theorem 5.5.6, the structure here will be exactly the same. Every R_A belongs to $\mathfrak{A}$ and is therefore a regular operator, self-adjoint in $\mathcal{H}$. However, an arbitrary symmetric element A of $\mathfrak{A}$ need *not* be essentially self-adjoint on $V^{\#}$ (a criterion is given in [27]).
- Lemma 5.5.10 was proved in Mathot [Mat75] for sequence spaces. Corollary 5.5.12 strengthens the results of Antoine–Mathot [27, Proposition 5.2].
- The contents of the Sections 5.5.5 and 5.5.6, including the example on sequence spaces, come again from the unpublished paper [28]. For all results concerning locally convex spaces, we refer to Köthe [Köt69].
- The case of chains of Banach spaces is treated in detail in Krein–Petunin [132]. For the classical Riesz-Thorin theorem, see any textbook on interpolation theory, for instance, Lions–Magenes [LM68] or Bergh–Löfström [BL76]. For the normal interpolation property, see Krein–Petunin [132].
- Corollary 5.5.25 exploits the results of Fillmore–Williams [94]).
- The space $\omega_d = \mathbb{C}^d$, the coordinate space of uncountable power d, has been studied in Friedrich–Lassner [101]. See also Köthe [Köt69, §23.8].
- Kürsten's domain is described in [134].
- The relation $\#_\alpha < \#_\beta$ has been proved by Garling [104]. See also Köthe [Köt69, §30, 10]. That a space $U\ell^2(s)$ is α-perfect if and only if $U\ell^2(s) = \ell^2(u)$, for some weight sequence u, follows from a theorem of Köthe–Toeplitz [131, §12, Satz 8].

Chapter 6
Partial *-Algebras of Operators in a PIP-Space

The family of operators on a PIP-space V is endowed with two, possibly different, *partial multiplications*, where *partial* means that the multiplication is not defined for any pair A, B of elements of $\mathrm{Op}(V)$ but only for certain couples. The two multiplications, to be called *strong* and *weak*, give rise to two different structures that coincide in certain situations. In this chapter we will discuss first the structure of $\mathrm{Op}(V)$ as partial *-algebra in the sense of [AIT02] and then the possibility of representing an abstract partial *-algebra into $\mathrm{Op}(V)$.

6.1 Basic Facts About Partial *-Algebras

For convenience, we collect here the basic definitions about partial *-algebras. A full treatment may be found in our monograph [AIT02].

A *partial *-algebra* is a complex vector space $\mathfrak{A}$, endowed with an involution $x \mapsto x^*$, that is, a conjugate linear bijection such that $x^{**} = x$, and a subset Γ of $\mathfrak{A} \times \mathfrak{A}$ such that:

(i) $(x, y) \in \Gamma$ if and only if $(y^*, x^*) \in \Gamma$;
(ii) if $(x, y) \in \Gamma$ and $(x, z) \in \Gamma$, then $(x, \lambda y + \mu z) \in \Gamma$ for all $\lambda, \mu \in \mathbb{C}$;
(iii) whenever $(x, y) \in \Gamma$, there exists an element $x \cdot y$ of $\mathfrak{A}$ with the usual properties of the multiplication:

$$x \cdot (y + \lambda z) = x \cdot y + \lambda (x \cdot z) \text{ and } (x \cdot y)^* = y^* \cdot x^* \text{ for } (x, y), (x, z) \in \Gamma \text{ and } \lambda \in \mathbb{C}.$$

Notice that the partial multiplication is *not* required to be associative (and often it is not). We shall always assume the partial *-algebra $\mathfrak{A}$ contains an identity 1, by which we mean that $1^* = 1, (1, x) \in \Gamma$ and $1 \cdot x = x \cdot 1 = x$, for every $x \in \mathfrak{A}$.

Given the defining set Γ, spaces of multipliers are defined in the obvious way. Whenever $(x, y) \in \Gamma$, we say that x is a *left multiplier* of y or y is a *right multiplier* of x, and write $x \in L(y)$ and $y \in R(x)$. By (ii), $L(x)$ and $R(x)$ are vector subspaces of $\mathfrak{A}$. For any subset $\mathfrak{N}$ of $\mathfrak{A}$, we write

J.-P. Antoine and C. Trapani, *Partial Inner Product Spaces:*
Theory and Applications, Lecture Notes in Mathematics 1986,
DOI 10.1007/978-3-642-05136-4_6, © Springer-Verlag Berlin Heidelberg 2009

$$L\mathfrak{N} = \bigcap_{x \in \mathfrak{N}} L(x), \quad R\mathfrak{N} = \bigcap_{x \in \mathfrak{N}} R(x),$$

and, of course, the involution exchanges the two:

$$(L\mathfrak{N})^* = R\mathfrak{N}^*, \quad (R\mathfrak{N})^* = L\mathfrak{N}^*.$$

Clearly all these multiplier spaces are vector subspaces of $\mathfrak{A}$, containing the identity 1.

The set $R\mathfrak{A} = \{y \in \mathfrak{A} : y \in R(x), \forall x \in \mathfrak{A}\}$ of the so-called right universal multipliers will play a significant role in the sequel.

The partial *-algebra is *commutative*, or *abelian*, if $L(x) = R(x)$, $\forall\, x \in \mathfrak{A}$, and then $x \cdot y = y \cdot x$, $\forall\, x \in L(y)$. In that case, we write simply for the multiplier spaces $L(x) = R(x) =: M(x)$, $L\mathfrak{N} = R\mathfrak{N} =: M\mathfrak{N}$ ($\mathfrak{N} \subset \mathfrak{A}$).

A partial *-subalgebra of $\mathfrak{A}$ is a subspace $\mathfrak{B}$ such that $x^* \in \mathfrak{B}$ whenever $x \in \mathfrak{B}$, and $x \cdot y \in \mathfrak{B}$ for all $x, y \in \mathfrak{B}$ such that $x \in L(y)$.

Notice that the multiplication is not required to be associative, but it must be distributive with respect to the addition by (iii). There are, however, several interesting examples where a weaker form of associativity holds true. Namely if $y \in R(x)$ implies $y \cdot z \in R(x)$ for every $z \in R\mathfrak{A}$ and

$$(x \cdot y) \cdot z = x \cdot (y \cdot z), \tag{6.1}$$

then we say that $\mathfrak{A}$ is *semi-associative*.

Remark 6.1.1. A crucial fact of partial *-algebras is that the couple of maps (L, R) defines a *Galois connection* on the complete lattice of all vector subspaces of $\mathfrak{A}$, ordered by inclusion (see Appendix A). This means that (i) both L and R reverse order; and (ii) both LR and RL are closures, i.e.,

$$\mathfrak{N} \subset LR\mathfrak{N} \quad \text{and} \quad LRL = L$$

$$\mathfrak{N} \subset RL\mathfrak{N} \quad \text{and} \quad RLR = R.$$

Let us denote by $\mathcal{F}^L$, resp. $\mathcal{F}^R$, the set of all LR-closed, resp. RL-closed, subspaces of $\mathfrak{A}$:

$$\mathcal{F}^L = \{\mathfrak{N} \subset \mathfrak{A} : \mathfrak{N} = LR\mathfrak{N}\},$$
$$\mathcal{F}^R = \{\mathfrak{N} \subset \mathfrak{A} : \mathfrak{N} = RL\mathfrak{N}\},$$

both ordered by inclusion. Then standard results from universal algebra yield the full multiplier structure of $\mathfrak{A}$. Indeed,

(1) $\mathcal{F}^L$ is a complete lattice with lattice operations

$$\mathfrak{M} \wedge \mathfrak{N} = \mathfrak{M} \cap \mathfrak{N}, \quad \mathfrak{M} \vee \mathfrak{N} = LR(\mathfrak{M} + \mathfrak{N}).$$

The largest element is $\mathfrak{A}$, the smallest $L\mathfrak{A}$.

(2) $\mathcal{F}^R$ is a complete lattice with lattice operations

$$\mathfrak{M} \wedge \mathfrak{N} = \mathfrak{M} \cap \mathfrak{N}, \quad \mathfrak{M} \vee \mathfrak{N} = RL(\mathfrak{M} + \mathfrak{N}).$$

The largest element is $\mathfrak{A}$, the smallest $R\mathfrak{A}$.
(3) Both $L : \mathcal{F}^R \to \mathcal{F}^L$ and $R : \mathcal{F}^L \to \mathcal{F}^R$ are lattice anti-isomorphisms:

$$L(\mathfrak{M} \wedge \mathfrak{N}) = L\mathfrak{M} \vee L\mathfrak{N}, \text{ etc.}$$

(4) The involution $\mathfrak{N} \leftrightarrow \mathfrak{N}^*$ is a lattice isomorphism between $\mathcal{F}^L$ and $\mathcal{F}^R$.

A *-*homomorphism* of a partial *-algebra $\mathfrak{A}$ into another one $\mathfrak{B}$ is a linear map $\sigma : \mathfrak{A} \to \mathfrak{B}$ such that (i) $\sigma(x^*) = \sigma(x)^*$ for each $x \in \mathfrak{A}$, and (ii) whenever $x \in L(y)$ in $\mathfrak{A}$, then $\sigma(x) \in L(\sigma(y))$ in $\mathfrak{B}$ and $\sigma(x) \cdot \sigma(y) = \sigma(x \cdot y)$. The map σ is a *-*isomorphism* if it is a bijection and $\sigma^{-1} : \mathfrak{B} \to \mathfrak{A}$ is also a *-homomorphism.

Remark 6.1.2. If σ is a *-homomorphism, the image $\sigma(\mathfrak{A})$ need not be a partial *-subalgebra of $\mathfrak{B}$. Indeed, there could be pairs $x, y \in \mathfrak{A}$ such that $x \notin L(y)$, but $\sigma(x) \in L(\sigma(y))$; then the product $\sigma(x) \cdot \sigma(y)$ is well-defined, but it might not belong to $\sigma(\mathfrak{A})$. This is the reason why when defining *-isomorphisms, one requires the inverse to be a *-homomorphism too.

Let $\mathcal{H}$ be a complex Hilbert space and $\mathcal{D}$ a dense subspace of $\mathcal{H}$. We denote by $\mathcal{L}^\dagger(\mathcal{D}, \mathcal{H})$ the set of all (closable) linear operators X such that $\mathcal{D}(X) = \mathcal{D}$, $\mathcal{D}(X^*) \supseteq \mathcal{D}$. The set $\mathcal{L}^\dagger(\mathcal{D}, \mathcal{H})$ is a partial *-algebra with respect to the following operations : the usual sum $X_1 + X_2$, the scalar multiplication λX, the involution $X \mapsto X^\dagger = X^* {\restriction} \mathcal{D}$ and the partial multiplication $X_1 \square X_2 = X_1^{\dagger *} X_2$, defined whenever X_1 is a weak left multiplier of X_2 (which is denoted by $X_1 \in L^w(X_2)$), that is, whenever $X_2 \mathcal{D} \subset \mathcal{D}(X_1^{\dagger *})$ and $X_1^* \mathcal{D} \subset \mathcal{D}(X_2^*)$. When we regard $\mathcal{L}^\dagger(\mathcal{D}, \mathcal{H})$ as a partial *-algebra with those operations, we denote it by $\mathcal{L}_w^\dagger(\mathcal{D}, \mathcal{H})$. Next, a subspace $\mathfrak{M}$ of $\mathcal{L}^\dagger(\mathcal{D}, \mathcal{H})$ is called a (weak) *partial O*-algebra* on $\mathcal{D}$ in $\mathcal{H}$ if it is stable under the weak multiplication $\square$ (in the sense that $X, Y \in \mathfrak{M}$ and $X \in L^w(Y)$ imply $X \square Y \in \mathfrak{M}$).
As for *-algebras, representations of a partial *-algebra $\mathfrak{A}$ are easily obtained by a (suitably generalized) GNS construction. The starting point is the notion of $\mathfrak{B}$-invariant positive sesquilinear form, where $\mathfrak{B} \subset R\mathfrak{A}$. These representations take their values in a weak partial *-algebra of operators. We refer to [AIT02] for a full discussion of this subject.

6.2 Op(V) as Partial *-Algebra

The set Op(V) of operators in a PIP-space is always a complex vector space with involution $A \mapsto A^\times$, where $A^\times$ is the adjoint of A as defined in Chapter 3. In this section we will discuss its structure of partial *-algebra. As we shall see, there are two natural ways of defining a partial multiplication: the first one, which we will call *strong*, is based on the possibility of factorizing the product via some intermediate assaying subspace (the product is then defined simply as a composition of maps); the second one, can be given in terms of the partial inner product itself and for this reason will be called *weak* multiplication. We shall see that these two notions coincide for indexed PIP-spaces.

6.2.1 The Composition of Maps as Partial Multiplication

Let us now consider the problem of multiplying two operators of $\mathrm{Op}(V)$. Let $A, B \in \mathrm{Op}(V)$ and assume that there exists an assaying subspace V_q such that $B : V^\# \to V_q$ and $A : V_q \to V$ (both continuous). Then the product AB can be defined by the ordinary composition $A \circ B$ of linear maps, as we did in Section 3.1.3. However, this definition could *a priori* depend on the choice of the assaying subspace V_q used to factorize the product (but it does not, as we shall see in a while). Indeed, if $V_{q'}$ is another assaying subspace such that $B : V^\# \to V_{q'}$ and $A : V_{q'} \to V$ continuously, It follows that B maps $V^\#$ continuously into V_q and $V_{q'}$, hence into their projective limit $(V_q \cap V_{q'})_{\mathrm{proj}}$, but *not* necessarily into $V_{q \wedge q'}$, since the Mackey topology might be strictly finer than the projective topology.

For similar reasons, even if BA and CA are both unambiguously defined, $(B + C)A$ need not be. Indeed, the first condition means there exist $q, r \in F$ such that the following factorizations are continuous (with respect to the respective Mackey topologies):

$$V^\# \xrightarrow{A} V_q \xrightarrow{B} V,$$

$$V^\# \xrightarrow{A} V_r \xrightarrow{C} V. \tag{6.2}$$

It follows that A maps $V^\#$ continuously into V_q and V_r, hence into their projective limit $(V_r \cap V_q)_{\mathrm{proj}}$, but, for the same reason as before, *not* necessarily into $V_{r \wedge q}$.

Thus, $\mathrm{Op}(V)$ is not, in general, a partial *-algebra with respect to $\circ$.

6.2.2 A Weak Multiplication in $\mathrm{Op}(V)$

The partial multiplication of operators of $\mathrm{Op}(V)$ is natural, since it makes use of the lattice of assaying subspaces. However, it exhibits some unpleasant features, as we have seen. There is, however, an alternative possible definition of the multiplication, suggested by the fact that a partial inner product is defined on V.

Definition 6.2.1. Let V be a nondegenerate PIP-space and $A, B \in \mathrm{Op}(V)$. Assume that the following conditions hold.

(i) $A^\times g \in \bigcap_{f \in V^\#} (Bf)^\#$, for every $g \in V^\#$.

(ii) There exists $X \in \mathrm{Op}(V)$ such that

$$\langle Bf | A^\times g \rangle = \langle Xf | g \rangle, \ \forall f, g \in V^\#.$$

Then we put $X = A \cdot B$ and call it the *weak product* of A and B.

Remark 6.2.2. It is clear that the operator $X = A \cdot B$ is uniquely determined by the conditions of the previous definition.

If condition (i) of Definition 6.2.1 is satisfied, then it is readily checked that one also has $B^\times f \in \bigcap_{g \in V^\#} (A^\times g)^\#$, for every $f \in V^\#$ and, by (ii),

$$\langle Bf | A^\times g \rangle = \langle f | X^\times g \rangle, \ \forall f, g \in V^\#.$$

Hence $B^\times \cdot A^\times$ is also well-defined.

As for the distributive property, let us assume that $B \cdot A$ and $C \cdot A$ are both well-defined. Then,

$$B^\times g \in \bigcap_{f \in V^\#} (Af)^\# \quad \text{and} \quad C^\times g \in \bigcap_{f \in V^\#} (Af)^\#, \ \forall g \in V^\#.$$

Hence, $(B+C)^\times g \in \bigcap_{f \in V^\#} (Af)^\#$. Moreover, there exist $X, Y \in \mathrm{Op}(V)$ such that

$$\langle Af | B^\times g \rangle = \langle Xf | g \rangle \quad \text{and} \quad \langle Af | C^\times g \rangle = \langle Yf | g \rangle, \ \forall f, g \in V^\#.$$

Hence,

$$\langle Af | (B + C)^\times g \rangle = \langle (X + Y)f | g \rangle, \ \forall f, g \in V^\#.$$

Therefore $(B + C) \cdot A$ is well-defined and $(B + C) \cdot A = B \cdot A + C \cdot A$. The other properties defining a partial *-algebra can be checked in similar way. Thus

Proposition 6.2.3. *Let V be a nondegenerate PIP-space. Then $\mathrm{Op}(V)$ is a partial *-algebra with respect to the weak multiplication.*

Proposition 6.2.4. *Let $A, B \in \mathrm{Op}(V)$ and assume that there exists an assaying subspace V_q such that*

$$B : V^\# \to V_q \quad \text{and} \quad A : V_q \to V, \ \text{both continuously.}$$

Then $A \cdot B$ is well-defined and $(A \cdot B)f = (A \circ B)f$, for every $f \in V^\#$.

Proof. Since $A : V_q \to V$ continuously, $A^\times : V^\# \to V_{\bar{q}}$ continuously, by duality; hence, condition (i) of Definition 6.2.1 is fulfilled. On the other hand, $A \circ B$ is continuous from $V^\#$ into V; thus $A \circ B \in \mathrm{Op}(V)$. Finally, one has

$$\langle Bf | A^\times g \rangle = \langle (A \circ B) f | g \rangle, \ \forall f, g \in V^\#.$$

This proves the statement.   ■

The following example shows that the existence of $A \cdot B$ does not imply the existence of $A \circ B$.

Example 6.2.5. Take ω, the space of all complex sequences. Let us fix an increasing unbounded sequence $a = (a_n)$ of positive real numbers and consider the subspaces of ω defined by

$$V = \ell^2(a) = \left\{ x = (x_n) \in \omega : \sum_n |x_n|^2 \, a_n^{-2} < \infty \right\}$$

and

$$W = \ell^2(a^{-1}) = \left\{ x = (x_n) \in \omega : \sum_n |x_n|^2 \, a_n^2 < \infty \right\}.$$

Then V is a PIP-space with compatibility

$$x \# y \ \Leftrightarrow \ (x, y) \in (V \times W) \cup (W \times V)$$

and partial inner product

$$\langle x | y \rangle = \sum_n \overline{x_n} \, y_n.$$

Clearly, V and $V^\# = W$ are the only assaying subspaces of V and $V^\# = W$. The operators

$$A : x \in V^\# \mapsto ax := (a_n x_n) \in V$$
$$B : x \in V^\# \mapsto a^2 x := (a_n^2 x_n) \in V$$

are symmetric elements of $\mathrm{Op}(V)$. The operator A does not map $V^\#$ into itself, hence there is no possibility of defining $A \circ A$. On the other hand, it is easy to see that $A \cdot A = B$.

Corollary 6.2.6. *Let $A, B \in \mathrm{Op}(V)$. Assume that there exist two assaying subspaces $V_q, V_{q'}$ such that*

$$B : V^\# \to V_q \ \text{ and } \ A : V_q \to V \ (\text{both continuously})$$

and

$$B : V^\# \to V_{q'} \quad and \quad A : V_{q'} \to V \ (both \ continuously).$$

Then,

$$A_q(Bf) = A_{q'}(Bf), \ \forall f \in V^\#,$$

where A_r denotes the restriction of A to V_r.

Proof. Indeed, both products coincide with $A \cdot B$.   ◼

As a consequence of Corollary 6.2.6, the composition $A \circ B$, when it exists, does not depend on the assaying subspace used as intermediate space for the factorization. Then it defines a partial multiplication, which we will call *strong* from now on and denote simply by AB.

6.2.3 The Case of an Indexed PIP-Space

For an indexed PIP-space V_I, the situation may improve. As defined in Chapter 2, V_I is said to be *projective* if, for every pair $r, q \in I$, the Mackey topology $\tau(V_{r \wedge q}, V_{\bar{r} \vee \bar{q}})$ on $V_{r \wedge q} = V_r \cap V_q$ coincides with the projective topology: $V_{r \wedge q} \simeq (V_r \cap V_q)_{\mathrm{proj}}$ (TVS isomorphism). This is the case, in particular, when V_I is a lattice of Hilbert spaces (type (H)) or of reflexive Banach spaces (type (B)). Then we have:

Proposition 6.2.7. *Let V_I be a projective indexed PIP-space. The following statements hold.*

(i) The domain $\mathcal{D}(A)$ of any operator $A \in \mathrm{Op}(V_I)$ is a vector subspace of V.
*(ii) $\mathrm{Op}(V_I)$ is a partial *-algebra, with respect to the strong partial multiplication.*

Proof. Given the above discussion and the results of Chapter 2, the proof is straightforward: the condition of projectivity implies that $V_{r \vee q} = V_r + V_q$, for all $r, q \in I$, from which the two statements follow immediately.   ◼

In the case of a projective indexed PIP-space V_I, one can also consider the algebraic properties of the partial *-algebra $\mathrm{Op}(V_I)$ and, in particular, identify its multiplier structure. The building blocks are the sets:

$$\mathcal{O}_{pq} = \{A \in \mathrm{Op}(V_I) : A_{pq} \text{ exists}\}. \tag{6.3}$$

Clearly we have

- $L\mathcal{O}_{pq} \equiv L_p := \{C \in \mathrm{Op}(V_I) : \exists s \text{ such that } C_{sp} \text{ exists}\}$
  $= \{C \in \mathrm{Op}(V_I) : p \in \mathsf{d}(C)\},$
- $R\mathcal{O}_{pq} \equiv R_q := \{B \in \mathrm{Op}(V_I) : \exists t \text{ such that } B_{qt} \text{ exists}\}$
  $= \{B \in \mathrm{Op}(V_I) : q \in \mathsf{i}(B)\},$

- $RLO_{pq} = RL_p = R_p \in \mathcal{F}^R$,
- $LRO_{pq} = LR_q = L_p \in \mathcal{F}^L$.

From this we deduce immediately, using the fact that L, R are lattice anti-isomorphisms:

$$L_p \wedge L_q = L_{p \vee q} , \quad L_p \vee L_q = L_{p \wedge q} ,$$

$$R_p \wedge R_q = R_{p \wedge q} , \quad R_p \vee R_q = R_{p \vee q} .$$

In particular, $q \leqslant q'$ implies $R_q \subseteq R_{q'}$ and $L_q \supseteq L_{q'}$. Thus $\mathcal{I}^L = \{L_p\}$ is a sublattice of $\mathcal{F}^L$, $\mathcal{I}^R = \{R_p\}$ is a sublattice of $\mathcal{F}^R$, and both are generating (except that they do not contain the extreme elements in general, see below). In addition $\mathcal{I}^L, \mathcal{I}^R$ consist of matching pairs (R_q, L_q). Indeed, given $A \in \mathrm{Op}(V_I)$, we may rewrite

$$\mathsf{d}(A) = \{q \in I : A \in L_q\}, \qquad \mathsf{i}(A) = \{p \in I : A \in R_p\}.$$

and therefore

$$A \in L(B) \quad \Leftrightarrow \quad \exists\, p \in I \text{ such that } A \in L_p, B \in R_p. \tag{6.4}$$

From (6.4), we deduce individual multiplier spaces:

$$L(A) = \bigvee\nolimits_{p \in \mathsf{i}(A)} L_p = L_{p_{\min}}, \quad R(A) = \bigvee\nolimits_{q \in \mathsf{d}(A)} R_q = R_{q_{\max}}.$$

Note that these two subsets do not belong to $\mathcal{I}^L$, resp. $\mathcal{I}^R$, in general, but to the complete lattice generated by the latter.

In the same way, we obtain

$$LOp(V_I) = \{C \in \mathrm{Op}(V_I) : \forall p, \exists\, s \text{ such that } C_{sp} \text{ exists}\} \tag{6.5}$$
$$= \{C \in \mathrm{Op}(V_I) : \mathsf{d}(C) = I\} = \bigvee\nolimits_{p \in I} L_p$$
$$ROp(V_I) = \{B \in \mathrm{Op}(V_I) : \forall q, \exists\, t \text{ such that } B_{qt} \text{ exists}\} \tag{6.6}$$
$$= \{B \in \mathrm{Op}(V_I) : \mathsf{i}(B) = I\} = \bigcap\nolimits_{q \in I} R_q.$$

Again, $LOp(V_I) \notin \mathcal{I}^L$, $ROp(V_I) \notin \mathcal{I}^R$.

The next step is to equip all these multiplier spaces with adequate topologies, but we will refrain from doing it, referring instead the reader to our monograph [AIT02].

6.3 Operators in a RHS

Given a rigged Hilbert space $\mathcal{D} \hookrightarrow \mathcal{H} \hookrightarrow \mathcal{D}^\times$, we denote by $\mathfrak{L}(\mathcal{D}, \mathcal{D}^\times)$ the set of all continuous linear maps from $\mathcal{D}[\mathsf{t}]$ into $\mathcal{D}^\times[\mathsf{t}^\times]$. The space $\mathfrak{L}(\mathcal{D}, \mathcal{D}^\times)$ carries a natural involution $A \mapsto A^\dagger$ defined by (remember the notational convention of Section 5.4.1)

$$\langle g|A^\dagger f\rangle = \overline{\langle f|Ag\rangle}, \ \forall f, g \in \mathcal{D}.$$

Furthermore, we denote by $\mathcal{L}^\dagger(\mathcal{D})$ the *-algebra of all closable operators A in $\mathcal{H}$ with the properties $D(A) = \mathcal{D}, D(A^*) \supseteq \mathcal{D}$ and both A and A^* leave $\mathcal{D}$ invariant (* denotes here the usual Hilbert adjoint). The map $A \mapsto A^* \!\restriction_\mathcal{D}$ defines an involution on $\mathcal{L}^\dagger(\mathcal{D})$, which coincides with † whenever $\mathcal{L}^\dagger(\mathcal{D}) \subset \mathfrak{L}(\mathcal{D}, \mathcal{D}^\times)$. This inclusion does not hold in general. But, for instance, when $\mathsf{t} = \mathsf{t}_{\mathcal{L}^\dagger(\mathcal{D})}$ is the so called *graph topology* defined by $\mathcal{L}^\dagger(\mathcal{D})$ on $\mathcal{D}$, then $\mathcal{L}^\dagger(\mathcal{D}) \subset \mathfrak{L}(\mathcal{D}, \mathcal{D}^\times)$. Under these circumstances, $(\mathfrak{L}(\mathcal{D}, \mathcal{D}^\times), \mathcal{L}^\dagger(\mathcal{D}))$ is a *quasi *-algebra* (see the Notes for the definition). Of course, instead of considering on $\mathcal{D}$ the initial topology t and on $\mathcal{D}^\times$ the strong dual topology $\mathsf{t}^\times$, one may also consider each of these spaces as endowed with their Mackey topologies, denoted as before by $\tau_\mathcal{D}, \tau_{\mathcal{D}^\times}$, respectively. The corresponding space of continuous linear maps of $\mathcal{D}[\tau_\mathcal{D}]$ into $\mathcal{D}^\times[\tau_{\mathcal{D}^\times}]$ will be denoted by $\mathfrak{L}_\tau(\mathcal{D}, \mathcal{D}^\times)$. In general, $\mathfrak{L}(\mathcal{D}, \mathcal{D}^\times) \subseteq \mathfrak{L}_\tau(\mathcal{D}, \mathcal{D}^\times)$ and $(\mathfrak{L}_\tau(\mathcal{D}, \mathcal{D}^\times), \mathcal{L}^\dagger(\mathcal{D}))$ is a quasi*-algebra. The general problem of multiplying operators acting on a rigged Hilbert space has already been discussed in our monograph [AIT02]. We give here a short account of those results, considering only the case where $\mathfrak{L}(\mathcal{D}, \mathcal{D}^\times) = \mathfrak{L}_\tau(\mathcal{D}, \mathcal{D}^\times)$. This equality is satisfied, for instance, if $\mathcal{D}[\mathsf{t}]$ is a reflexive space.

6.3.1 The Multiplication Problem

Let $\mathcal{E}, \mathcal{F}$ be interspaces and $\mathfrak{L}(\mathcal{E}, \mathcal{F})$ the linear space of all continuous linear maps from $\mathcal{E}$ into $\mathcal{F}$. Let us define

$$C(\mathcal{E}, \mathcal{F}) = \left\{ A \in \mathfrak{L}(\mathcal{D}, \mathcal{D}^\times) : A = \widetilde{A} \!\restriction \mathcal{D} \text{ for some } \widetilde{A} \in \mathfrak{L}(\mathcal{E}, \mathcal{F}) \right\}.$$

We remind the reader that each interspace (including $\mathcal{D}$ and $\mathcal{D}^\times$) is endowed with its Mackey topology. It is not difficult to prove that $\mathcal{L}^\dagger(\mathcal{D}) = C(\mathcal{D}, \mathcal{D}) \cap C(\mathcal{D}^\times, \mathcal{D}^\times)$.

Let now $A, B \in \mathfrak{L}(\mathcal{D}, \mathcal{D}^\times)$ and assume that there exists an interspace $\mathcal{E}$ such that $B \in C(\mathcal{D}, \mathcal{E})$ and $A \in C(\mathcal{E}, \mathcal{D}^\times)$; it would then be natural to define

$$A \cdot Bf = \widetilde{A}(Bf), \quad f \in \mathcal{D}.$$

However, this product is not well-defined, because it may depend on the choice of the interspace $\mathcal{E}$. This dependence is shown explicitly in the following examples. Once again, the pathology is due to the fact that $\mathcal{D}$ is not necessarily dense in the intersection $\mathcal{E} \cap \mathcal{F}$ of two interspaces $\mathcal{E}, \mathcal{F}$ endowed with the projective topology $\tau_\mathcal{E} \wedge \tau_\mathcal{F}$, a problem already discussed in Section 5.4.

Example 6.3.1. Let $\mathcal{H}$ be a separable Hilbert space. Then $\mathcal{H}$ admits a countable orthonormal basis $\{\phi_n\}$. Let $\mathcal{D} \subset \mathcal{H}$ be the subspace of $\mathcal{H}$ defined as follows

$$\mathcal{D} = \left\{ \phi = \sum_{n=1}^{\infty} x_n \phi_n : \sum_{n=1}^{\infty} |x_n|^2 n^{2k} < \infty, \text{ for every } k \in \mathbb{N} \right\}.$$

We endow $\mathcal{D}$ with the locally convex Hausdorff topology defined by the family of seminorms:

$$p_k(\phi) = \left(\sum_{n=1}^{\infty} |x_n|^2 n^{2k} \right)^{1/2}, \quad k \in \mathbb{N},$$

and denote by $\mathcal{D}^{\times}$ the conjugate dual space of $\mathcal{D}$. Let X and Y be the operators defined on $\mathcal{D}$ by

$$X\phi = \sum_{n=1}^{\infty} \langle -2n\phi_{2n} + (2n+1)\phi_{2n+1} | \phi \rangle (-2n\phi_{2n} + (2n+1)\phi_{2n+1}),$$

$$Y\phi = \sum_{n=1}^{\infty} \langle (2n-1)\phi_{2n-1} - 2n\phi_{2n} | \phi \rangle ((2n-1)\phi_{2n-1} - 2n\phi_{2n}).$$

It is easily seen that both X and Y belong to $\mathcal{L}^{\dagger}(\mathcal{D})$ and that they are essentially self-adjoint. Hence the domains $D(\overline{X})$ and $D(\overline{Y})$, endowed with the norms

$$\|\phi\|_{\mathcal{E}} = (\|\overline{X}\phi\|^2 + \|\phi\|^2)^{1/2}, \quad \|\phi\|_{\mathcal{F}} = (\|\overline{Y}\phi\|^2 + \|\phi\|^2)^{1/2},$$

are Hilbert spaces, which we will denote by $\mathcal{E}$ and $\mathcal{F}$, respectively. We consider now the Cauchy sequence (ψ_k) in $\mathcal{H}$

$$\psi_k = \sum_{n=1}^{k} n^{-1} \phi_n,$$

and we denote by $f = \sum_{n=1}^{\infty} n^{-1} \phi_n \in \mathcal{H}$ its limit. Notice that $\psi_k \in \mathcal{D}$, for every $k \in \mathbb{N}$, while $f \notin \mathcal{D}$. By applying X and Y to ψ_k, we get $X\psi_{2k+1} = Y\psi_{2k} = 0$. This implies that the subsequences (ψ_{2k+1}) and (ψ_{2k}) are Cauchy in $\mathcal{E}$ and in $\mathcal{F}$, respectively. Therefore, their common limit f belongs to $\mathcal{E} \cap \mathcal{F}$. Let $B : \mathcal{D} \to \mathcal{E} \cap \mathcal{F}$ be the operator defined by

$$B\phi = \langle \phi | \phi_1 \rangle f, \quad \phi \in \mathcal{D}.$$

Then B is a rank one operator. Moreover, $B \in \mathcal{C}(\mathcal{D}, \mathcal{E}) \cap \mathcal{C}(\mathcal{D}, \mathcal{F})$. Next, we consider the operator $\tilde{A} : \mathcal{E} \to \mathcal{D}^{\times}$, defined by

$$\tilde{A}\left(\sum_{n=1}^{\infty} x_n \phi_n\right) = \left(x_1 + \sum_{n=1}^{\infty}(-2nx_{2n} + (2n+1)x_{2n+1})\right)\phi_1.$$

Simple estimations show that $\tilde{A} \in \mathcal{L}(\mathcal{E}, \mathcal{D}^{\times})$. Analogously, the operator $\hat{A} \in \mathcal{L}(\mathcal{F}, \mathcal{D}^{\times})$ defined by

$$\hat{A}\left(\sum_{n=1}^{\infty} x_n \phi_n\right) = \sum_{n=1}^{\infty}((2n-1)x_{2n-1} - 2nx_{2n})\phi_1.$$

has the property $\tilde{A}\phi = \hat{A}\phi$, for every $\phi \in \mathcal{D}$, thus both $\tilde{A}$ and $\hat{A}$ are extensions of A. On the other hand,

$$\tilde{A}(B\phi) = \langle \phi | \phi_1 \rangle \phi_1, \quad \hat{A}(B\phi) = 0, \quad \forall \phi \in \mathcal{D}.$$

Hence, $\tilde{A} \cdot B \neq \hat{A} \cdot B$, which proves that the product depends on the chosen extension of A.

Example 6.3.2. Let us consider again the rigged Hilbert space of Example 5.4.3. Let $\eta_1, \eta_2 \in \mathcal{S}(\mathbb{R})$ be fixed functions with the properties

$$\eta_1 \geqslant 0, \ \mathrm{supp}(\eta_1) \subset (0,1), \ \int_0^1 \eta_1(t)dt = 1$$

$$\mathrm{supp}(\eta_2) \subset (-1,1), \ \eta_2(0) = 1.$$

Define

$$f_j(t) = j\eta_1(jt),$$
$$g_j(t) = f_j(-t),$$
$$h_j = f_j * (\chi_{[2/j,1]} \cdot \eta_2) \quad (h_1 = 0)$$
$$k_j = f_j * (\chi_{[-2/j,1]} \cdot \eta_2)$$
$$k = \chi_{[0,1]} \cdot \eta_2,$$

where as usual χ_E denotes the characteristic function of the set E. These test functions satisfy the following statements:

$$\lim_{j \to \infty} f_j = \delta \quad \text{in } X,$$

$$\lim_{j \to \infty} g_j = \delta \quad \text{in } X^{\times},$$

$$\lim_{j \to \infty} h_j = k \quad \text{in } X,$$

$$\lim_{j \to \infty} k_j = k \quad \text{in } X^\times,$$

$$g_j \cdot h_j = 0$$

$$\lim_{j \to \infty} \langle f_j | k_j \rangle = 1.$$

The properties listed above imply that the operators

$$T_1 : \varphi \in \mathcal{S}(\mathbb{R}) \mapsto \langle \delta, \varphi \rangle \delta \in \mathcal{S}^\times(\mathbb{R})$$

$$T_2 : \varphi \in \mathcal{S}(\mathbb{R}) \mapsto \langle k, \varphi \rangle k \in \mathcal{S}^\times(\mathbb{R}),$$

which both belong to $\mathcal{L}(\mathcal{S}(\mathbb{R}), \mathcal{S}^\times(\mathbb{R}))$, satisfy

(i) T_1 maps continuously $\mathcal{S}(\mathbb{R})$ into X and also into $X^\times$, these latter spaces carrying the norm topology defined earlier;
(ii) T_2 has continuous extensions to both X and $X^\times$.

One can compute $T_2 \cdot T_1$ taking X as intermediate space for the factorization. In this case,

$$T_2 \cdot T_1(\varphi) = \lim_{j \to \infty} \langle \delta, \varphi \rangle \langle k_j, f_j \rangle k = \varphi(0)k.$$

If we take $X^\times$, instead,

$$T_2 \cdot T_1(\varphi) = \lim_{j \to \infty} \langle \delta, \varphi \rangle \langle g_j, h_j \rangle k = 0.$$

Hence the multiplication is not well defined.

But there is more. Namely, there exist two linear operators S, T mapping the Schwartz space $\mathcal{S}(\mathbb{R})$ into its (conjugate) dual $\mathcal{S}^\times(\mathbb{R})$ in such a way that *every* operator R mapping continuously $\mathcal{S}(\mathbb{R})$ into $\mathcal{S}^\times(\mathbb{R})$ may be obtained as a factorization product $S \cdot T$, in the sense that there exists a Hilbert space $\mathcal{H}_R$ such that $\mathcal{S}(\mathbb{R}) \hookrightarrow \mathcal{H}_R \hookrightarrow \mathcal{S}^\times(\mathbb{R})$, the embeddings being continuous; $T : \mathcal{S}(\mathbb{R}) \to \mathcal{H}_R$, S has a continuous extension $\widetilde{S} : \mathcal{H}_R \to \mathcal{S}^\times(\mathbb{R})$ and $\widetilde{S}(T\xi) = R\xi$, for every $\xi \in \mathcal{S}(\mathbb{R})$.

In the light of the preceding discussion, we can conclude that an unambiguous definition of the multiplication can only be given if it is possible to select a family $\mathfrak{L}_0$ of interspaces with the property that the intersection of any two of them is an interspace (not necessarily belonging to $\mathfrak{L}_0$!).

Definition 6.3.3. A family $\mathfrak{L}_0$ of interspaces in the rigged Hilbert space $(\mathcal{D}[t], \mathcal{H}, \mathcal{D}^\times[t^\times])$ is said to be *tight* (around $\mathcal{D}$) if $\mathcal{E} \cap \mathcal{F}$ is an interspace for any pair of interspaces $\mathcal{E}, \mathcal{F} \in \mathfrak{L}_0$.

Starting from a tight family of interspaces, the partial multiplication in $\mathfrak{L}(\mathcal{D}, \mathcal{D}^\times)$ can be defined as follows.

Definition 6.3.4. Let $\mathfrak{L}_0$ be a tight family of interspaces in the rigged Hilbert space $(\mathcal{D}[t], \mathcal{H}, \mathcal{D}^\times[t^\times])$. The product $A \cdot B$ of two elements of $\mathcal{L}(\mathcal{D}, \mathcal{D}^\times)$ is defined, with respect to $\mathfrak{L}_0$, if there exist three interspaces $\mathcal{E}, \mathcal{F}, \mathcal{G} \in \mathfrak{L}_0$ such that $A \in \mathcal{C}(\mathcal{F}, \mathcal{G})$ and $B \in \mathcal{C}(\mathcal{E}, \mathcal{F})$. In this case the multiplication $A \cdot B$ is defined by:

$$A \cdot B = \left(\tilde{A}\tilde{B} \right) \restriction \mathcal{D}$$

or, equivalently, by:

$$A \cdot Bf = \tilde{A}\tilde{B}f, \quad f \in \mathcal{D}.$$

where $\tilde{A}$ (resp., $\tilde{B}$) denotes the extension of A (resp., B) to $\mathcal{E}$ (resp., $\mathcal{F}$).

This definition does not depend on the particular choice of the interspaces $\mathcal{E}, \mathcal{F}, \mathcal{G} \in \mathfrak{L}_0$, but may depend, of course, on $\mathfrak{L}_0$.

Obviously, we may suppose that $\mathcal{E} = \mathcal{D}$ and $\mathcal{G} = \mathcal{D}^\times$. With this choice, for the product $A \cdot B$ to make sense, we have only to require the existence of one interspace $\mathcal{F}$ such that $A \in \mathcal{C}(\mathcal{F}, \mathcal{D}^\times)$ and $B \in \mathcal{C}(\mathcal{D}, \mathcal{F})$. The price is, of course, a loss of information on the range of $A \cdot B$.

In a similar way, the product $A_n \cdot A_{n-1} \cdot \ldots \cdot A_1$ is defined, with respect to $\mathfrak{L}_0$, if there are interspaces $\mathcal{E}_0, \mathcal{E}_1 \ldots \mathcal{E}_n$ in $\mathfrak{L}_0$ such that $A_j \in \mathcal{C}(\mathcal{E}_{j-1}, \mathcal{E}_j)$. It is clear from this definition that, if $A \cdot B \cdot C$ exists in that sense, then $(A \cdot B) \cdot C$ and $A \cdot (B \cdot C)$ also exist and

$$(A \cdot B) \cdot C = A \cdot (B \cdot C) = A \cdot B \cdot C.$$

The converse statement is not true, i.e., if both $(A \cdot B) \cdot C$ and $A \cdot (B \cdot C)$ exist, they are not necessarily equal.

A natural question arises now: Given a tight family of interspaces around $\mathcal{D}$, is $\mathcal{L}(\mathcal{D}, \mathcal{D}^\times)$ a partial *-algebra with respect to the multiplication defined above? The answer is in general negative. In order to get this result, the family of interspaces must be closed under the operation of taking duals and under finite intersections. This justifies the definition of *multiplication framework* as a family $\mathfrak{L}$ of interspaces satisfying

(i) $\mathcal{D} \in \mathfrak{L}$;
(ii) for every $\mathcal{E} \in \mathfrak{L}$, its conjugate dual $\mathcal{E}^\times$ also belongs to $\mathfrak{L}$;
(iii) for every $\mathcal{E}, \mathcal{F} \in \mathfrak{L}$, $\mathcal{E} \cap \mathcal{F} \in \mathfrak{L}$.

These conditions are exactly the same that we used in Section 5.4.1 for constructing a PIP-space from a rigged Hilbert space. In that context, we called a family with these properties a *generating family* of interspaces. In this chapter we keep the name *multiplication framework* to point out their role in the problem of multiplying operators. Essentially the same conditions were also needed in Section 5.5.2 for constructing a LHS from a family or an algebra of unbounded operators.

Remark 6.3.5. It is clear that the family $\{\mathcal{D}, \mathcal{D}^\times\}$ is always a multiplication framework. So the set of multiplication frameworks in a rigged Hilbert space is always non empty. The family $\{\mathcal{D}, \mathcal{D}^\times\}$ gives rise to the simplest and poorest lattice of multipliers consisting only of $\mathfrak{L}(\mathcal{D}, \mathcal{D}^\times)$, $\mathfrak{L}(\mathcal{D}, \mathcal{D})$, $\mathfrak{L}(\mathcal{D}^\times, \mathcal{D}^\times)$ and $\mathcal{L}^\dagger(\mathcal{D})$. On the other hand, a maximal multiplication framework need not exist.

Remark 6.3.6. A generating family of interspaces or, in other terms, a multiplication framework, is an involutive lattice of interspaces, under the operations of intersection and duality. So we have recovered exactly the notion of projective indexed PIP-space (see Section 2.4). However, the motivations of the two approaches are totally different. In the PIP-space set-up, one starts with a binary relation on a vector space V (the compatibility) and a partial inner product defined on compatible pairs. From this, one generates, by intersection and duality, a complete involutive lattice of subspaces, each subspace being equipped with its Mackey topology. Then, one selects an involutive sublattice, for which the condition of projectivity guarantees that no topological pathology arises. Finally, a generating family is a subset of this sublattice that generates the latter. When this is done, operators and their multiplication rule follow naturally.

Here, on the other hand, one starts from the other end, with only the rigged Hilbert space $(\mathcal{D}[t], \mathcal{H}, \mathcal{D}^\times[t^\times])$, and one tries to define a suitable notion of product on operators from $\mathfrak{L}(\mathcal{D}, \mathcal{D}^\times)$, which is by necessity only partially defined. The multiplication framework is nothing but an involutive lattice of interspaces that solves that problem, so that one gets exactly the same structure as in the PIP-space set-up.

Let us now consider whether a multiplication framework can be generated starting from a given tight family $\mathfrak{L}_0$ of interspaces around $\mathcal{D}$. The natural procedure would consist in taking, together with $\mathfrak{L}_0$, all possible finite intersections of elements of $\mathfrak{L}_0$. Nevertheless, this would not produce, as a result, a multiplication framework, since $\mathcal{D}$ need not be dense in these finite intersections of interspaces, endowed with their projective topology. For this reason, we can call *generating* a tight family of interspaces $\mathfrak{L}_0$, closed under duality and enjoying the following property:

(g) $\mathcal{D}$ is dense in $\mathcal{E}_1 \cap \cdots \cap \mathcal{E}_n$, endowed with its projective topology, for any finite set $\{\mathcal{E}_1, \ldots, \mathcal{E}_n\}$ of elements of $\mathfrak{L}_0$.

The notion of *generating* family is useful, since it is sometimes reasonably easy to check.

Example 6.3.7. Let us consider the rigged Hilbert space:

$$\mathcal{S}(\mathbb{R}^n) \hookrightarrow L^2(\mathbb{R}^n) \hookrightarrow \mathcal{S}^\times(\mathbb{R}^n).$$

The family of interspaces $\{L^p(\mathbb{R}^n),\ p \in (1,+\infty)\}$ is tight and generates a multiplication framework (i.e., it is generating in the sense discussed above). Also the family of Sobolev spaces $\{W^{k,p}(\mathbb{R}^n),\ k \in \mathbb{Z}, p \in (1,+\infty)\}$ generates a multiplication framework.

In conclusion, we may state the following

Theorem 6.3.8. *Let $\mathfrak{L}$ be a multiplication framework in the rigged Hilbert space $(\mathcal{D}, \mathcal{H}, \mathcal{D}^\times)$. Then $\mathfrak{L}(\mathcal{D}, \mathcal{D}^\times)$, with the multiplication defined above, is a partial *-algebra.*

Let now $\mathcal{D}$ be a reflexive Fréchet domain. Then the definition of multiplication of two operators $A, B \in \mathfrak{L}(\mathcal{D}, \mathcal{D}^\times)$ given above can be formulated in the same terms as those used for defining the weak multiplication in $\mathrm{Op}(V)$.

Let indeed $A, B \in \mathfrak{L}(\mathcal{D}, \mathcal{D}^\times)$. Assume that there exists an interspace $\mathcal{E}$ such that $B \in \mathcal{C}(\mathcal{D}, \mathcal{E}^\times)$ and $A^\dagger \in \mathcal{C}(\mathcal{D}, \mathcal{E})$. Then the sesquilinear form

$$\gamma_{A,B}(\phi, \psi) = \langle A^\dagger \phi | B \psi \rangle, \quad \phi, \psi \in \mathcal{D}$$

is jointly continuous; hence there exists an operator $C \in \mathfrak{L}(\mathcal{D}, \mathcal{D}^\times)$ such that

$$\langle \phi | C \psi \rangle = \langle A^\dagger \phi | B \psi \rangle, \ \forall \phi, \psi \in \mathcal{D}.$$

In this case also, C may depend on the choice of $\mathcal{E}$, unless $\mathcal{E}$ is not chosen in a multiplication framework $\mathfrak{L}$.

Thus, if $\mathfrak{L}$ is a multiplication framework, for any pair $A, B \in \mathfrak{L}(\mathcal{D}, \mathcal{D}^\times)$ for which there exists an interspace $\mathcal{E} \in \mathfrak{L}$ such that $B \in \mathcal{C}(\mathcal{D}, \mathcal{E}^\times)$ and $A^\dagger \in \mathcal{C}(\mathcal{D}, \mathcal{E})$, one can put $A \cdot B := C$ and so the natural equality

$$\langle \phi | (A \cdot B) \psi \rangle = \langle A^\dagger \phi | B \psi \rangle, \ \forall \phi, \psi \in \mathcal{D}$$

holds. With this multiplication, $\mathfrak{L}(\mathcal{D}, \mathcal{D}^\times)$ is a partial *-algebra.

Moreover, if $A \cdot B$ is well-defined, then a simple duality argument shows that $A \in \mathcal{C}(\mathcal{E}^\times, \mathcal{D}^\times)$, hence the product $A \cdot B$ is also well-defined.

Theorem 6.3.9. *Let $\mathfrak{L}$ be a multiplication framework in the rigged Hilbert space $(\mathcal{D}, \mathcal{H}, \mathcal{D}^\times)$, where $\mathcal{D}$ is a reflexive Fréchet domain. Then the multiplications $\cdot$ and $\cdot$ coincide and $\mathfrak{L}(\mathcal{D}, \mathcal{D}^\times)$ is a partial *-algebra.*

Remark 6.3.10. This result is not surprising. Indeed, the family of interspaces of a multiplication framework makes $\mathcal{D}^\times$ into an indexed PIP-space and the family of operators on an indexed PIP-space always form a partial *-algebra. Hence, Theorem 6.3.9 is only a special case of Proposition 6.2.3 and of Proposition 6.2.7.

6.3.2 Differential Operators

We describe here briefly the quasi *-algebra of differential operators with coefficients in $\mathcal{S}^\times(\mathbb{R})$ and we will give some sufficient conditions in order that the product of two operators of this kind be well-defined.

We maintain the notation of Section 6.3 and, for shortness we put $\mathcal{S} := \mathcal{S}(\mathbb{R})$ and $\mathcal{S}^\times := \mathcal{S}^\times(\mathbb{R})$. First we notice that $(\mathcal{S}^\times, \mathcal{S})$ is a quasi *-algebra. Indeed, if $F \in \mathcal{S}^\times$ and $\phi \in \mathcal{S}$, then the product $F\phi$ defined by

$$\langle \psi | F\phi \rangle = \langle \phi^* \psi | F \rangle, \ \psi \in \mathcal{S}$$

is an element of $\mathcal{S}^\times$ (ϕ^* denotes the complex conjugate of $\phi \in \mathcal{S}$).

Moreover, the map

$$L_F : \phi \in \mathcal{S} \mapsto F\phi \in \mathcal{S}^\times$$

is continuous; thus $L_F \in \mathcal{L}(\mathcal{S}, \mathcal{S}^\times)$.

On the other hand the operator $P = i\frac{d}{dx}$ (and all its powers) belongs to $\mathcal{L}^\dagger(\mathcal{S})$. Then an expression of the form

$$\sum_{k=0}^n F_k P^k, \quad F_k \in \mathcal{S}^\times, n \in \mathbb{N}$$

defines an operator of $\mathcal{L}(\mathcal{S}, \mathcal{S}^\times)$ that we call a *differential operator* with coefficients in $\mathcal{S}^\times$. We put

$$\mathfrak{A} = \left\{ \sum_{k=0}^n F_k P^k; \ F_k \in \mathcal{S}^\times, n \in \mathbb{N} \right\}$$

and

$$\mathfrak{A}_0 = \left\{ \sum_{j=0}^m f_j P^j; \ f_j \in \mathcal{S}, m \in \mathbb{N} \right\}.$$

Then, making use of the identity (the so-called canonical commutation relation)

$$P(xf) - x(Pf) = if, \ \forall f \in \mathcal{S},$$

it is not difficult to prove that $(\mathfrak{A}, \mathfrak{A}_0)$ is a quasi *-algebra (the quasi *-algebra of differential operators with coefficients in $\mathcal{S}^\times$).

Now let us be given $X, Y \in \mathfrak{A}$. Under which conditions is $X \cdot Y$ well-defined? In order to discuss this question, following the path of Section 6.3.1, we have to choose a suitable multiplication framework between $\mathcal{S}$ and $\mathcal{S}^\times$. Since only integer powers of P are involved, it is reasonable to consider, at first, the multiplication framework generated by the Sobolev spaces $W^{k,p}(\mathbb{R})$, $k \in \mathbb{Z}$, $1 < p < \infty$, defined in (5.20) (these spaces also will be indicated as $W^{k,p}$ in what follows).

Notice that $P^n : W^{m,p} \to W^{m-n,p}$ for each $n \in \mathbb{N}$, $m \in \mathbb{Z}$ and $1 < p < \infty$. So, if $F_h \in W^{r,q}$ with $r + m - n \geq 0$ and $\frac{1}{p} + \frac{1}{q} = \frac{1}{s}$, then $X = \sum_{h=0}^n F_h P^h$ maps $W^{m,p}$ into $W^{l,s}$ with $l = \min\{m - n, r\}$.

Analogously, if $Y = \sum_{j=0}^{n'} G_j P^j$ with $G_j \in W^{r',q'}$ with $r' + m' - n' \geq 0$ and $\frac{1}{p'} + \frac{1}{q'} = \frac{1}{s'}$, then Y maps $W^{m',p'}$ into $W^{l',s'}$ with $l' = \min\{m' - n', r'\}$. It follows that if $r + l' - n \geq 0$ and $\frac{1}{q} + \frac{1}{s'} = \frac{1}{s''}$, then $X \cdot Y$ is well-defined and maps $W^{m',p'}$ into $W^{l,s''}$. So, in conclusion,

Proposition 6.3.11. *Let* $X = \sum_{h=0}^n F_h P^h$ *and* $Y = \sum_{j=0}^{n'} G_j P^j$ *be two differential operators, with* $F_h \in W^{r,q}$, $h = 0, \ldots n$, *and* $G_j \in W^{r',q'}$, $j = 0, \ldots n'$. *Then their product* $X \cdot Y$ *is well-defined in* $\mathfrak{A}$ *and its coefficients belong to a suitable space* $W^{k,p}$.

6.3.3 Multiplication of Distributions

The problem of multiplying distributions dates back to the early times of the theory. L. Schwartz himself, in fact, proved already in the 1950s that the Dirac δ distribution cannot be multiplied by itself. Since then, several approaches have been developed for defining the multiplication of distributions both with heuristic methods or within a rigorous mathematical framework. Schwartz's result essentially means that this multiplication can only be partial. The so called *duality method* provides a rigorous way for considering this problem. In this context the problem can also be studied making use of the formalism of interspaces developed in Section 6.3.1. Indeed, this can be done by identifying distributions with certain multiplication operators acting in the rigged Hilbert space of distributions itself. This can be done abstractly, i.e., without making reference to specific test function spaces, such as $\mathcal{D}(\mathbb{R}^n)$ or $\mathcal{S}(\mathbb{R}^n)$. The starting point is a dense domain in a Hilbert space, which is at the same time a commutative *-algebra satisfying additional topological requirements.

Let A be a self-adjoint operator in the Hilbert space $\mathcal{H}$ and put, as usual, $R_A = (1 + A^2)^{1/2}$. Let us consider a subspace $\mathcal{D} \subset \mathcal{D}^\infty(A)$ endowed with a locally convex topology t which makes of $\mathcal{D}$ a semireflexive space and satisfies the following properties:

(d_1) The topology t of $\mathcal{D}$ is finer than the graph topology induced from $\mathcal{D}^\infty(A)$;

(d_2) Both A and R_A map $\mathcal{D}[t]$ continuously into itself;

(d_3) For all $n \in \mathbb{N}$, $\mathcal{D}$ is a core for R_A^n, that is, $\overline{R_A^n \restriction \mathcal{D}} = R_A^n$.

If $\mathcal{D}^\times$ denotes the conjugate dual of $\mathcal{D}[t]$, we have the following situation

$$\mathcal{D} \hookrightarrow \mathcal{D}^\infty(A) \hookrightarrow \mathcal{H} \hookrightarrow \mathcal{D}_{\overline{\infty}}(A) \hookrightarrow \mathcal{D}^\times,$$

where all the embeddings are continuous with dense range. Thus, if $\{G_\alpha\}$ is a multiplication framework between $\mathcal{D}^\infty(A)$ and $\mathcal{D}_{\overline{\infty}}(A)$, it is one between $\mathcal{D}$ and $\mathcal{D}^\times$ also.

Let $\mathcal{D}$ be a domain satisfying the conditions (d_1)-(d_3) above with respect to a fixed self-adjoint operator A. We assume, in addition, that $\mathcal{D}$ is a *generalized* commutative Hilbert *-algebra, in the sense that $\mathcal{D}$ is a commutative *-algebra with respect to the involution $\phi \mapsto \phi^*$ and the multiplication $(\phi, \psi) \mapsto \phi\psi \, (= \psi\phi)$ and the following conditions hold:

(h_1) $\langle \psi | \phi \rangle = \langle \phi^* | \psi^* \rangle$, $\forall\, \phi, \psi \in \mathcal{D}$;
(h_2) $\langle \chi | \phi\psi \rangle = \langle \phi^*\chi | \psi \rangle$, $\forall\, \phi, \psi, \chi \in \mathcal{D}$;
(h_3) The multiplication $(\phi, \psi) \mapsto \phi\psi \, (= \psi\phi)$ is *jointly* continuous with respect to the topology t of $\mathcal{D}$;
(h_4) The involution $\phi \mapsto \phi^*$ is continuous in $\mathcal{D}[\mathsf{t}]$.
(h_5) $\mathcal{D} \cdot \mathcal{D}$ is dense in $\mathcal{D}[\mathsf{t}]$.

Proposition 6.3.12. *Let $\mathcal{D}^\times$ be the conjugate dual of $\mathcal{D}$. If we define the multiplication of an element $F \in \mathcal{D}^\times$ and an element $\phi \in \mathcal{D}$ by*

$$\langle \psi | F\, \phi \rangle = \langle \psi | \phi\, F \rangle := \langle \phi^*\, \psi | F \rangle, \ \forall\, \psi \in \mathcal{D},$$

*then $(\mathcal{D}^\times[\mathsf{t}^\times], \mathcal{D})$ is a locally convex quasi *-algebra.*

From these facts it follows that, to each element F of $\mathcal{D}^\times$, we can associate an operator L_F of multiplication on $\mathcal{D}$ defined by

$$L_F : \phi \in \mathcal{D} \mapsto F\phi \in \mathcal{D}^\times.$$

This is a continuous linear map of $\mathcal{D}$ into $\mathcal{D}^\times$. Therefore, the problem of multiplying two distributions F, G can be formulated in terms of the multiplication of the corresponding operators L_F and L_G. The multiplication of operators of this kind can then be studied in the terms proposed in Section 6.3.1. Nevertheless, even though the product $L_F \cdot L_G$ exists in $\mathfrak{L}(\mathcal{D}, \mathcal{D}^\times)$, it is not necessarily an operator of multiplication by some distribution. For this to happen, additional conditions must be added.

A net (η_ϵ) of elements of $\mathcal{D}$ is called an *approximate identity* of $\mathcal{D}$ if

$$\mathsf{t}-\lim_\epsilon \eta_\epsilon \phi = \phi, \ \forall\, \phi \in \mathcal{D}.$$

Then, using this notion, one may prove the following

Proposition 6.3.13. *Assume that $\mathcal{D}$ has an approximate identity and let $X \in \mathfrak{L}(\mathcal{D}, \mathcal{D}^\times)$. Then $X = L_V$ for some $V \in \mathcal{D}^\times$ if, and only if, the following two conditions are fulfilled* *(i)* $X(\phi\psi) = \phi X\psi, \quad \forall \phi, \psi \in \mathcal{D}$;

(ii) There exists a t-*continuous seminorm p in $\mathcal{D}$ such that*

$$|\langle \psi | X \phi \rangle| \leqslant p(\psi^* \phi), \ \forall \phi, \psi \in \mathcal{D}.$$

Let now $\mathcal{T}$ be a multiplication framework satisfying the following properties:

(a₁) If $\mathcal{F} \in \mathcal{T}$, the involution $\phi \mapsto \phi^*$ is continuous from $\mathcal{F}$ into itself.

(a₂) If $\mathcal{F} \in \mathcal{T}$ and $\phi \in \mathcal{D}$, then $\phi\mathcal{F} \subset \mathcal{F}$, and the map $T_\phi : \mathcal{F} \to \mathcal{F}$, defined by:

$$u \in \mathcal{F} \mapsto \phi u \in \mathcal{F}$$

is continuous in $\mathcal{F}$.

(a₃) If $\mathcal{F} \in \mathcal{T}$, $u \in \mathcal{F}$, the map $L_u : \phi \in \mathcal{D} \mapsto \phi u \in \mathcal{F}$ is continuous from $\mathcal{D}$ into $\mathcal{F}$.

The multiplication induced by $\mathcal{T}$ is *commutative*, i.e, if $u, v \in \mathcal{D}^\times$ and $L_u \circ L_v$ is well-defined with respect to $\mathcal{T}$, then also $L_v \circ L_u$ is well-defined and $L_u \circ L_v = L_v \circ L_u$.

Proposition 6.3.14. *Let $\mathcal{T}$ be a multiplication framework satisfying the properties* (a₂)*,* (a₃) *and $\mathcal{E}$, $\mathcal{F}$ two interspaces of $\mathcal{T}$. For fixed $v \in \mathcal{E}$ and $u \in \mathcal{D}^\times$, let L_v, L_u be the respective multiplication maps from $\mathcal{D}$ into $\mathcal{D}^\times$. If L_u has a continuous extension, $\widetilde{L}_u := \mathcal{E} \to \mathcal{F} \in \mathcal{T}$, then there exists $w \in \mathcal{F}$ such that $L_w = \widetilde{L}_u \cdot L_v : \mathcal{D} \to \mathcal{F}$.*

The next corollary summarizes the previous discussion.

Corollary 6.3.15. *Let $\mathcal{T}$ be a multiplication framework satisfying the properties* (a₁)-(a₃)*, and let $v \in \mathcal{E}$ and $u \in \mathcal{D}^\times$, with L_v, L_u the corresponding multiplication maps. If $L_u \in \mathcal{C}(\mathcal{E}, \mathcal{F})$, then it is possible to define a product $w = u \cdot v$ with $w \in \mathcal{F}$. The multiplication defined in this way is commutative.*

A special case of Corollary 6.3.15 gives the more familiar *duality method* for multiplying distributions.

Corollary 6.3.16. *Let $\mathcal{T}$ be a multiplication framework satisfying the properties* (a₁)-(a₃) *and $u, v \in \mathcal{D}^\times$. If there exists an interspace $\mathcal{M} \in \mathcal{T}$ such that $u \in \mathcal{M}$ and $v \in \mathcal{M}^\times$, then $L_u \circ L_v$ is well defined as a multiplication map, i.e, there exists $w \in \mathcal{D}^\times$ such that $L_w = L_u \circ L_v$. This product is commutative.*

Let now $u \in \mathcal{D}^\times$ and $\mathcal{T}$ be a multiplication framework. In general, u need not belong to any proper interspace $\mathcal{F} \in \mathcal{T}$. This is, however, needed to apply Corollaries 6.3.15 and 6.3.16. For this reason we put:

$$\mathcal{D}_{\mathcal{T}}^\times = \bigcup \left\{ \mathcal{F} : \mathcal{F} \in \mathcal{T}, \mathcal{D} \hookrightarrow \mathcal{F} \hookrightarrow \mathcal{D}^\times \right\}.$$

At the light of the previous discussion, we get

Proposition 6.3.17. *Let $\mathcal{T}$ be a multiplication framework satisfying the properties* (a$_1$)-(a$_3$). *Then $\mathcal{D}_{\mathcal{T}}^{\times}$ can be identified with a commutative partial *-algebra*

$$\mathfrak{A} = \{L_u, \, u \in \mathcal{D}_{\mathcal{T}}^{\times} \subset \mathfrak{L}(\mathcal{D}, \mathcal{D}^{\times})\},$$

where the multiplication is that defined in $\mathfrak{L}(\mathcal{D}, \mathcal{D}^{\times})$ by the multiplication framework $\mathcal{T}$.

Remark 6.3.18. We notice that in the Propositions given above, the assumption that $\mathcal{T}$ is a multiplication framework having the properties (a$_1$)-(a$_3$), can be replaced with the requirement that $\mathcal{T}$ is a generating family of interspaces enjoying the same properties. This is due to the fact that, if $\mathcal{T}$ is a generating family of interspaces with the properties (a$_1$)-(a$_3$), then the generated multiplication framework $\widehat{\mathcal{T}}$ has again the properties (a$_1$)-(a$_3$).

Example 6.3.19. A tempered distribution u belongs to the Bessel potential space (see Section 5.4.2), $L^{s,p}$ if and only if $\mathcal{F}^{-1}\left(\left(1+|\xi|^2\right)^{-s/2}\mathcal{F}u\right) \in L^p(\mathbb{R})$. These spaces generate a multiplication framework $\mathcal{T}$ in the rigged Hilbert space corresponding to $A = -i\frac{d}{dx}$ defined on $W^{1,2}(\mathbb{R})$ and it has the properties (a$_1$)-(a$_3$). Then they can be used to reformulate the above abstract results in the concrete case of tempered distributions. So, for instance, an immediate application of Corollary 6.3.15 yields:

Proposition 6.3.20. *Let $v \in L^{t,q}(\mathbb{R})$ and $u \in \mathcal{S}^{\times}$. Assume that there exist $s, t \in \mathbb{R}$ and $p, q \in [1, \infty[$ such that $L_u \in \mathcal{C}(L^{s,p}(\mathbb{R}), L^{t,q}(\mathbb{R}))$ Then the product $w = u \cdot v$ exists and $w \in L^{t,q}(\mathbb{R})$.*

As a consequence of Proposition 6.3.17, the set $\mathcal{S}_{\mathcal{T}}^{\times}$ of tempered distributions belonging to some Bessel potential space $L^{s,p}(\mathbb{R})$, is a partial *-algebra with respect to the partial multiplication inherited by $\mathcal{L}(\mathcal{S}, \mathcal{S}^{\times})$.

6.4 Representations of Partial *-Algebras

As we have seen in Section 6.2.2, $\mathrm{Op}(V)$ is a partial *-algebra with respect to the weak multiplication $..$. It is therefore natural to ask the question of the existence of *-representations of an abstract partial *-algebra $\mathfrak{A}$, i.e., a *-homomorphism of $\mathfrak{A}$ into $\mathrm{Op}(V)$. In the theory of *-algebras, a canonical way to construct a *-representation of a given *-algebra $\mathfrak{A}$ is provided by the Gel'fand–Naĭmark–Segal (GNS) construction, which has a positive linear functional ω as starting point. This construction can be extended in several ways to partial *-algebras, but in most cases one has to deal with positive sesquilinear forms defined on $\mathfrak{A} \times \mathfrak{A}$ rather than with positive linear functionals. This choice recommends itself when one wants to get a *-representation living in a Hilbert space. Since the GNS representation in very relevant for applications, we will study in this section several possible approaches to it; but, of course, we will enlarge our set-up, considering *-representations taking values in $\mathrm{Op}(V)$.

6.4.1 A GNS Construction for $\mathfrak{B}$-Weights on a Partial *-Algebra

In this section we will deal with the question of GNS-representation in a partial *-algebra, but with a different set-up. We will start with a possibly non everywhere defined sesquilinear form fulfilling certain properties of invariance.

Let $\mathfrak{A}$ be a partial *-algebra and Ω a positive sesquilinear form defined on a domain $\Gamma_\Omega \subseteq \mathfrak{A} \times \mathfrak{A}$, by which we mean the following properties:[1]

(d$_1$) Γ_Ω preserves linearity : if $(x, y) \in \Gamma_\Omega$ and $(x, z) \in \Gamma_\Omega$, then $(x, y + \lambda z) \in \Gamma_\Omega \forall \lambda \in \mathbb{C}$;
(d$_2$) Γ_Ω is symmetric, i.e., if $(x, y) \in \Gamma_\Omega$, then $(y, x) \in \Gamma_\Omega$;
(d$_3$) Ω is Hermitian, i.e., $\Omega(x, y) = \overline{\Omega(y, x)}, \forall (x, y) \in \Gamma_\Omega$;
(d$_4$) Ω is positive, i.e., if $(x, x) \in \Gamma_\Omega$, then $\Omega(x, x) \geqslant 0$.

If $\mathfrak{M}$ is any subset of $\mathfrak{A}$, we denote by $\mathfrak{M}_\Omega$ the following subspace of $\mathfrak{A}$:

$$\mathfrak{M}_\Omega = \{x \in \mathfrak{A} : (x, b) \in \Gamma_\Omega, \forall b \in \mathfrak{M}\}.$$

Definition 6.4.1. The positive sesquilinear form Ω is called a $\mathfrak{B}$-*weight* on the partial *-algebra $\mathfrak{A}$ if there exists a subspace $\mathfrak{B} \subseteq R\mathfrak{A} \subseteq \mathfrak{A}$ such that

(i) $\mathfrak{B} \times \mathfrak{B} \subseteq \Gamma_\Omega$;
(ii) If $x \in \mathfrak{A}$ and $b \in \mathfrak{B}$, then $(xb, c) \in \Gamma_\Omega, \forall c \in \mathfrak{B}$;
(iii) $\Omega(xb_1, b_2) = \Omega(b_1, x^*b_2), \forall x \in \mathfrak{A}, b_1, b_2 \in \mathfrak{B}$;
(iv) If $x_1 \in L(x_2)$, then $(x_1^*b_1, x_2b_2) \in \Gamma_\Omega, \forall b_1, b_2 \in \mathfrak{B}$ and

$$\Omega(x_1^*b_1, x_2b_2) = \Omega(b_1, (x_1 x_2)b_2);$$

(v) If $x \in \mathfrak{B}$ and $\Omega(b, x) = 0, \forall b \in \mathfrak{B}$, then $\Omega(y, x) = 0, \forall y \in \mathfrak{B}_\Omega$.

If $\mathfrak{A}$ possesses a unit e we say that a $\mathfrak{B}$-weight Ω is a $\mathfrak{B}$-*state* if Ω is normalized, i.e., if $e \in \mathfrak{B}$ and $\Omega(e, e) = 1$.

If $\mathfrak{A}$ has a unit, then (i) and (ii) imply that $\mathfrak{B} \times \mathfrak{A} \subseteq \Gamma_\Omega$ and $\mathfrak{B}_\Omega = \mathfrak{A}$. We notice that, since Ω is Hermitian and positive, a Schwarz inequality holds in $\mathfrak{B}$, namely,

$$|\Omega(x, y)|^2 \leqslant \Omega(x, x)\,\Omega(y, y), \ \forall x, y \in \mathfrak{B}.$$

---

[1] In accordance with the literature, we take $\Omega(x, y)$ to be linear in x and antilinear in y.

Remark 6.4.2. Conditions (iii) and (iv) of the definition above take into account the possible lack of associativity in a partial *-algebra.

Condition (v) looks like separateness of dual pairs. It will in fact enable us to recover the density of our representation space in a bigger one, as we shall see below. If Ω is everywhere defined, i.e., $\Gamma_\Omega = \mathfrak{A} \times \mathfrak{A}$, and $\mathfrak{B} \subset R\mathfrak{A}$, Definition 6.4.1 reduces to that of $\mathfrak{B}$-invariant positive sesquilinear form given in [21]; such forms are defined by obvious modifications of (iii), (iv) and (v).

Finally, we notice that in many instances, instead of (i) and (ii) of the previous definition, the stronger condition $\mathfrak{B} \times \mathfrak{A} \subseteq \Gamma_\Omega$ holds true (i.e., $\mathfrak{B}_\Omega = \mathfrak{A}$). However, we prefer a more general formulation which allows also weights on *-algebras to fit into our framework.

Example 6.4.3. Let $\mathfrak{A}$ be a *-algebra and ω a weight (in the usual sense) on $\mathfrak{A}$, i.e., ω is a function from $\mathfrak{A}_+ = \{\sum_{k=1}^n x_k^* x_k; \, x \in \mathfrak{A}\}$ (the positive cone of $\mathfrak{A}$) into $[0, \infty]$ satisfying

$$\omega(x + y) = \omega(x) + \omega(y), \quad x, y \in \mathfrak{A}_+,$$
$$\omega(\alpha x) = \alpha \omega(x), \quad \alpha \in \mathbb{R}_+, x \in \mathfrak{A}_+,$$

where we assume $0.\infty = 0$.

Define the set

$$\mathcal{L}_\omega^o = \{x \in \mathfrak{A} : \omega(x^* x) < \infty\}.$$

Assume that $\mathcal{L}_\omega^o$ is a left ideal in $\mathfrak{A}$ (if $\mathfrak{A}$ is a C*-algebra, this is automatically true). Let $\mathcal{D}_\omega$ be the subspace of $\mathfrak{A}$ generated by the set $\{x^* x : x \in \mathcal{L}_\omega^o\}$. Since $\mathcal{L}_\omega^o$ is a left ideal of $\mathfrak{A}$, $\mathcal{D}_\omega$ coincides with the linear span of $\{y^* x : x, y \in \mathcal{L}_\omega^o\}$. Then every element of the form $\sum_{k=1}^m \lambda_k y_k^* x_k$, with $\lambda_k \in \mathbb{C}$, $x_k, y_k \in \mathcal{L}_\omega^o$, can be represented as $\sum_{h=1}^n \mu_h x^* x$. Then ω extends, in an obvious manner, to a linear functional on $\mathcal{D}_\omega$ which we denote with the same symbol. Now let us set $\Gamma_\Omega = \mathcal{L}_\omega^o \times \mathcal{L}_\omega^o$ and

$$\Omega(x, y) = \omega(y^* x), \quad \text{for } (x, y) \in \Gamma_\Omega.$$

It is easily seen that Ω and Γ_Ω satisfy Conditions (d_1)-(d_4). Moreover, if we take $\mathfrak{B} = \mathcal{L}_\omega^o$, taking into account that $\mathcal{L}_\omega$ is a left ideal of $\mathfrak{A}$, it is readily checked that Ω and Γ_Ω satisfy also Conditions (i)-(iv) of Definition 6.4.1. Condition (v) also holds true. In this case, in fact, $\mathfrak{B}_\Omega = \mathcal{L}_\omega$ and then the result is obtained making use of the Schwarz inequality on $\mathcal{L}_\omega$. Therefore the sesquilinear form associated to a weight ω on a *-algebra is a $\mathcal{L}_\omega$-weight in the sense of Definition 6.4.1.

Now we will show that a $\mathfrak{B}$-weight generates on the partial *-algebra $\mathfrak{A}$ a natural structure of PIP-space.

Let $\mathfrak{A}, \mathfrak{B}, \Omega$ and $\mathfrak{B}_\Omega$ be as above. It is readily checked that $\mathfrak{B}_\Omega$ is a vector space and $\mathfrak{B} \subseteq \mathfrak{B}_\Omega$. We can introduce a compatibility in $\mathfrak{B}_\Omega$ by

$$x \# y \quad \text{if and only if} \quad (x, y) \in \Gamma_\Omega.$$

Because of Condition (d_1), $\#$ is a linear compatibility; therefore, for any subset $\mathcal{C} \subseteq \mathfrak{B}_\Omega$, $\mathcal{C}^\#$ is a vector subspace of $\mathfrak{B}_\Omega$. Thus $\mathfrak{B}_\Omega = \mathfrak{B}^\#$ and therefore

$$\mathfrak{B} \subseteq \mathfrak{B}_\Omega^\# \subseteq \mathfrak{B}_\Omega \; (= \mathfrak{B}_\Omega^{\#\#} = \mathfrak{B}^\#).$$

The partial inner product is then defined on compatible pairs (x, y) by $\langle x | y \rangle = \Omega(x, y)$.[2] Clearly, this PIP-space is degenerate, in general. However it is not difficult to remove this degeneracy. To do this, let us introduce the following vector spaces:

$$(\mathfrak{B}_\Omega)^\perp = \{x \in \mathfrak{B}_\Omega^\# : \Omega(x, y) = 0, \forall y \in \mathfrak{B}_\Omega\},$$
$$(\mathfrak{B}_\Omega^\#)^\perp = \{x \in \mathfrak{B}_\Omega : \Omega(x, y) = 0, \forall y \in \mathfrak{B}_\Omega^\#\},$$
$$\mathfrak{B}^\perp = \{x \in \mathfrak{B}_\Omega : \Omega(x, y) = 0, \forall y \in \mathfrak{B} \}.$$

Clearly one has $(\mathfrak{B}_\Omega)^\perp \subseteq (\mathfrak{B}_\Omega^\#)^\perp \subseteq \mathfrak{B}^\perp$.

Lemma 6.4.4. *If* $x \in \mathfrak{B} \cap (\mathfrak{B}_\Omega)^\perp$ *and* $a \in \mathfrak{A}$, *then* $ax \in (\mathfrak{B}_\Omega)^\perp$.

Proof. Let $x \in \mathfrak{B} \cap (\mathfrak{B}_\Omega)^\perp$ and $x' \in \mathfrak{B}$, then, by (ii) and (iii) of Definition 6.4.1, we have first that both ax and $a^*x' \in \mathfrak{B}_\Omega$, and

$$\Omega(x', ax) = \Omega(a^*x', x) = 0$$

since $x \in (\mathfrak{B}_\Omega)^\perp$.   ∎

Let now $\widehat{\mathfrak{A}} = \mathfrak{B}_\Omega/(\mathfrak{B}_\Omega)^\perp$ and $\widehat{\mathfrak{B}} = \mathfrak{B}/(\mathfrak{B} \cap (\mathfrak{B}_\Omega)^\perp)$. Elements of these sets will be denoted respectively by $\lambda_\Omega(b), \mu_\Omega(a)$.

Lemma 6.4.5. $\widehat{\mathfrak{B}}$ *can be identified with a subspace of* $\widehat{\mathfrak{A}}$.

Proof. The map $i : \lambda_\Omega(b) \in \widehat{\mathfrak{B}} \mapsto \mu_\Omega(b) \in \widehat{\mathfrak{A}}$ is well-defined and linear. We get $\operatorname{Ker} i = \{0\}$. Indeed, if $\mu_\Omega(b) = 0$, then $b \in \mathfrak{B} \cap (\mathfrak{B}_\Omega)^\perp$, so that $\lambda_\Omega(b) = 0$.   ∎

Because of this identification, we will indicate the rest classes with the same symbol $\lambda_\Omega(a)$, whenever this will not create confusion.

Remark 6.4.6. The identification stated in Lemma 6.4.5 (which by the way remains valid if we replace $(\mathfrak{B}_\Omega)^\perp$ with an arbitrary subspace of $\mathfrak{B}_\Omega$) is crucial for our purposes. We wish to build a GNS-representation of $\mathfrak{A}$ just

---

[2] According to our convention for sesquilinear forms, this definition implies that the partial inner product $\langle x | y \rangle$ is linear in x and antilinear in y. Of course, this does not change anything in the arguments.

by generalizing the usual procedure: the representation will be defined as multiplication of elements of $\widehat{\mathfrak{A}}$ by elements of $\widehat{\mathfrak{B}}$. Lemma 6.4.5 makes this possible.

Lemma 6.4.7. *(i) The* PIP-*space* $(\widehat{\mathfrak{A}}, \widehat{\#}, \langle \cdot | \cdot \rangle)$ *is well-defined:*

$$\lambda_\Omega(a) \,\widehat{\#}\, \lambda_\Omega(b) \quad \text{if and only if } a \# b \text{ and } \langle \lambda_\Omega(a) | \lambda_\Omega(b) \rangle = \langle a | b \rangle = \Omega(a, b).$$

(ii) The PIP-*space* $(\widehat{\mathfrak{A}}, \widehat{\#}, \langle \cdot | \cdot \rangle)$ *is nondegenerate if and only if*

$$(\mathfrak{B}_\Omega)^\perp = (\mathfrak{B}_\Omega^\#)^\perp. \tag{6.7}$$

(iii) $(\widehat{\mathfrak{B}}, \widehat{\mathfrak{A}})$ *is a dual pair if and only if*

$$(\mathfrak{B}_\Omega)^\perp = (\mathfrak{B}_\Omega^\#)^\perp = \mathfrak{B}^\perp. \tag{6.8}$$

Proof. (i) Indeed, by the linearity of $\#$, we have

$$a \# b \iff (a + n) \,\#\, (b + n'), \ \forall n, n' \in (\mathfrak{B}_\Omega)^\perp,$$

and, by the definition of $(\mathfrak{B}_\Omega^\#)^\perp$,

$$\Omega(a, b) = \Omega(a + n, b + n'), \quad \forall n, n' \in (\mathfrak{B}_\Omega)^\perp.$$

(ii) $(\widehat{\mathfrak{A}}, \widehat{\#}, \langle \cdot, \cdot \rangle)$ is nondegenerate if and only if $(\widehat{\mathfrak{B}}^{\#})^\perp = \{0\}$, where $\widehat{\mathfrak{B}} = \lambda_\Omega(\mathfrak{B})$. This is clearly equivalent to say that $(\mathfrak{B}_\Omega)^\perp = (\mathfrak{B}_\Omega^\#)^\perp$.

(iii) $(\widehat{\mathfrak{B}}, \widehat{\mathfrak{A}})$ is a dual pair if and only if also $\widehat{\mathfrak{B}}^\perp = \{0\}$. But $\widehat{\mathfrak{B}}^\perp = \lambda_\Omega(\mathfrak{B}^\perp)$ and this is the null subspace if and only if $\mathfrak{B}^\perp = (\mathfrak{B}_\Omega)^\perp$. ∎

Remark 6.4.8. (ii) implies (iii), in general.

Remark 6.4.9. For the new compatibility $\widehat{\#}$, one has $\widehat{\mathfrak{A}}^{\widehat{\#}} = \lambda_\Omega(\mathfrak{B}_\Omega^\#)$. Indeed,

$$\lambda_\Omega(a) \in \widehat{\mathfrak{A}}^{\widehat{\#}} \iff \lambda_\Omega(a) \widehat{\#} \lambda_\Omega(b), \forall \lambda_\Omega(b) \in \widehat{\mathfrak{A}} \iff a \# b, \forall b \in \mathfrak{B}_\Omega$$
$$\iff a \in \mathfrak{B}_\Omega^\# \iff \lambda_\Omega(a) \in \lambda_\Omega(\mathfrak{B}_\Omega^\#).$$

Furthermore, if $\langle \widehat{\mathfrak{B}}, \widehat{\mathfrak{A}} \rangle$ is a dual pair, then $\widehat{\mathfrak{B}}$ is Mackey dense in $\mathfrak{B}_\Omega^\#$. Indeed, let $\lambda_\Omega(a)$ be such that $\langle \lambda_\Omega(b) | \lambda_\Omega(a) \rangle = 0, \forall \lambda_\Omega(b) \in \widehat{\mathfrak{B}}$. Then $\Omega(b, a) = 0$, $\forall b \in \mathfrak{B}$, i.e., $a \in \mathfrak{B}^\perp = (\mathfrak{B}_\Omega)^\perp$ and so $\lambda_\Omega(a) = 0$.

The PIP-space constructed in this way does not possess, in general, a central Hilbert space. As remarked in Section 2.5, the set

$$\Lambda = \{\lambda_\Omega(a) \in \widehat{\mathfrak{A}} : a \# a\}$$

is in general neither a pre-Hilbert space nor complete. We shall see an instance of this situation in Example 6.4.12 below.

We can now prove a generalization of the GNS-theorem to the present situation.

Proposition 6.4.10. *Let $\mathfrak{A}$ be a partial *-algebra and Ω be a $\mathfrak{B}$-weight on $\mathfrak{A}$, with the property that $(\mathfrak{B}_\Omega)^\perp = (\mathfrak{B}_\Omega^\#)^\perp$. Then there exist a nondegenerate PIP-space $(\widehat{\mathfrak{A}}, \#)$ and a linear map π from $\mathfrak{A}$ into $\mathrm{Op}(\widehat{\mathfrak{A}})$, which is a *-homomorphism from $\mathfrak{A}$ into $\mathrm{Op}(\widehat{\mathfrak{A}})$ endowed with the weak multiplication $\boldsymbol{.}$. Moreover, if $\mathfrak{A}$ has a unit e and $e \in \mathfrak{B}$, then π is cyclic in the sense that $\pi(\mathfrak{A})[\lambda_\Omega(e)] = \widehat{\mathfrak{A}}$.*

Proof. For $a \in \mathfrak{A}$, set

$$\pi(a)\lambda_\Omega(b) = \lambda_\Omega(ab), \forall\, b \in \mathfrak{B}. \qquad (6.9)$$

The map $\pi(a)$ is well-defined. In fact if $\lambda_\Omega(b) = \lambda_\Omega(b')$, then $b-b' \in (\mathfrak{B}_\Omega)^\perp \cap \mathfrak{B}$ so that $a(b - b') \in (\mathfrak{B}_\Omega)^\perp$ or, equivalently, $\lambda_\Omega(ab) = \lambda_\Omega(ab')$.

Now, in order to show that $\pi(a) \in \mathrm{Op}(\widehat{\mathfrak{A}})$, we need only to prove that $\pi(a)$ admits an adjoint; this is, in fact, equivalent to the Mackey continuity of the operator. But this follows easily from (iii) of Definition 6.4.1.

On the other hand, conditions (ii) and (iv) of the same definition imply that π is a *-homomorphism. Assume indeed that $x_1 \in L(x_2)$; then, for every $b_1, b_2 \in \mathfrak{B}$,

$$\begin{aligned}
\langle \pi(x_1 x_2)\lambda_\Omega(b_1) | \lambda_\Omega(b_2) \rangle &= \Omega((x_1 x_2)b_1, b_2) = \Omega(x_2 b_1, x_1^* b_2) \\
&= \langle \pi(x_2)\lambda_\Omega(b_1) | \pi(x_1^*)\lambda_\Omega(b_2) \rangle.
\end{aligned}$$

The equality between the first and the fourth term implies that the weak product $\pi(x_1) \circ \pi(x_2)$ is well-defined and equals $\pi(x_1 x_2)$. Similarly, $\pi(x^*) = \pi(x)^\times$ follows from (ii). Finally, if $e \in \mathfrak{B}$, the vector $\lambda_\Omega(e)$ is evidently cyclic in the sense said above.   ∎

Let us give two concrete examples of the generalized GNS-construction just described. In both cases, $\mathfrak{B}^\perp = \{0\}$. Hence the condition $(\mathfrak{B}_\Omega)^\perp = (\mathfrak{B}_\Omega^\#)^\perp$ is certainly satisfied.

Example 6.4.11. Let $\mathfrak{A} = \mathcal{B}(\mathcal{H})$, the *-algebra of bounded operators in a separable Hilbert space $\mathcal{H}$, and $\mathcal{B}(\mathcal{H})_+$ the positive cone of $\mathcal{B}(\mathcal{H})$. For $A \in \mathcal{B}(\mathcal{H})_+$, let us define

$$\omega(A) = \mathrm{Tr}(A).$$

Then

$$\mathcal{L}_\omega = \{A \in \mathfrak{A} : \mathrm{Tr}(A^* A) < \infty\} = \mathcal{C}^2(\mathcal{H}),$$

where $\mathcal{C}^2(\mathcal{H})$ is the *-ideal of Hilbert-Schmidt operators in $\mathcal{H}$.

As shown in Example 6.4.3, the sesquilinear form $\Omega(A, B) = \mathrm{Tr}(B^* A)$, defined on $\mathcal{C}^2(\mathcal{H}) \times \mathcal{C}^2(\mathcal{H})$, is a $\mathcal{C}^2(\mathcal{H})$-weight in our sense.

In this case the GNS-representation of $\mathcal{B}(\mathcal{H})$ generated by ω acts on the Hilbert space $\mathcal{H}_1 = \mathcal{C}^2(\mathcal{H})$ and it is defined by

$$\pi_\omega(A)\,B = AB, \quad \text{for } B \in \mathcal{C}^2(\mathcal{H}).$$

But in our approach other possibilities are allowed. In fact, if we define

$$\Gamma_\Omega = \{(A, B) \in \mathcal{B}(\mathcal{H}) \times \mathcal{B}(\mathcal{H}) : |\mathrm{Tr}(B^*A)| < \infty\}$$

and

$$\Omega(A, B) = \mathrm{Tr}(B^*A), \quad \text{for } A, B \in \Gamma_\Omega,$$

then Ω and Γ_Ω satisfy once again the conditions (d_1)-(d_4).

As for the choice of $\mathfrak{B}$, condition (i) of Definition 6.4.1 forces us to take $\mathfrak{B} \subset \mathcal{C}^2(\mathcal{H})$. Let, for instance, $\mathfrak{B} = \mathcal{C}^1(\mathcal{H})$, the trace class operators; then $\mathfrak{B}_\Omega = \mathcal{B}(\mathcal{H})$. As it is easily seen, also in this case conditions (ii)-(v) of Definition 6.4.1 hold true.

The difference with the previous situation is that in the present case we get really a (nondegenerate) PIP-space, whose extremal spaces are $\mathfrak{B} = \mathcal{C}^1(\mathcal{H})$ and $\mathfrak{B}_\Omega = \mathcal{B}(\mathcal{H})$. As we have seen, the PIP-space we have obtained admits a central Hilbert space, namely $\mathcal{H}_1 = \mathcal{C}^2(\mathcal{H})$. In terms of domains of definition, the first case corresponds to $\mathcal{L}_\omega \times \mathcal{L}_\omega$, whereas the second one uses the larger set Γ_Ω.

Example 6.4.12. Let (X, μ) be a measure space such that $\mu(X) = \infty$ and μ has no atoms; as usual we denote by $L^p(X, d\mu)$ the Banach space of (equivalence classes of) measurable functions $f : X \to \mathbb{C}$ such that $\|f\|_p = \left(\int_X |f|^p \, d\mu\right)^{1/p} < \infty$. As it is well-known (see Example 4.4.4), the spaces $L^p(X, d\mu)$ are not comparable, but the spaces $L^1(X, d\mu) \cap L^p(X, d\mu), 1 \leqslant p \leqslant \infty$, form a chain and $L^1(X, d\mu)$ is an associative partial *-algebra if we define

$$f \in L(g) = R(g) \iff \exists p \in \omega(f), q \in \omega(g) \text{ such that } 1/p + 1/q = 1,$$

where

$$\omega(f) = \{q \in [1, \infty) : \|f\|_q < \infty\}.$$

Let now $w \in L^r(X, d\mu), w > 0$, and define

$$\Gamma_\Omega = \{(f, g) \in L^1(X, d\mu) \times L^1(X, d\mu) : \exists p \in \omega(f), q \in \omega(g)$$
$$\text{such that } 1/p + 1/q + 1/r = 1\}. \tag{6.10}$$

For such pairs define

$$\Omega(f, g) = \int_X f(x)\,\overline{g(x)}\,w(x)\,d\mu(x).$$

It is readily checked that both Γ_Ω and Ω satisfy all requirements of Definition 6.4.1 with $\mathfrak{B}_\Omega = L^1(X, d\mu)$ and $\mathfrak{B} = L^\infty(X, d\mu) \cap L^1(X, d\mu)$. So the previous statements apply. It is worth remarking that the compatibility $\#_\Omega$ defined by Ω is different from the natural one, which makes $L^1(X, dm)$ into a PIP-space (unless there exists $a, b > 0$ such that $a \leqslant w(x) \leqslant b, \forall x$). The latter is defined by $f \# g$ if, and only if, $f \in L(g)$. The representation π is defined just by $\pi(f)g = fg$ (taking into account that in this case $(\mathfrak{B}_\Omega)^\perp = (\mathfrak{B}_\Omega^{\#_\Omega})^\perp = \mathfrak{B}^\perp = \{0\}$). Finally notice that there is no central Hilbert space between $V = L^1(X, d\mu)$ and $V^{\#_\Omega} = L^\infty(X, d\mu) \cap L^1(X, d\mu)$; indeed $\Lambda = L^1(X, d\mu) \cap L^2(X, d\mu)$ is not a Hilbert space.

There are very natural instances where the situation described so far simplifies. This is the case when we start from a linear functional on $\mathfrak{A}$ which is positive 'as much as it can', in the following sense.

Let $\mathfrak{A}$ be a partial *-algebra and ω a linear functional on $\mathfrak{A}$ such that

(1) $\omega(x^*x) \geqslant 0$ whenever $x \in R(x^*)$;
(2) $\omega(y^*x) = \overline{\omega(x^*y)}$, $\forall x, y$ such that $x \in R(y^*)$.

If (1) and (2) are satisfied, we say that ω is *positive*.

Remark 6.4.13. This notion reduces to the usual one if $\mathfrak{A}$ is a *-algebra and, in that case, (1) implies (2).

If ω is a positive linear functional and $x \in R(y^*)$, we define

$$\Omega(x, y) = \omega(y^*x).$$

Then Ω is a positive sesquilinear form defined on the domain

$$\Gamma_\Omega = \{(x, y) \in \mathfrak{A} \times \mathfrak{A} : x \in R(y^*)\}.$$

It is easy to verify that Ω and Γ_Ω fulfill the conditions (d_1)-(d_4) above.

Now we ask whether Ω is a $\mathfrak{B}$-weight in the sense of Definition 6.4.1 for some $\mathfrak{B} \subset R(\mathfrak{A})$. It is easily seen that conditions (i) and (ii) are fulfilled for any $\mathfrak{B} \subset R(\mathfrak{A})$. This is no longer true for (iii), (iv) and (v), however. For a general partial *-algebra $\mathfrak{A}$, we have then to select a subspace $\mathfrak{B} \subset R(\mathfrak{A})$ and proceed as above.

We will now consider a special situation, namely, we assume that $\mathfrak{A}$ is a semi-associative partial *-algebra. Examples of such objects are provided by quasi *-algebras and by $\mathcal{L}^\dagger(\mathcal{D}, \mathcal{H})$ for self-adjoint domains. In this case we get the following result.

Proposition 6.4.14. *Let $\mathfrak{A}$ be a semi-associative partial *-algebra and ω a positive linear functional on $\mathfrak{A}$. Let*

$$\Gamma_\Omega = \{(x, y) \in \mathfrak{A} \times \mathfrak{A} : x \in R(y^*)\}.$$

and define $\Omega(x, y) = \omega(y^*x)$, $(x, y) \in \Gamma_\Omega$. *Then* Ω *is a* $\mathfrak{B}$-*weight on* $\mathfrak{A}$ *for any subspace* $\mathfrak{B} \subset R(\mathfrak{A})$ *for which the following condition holds :*

$$\text{If } x \in \mathfrak{B} \quad \text{and } \Omega(x, x) = 0, \text{ then } \Omega(y, x) = 0, \forall y \in \mathfrak{A}.$$

Proof. We need only to show that (iii) and (iv) of Definition 6.4.1 hold true. The semi-associativity of $\mathfrak{A}$ will play a crucial role.
As for (iii) :

$$\Omega(xb_1, b_2) = \omega(b_2^*(xb_1)) = \omega((b_2^*x)b_1)) = \Omega(b_1, x^*b_2), \forall\, x \in \mathfrak{A}, b_1, b_2 \in \mathfrak{B}.$$

As for (iv) :
Let $x_1 \in L(x_2)$: if $b_2 \in \mathfrak{B}$, then $x_2b_2 \in R(x_1)$ and $(x_1x_2)b_2 = x_1(x_2b_2)$. Now, for $b_1 \in \mathfrak{B}$, we have $x_1x_2 \in R(b_1^*)$, so that $(x_1x_2)b_2 \in R(b_1^*)$ and

$$(b_1^*(x_1x_2))b_2 = b_1^*((x_1x_2)b_2) = b_1^*(x_1(x_2b_2)).$$

Applying the semi-associativity from the left-hand side, this also implies that $b_1^*x_1 \in L(x_2b_2)$. Then finally

$$\Omega(x_1^*b_1, x_2b_2) = \omega((x_2b_2)^*x_1^*b_1) = \omega(b_2^*(x_2^*x_1^*)b_1) = \Omega(b_1, (x_1x_2)b_2).$$

∎

6.4.2 Weights on Partial *-Algebras: An Alternative Approach

As mentioned in Example 6.4.3, a weight on a *-algebra $\mathfrak{A}$ is an $\mathcal{L}_\omega$-weight in the sense of Section 6.4.1. We will show this as a simple consequence of some facts that we will discuss here. We will, in fact, introduce an alternative definition of weight, closer to the usual one. It turns out that this new class is contained in that considered in Definition 6.4.1 and it presents an interesting feature, namely, the GNS-representation built from one of them gives rise to operators in a Hilbert space.

In a partial *-algebra, we do not have a notion of positive elements (x^*x need not be defined for arbitrary x). Their role will be played by diagonal elements of $\mathfrak{A} \times \mathfrak{A}$ and a weight will be for us nothing else than a non-negative function of them. But such a function need not extend to a larger domain without additional assumptions.

Definition 6.4.15. Let $\mathfrak{A}$ be a partial *-algebra and

$$\mathcal{K} = \{(x, y) \in \mathfrak{A} \times \mathfrak{A} : x = y\}.$$

A *weight* on $\mathfrak{A}$ is a map $\Omega : \mathcal{K} \to [0, \infty]$ satisfying the following conditions:

(w_1) $\Omega(\lambda x, \lambda x) = |\lambda|^2 \,\Omega(x, x)$;

(w_2) $\Omega(x + y, x + y) + \Omega(x - y, x - y) = 2\Omega(x, x) + 2\Omega(y, y)$;

(w_3) If $\Omega(x, x) < \infty$ and $y \in L(x)$, then $\Omega(yx, yx) < \infty$.

Let

$$\mathcal{L}_\Omega = \{x \in \mathfrak{A} : \Omega(x, x) < \infty\}.$$

Condition (w_2) implies that $\Omega(x + y, x + y) \leqslant 2\Omega(x, x) + 2\Omega(y, y)$; together with ($w_1$), this shows that $\mathcal{L}_\Omega$ is a complex vector space.

Condition (w_3), which mimicks the usual ideal property, allows us to extend, formally, Ω to $\mathfrak{A} \times \mathfrak{A}$ by polarization (for simplicity we denote this extension by the same symbol):

$$\Omega(x, y) = \frac{1}{4} \sum_{k=0^3} i^k \Omega(x + i^k y, x + i^k y).$$

Then condition (w_2) implies that (the extended) Ω is a sesquilinear form [RN56, Nr.87]) on $\mathfrak{A} \times \mathfrak{A}$. The largest set where it can be correctly defined is

$$\Gamma_\Omega = \{(x, y) \in \mathfrak{A} \times \mathfrak{A} : x \pm y \in \mathcal{L}_\Omega \text{ and } x \pm iy \in \mathcal{L}_\Omega\}.$$

It is easily seen that $\Gamma_\Omega = \mathcal{L}_\Omega \times \mathcal{L}_\Omega$.

Simple calculations show that Ω and Γ_Ω satisfy conditions (d_1)-(d_4) of Section 6.4.1. Let now $\mathfrak{B}$ be a vector subspace of $\mathcal{L}_\Omega$. For all $b \in \mathfrak{B}$ and $x \in \mathfrak{A}$, from (w_3) we get $xb \in \mathcal{L}_\Omega$ and therefore $(xb, c) \in \Gamma_\Omega, \forall c \in \mathfrak{A}$. Hence (i) and (ii) of Definition 6.4.1 are fulfilled.

However conditions (iii), (iv) and (v) of Definition 6.4.1 do not depend on (w_1)-(w_3). This fact suggests the following definition analogous to that of invariant p.s. form given in [21]. More generally, *-representations of partial *-algebras in Hilbert space can be constructed through the notion of *biweight*, which is somewhat more restrictive of that of $\mathfrak{B}$-weight, but, in a sense, more natural. For more details we refer to our monograph [AIT02].

Definition 6.4.16. Let Ω be a weight on the partial *-algebra $\mathfrak{A}$ and $\mathfrak{B}$ a vector subspace of $\mathcal{L}_\Omega \cap R\mathfrak{A}$. We say that Ω is a $\mathfrak{B}$-*invariant weight* if

(i) $\Omega(xb_1, b_2) = \Omega(b_1, x^* b_2), \forall\, x \in \mathfrak{A}, b_1, b_2 \in \mathfrak{B}$.

(ii) If $x_1 \in L(x_2)$, then $(x_1^* b_1, x_2 b_2) \in \Gamma_\Omega, \forall\, b_1, b_2 \in \mathfrak{B}$ and

$$\Omega(x_1^* b_1, x_2 b_2) = \Omega(b_1, (x_1 x_2) b_2).$$

(iii) If $x \in \mathfrak{B}$ and $\Omega(b, x) = 0, \forall b \in \mathfrak{B}$ or, equivalently, if $\Omega(x, x) = 0$, then $\Omega(y, x) = 0, \forall\, y \in \mathcal{L}_\Omega$.

It is clear at this point that if Ω is a $\mathfrak{B}$-invariant weight, then it is also a $\mathfrak{B}$-weight in the sense of Definition 6.4.1. The converse, however, is false in general. Indeed, the form Ω corresponding to a $\mathfrak{B}$-weight is defined on the widest possible domain, so that $\Gamma_\Omega \supset \mathcal{L}_\Omega \times \mathcal{L}_\Omega$. On the other hand, in the case of a $\mathfrak{B}$-invariant weight, all elements are forced to be self-compatible (in the sense of Section 6.4.1), and this restricts the domain to $\Gamma_\Omega = \mathcal{L}_\Omega \times \mathcal{L}_\Omega$. An explicit counterexample is given by the two variants of Example 6.4.11. Another one is implicit in Example 6.4.12. To get a $\mathfrak{B}$-invariant weight, one should define the form Ω on $\mathcal{L}_\Omega \times \mathcal{L}_\Omega$, where

$$\mathcal{L}_\Omega := \{ f \in L^1(X, dx) : \int_X |f(x)|^2 \, w(x) dx < \infty \} = L^1(X, dx) \cap L^2(X, w \, dx),$$

and $\mathcal{L}_\Omega \times \mathcal{L}_\Omega$ is strictly smaller than the domain Γ_Ω given in Eq.(6.10).

Since every $\mathfrak{B}$-invariant weight Ω is a $\mathfrak{B}$-weight, we can build the corresponding GNS-representation as in Proposition 6.4.10. However, since in this case $\mathfrak{B}_\Omega = \mathcal{L}_\Omega$, then $\Omega(x, x)$ is well-defined and finite on the whole PIP-space of the representation; in other words the PIP-space is just a pre-Hilbert space. Then we can formulate the following variant of the GNS-theorem.

Proposition 6.4.17. *Let Ω be a $\mathfrak{B}$-invariant weight on the partial *-algebra $\mathfrak{A}$. Then there exist a pre-Hilbert space $\mathcal{D}$, whose norm-completion will be denoted by $\mathcal{H}$, and a linear map π from $\mathfrak{A}$ into the weak partial *-algebra $\mathcal{L}^\dagger(\mathcal{D}, \mathcal{H})$ such that:*

*(i) π is a *-homomorphism of partial *-algebras;*
(ii) if $\mathfrak{A}$ has a unit $e \in \mathfrak{B}, \pi$ is cyclic.

Remark 6.4.18. Let $\mathfrak{A}$ be a *-algebra and ω a weight (in the usual sense) on $\mathfrak{A}$. With the same notations as in Example 6.4.3, if we define

$$\Omega(x, x) = \omega(x^* x), \ \forall \, x \in \mathfrak{A},$$

then it is straightforward to prove that Ω defines a $\mathcal{L}_\omega$-invariant weight. So, both $\mathfrak{B}$-weights and $\mathfrak{B}$-invariant weights generalize the usual definition.

6.4.3 Examples: Graded Partial *-Algebras

Besides the standard examples of spaces of bounded operators or integrable functions discussed so far, a natural class of examples of weights on partial *-algebras arises in the context of graded *-algebras with infinite gradation, such as the Borchers algebra (see Section 7.3.2) or tensor algebras. In all cases, a generic element of the algebra is represented by an infinite sequence, and a special role is played by those elements corresponding to finite sequences. For the clarity of the discussion, we will distinguish two different types of examples.

6.4.3.1 Tensor Algebras

Let $\mathcal{U}$ be a *-vector space, that is, a vector space with an involution. Then $\mathcal{T}_m = \mathcal{U}^{\otimes m} (m > 1)$ is also a *-vector space with the usual operations and the involution defined by

$$x_m = (u_1, \ldots, u_m) \mapsto x_m^* = (u_m^*, \ldots, u_1^*).$$

For $m = 0$, we set $\mathcal{T}_0 = \mathbb{C}$.

Let now $\mathcal{T}$ be the set of all sequences $x = (x_0, x_1, \ldots, x_m, \ldots), x_m \in \mathcal{T}_m$. In $\mathcal{T}$ an involution and a multiplication can be defined in the following way:

$$(x^*)_m = x_m^* \, ,$$

$$(xy)_m = \sum_{i=0}^{m} x_i y_{m-i} \, .$$

Together with the usual addition and multiplication by scalars, these operations make $\mathcal{T}$ into a *-algebra. The product $x_i y_j$ of $x_i = (u_1, \ldots, u_i) \in \mathcal{T}_i$ and $y_i = (v_1, \ldots, v_j) \in \mathcal{T}_j$ is defined as $x_i y_j = (u_1, \ldots, u_i, v_1, \ldots, v_j) \in \mathcal{T}_{i+j}$. We denote by $\mathcal{T}^F$ the *-subalgebra of all elements $x \in \mathcal{T}$ which have finite length (i.e., $x_m = 0$ for m large enough). $\mathcal{T}^F$, and thus also $\mathcal{T}$, has a unit $e = (1, 0, 0, \ldots)$.

Let now ω be a state on $\mathcal{T}^F$, i.e., a positive, normalized linear functional. As usual, we can define from ω a sesquilinear form Ω on $\mathcal{T}^F \times \mathcal{T}^F$ by

$$\Omega(x, y) = \omega(x^* y).$$

We will now sketch how to build up, from the spaces $\mathcal{T}, \mathcal{T}^F$ and the state ω, partial *-algebras and weights that fit into the framework developed in Sections 6.4.1 and 6.4.2.

We will first extend ω (or Ω) to certain elements of $\mathcal{T}$. In order to do this, we will cut-off elements of $\mathcal{T}$ in a natural way. In fact, to any $x \in \mathcal{T}$ we can associate the element $x^{[M]} \in \mathcal{T}^F$ defined as

$$x_m^{[M]} = \begin{cases} x_m, & \text{if } m \leqslant M, \\ 0 \, , & \text{if } m > M. \end{cases}$$

Then an extension of ω (denoted by the same symbol) can be defined for all those $x \in \mathcal{T}$ such that

$$\lim_{M \to \infty} |\omega(x^{[M]})| < \infty.$$

In a similar way, Ω can be extended to those pairs of elements $(x, y) \in \mathcal{T} \times \mathcal{T}$ for which the limit $\lim_{M \to \infty} \Omega(x^{[M]}, y^{[M]})$ exists.

Remark 6.4.19. Even if $\omega(x^*y)$ and $\Omega(x, y)$ exist, we cannot conclude that

$$\lim_{M \to \infty} \omega((x^*y)^{[M]}) = \lim_{M \to \infty} \Omega(x^{[M]}, y^{[M]}).$$

This is one of the reasons why we have to select appropriate subsets of $\mathcal{T}$ where our GNS construction will be possible.

Example 6.4.20. Let

$$\mathfrak{A} = \left\{ x \in \mathcal{T} : \lim_{M \to \infty} \omega((b_1^* x b_2)^{[M]}) \text{ exists } \forall b_1, b_2 \in \mathcal{T}^F \right\}.$$

From now on we set $\mathfrak{B} = \mathcal{T}^F$.

Let $\Gamma = \{(x_1, x_2) \in \mathfrak{A} \times \mathfrak{A} : x_1 x_2 \in \mathfrak{A}\}$. Then it is readily checked that $(\mathfrak{A}, \Gamma, \diamond)$ is a partial *-algebra if the partial multiplication is defined as

$$x_1 \diamond x_2 = x_1 x_2, \quad \text{for } (x_1, x_2) \in \Gamma.$$

It follows from the very definition that $\mathfrak{B} \subset R\mathfrak{A} \cap L\mathfrak{A}$.

Let now Γ_Ω be the set of pairs $(x_1, x_2) \in \mathfrak{A} \times \mathfrak{A}$ such that

$$\lim_{M \to \infty} \omega((b_1^* x_1)^{[M]}(x_2 b_2)^{[M]}) = \lim_{M \to \infty} \omega((b_1^* x_1 x_2 b_2)^{[M]}), \ \forall b_1, b_2 \in \mathfrak{B}.$$
$$(6.11)$$

Remark 6.4.21. It is easily seen that, if $b_1, b_2 \in \mathcal{T}^F$ and $x \in \mathcal{T}$, then there exists $M \in \mathbb{N}$ such that

$$\omega((b_1^* x b_2)^{[M]}) = \omega((b_1^* x^{[M]} b_2)).$$

This relation may also be applied to (6.11), which thus becomes

$$\lim_{M \to \infty} \omega((b_1^* x_1^{[M]} x_2^{[M]} b_2) = \lim_{M \to \infty} \omega((b_1^* (x_1 x_2)^{[M]} b_2), \ \forall b_1, b_2 \in \mathfrak{B}.$$

Now, according to the previous discussion, we can define an extension of Ω (we keep the same notation) on Γ_Ω by

$$\Omega(x_1, x_2) = \lim_{M \to \infty} \Omega(x_1^{[M]}, x_2^{[M]}), \ \forall (x_1, x_2) \in \Gamma_\Omega.$$

Conditions (d_1)-(d_4) of Section 6.4.1 are then fulfilled. Concerning (i)-(iv) of Definition 6.4.1, we just mention the fact that (iv) is fulfilled because it is included in the definition of Γ_Ω. Condition (v) is also satisfied. In fact, since $\mathfrak{A}$ has a unit, $\mathfrak{B}_\Omega = \mathfrak{A}$; then if $x \in \mathfrak{B}$ and $\Omega(b, x) = 0, \ \forall b \in \mathfrak{B}$, one has

$$\Omega(y, x) = \lim_{M \to \infty} \Omega(y, x^{[M]}) = 0.$$

In conclusion, Ω is a $\mathfrak{B}$ -weight on $\mathfrak{A}$.

Remark 6.4.22. Notice that we have not imposed any condition on Ω: it is the adequate definition of $\mathfrak{A} \subset \mathcal{T}$ which ensures that Ω is a $\mathfrak{B}$-weight on the partial *-algebra $\mathfrak{A}$.

Example 6.4.23. We may obtain an example of the situation described in Section 6.4.2 if we restrict further the definition of the partial *-algebra $\mathfrak{A}$ of the previous example.

Let

$$\mathfrak{A}_1 = \left\{ x \in \mathfrak{A} : \lim_{M \to \infty} \omega(b^* x^{*[M]} x^{[M]} b) \text{ exists}, \ \forall b \in \mathfrak{B} \right\}.$$

As usual, the above limit may be finite or infinite. Now, define Ω according to (6.4.3.1) and set

$$\mathcal{L}_\Omega = \left\{ x \in \mathfrak{A}_1 : \lim_{M \to \infty} \omega(b^* x^{*[M]} x^{[M]} b) < \infty, \ \forall b \in \mathfrak{B} \right\}.$$

In this case, Ω is a $\mathfrak{B}$-invariant weight on $\mathfrak{A}_1$ and thus Proposition 6.4.17 applies. The GNS representation corresponding to Ω indeed lives in a Hilbert space.

6.4.3.2 A Fock Space Construction

A second example, rather similar to the first one, is obtained by a construction analogous to the occupation number representation of Fock space. Let $\mathcal{U}$ be a (pre)-Hilbert space and $\{\phi_i\}$ a fixed orthonormal basis in $\mathcal{U}$. Then every vector $\phi \in \mathcal{U}, \phi = \sum_j c_j \phi_j$ may be identified with the sequence of its Fourier coefficients $\{c_j\} \in \ell^2$. Now we define $\mathcal{T}$ as before, but change the definition of $\mathcal{T}^F$ by introducing two cutoffs: M for m and N for j. Replacing everywhere $x^{[M]}$ by $x^{[M,N]}$ and $\lim_{M \to \infty}$ by $\lim_{M,N \to \infty}$, all the arguments go through.

This example is directly applicable to quantum field theory (Section 7.3). Let $\mathcal{U} = \mathcal{H}^{(+)} \oplus \mathcal{H}^{(-)}$ where $\mathcal{H}^{(+)}$ (resp $\mathcal{H}^{(-)}$) is the space of positive (resp. negative) energy solutions of the Klein-Gordon equation. Choose a fixed basis $\{\phi_j^+\} \subset \mathcal{H}^{(+)}$ and the corresponding basis $\{\phi_j^-\} \subset \mathcal{H}^{(-)}$, with $\phi_j^- = \overline{\phi_j^+}$. Then each basis vector $\phi_j^\pm$ is represented by the corresponding basis sequence $\theta_j^\pm = \{0, \ldots, 1, \ldots\}$, where the nonzero element corresponds to the vector $\phi_j^\pm$. Together with the unit element $1 \in \mathcal{T}_0$, the elements $\theta_j^\pm \in \mathcal{T}^F \cap \mathcal{T}_0$ generate $\mathcal{T}^F$ by sums and products. In this language, the Fock state ω_F is defined by the following relations:

$$\omega_F(\theta_{j_1}^\pm \theta_{j_2}^\pm \cdots \theta_{j_k}^\pm) = 0, \text{ unless } k = 2l \text{ and the product contains}$$
$$l \text{ factors } \theta^+ \text{ and } l \text{ factors } \theta^-, \tag{6.12}$$
$$\omega_F(b_1 \theta_m^- \theta_n^+ b_2) = \omega_F(b_1 \theta_n^+ \theta_m^- b_2) + \langle \phi_m | \phi_n \rangle \, \omega_F(b_1 b_2), \ \forall b_1, b_2 \in \mathcal{T}^F, \tag{6.13}$$

$$\omega_F(b_1\theta_m^-) = 0, \; \forall \, b_1 \in \mathcal{T}^F, \tag{6.14}$$

$$\omega_F(1) = 1. \tag{6.15}$$

Then the Fock state ω_F generates, by the GNS construction, the usual Fock representation of the CCR. In the same way, the Araki-Woods representation is obtained from (6.12), (6.13), (6.15) and

$$\omega_{AW}(\theta_{m_1}^+\theta_{m_2}^+\cdots\theta_{m_s}^+\theta_{n_s}^-\theta_{n_{s-1}}^-\cdots\theta_{n_1}^-) = \sum_P \langle\phi_{n_1}|B^2\phi_{m_1}\rangle\langle\phi_{n_2}|B^2\phi_{m_2}\rangle\cdots\langle\phi_{n_s}|B^2\phi_{m_s}\rangle, \tag{6.16}$$

where $\sum_P$ denotes the sum over all permutations of the set $\{m_1, m_2, \ldots, m_s\}$ and B^2 is a positive bounded operator on $\mathcal{H}^{(+)}$.

Notes to Chapter 6

Section 6.1. The theory of partial *-algebras, originally introduced by Antoine–Karwowski in [24,25], has been extensively studied by A. Inoue and the authors. A complete overview can be found in [AIT02].

- Concerning Remark 6.1.2, see the discussion in Antoine–Mathot [29], Sec.3.

Section 6.2. The problem of multiplying operators on a PIP-space was already considered in the early papers of Antoine–Grossmann [17,18].

- The notion of weak multiplication in $\mathrm{Op}(V)$ comes from a definition given by Kürsten in [136], but it has never been explored so far in detail. So the contents of Section 6.2.2 are essentially new. Example 6.2.5 is due to Kürsten [136].

Section 6.3. As defined by Lassner [138, 139], a quasi *-algebra is a pair $(\mathfrak{A}, \mathfrak{A}_0)$, where $\mathfrak{A}$ is a partial *-algebra and $\mathfrak{A}_0 \subset \mathfrak{A}$ is a *-algebra, such that $(x, y) \in \Gamma$ if and only if $x \in \mathfrak{A}_0$ or $y \in \mathfrak{A}_0$ and $\mathfrak{A}$ is an $\mathfrak{A}_0$-bimodule.

- The space $\mathcal{C}(\mathcal{E}, \mathcal{F})$ has been introduced by Kürsten [135], who was the first to consider the multiplication problem. The analysis presented here comes essentially from Trapani–Tschinke [183]. Examples and counterexamples, in particular Examples 6.3.1 and 6.3.2, are also due to Kürsten [135, 136]. An application to the problem of multiplication of distributions has been considered by Trapani–Tschinke in [184]. The counterexample given after Example 6.3.2 is due to Kürsten–Läuter [137], namely, there exist two linear operators $S, T : \mathcal{S} \to \mathcal{S}^\times$ such that every operator $R : \mathcal{S} \to \mathcal{S}^\times$ may be factorized as $\tilde{S} \cdot T$, where $\tilde{S}$ is an appropriate extension of S.

- The partial multiplication introduced in Definition 6.3.4 is due to Kürsten [135]. The independence of the product on the particular choice of the interspaces $\mathcal{E}, \mathcal{F}, \mathcal{G} \in \mathfrak{L}_0$ is Proposition 3.2 of that paper, while the pathology concerning associativity is explained in Proposition 3.8.
- The definition of multiplication framework was given in [77].
- The quasi *-algebra of differential operators has been described by Russo–Trapani [173].
- The canonical commutation relations on an interval and the corresponding quasi *-algebra have been studied by Lassner–Lassner–Trapani [140].
- The result of Schwartz may be found in his textbook [Sch57]. An overview of the different methods for defining a multiplication of distributions may be found in the monograph of Oberguggenberger [Obe92].

Section 6.4. Representations of partial *-algebras in Hilbert space have been studied in [AIT02].

- For the standard notion of weight on a *-algebra, we refer to the textbooks of Bratteli–Robinson [BR79] and Strătilă–Zsidó [SZ79].
- The notion of $\mathfrak{B}$-weight first appeared in Antoine–Soulet–Trapani [30]. The proof of the GNS construction contained, however, some gaps pointed out by Kürsten. See also [136, Wag97].
- The Borchers algebra, introduced in [51, 52], is described in detail in Section 7.3.2.
- For the occupation number representation of Fock space and, more generally, for quantum field theory, we refer to the textbook of Schweber [Sch62].
- The Araki–Woods representation is described in [32]. See also Chaiken [57].

Chapter 7
Applications in Mathematical Physics

It turns out that PIP-space methods have many applications in physics, although they are seldom mentioned as such. To draw on a literary analogy, like Molière's Monsieur Jourdain speaking in prose without knowing so, many authors have been using PIP-space language without realizing it. In particular, chains or lattices of Hilbert spaces are quite common in many fields of mathematical physics. Some of these applications will be discussed at length in this chapter. To mention a few examples: quantum mechanics, in particular singular interactions (Section 7.1.3), scattering theory (Section 7.2), quantum field theory (Section 7.3), representations of Lie groups (Section 7.4), etc.

7.1 Quantum Mechanics

The mathematical description of a quantum system has evolved considerably since the creation of quantum mechanics in the 1920s. The whole edifice rests on three basic principles:

(i) The *superposition principle*, which implies that the set of states of the system has a linear structure;
(ii) The notion of *transition amplitude*, given by an inner product: $A(\psi_1 \to \psi_2) = \langle \psi_2 | \psi_1 \rangle$. The latter in turn yields transition probabilities by $P(\psi_1 \to \psi_2) = |\langle \psi_2 | \psi_1 \rangle|^2$.
(iii) Then the *probabilistic interpretation* requires that $\langle \psi | \psi \rangle = \|\psi\|^2 > 0$, whenever $\psi \neq 0$.

Combining these basic principles implies that the set of states of the system is a positive definite inner product space Φ, that is, a pre-Hilbert space. On this basis, Dirac developed a formalism for quantum physics with great computational capacity and broad predictive power. The essential features of Dirac's formalism are the following:

(i) Physical observables are represented by linear operators in the space Φ and these operators form an algebra. Therefore, it makes sense to arbitrarily add and multiply operators to form new operators.

J.-P. Antoine and C. Trapani, *Partial Inner Product Spaces:*
Theory and Applications, Lecture Notes in Mathematics 1986,
DOI 10.1007/978-3-642-05136-4_7, © Springer-Verlag Berlin Heidelberg 2009

(ii) For a given quantum physical system, there exist complete systems of commuting observables (CSCO) in the algebra of observables. The system of eigenvectors for a chosen CSCO provides a basis for the space Φ, i.e., every vector $\phi \in \Phi$ can be expanded into the eigenvectors of the CSCO.

For instance, let H, $\boldsymbol{J}^2$ and J_3 be such a CSCO for a spherically symmetric Hamiltonian H (where the $J_i, i = 1, 2, 3$, are the angular momentum operators and $\boldsymbol{J}^2 = J_1^2 + J_2^2 + J_3^2$). This CSCO has common eigenvectors $|Ejj_3\rangle$:

$$H|Ejj_3\rangle = E|Ejj_3\rangle, \; E \in \mathrm{sp}\, H \subset \mathbb{R},$$
$$J^2|Ejj_3\rangle = j(j+1)|Ejj_3\rangle, \; j = 0, 1, 2, \ldots,$$
$$J_3|Ejj_3\rangle = j_3|Ejj_3\rangle, \; j_3 = -j, -j+1, \ldots, j.$$

The eigenvalues E may be discrete or continuous and every $\phi \in \Phi$ can be expanded as

$$\phi = \sum_{E_n j j_3} |E_n j j_3\rangle\langle E_n j j_3|\phi\rangle + \sum_{j j_3} \int_0^\infty dE |Ejj_3\rangle\langle Ejj_3|\phi\rangle. \tag{7.1}$$

For discrete E_n, the eigenvectors $|E_n j j_3\rangle$ satisfy the orthogonality conditions

$$\langle E_{n'} j' j_3'|E_n j j_3\rangle = \delta_{n'n}\delta_{j'j}\delta_{j_3' j_3}. \tag{7.2}$$

For continuous E, the $|Ejj_3\rangle$ are the so-called Dirac kets, which fulfill the Dirac orthogonality condition

$$\langle E' j' j_3'|Ejj_3\rangle = \delta(E' - E)\delta_{j'j}\delta_{j_3' j_3}. \tag{7.3}$$

Clearly such vectors cannot belong to the pre-Hilbert space Φ, nor to its completion $\mathcal{H}$. Thus Dirac's formalism, while extremely practical and used by physicists on a daily basis, is not mathematically well-defined.

For that reason, von Neumann formulated a rigorous version of quantum mechanics, in a pure Hilbert space language. His formulation consists in the following two axioms:

(i) *Pure states* are represented by rays (i.e., one-dimensional subspaces) in a Hilbert space $\mathcal{H}$;
(ii) *Observables* are represented by self-adjoint operators in $\mathcal{H}$.

One even adds sometimes that every self-adjoint operator in $\mathcal{H}$ represents an observable, but this seems unrealistic. This formulation is well-defined mathematically, but too restrictive. Nonnormalizable eigenvectors, corresponding to points of a continuous spectrum, cannot belong to $\mathcal{H}$, yet they are extremely useful (plane waves, for instance). Observables may be unbounded, so that domain considerations must be taken into account. In particular, unbounded operators may not always be multiplied.

A more general approach is provided by the abstract algebraic version of the theory developed in the 1960s. Here observables are realized by self-adjoint elements of a C*-algebra $\mathfrak{A}$ and states by (normalized) continuous linear functionals on $\mathfrak{A}$. In this framework, a concrete Hilbert space representation is obtained via the Gel'fand–Naĭmark–Segal (GNS) construction. Thus we are back to the traditional approach, with observables constituting an algebra, but represented by *bounded* operators.

7.1.1 Rigorous Formulation of the Dirac Formalism

(i) The Rigged Hilbert Space approach

Although standard, the traditional Hilbert space approach has difficulties. Unbounded operators are often more natural than bounded ones (e.g. representatives of a Lie algebra, such as symmetry generators), but then one may have domain problems. Also not all self-adjoint operators can be interpreted as physical observables. Neither do all states play the same role. Indeed, there are "*physical*" states, that can actually be prepared, and "*generalized*" states, associated with quantum measurements.

A first way to overcome these obstacles is to enlarge the mathematical framework from a Hilbert space to a *rigged Hilbert space (RHS)* (see Section 5.4). Assume all "relevant" observables have a common, dense, invariant domain $\mathcal{D} \subset \mathcal{H}$. Then one gets a RHS by putting on $\mathcal{D}$ a locally convex topology t_Φ, that is finer than the norm topology inherited from $\mathcal{H}$ and makes all these observables into continuous operators from $\Phi := \mathcal{D}[t_\Phi]$ into itself:

$$\Phi \hookrightarrow \mathcal{H} \hookrightarrow \Phi^\times \qquad (7.4)$$

Then one interprets Φ as the set of all physical states and $\Phi^\times$ as that of the generalized states associated to measurement devices.

The problem, of course, is how to build Φ. A solution is to start from a distinguished set $\mathcal{O}$ of *labeled* observables, which have both a physical interpretation (how does one measure it?) *and* a mathematical definition (as a self-adjoint operator in $\mathcal{H}$). The elements of $\mathcal{O}$, which characterize the system (physics): position, momentum, energy (Hamiltonian), ..., are supposed to have a common, dense, invariant domain $\mathcal{D}$ in $\mathcal{H}$ (mathematics). If one equips $\mathcal{D}$ with a suitable (intrinsic) topology (usually the graph topology $t_\mathcal{O}$ defined by the family $\mathcal{O}$ itself, as defined Section 5.5.2), one obtains a RHS (7.4) defined by the system.

The simplest example in nonrelativistic quantum mechanics is that of a particle, either free or in a nice potential v. The labeled observables are position $\mathbf{q}$, momentum $\mathbf{p}$ and energy $H = -\mathbf{p}^2/2m + v$. The corresponding RHS is the familiar Schwartz triplet $\mathcal{S}(\mathbb{R}^3) \hookrightarrow L^2(\mathbb{R}^3) \hookrightarrow \mathcal{S}^\times(\mathbb{R}^3)$.

In the RHS (7.4), the space, Φ is required, in general, to fulfill the following conditions:

(1) Φ should be *complete* with respect to t_Φ.
(2) Φ should be *reflexive*, that is, $(\Phi^\times)^\times \simeq \Phi$.

In most cases, Φ can be obtained as the intersection of a *countable* family of Hilbert spaces, that is, a countably Hilbert space, $\Phi = \cap_{n \in \mathbb{N}} \mathcal{H}_n$. It is then a Fréchet space.

(3) Φ should be *nuclear*. In the case where $\Phi = \cap_{n \in \mathbb{N}} \mathcal{H}_n$, this means that, for each n, there is an $m > n$ such that the embedding $\mathcal{H}_m \to \mathcal{H}_n$ is a Hilbert-Schmidt operator.

The motivation for the completion property (1) and the reflexivity property (2) is that of symmetry, namely, there should be only *two* spaces at hand, in addition to $\mathcal{H}$, with complete symmetry between the two. Indeed, if Φ is not complete, we can build its completion $\tilde{\Phi}$. However, the latter need *not* be included in $\mathcal{H}$, although its conjugate dual $\tilde{\Phi}^\times$ still coincides with $\Phi^\times$. Thus, if $\tilde{\Phi} \subset \mathcal{H}$, and only then, we can replace Φ by its completion – or simply assume from the beginning that Φ is complete. Note that the condition $\tilde{\Phi} \subset \mathcal{H}$ is always satisfied if the topology of Φ is the graph topology t_o. Similarly, should Φ be nonreflexive, we can build its bidual $\Phi^{\times\times}$ and we have $\Phi \subset \Phi^{\times\times}$ (they coincide as vector spaces if Φ is semireflexive). But then, the bidual $\Phi^{\times\times}$ is necessarily nonreflexive also and we get a whole family of spaces by taking successive conjugate duals. Thus in all cases, there arises at least a third space and the physical interpretation in terms of states is lost (or at least becomes ambiguous).

As for the nuclearity property (3), the justification is that it allows one to exploit the *nuclear spectral theorem* of Gel'fand and Maurin, which says the following. Let A be a closed linear operator in $\mathcal{H}$, which maps Φ into itself continuously with respect to t_Φ. Then A may be transposed by duality to a linear operator $A^\times : \Phi^\times \to \Phi^\times$, which is an extension of $A^\dagger := A^* \upharpoonright \Phi$, where A^* is the usual Hilbert space adjoint operator, namely,

$$A^\times F(\phi) = F(A\phi), \text{ for all } \phi \in \Phi \text{ and for all } F \in \Phi^\times, \qquad (7.5)$$

which we also write, using the notation of Section 5.4.1,

$$\langle \phi | A^\times F \rangle = \langle A\phi | F \rangle, \ \forall \phi \in \Phi, F \in \Phi^\times. \qquad (7.6)$$

For such an operator, the vector $\xi_\lambda \in \Phi^\times$ is called a *generalized eigenvector* of A, with eigenvalue $\lambda \in \mathbb{C}$, if it satisfies

$$\langle \phi | A^\times \xi_\lambda \rangle := A^\times \xi_\lambda(\phi) = \overline{\lambda} \, \xi_\lambda(\phi) \equiv \overline{\lambda} \langle \phi | \xi_\lambda \rangle, \text{ for all } \phi \in \Phi. \qquad (7.7)$$

This equality can also be written in the Dirac notation as

$$A^\times |\xi_\lambda\rangle = \bar{\lambda}|\xi_\lambda\rangle, \quad |\xi_\lambda\rangle \in \Phi^\times. \tag{7.8}$$

Now assume that A has a self-adjoint extension A_0 in $\mathcal{H}$ with a nondegenerate spectrum, and that Φ is nuclear and complete. In this case, $\widehat{A} := A^\times$ is an extension of both A and A_0. Then the nuclear spectral theorem asserts that A (or $\widehat{A}$) possesses a complete orthonormal set of generalized eigenvectors $\xi_\lambda \in \Phi^\times$, $\lambda \in \mathbb{R}$. This means that, for any two $\phi, \psi \in \Phi$, one has

$$\langle \phi | \psi \rangle = \int_\mathbb{R} \xi_\lambda(\phi)\, \overline{\xi_\lambda(\psi)}\, d\mu(\lambda)$$

$$\equiv \int_\mathbb{R} \langle \phi|\xi_\lambda\rangle \langle \xi_\lambda|\psi\rangle\, d\mu(\lambda)(ii)) \tag{7.9}$$

for some measure μ on $\mathbb{R}$. For quantum mechanical operators A, the measure μ may be split into a discrete and an absolutely continuous part such that (7.9) can be written as

$$\langle\phi|\psi\rangle = \sum_i \langle\phi|\lambda_i\rangle\langle\lambda_i|\psi\rangle + \int \langle\phi|\lambda_\rho\rangle\langle\lambda_\rho|\psi\rangle\rho(\lambda)d\lambda, \tag{7.10}$$

where the $\{\lambda_i\}$ are the discrete eigenvalues of A in $\mathcal{H}$, $|\xi_\lambda\rangle\langle\xi_\lambda|d\mu(\lambda) = |\lambda_\rho\rangle\langle\lambda_\rho|\rho(\lambda)d\lambda$, where $\rho(\lambda)$ is a non-negative integrable function and the integral extends over the absolutely continuous Hilbert space spectrum of A. Thus the Dirac kets are $|\lambda\rangle = |\lambda_\rho\rangle\sqrt{\rho(\lambda)}$.

The net result of this theorem is to put the eigenvalues and the points of the continuous spectrum of A on the same footing – exactly what is usually assumed in the Dirac formulation of quantum mechanics. Indeed, using Dirac's notation, (7.9) and (7.10) may be written as a decomposition of the identity:

$$I = \int_\mathbb{R} |\xi_\lambda\rangle\langle\xi_\lambda|\, d\mu(\lambda) = \sum_i |\lambda_i\rangle\langle\lambda_i| + \int_\mathbb{R} d\lambda\, |\lambda\rangle\langle\lambda|, \quad \langle\lambda|\lambda'\rangle = \delta(\lambda - \lambda'), \tag{7.11}$$

with the proviso that this quantity makes sense only between two vectors of Φ. In other words, I must be understood as the (linear) embedding of Φ into $\Phi^\times$ or, equivalently as a sesquilinear form on $\Phi \times \Phi$.

Actually the symbol $|\xi_\lambda\rangle\langle\xi_\lambda|$ in (7.11) may be interpreted as a genuine projection operator from Φ onto the λ-component in the decomposition, if one combines von Neumann's direct integral approach with the nuclear spectral theorem. According to von Neumann, the self-adjoint operator A determines a decomposition of $\mathcal{H}$ into a direct integral of one-dimensional spaces $\mathcal{H}(\lambda)$:

$$\mathcal{H} \simeq \int_\mathbb{R}^\oplus \mathcal{H}(\lambda)\, d\mu(\lambda), \tag{7.12}$$

which "diagonalizes" A:

$$f \sim \{f(\lambda)\}, \ f(\lambda) \in \mathcal{H}(\lambda), \ \text{with} \ \|f\|^2 = \int_{\mathbb{R}} |f(\lambda)|^2 \, d\mu(\lambda), \qquad (7.13)$$

$$Af \sim \{\lambda f(\lambda)\}. \qquad (7.14)$$

As already mentioned, the difficulty with this formulation is that $\mathcal{H}(\lambda)$ is *not* a subspace of $\mathcal{H}$ if λ is a point of μ-measure zero (for instance, if $f(\lambda_0) \neq 0$ and $f(\lambda) = 0$ for $\lambda \neq \lambda_0$, then $\|f\|^2 = 0$, i.e., $f = 0$ as a vector of $\mathcal{H}$; in other words, $f \notin \mathcal{H}$). This is why there are no true eigenvectors associated to the points of the continuous spectrum.

However, if the space Φ in (7.4) is nuclear, then the map $\tau_\lambda : \phi \mapsto \phi(\lambda)$, $\phi \in \Phi$, $\phi(\lambda) \in \mathcal{H}(\lambda)$, is continuous and nuclear for μ-almost all λ. Therefore, one may write

$$\tau_\lambda \phi = \phi(\lambda) = \langle \phi | \xi_\lambda \rangle h(\lambda), \ \text{where} \ \xi_\lambda \in \Phi^\times, \ h(\lambda) \in \mathcal{H}(\lambda). \qquad (7.15)$$

Then the dual mapping $\tau_\lambda^\times : \mathcal{H}(\lambda) \to \Phi^\times$ is continuous as well and allows us to identify each vector $\xi \in \mathcal{H}(\lambda)$ with a functional $\tilde{\xi} = \tau_\lambda^\times \xi \in \Phi^\times$. Finally, the combined map $\chi_\lambda = \tau_\lambda^\times \tau_\lambda$, which is a nuclear operator mapping Φ into $\Phi^\times$, acts as a projection operator onto the eigensubspace $\Phi_\lambda^\times$ corresponding to the eigenvalue λ. This is a so-called *eigenoperator* associated with A, that is, a positive mapping $\chi_\lambda : \Phi \to \Phi^\times$ such that

$$\chi_\lambda A = \widehat{A} \chi_\lambda = \lambda \chi_\lambda.$$

If the spectrum of the self-adjoint operator A_0 has nontrivial multiplicity, i.e., $\dim \mathcal{H}(\lambda) > 1$, as in the case of a spherically symmetric Hamiltonian described above, then the whole machinery still goes through. The map τ_λ of (7.15) reads as

$$\tau_\lambda \phi = \phi(\lambda) = \sum_n \langle \phi | \xi_{\lambda,n} \rangle h_n(\lambda), \qquad (7.16)$$

where $\xi_{\lambda,n} \in \Phi^\times$ and $\{h_n(\lambda), \ n = 1, 2, \ldots \dim \mathcal{H}(\lambda)\}$ is a basis of $\mathcal{H}(\lambda)$. Thus the expansion (7.11) becomes

$$I = \int_{\mathbb{R}} \sum_n |\xi_{\lambda,n}\rangle\langle\xi_{\lambda,n}| \, d\mu(\lambda)$$

$$= \sum_{i,n} |\lambda_{i,n}\rangle\langle\lambda_{i,n}| + \int_{\mathbb{R}} d\lambda \sum_n |\lambda, n\rangle\langle\lambda, n|, \quad \langle\lambda, n|\lambda', n'\rangle = \delta(\lambda - \lambda')\delta_{n,n'}.$$

$$(7.17)$$

Yet a word of caution is necessary here. If one is interested only in the spectral properties of A, one may require that the spectrum of $\widehat{A}$ in $\Phi^\times$

consists exactly of the points of the spectrum of A in $\mathcal{H}$. If this is the case, one says that (7.4) is a *tight rigging* for A. Tight riggings are by no means guaranteed for a given operator A, as can be seen from the sufficient conditions given in the literature. On the other hand, there are important cases where one actually needs generalized eigenvalues that do *not* belong to the Hilbert space spectrum of A. Scattering theory is a major example, where resonances are associated with complex eigenvalues of the Hamiltonian, with Gamow vectors as generalized eigenvectors. As operators in the Hilbert space, these Hamiltonians are self-adjoint and as such their Hilbert space spectra are real. Therefore, a tight rigging would not permit a description of resonance states by complex eigenvectors. However, in the more general case, it is possible to construct rigged Hilbert spaces such that self-adjoint Hamiltonians have complex generalized eigenvalues.

Before closing this section, we ought to mention a slight variant of the RHS approach described so far. The idea is to consider *simultaneously* two (barely) different triplets, namely,

$$\Phi \hookrightarrow \mathcal{H} \hookrightarrow \Phi^{\times} \quad \text{and} \quad \Phi \hookrightarrow \mathcal{H} \hookrightarrow \Phi', \tag{7.18}$$

where Φ' is the usual dual, the space of continuous *linear* functionals on Φ. The interpretation is then that the Dirac kets are vectors in $\Phi^{\times}$, whereas the Dirac bras are vectors in Φ', the vectors of Φ being still interpreted as physically realizable states. This is somewhat more complicated, but has the advantage of keeping the bijection between bras and kets, inherent to Dirac's formalism. Otherwise, there is not much difference with the conventional RHS formalism, based on (7.4) only.

(ii) Analyticity/trajectory spaces

The analyticity/trajectory spaces described in Section 4.5 were specifically designed for obtaining a rigorous formulation of Dirac's bra-and-ket formalism of quantum mechanics. The original treatment being quite technical (and in fact rather unknown among quantum physicists), we will only sketch the basic features of this approach.

The framework is a triplet as in (4.22), $\mathcal{S}_{\mathcal{H},A} \subset \mathcal{H} \subset \mathcal{T}_{\mathcal{H},A}$, such that the trajectory space $\mathcal{T}_{\mathcal{H},A}$ is nuclear. The operator A has to be chosen in such a way that $\mathcal{T}_{\mathcal{H},A}$ is best adapted to the quantum system under consideration. Nuclearity implies that A has a discrete spectrum $\{\lambda_d, \ d \in \mathbb{D}\}$, where $\mathbb{D}$ denotes a countable index set. The eigenvalues λ_d are non-negative and satisfy the condition $\sum_{d \in \mathbb{D}} e^{-\lambda_d t} < \infty$, $\forall t > 0$ (counting multiplicity). The corresponding eigenvectors are denoted by $v_d : A v_d = \lambda_d v_d$, $\forall d \in \mathbb{D}$.

In these terms, the key ingredients of Dirac's formalism are defined as follows.

(i) Ket vectors:

The elements of $\mathcal{T}_{\mathcal{H},A}$ are taken as ket vectors and denoted by $|F\rangle$, that is, the ket $|F\rangle$ represents the trajectory $t \mapsto |F\rangle(t)$, $t > 0$, satisfying the relation (4.21). A ket $|F\rangle$ is called *normalizable* if $|F\rangle(0) := \lim_{t\to 0} |F\rangle(t)$ exists in $\mathcal{H}$. It is called a *test ket* if $|F\rangle(0) \in \mathcal{S}_{\mathcal{H},A}$.

(ii) Bra vectors:

In order to define bra vectors following closely the original Dirac notation, it is simpler to consider the dual $\mathcal{H}'$ of $\mathcal{H}$ rather than the conjugate dual $\mathcal{H}^\times$. Denote by $f \mapsto f'$ the *antilinear* isometric isomorphism from $\mathcal{H}$ onto $\mathcal{H}'$, so that one has $\langle f'|g'\rangle_{\mathcal{H}'} = \langle g|f\rangle_{\mathcal{H}}$. The vectors $\{v'_d, d \in \mathbb{D}\}$ form an orthonormal basis in $\mathcal{H}'$. Define the operator

$$A'f' := (Af)' = \sum_{d\in\mathbb{D}} \lambda_d \langle v'_d|f'\rangle_{\mathcal{H}'} v'_d$$

on the domain $D(A') = \{f' \in \mathcal{H}' : f \in D(A)\}$. Then the *bra vectors* are taken as the elements of the trajectory space $\mathcal{T}_{\mathcal{H}',A'}$ and denoted by $\langle B|$. Normalizable bras are defined as the corresponding kets. Clearly there is an antilinear bijection $\langle B| \leftrightarrow |B\rangle$ between bras and kets.

(iii) Bracket:

Given a bra $\langle B| \in \mathcal{T}_{\mathcal{H}',A'}$ and a ket $|F\rangle \in \mathcal{T}_{\mathcal{H},A}$, their bracket is taken as the complex valued function on $(0,\infty)$ defined by

$$\langle B|F\rangle : t \mapsto \langle B|\{|F\rangle(t)\}, \; t > 0. \tag{7.19}$$

This quantity is well-defined, since $|F\rangle(t) \in \mathcal{S}_{\mathcal{H},A}$ for all $t > 0$. The bracket (7.19) can also be expressed in terms of sequences

$$\langle B|F\rangle(t) = \sum_{d\in\mathbb{D}} e^{-\lambda_d} \xi_d \zeta_d \,,$$

where (ξ_d) and (ζ_d) are the coefficients in the expansion of $\langle B|$ and $|F\rangle$, respectively, in terms of the basis $\{v_d\}$. The bracket can also be extended to a holomorphic function on the open right half $t > 0$ of the complex plane $\{z = t + iy\}$.

Armed with these notions, one can then define linear operators and reconstruct the whole quantum mechanics. The key tool for reproducing Dirac's formalism is the notion of *Dirac basis*, a generalization of the usual notion that encompasses both basis vectors indexed by a discrete parameter (as in Hilbert space) and those indexed by a continuous parameter (with orthogonality in terms of the Dirac δ measure). The development is rather complex, so we refer the interested reader to the books mentioned in the Notes.

(iii) Beyond the RHS: a PIP-space approach

As in the standard approach, there is a corresponding abstract version of the RHS theory, in which the observable algebra $\mathfrak{A}$ is assumed to be a *-algebra of unbounded operators*. The mathematical technology is available thanks to the work of many authors, including the GNS construction and the description of *-automorphism groups and derivations of $\mathfrak{A}$. However, more difficulties may arise. Indeed, it is not always possible, or convenient, to find an *invariant* common dense domain for all relevant observables of the system. However, if one drops the requirement of invariance of the domain, the product of two such operators A, B need no longer be defined. Namely, AB makes sense only if the range of B is contained in the domain of A. This suggests to extend one step further the description of $\mathfrak{A}$, and take it as a *partial *-algebra of closable operators* on $\mathcal{H}$. Once again, the mathematical technology is available, including the GNS construction and the notions of *-automorphism groups and derivations. In fact most concepts familiar in the theory of C*-algebras extend to this wider framework, but at the price of severe technical complications. Thus this approach, while intellectually satisfying, is not directly applicable for quantum mechanics, we need something simpler and more natural. Our answer to that query is, of course, the notion of PIP-space. As for (ii) above, it is worth recalling that a trajectory space is a special kind of PIP-space, as mentioned in Section 4.5.

Instead of the RHS (7.4) (which is already a PIP-space), we will state the following axioms.

(i) The set of states is a PIP-space V_I with nondegenerate, positive definite, partial inner product and a central Hilbert space $\mathcal{H}_0$, to be identified with the Hilbert space of von Neumann's formulation. The subspace $V^{\#}$ is the set of physically realizable states. Transition amplitudes, whenever possible, are given by the partial inner product between two compatible states: $A(\psi_1 \rightarrow \psi_2) = \langle \psi_2 | \psi_1 \rangle$.

(ii) Observables of the system are represented by symmetric operators in $\mathrm{Op}(V_I)$. In particular, questions (that is, yes-no experiments) correspond to orthogonal projections in V_I.

Next, we want to apply the nuclear spectral theorem. The simplest way is to require that V_I be a LHS with the property that, for each $n \in I$, there is an $m > n$ such that the embedding $\mathcal{H}_m \rightarrow \mathcal{H}_n$ is a Hilbert-Schmidt operator. Indeed, in this case, the definition of nuclearity is a PIP-space-type statement. Similarly, the eigenoperators (spectral projections) $\chi_\lambda = \tau_\lambda^\times \tau_\lambda$ may be given a PIP-space formulation. Of course, an alternative could be to require that the triplet $V^{\#} \subset \mathcal{H}_0 \subset V$ obey the restrictions imposed on the RHS (7.4), but this is somewhat artificial and contrary to the PIP-space spirit.

As for the probabilistic interpretation, the mean value of an observable $A \in \mathrm{Op}(V_I)$, namely, $\langle A \rangle_\phi := \langle \phi | A\phi \rangle$, is obviously well-defined for any $\phi \in V^\#$. However, the variance

$$\sigma_\phi(A) = \left\{ \langle A^2 \rangle_\phi - \langle A \rangle_\phi^2 \right\}^{1/2}$$

is well-defined if and only if the product A^2 exists. Clearly more interpretational work is needed here.

7.1.2 Symmetries in Quantum Mechanics

An important aspect of quantum theory is the realization of symmetries of a given system. Thus it is worthwhile to briefly describe how one proceeds in the successive formulations of quantum mechanics presented in the previous sections. In the standard Hilbert space approach, a *symmetry* is defined as a bijection between states that preserves the absolute values of all transition amplitudes. According to Wigner, a symmetry τ is realized by a unitary or an anti-unitary operator in $\mathcal{H}$. Then, if the system admits a *symmetry group* $\{\tau_g, g \in G\}$, with G a Lie group, the latter is realized by a strongly continuous unitary (projective) representation $U(g)$ of G in $\mathcal{H}$ (Wigner–Bargmann). In the algebraic version of the theory, a symmetry τ is realized by a *-automorphism σ of $\mathfrak{A}$ and a symmetry group $\{\tau_g, g \in G\}$ by a continuous *-automorphism group $\{\sigma_g, g \in G\}$ of $\mathfrak{A}$.

In the RHS approach, consistency requires that U maps physical states into physical states and similarly for the measuring devices. Thus one should have two other realizations of U, in addition to U itself, which acts in $\mathcal{H}$, namely:

- U_Φ acting in Φ (*active* point of view);
- $\widehat{U}_\Phi$ acting in $\Phi^\times$ (*passive* point of view).[1]

The equivalence of the two points of view is manifested by the requirement that U_Φ and $\widehat{U}_\Phi$ are contragredient of each other, that is,

$$\langle \widehat{U}_\Phi(g)F | \phi \rangle = \langle F | U_\Phi(g^{-1})\phi \rangle, \, \forall g \in G, \phi \in \Phi, F \in \Phi^\times,$$

or, equivalently,

$$\langle \widehat{U}_\Phi(g)F | U_\Phi(g)\phi \rangle = \langle F | \phi \rangle, \, \forall g \in G, \phi \in \Phi, F \in \Phi^\times.$$

---

[1] It would be more natural to write $U_\Phi^\times$, but this would create an obvious confusion with the PIP-space notation of the adjoint.

This corresponds to the unitarity of U acting in $\mathcal{H}$:

$$\langle U(g)f|U(g)h\rangle = \langle f|h\rangle, \; \forall\, g \in G, f, h \in \mathcal{H}.$$

Thus the definition implies that $\widehat{U}_\Phi$ is an extension of both U_Φ and U, as it should in view of (7.4).

As far as continuity properties are concerned, the following is known. Assume Φ is a reflexive Fréchet space, for instance, the Schwartz space $\mathcal{S}(\mathbb{R}^n)$. If U_Φ is continuous in Φ, i.e., the map $g \mapsto U_\Phi(g)\phi$ is continuous from G to Φ, $\forall \phi \in \Phi$, then $\widehat{U}_\Phi$ is automatically continuous in $\Phi^\times$.

Turning now to the PIP-space version, suppose the system has a symmetry group G. Thus, there exists a unitary representation U_{00} of G in $\mathcal{H}_0$:

$$U_{00}U_{00}^\times = U_{00}^\times U_{00} = 1_{00}, \text{ the identity operator in } \mathcal{H}_0.$$

In virtue of the conservation of transition amplitudes, if $\psi_2 \,\#\, \psi_1$, one must have $U(g)\psi_2 \,\#\, U(g)\psi_1$, $\forall\, g \in G$, and

$$\langle U(g)\psi_2|U(g)\psi_1\rangle = \langle \psi_2|\psi_1\rangle.$$

Since $U(g)^\times = U(g^{-1})$, this implies that $U(g)$ must be a unitary homomorphism, $\forall\, g \in G$. Therefore, U_{00} must extend to a unitary representation U in V_I, in the sense of Definition 3.3.17.

Take again a nonrelativistic particle in a nice rotation invariant potential v, discussed in Example 3.3.19 (1). The system admits $G = \mathrm{SO}(3)$ as symmetry group[2] and the representation U_{00} is the natural representation of $\mathrm{SO}(3)$ in $L^2(\mathbb{R}^3)$:

$$[U_{00}(R)\psi](\mathbf{x}) = \psi\left(R^{-1}\mathbf{x}\right), \quad R \in \mathrm{SO}(3).$$

If we take as PIP-space V_I the scale built on the powers of the Hamiltonian, we do get a unitary representation of $\mathrm{SO}(3)$ by totally regular isomorphisms of V_I. But if we take instead the lattice $\{L^2(r)\}$ of weighted L^2 spaces, a given assaying subspace $L^2(r)$ is invariant under U only if the weight function r is itself rotation invariant, yet we still have a unitary representation in V_I.

This solves the problem of describing a symmetry under a Lie group G. But a quantum system has in general two types of (labeled) observables, some which can be derived from the symmetry group G, as representatives of elements of the Lie algebra $\mathfrak{g}$ or its universal enveloping algebra $\mathfrak{U}(\mathfrak{g})$,[3] and some who don't. A global solution may be obtained as follows.

---

[2] In fact, the full symmetry group of the system may be larger that $\mathrm{SO}(3)$. For instance, the Coulomb potential admits $\mathrm{SO}(4)$ as symmetry group for its bound states; this explains the additional degeneracy of the spectrum, manifested in the well-known Balmer formula.

[3] Roughly speaking, the universal enveloping algebra $\mathfrak{U}(\mathfrak{g})$ consists of all polynomials in the elements of $\mathfrak{g}$, modulo the commutation relations.

Given the representation U of G in $\mathcal{H}$, a vector $\psi \in \mathcal{H}$ is called a C^∞-*vector* for U (resp. an *analytic* vector) if the map $g \mapsto U(g)\psi$ of G into $\mathcal{H}$ is C^∞ (resp. analytic). Notice that an equivalent definition of analytic vectors was given in Section 4.5. The set $\mathcal{H}^\infty$ of all C^∞-vectors is dense in $\mathcal{H}$, and so is the set of analytic vectors. In order to construct C^∞-vectors, one considers the so-called $G\mathring{a}rding$ *domain* $\mathcal{H}^G$ of the representation U. Let $\mathcal{D}(G)$ denote the space of C^∞ functions on G with compact support, with its Schwartz nuclear topology. To each $f \in \mathcal{D}(G)$, one associates the operator $\widetilde{U}(f)\psi$ defined by the relation

$$\widetilde{U}(f)\psi = \int_G U(g)\psi f(g)\, dg \quad (dg = \text{left-invariant Haar measure on } G).$$

Every such vector is a C^∞-vector for U. Then the $G\mathring{a}rding$ domain $\mathcal{H}^G$ is the space of finite linear combinations of vectors of the form $\widetilde{U}(f)\psi$, $f \in \mathcal{D}(G)$ and one has $\mathcal{H}^G \subset \mathcal{H}^\infty$. The domain $\mathcal{H}^G$ is dense in $\mathcal{H}$, stable under $U(g)$, $\forall g \in G$, contained in the domain the representatives of all elements of the Lie algebra $\mathfrak{g}$ of G and stable under them. Hence every element T of the universal enveloping algebra $\mathfrak{U}(\mathfrak{g})$ of $\mathfrak{g}$ is represented in $\mathcal{H}^G$ by an operator $\widetilde{U}(T)$ defined by the relation

$$\widetilde{U}(T)\widetilde{U}(f)\psi = \widetilde{U}(Tf)\psi, \ f \in \mathcal{D}(G), \ \psi \in \mathcal{H}$$

(this makes sense, since elements of $\mathfrak{U}(\mathfrak{g})$ act on $\mathcal{D}(G)$ as differential operators). In addition, for most operators of physical interest (elements of $\mathfrak{g}$, symmetric elements of $\mathfrak{U}(\mathfrak{g})$), the operator $\widetilde{U}(T)$ is essentially self-adjoint on $\mathcal{H}^G$, thus a potential labeled observable. The problem is that $\mathcal{H}^G$ is not a nuclear space.

To circumvent this difficulty, one proceeds in two steps. First, one realizes that the enveloping algebra $\mathfrak{U}(\mathfrak{g})$ may be identified with the space $\mathcal{E}'_e$ of distributions on G with support reduced to the unit element e. Then one extends the representation U to the whole space $\mathcal{E}'_G$ of distributions of compact support (which is a *-algebra under convolution) as follows. Assume $\psi \in \mathcal{H}^\infty$ is a C^∞-vector for U. For any $T \in \mathcal{E}'_G$, one defines the operator $\widetilde{U}(T)$ by the equivalent relations[4]

$$\widetilde{U}(T)\psi = T(U(\cdot)\psi) \equiv \int_G U(g)\psi\, dT(g), \ T \in \mathcal{E}'_G,$$

$$\langle \phi | \widetilde{U}(T)\psi \rangle = T(\langle \phi | U(\cdot)\psi \rangle) \equiv \int_G \langle \phi | U(g)\psi \rangle\, dT(g), \ \phi \in \mathcal{H}.$$

---

[4] For simplicity, we denote here by $T(f)$ the action of the distribution T on the test function f.

Then one can show that $T \mapsto \widetilde{U}(T)$ is a representation of the *-algebra $\mathcal{E}'_G$ by operators in $\mathcal{H}^\infty$, since $\widetilde{U}(T)\,\widetilde{U}(S) = \widetilde{U}(T*S)$, with all the needed properties. In particular, $\widetilde{U}(\delta_g) = U(g)$, i.e., $\widetilde{U}$ is an extension of U.

Using that result, the construction of a RHS adapted to the symmetry under G runs as follows. Let $\Psi \subset \mathcal{H}$ a dense subspace with a nuclear topology. Consider the projective tensor product[5] $\mathcal{D}(G) \otimes_\pi \Psi$ and the continuous map $u : \mathcal{D}(G) \otimes_\pi \Psi \to \mathcal{H}$ defined by $u(f \otimes \psi) = \widetilde{U}(f)\psi$. Let N be the kernel of u, i.e., the linear span of $\{f \otimes \psi : \widetilde{U}(f)\psi = 0\}$. N is closed in $\mathcal{D}(G) \otimes_\pi \Psi$, hence one gets an isomorphism $[u] : \mathcal{D}(G) \otimes_\pi \Psi/N \to \Theta := \operatorname{Ran} u \subset \mathcal{H}$. Finally, denote by Φ the completion of Θ. Then one can show

Proposition 7.1.1.   (i) Φ is a complete nuclear space, dense in $\mathcal{H}$, with continuous embedding. Thus $\Phi \subset \mathcal{H} \subset \Phi^\times$ is a RHS.
(ii) For any $T \in \mathcal{E}'_G$, $\widetilde{U}(T)$ maps Φ continuously into itself, thus $T \mapsto \widetilde{U}(T)$ is a continuous representation of $\mathcal{E}'_G$ in Φ.
(iii) In particular, $g \mapsto U(g)$ is a continuous representation of G in Φ.

Further details may be found in the literature quoted in the Notes to this chapter. We may notice that, in many cases, one could replace the distribution space $\mathcal{D}(G)$ by that of tempered distributions, $\mathcal{S}(G)$. Since the latter may be represented as the end space of a Hilbert scale, and similarly for Ψ, the PIP-space version of the construction is immediate.

7.1.3 Singular Interactions

We have given in Section 3.3.5 a number of results concerning the existence of self-adjoint restrictions of symmetric operators in a PIP-space, in particular the Generalized KLMN Theorems 3.3.27 and 3.3.28 . As said there, the key fact is the existence of a suitable invertible representative, for which Proposition 3.3.31 gives a handy criterion. In order to treat singular interactions in quantum mechanics, we will consider now the case where T is given as a multiplication operator by a function $t(p)$ on a measure space and U is given as a kernel $U(p, p')$ on the same space, as one has for the Schrödinger equation in momentum representation. Hence the resolvent $R(\lambda)$ of T is the multiplication operator by the function $(t(p) - \lambda)^{-1}$. Then Theorem 3.3.27 takes the following form.

Theorem 7.1.2. Take $V = L^1_{\mathrm{loc}}(\mathbb{R}^n, dp)$ and let T be multiplication operator by the real-valued and locally bounded function $t(p)$. Assume that the range of $t(p)$ has a gap, i.e., that there exists a $\lambda \in \mathbb{R}$ such that $(t(p) - \lambda)^{-1}$ is a

---

[5] This means, with the finest topology on the tensor product such that all maps $(f, \psi) \mapsto f \otimes \psi$ are continuous.

bounded function. Assume that the kernel $U(p, p')(p, p' \in \mathbb{R}^n)$ is Hermitian and such that, for some $r < 0$, the kernel

$$K(p, p') = |t(p) - \lambda|^{-(r/2)-1} U(p, p')|t(p') - \lambda|^{(r/2)} \qquad (7.20)$$

defines in the Hilbert space $\mathcal{H}_0 = L^2(\mathbb{R}^n, dp)$ an operator of bound norm <1. Then $H = T + U$ has a self-adjoint restriction H in $\mathcal{H}_0$, and all the other conclusions of Theorem 3.3.27 hold. In particular, the domain of H is $\mathcal{D}(\mathsf{H}) = \{f \in V : Hf \in \mathcal{H}_0\}$.

A beautiful application of this theorem is the description of quantum mechanical systems with local, many-center Hamiltonians. Let $V = L^1_{\text{loc}}(\mathbb{R}^n, dp)$ and let T be multiplication operator by the positive unbounded function $t(p)$, so that $\lambda = -1$ is a real point in the resolvent set of T (typically $t(p) = p^2$, that is, T is the free Hamiltonian). Then consider the scale built on the powers of $T + 1$,

$$\mathcal{H}_r(\mathbb{R}^n) := \{\phi \in L^1_{\text{loc}}(\mathbb{R}^n) : \|\phi\|^2_r := \int_{\mathbb{R}^n} \left(t(p) + 1\right)^r |\phi(p)|^2 \, dp < \infty\}. \quad (7.21)$$

In particular, we will use the central part of the scale (7.21), namely,

$$\mathcal{H}_2 \hookrightarrow \mathcal{H}_1 \hookrightarrow \mathcal{H}_0 \hookrightarrow \mathcal{H}_{\bar{1}} \hookrightarrow \mathcal{H}_{\bar{2}} , \qquad (7.22)$$

where as usual $\mathcal{H}_{\bar{r}} = \mathcal{H}_{-r}$. The free resolvent is the operator $R(E) = (T-E)^{-1}$, where E belongs to the resolvent set $\rho(T) = \mathbb{C} \setminus [0, \infty)$. Then:

(i) The operator $R(E)$ satisfies the identities

$$R(E) - R(E') = (E - E')R(E)R(E') \qquad (7.23)$$

$$\frac{d}{dE}R(E) = R(E)^2. \qquad (7.24)$$

(ii) $R(E) : \mathcal{H}_r \to \mathcal{H}_{r+2}$ is bounded with bounded inverse, and similarly for $R(E)^{1/2} : \mathcal{H}_r \to \mathcal{H}_{r+1}$. Therefore, $\langle \phi | R(E)\psi \rangle$ is well-defined for $\phi, \psi \in \mathcal{H}_{\bar{1}}$, and $\langle \phi | R(E)^2 \psi \rangle$ is well-defined for $\phi, \psi \in \mathcal{H}_{\bar{2}}$.

We want to define properly the Hamiltonian of a system with singular interactions, such as point interactions (δ-potential) or dipole interactions (δ'-potential). To that effect, notice that the result depends on the dimension n. Define the exponential functions as

$$e_n^x(p) = (2\pi)^{-n/2} e^{ix \cdot p}, \ x, p \in \mathbb{R}^n.$$

Then one verifies immediately that

$$e_1^x \in \mathcal{H}_{\bar{1}}(\mathbb{R}),$$

$$e_2^x \notin \mathcal{H}_{\overline{1}}(\mathbb{R}^2), \text{ but } e_2^x \in \mathcal{H}_{\overline{2}}(\mathbb{R}^2),$$
$$e_3^x \notin \mathcal{H}_{\overline{1}}(\mathbb{R}^3), \text{ but } e_3^x \in \mathcal{H}_{\overline{3}}(\mathbb{R}^3),$$
$$e_n^x \notin \mathcal{H}_{\overline{2}}(\mathbb{R}^n) \text{ for } n \geqslant 4.$$

This means that one cannot define a point interaction in dimension larger than 3. As for the dipole interaction in one dimension, we have $(2\pi)^{-1/2} p e^{ix \cdot p} \in \mathcal{H}_{\overline{2}}(\mathbb{R}^2)$.

Consider now dyadic operators of the form $|\psi'\rangle\langle\psi''|$, $\psi', \psi'' \in \mathcal{H}_{\overline{2}}$ (see Section 3.1.4(ii)). More generally, given two n-tuples $\boldsymbol{\psi}' = \{\psi_1', \ldots, \psi_n'\}$, $\boldsymbol{\psi}'' = \{\psi_1'', \ldots, \psi_n''\}$ of vectors from $\mathcal{H}_{\overline{2}}$ and an arbitrary $n \times n$ matrix $\mathbf{B} = [B_{ij}]$, one defines the operator

$$|\boldsymbol{\psi}'\rangle\mathbf{B}\langle\boldsymbol{\psi}''| := \sum_{i,j=1}^{n} B_{ij} |\psi_i'\rangle\langle\psi_j''|. \tag{7.25}$$

Using this notation, we can define the Hamiltonian H as T perturbed by a dyadic of the form (7.25). The first result covers the case of a mildly singular perturbation.

Proposition 7.1.3. *Let* $\boldsymbol{\Phi} = \{\phi_1, \ldots, \phi_n\}, \phi_k \in \mathcal{H}_{\overline{1}}$, *and let* $\mathbf{B}$ *be an invertible* $n \times n$ *matrix. Then the natural restriction of* $H = T - |\boldsymbol{\Phi}\rangle\mathbf{B}\langle\boldsymbol{\Phi}|$ *is a closed operator in* $\mathcal{H}_0$. *The resolvent of this operator is*

$$R(E) - R(E)|\boldsymbol{\Phi}\rangle\Gamma(E)^{-1}\langle\boldsymbol{\Phi}|R(E),$$

where $\Gamma(E) = \langle\boldsymbol{\Phi}|R(E)\boldsymbol{\Phi}\rangle - \mathbf{B}^{-1}$. *If* $\mathbf{B}$ *is Hermitian, then* H *is self-adjoint.*

By 'natural restriction', we mean, of course, restriction in PIP-space-sense, as in Theorems 3.3.27 and 7.1.2.

In the case of a strongly singular perturbation, that is, when $\phi_k \in \mathcal{H}_{\overline{2}}$, but $\phi_k \notin \mathcal{H}_{\overline{1}}$, the natural restriction of $T - |\boldsymbol{\Phi}\rangle\mathbf{B}\langle\boldsymbol{\Phi}|$ is *not* self-adjoint, but it admits a family of self-adjoint extensions. In order to tackle this case, we have to let $\boldsymbol{\Phi}$ vary over all invertible $n \times n$ matrices and thus define a family $\{H_{\Gamma}^{\boldsymbol{\Phi}}\}$ of closed operators, indexed by n^2 complex parameters.

Assume that $\boldsymbol{\Psi} = \{\psi_1, \ldots, \psi_n\}, \psi_k \in \mathcal{H}_{\overline{2}}$, where the vectors $\psi_k, k = 1, \ldots, n$, are such that no nontrivial linear combination $\sum_1^n \lambda_k \psi_k$ belongs to $\mathcal{H}_0$ (we call this 'Condition (a)'). This is the case for the exponential functions described above, corresponding to point interactions in dimension less than 3 and to the dipole interaction in dimension 1. Then $\boldsymbol{\Psi}$ determines a family of n^2 functions $\Gamma(E)_{jk}$, each analytic in (the connected components of) $\rho(T)$ and determined up to an additive constant. These functions are defined as any solution of the equation

$$\frac{d}{dE}\Gamma_{jk}(E) = \langle\psi_j|R(E)^2\psi_k\rangle, \quad E \in \rho(T), 1 \leqslant j, k \leqslant n. \tag{7.26}$$

In shorthand notation, as before,

$$\frac{d}{dE}\Gamma(E) = \langle\psi|R(E)^2\psi\rangle, \ E \in \rho(T). \tag{7.27}$$

Explicit solutions of Eq. (7.27) may be found for the exponential functions corresponding to point and dipole interactions. In this language, the central results is given as follows.

Theorem 7.1.4. *Let Ψ be a n-tuple of vectors from $\mathcal{H}_{\overline{2}}$ satisfying Condition (a). Let $\Gamma(E)$ be any solution of Eq. (7.27). Consider in $\mathcal{H}_0$ the family of operators $Q_\Gamma^\Phi(E)$ defined by*

$$Q_\Gamma^\Phi(E) = R(E) - R(E)|\Psi\rangle\Gamma(E)^{-1}\langle\Psi|R(E), \tag{7.28}$$

where $\Gamma(E)^{-1}$ is the matrix inverse of $\Gamma(E)$ and $R(E) = (T - E)^{-1}$. Then $Q_\Gamma^\Phi(E)$ is the resolvent family of a closed, densely defined operator $\{H_\Gamma^\Phi\}$,

$$Q_\Gamma^\Phi(E) = (H_\Gamma^\Phi - E)^{-1}.$$

This theorem defines the Hamiltonian only indirectly, through its resolvent. In particular, it does not allow to write $H_\Gamma^\Phi = T + V$, unless the perturbation is mildly singular, that is, $\psi_k \in \mathcal{H}_{\overline{1}}$ for all k. Yet this is sufficient for obtaining a complete description of the spectral properties of the Hamiltonian: eigenvalues and bound states, resonances, continuous spectrum.

The analysis made so far covers the case of a *finite* number of point or dipole interactions. The generalization to infinitely many centers may be obtained by similar methods. This type of analysis applies to many situations of condensed matter physics (crystals, straight polymers, monomolecular layers, gratings (linear interferometers), ...). In the case of a periodic interaction, as for the Kronig-Penney crystal model (in dimension 1, 2, or 3), one needs a slight modification of the PIP-space used. Let Λ be a lattice in momentum space (the reciprocal lattice) and let $k \in \mathbb{R}^3$. Then one defines $\mathcal{H}_r$ as the Hilbert space of complex-valued functions ψ on the displaced lattice $k + \Lambda$ such that

$$\sum_{p \in k+\Lambda} (t(p) + 1)^r |\psi(p)|^2 < \infty.$$

Then all the analysis may be repeated. Of course, k and $k + K$, $K \in \Lambda$, yield the same space $\mathcal{H}_r$.

For further details on these results, we refer to the literature quoted in the Notes.

7.2 Quantum Scattering Theory

7.2.1 Phase Space Analysis of Scattering

In scattering theory, one uses frequently scales of Hilbert spaces built on the powers of $(1 + |\mathbf{x}|^2)$ or $(1 + |\mathbf{p}|^2)$, and lattices obtained by combinations of both. These two operators do not commute and their domains are not included into one another. Thus, in order to characterize the underlying LHS structure, we expand slightly the analysis of Section 5.3 by considering two noncomparable subspaces $\mathcal{H}_1, \mathcal{H}_2$ of $\mathcal{H}_0$. For $j = 1, 2$, we denote by A_j, B_j, respectively, the interpolating operators in the triplet $\mathcal{H}_1 \cap \mathcal{H}_2 \hookrightarrow \mathcal{H}_j \hookrightarrow \mathcal{H}_0$. The situation so obtained is depicted in Fig. 7.1. Then, proceeding as in Section 5.3, we obtain the following results:

(i) Using again the uniqueness of the interpolation operator, we get

$$(B_1^{1/2})^* A_1 B_1^{1/2} = (B_2^{1/2})^* A_2 B_2^{1/2} = A_1 \dotplus A_2,$$

where $(B_j^{1/2})^*$ is the adjoint of $B_j^{1/2}$ in $\mathcal{H}_0$. This operator is self-adjoint, with form domain $\mathcal{H}_1 \cap \mathcal{H}_2$ and defines an interpolating scale $\{Q((A_1 \dotplus A_2)^\alpha), 0 \leqslant \alpha \leqslant 1\}$, between $\mathcal{H}_1 \cap \mathcal{H}_2$ and $\mathcal{H}_0$.

(ii) Let $\{Q(A_j^\alpha), j = 1, 2, 0 \leqslant \alpha \leqslant 1\}$ be the scale interpolating between $\mathcal{H}_j$ and $\mathcal{H}_0$. Then these two scales generate again, by intersection and vector sum, a genuine lattice, and one has

$$Q(A_1^\alpha) \cap Q(A_2^\beta) = Q(A_1^\alpha \dotplus A_2^\beta), \ 0 \leqslant \alpha, \beta \leqslant 1.$$

(iii) The spaces defined in (i) and (ii) are related by the following inclusion relations, which follow from the interpolation property;

$$Q((A_1 \dotplus A_2)^\alpha) \subseteq Q(A_1^\alpha) \cap Q(A_2^\alpha), \ 0 \leqslant \alpha \leqslant 1. \tag{7.29}$$

Equality holds if A_1 and A_2 commute.

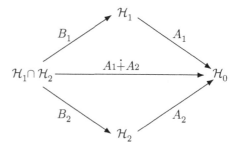

Fig. 7.1 Interpolation between two noncomparable subspaces

Coming back to scattering theory, we choose $\mathcal{H}_0 = L^2(\mathbb{R}^3, d\mathbf{x})$, and define $A_1 = (1 + |\mathbf{x}|^2)$, $A_2 = (1 + |\mathbf{p}|^2)$, which gives $B_1 = 1 + (1 + |\mathbf{x}|^2)^{-1}(1 + |\mathbf{p}|^2)$ and $B_2 = 1 + (1 + |\mathbf{p}|^2)^{-1}(1 + |\mathbf{x}|^2)$. The operators $A_1^{1/2} A_2^{-1/2}$ and $A_2^{1/2} A_1^{-1/2}$ being both unbounded, Lemma 5.3.1 implies that $\mathcal{H}_1$ and $\mathcal{H}_2$ are indeed noncomparable. Moreover, since A_1 and A_1 do not commute, iterating the construction above leads to a genuine LHS.

This example, or variants thereof, constitutes an essential tool for scattering theory. It contains the Sobolev spaces (the scale built on A_2), the weighted spaces L_s^2 (the scale built on A_1), and spaces of mixed type. In particular, operators of the form $f(\mathbf{x})g(\mathbf{p})$, for suitable functions f, g, play an essential role in the so-called phase space approach to scattering theory and they may be controlled by the LHS just described. For instance, their trace ideal properties may be derived in this way and they are used for proving the absence of singular continuous spectrum by the limiting absorption principle.

On the other hand, the standard examples of commuting operators are those that depend on different variables. Take again $\mathcal{H}_0 = L^2(\mathbb{R}^3, d\mathbf{x})$ with $A_j = (1 + x_j^2), j = 1, 2, 3$. Each of these generates a scale $Q(A_j^\alpha)$ and those scales generate a lattice. Then (7.29) gives, for each $0 \leqslant \alpha \leqslant 1$:

$$Q((A_1 \dotplus A_2 \dotplus A_3)^\alpha) = \bigcap_{j=1}^{3} Q(A_j^\alpha),$$

i.e.,

$$L^2(\mathbb{R}^3, (1 + |\mathbf{x}|^2)^\alpha d\mathbf{x}) = \bigcap_{j=1}^{3} L^2(\mathbb{R}^3, (1 + x_j^2)^\alpha d\mathbf{x}).$$

An interesting generalization of this construction arises for a Schrödinger Hamiltonian $H = H_0 + V(x)$ (the discussion extends to the multiparticle case, thus in $L^2(\mathbb{R}^{3d})$, but for simplicity we restrict ourselves to the one-particle case; see [126] for the general case). In the discussion above, the operator A_2 is essentially the nonrelativistic free Hamiltonian H_0. Assume V to be a Kato potential, that is, $D(H_0) \subset D(V)$ and there exist constants $a < 1$, $b < \infty$, such that, for all $u \in D(H_0)$, $\|Vu\| \leqslant a\|H_0u\| + b\|u\|$. Then the Hamiltonian H has domain $D(H) = D(H_0)$ and is self-adjoint.

Next, given a multi-index $n = (n_1, n_2, n_3)$, n_i integer $\geqslant 0$, define the monomial $x^n := x_1^{n_1} x_2^{n_2} x_3^{n_3}$, considered as a self-adjoint operator of multiplication in $L^2(\mathbb{R}^3)$. We write $|n| = n_1 + n_2 + n_3$ and $k \leqslant n$, meaning $k_j \leqslant n_j$, $j = 1, 2, 3$. Let us define the subspace D_n and the norm $\|\cdot\|_n$ on it as follows:

$$D_n = \bigcap_{k \leqslant n,\, m \leqslant |n| - |k|} D(x^k H^m), \quad \|u\|_n = \sup_{k \leqslant n,\, m \leqslant |n| - |k|} \|x^k H^m u\| \quad (m \geqslant 0,\, \text{integer}).$$

$$\text{(7.30)}$$

Then the main results are the following:

(i) Every subspace D_n is invariant under the unitary group e^{-iHt}.
(ii) For every $u \in D_n$, $e^{-iHt}u$ is $\| \cdot \|_n$-continuous and there is a constant c_n such that

$$\|e^{-iHt}u\|_n \leqslant c_n(1 + |t|)^{|n|})\|u\|_n.$$

Of course, D_n may be reduced to $\{0\}$ for n large enough, but this cannot happen if the potential V is a C^∞ function on some open subset of $\mathbb{R}^3$, in which case $D_\infty =: \bigcap_n D_n$ is dense in $L^2(\mathbb{R}^3)$. More interesting, if V is a bounded C^∞ function with bounded derivatives, then one can replace H by H_0 in the definitions (7.30), so that $D_\infty = \mathcal{S}(\mathbb{R}^3)$.

As they stand in (7.30), the subspaces D_n are only normed spaces and they are not comparable. However, it is trivial to build Hilbert spaces out of them, simply by replacing the 'sup' by a sum and the norm by a squared norm. Then we get a Hilbert norm, indeed a graph norm, so that D_n becomes a Hilbert space. Since the resulting spaces are still not comparable in general, one gets in this way a genuine lattice, that is, a LHS, with extreme spaces $\mathcal{S}$ and $\mathcal{S}^\times$ in the special case of a nice potential V. The construction of this LHS may be further restricted to get a Hilbert chain, as has been done, for instance, in the so-called Hermite representation of $\mathcal{S}$ or the construction of Grossmann's Hilbert spaces of type S (Section 5.4.3). Each of these spaces is in fact a LHS in which every assaying subspace is Fourier invariant. In other words, the Fourier operator is totally regular and an isomorphism. One interesting point of the analysis above is that the time evolution operator e^{-iHt} has the same properties.

7.2.2 A LHS of Analytic Functions for Scattering Theory

In this section, we shall indicate how the LHS of analytic functions described in Section 4.6.3 simplifies considerably the Weinberg-van Winter formulation of scattering theory. Furthermore, it is a crucial tool for proving that the latter is a particular case of the complex scaling method (CSM), by now a standard approach to quantum scattering theory.

Take again the Hamiltonian $H = H_0 + V$, where H_0 is the unperturbed Hamiltonian and V is the interaction. H_0 may be the free Hamiltonian, but not necessarily so. Define the resolvents $R(z) = (H - z)^{-1}$, $R_0(z) = (H_0 - z)^{-1}$, which are integral operators, since H, H_0 are differential operators.

The idea of van Winter is to formulate the Weinberg-van Winter (WVW) integral equations

$$R(z) = R_0(z) - R_0(z)VR_0(z) \tag{7.31}$$

in an appropriate Hilbert space, so that they become Fredholm equations, with a Hilbert-Schmidt kernel, uniquely solvable by standard methods. For simplicity, we consider only two-body scattering, with angular variables omitted, but this is not a serious restriction. The general case may be found in the literature.

The essence of van Winter's approach is to formulate the entire theory of scattering in the space $G(a,b)$ defined in Section 4.6.3, with the following steps (from now on, we take $-\frac{\pi}{2} < a < b < \frac{\pi}{2}$):

(1) The interaction V is defined as $v(re^{i\varphi}) \in G(-a,a)$ if it is local, or as $\mathsf{V} \in \mathfrak{K}$ if it is nonlocal.
(2) The Hamiltonian $H = H_0 + \mathsf{V}$ and the resolvents $R(z) = (H - z)^{-1}$, $R_0(z) = (H_0 - z)^{-1}$ are defined as operators in $G(a,b)$.
(3) The WVW integral equations (7.31) are interpreted as integral equations in $G(a,b)$.

Then it is proven that $R_0(z)\mathsf{V}$ belongs to $\mathfrak{K}$, hence is Hilbert-Schmidt, so that the WVW equation is a Fredholm equation in $G(a,b)$, which can be solved uniquely.

We shall now show that the formalism sketched above simplifies considerably if one uses the LHS language, namely, by considering the spaces $\widetilde{G}(a,b)$ and the corresponding LHS depicted in Figure 4.4. In particular, this approach clarifies the connection with the complex scaling method.

The dilation analyticity or complex scaling method (CSM) is by now a well established tool in scattering theory, including for numerical work in atomic and molecular systems. Its popularity rests largely on the fact that it turns a resonance into a discrete eigenvalue of a non-self-adjoint operator, thus allowing a complete disentangling of the spectrum of the Hamiltonian. The WVW method is obviously related to this one, but the exact link has not been identified so far. Our aim in the rest of this section is to show, in a precise mathematical way, that the WVW formalism is a particular case of CSM.

The starting point of CSM is the familiar unitary representation of the dilation group in $L^2(\mathbb{R}^3)$ (we again omit the angles):

$$(U(\rho)f)(r) = \rho^{3/2} f(\rho r), \ \rho > 0, \ f \in L^2(\mathbb{R}^3). \tag{7.32}$$

The basic notion of dilation analyticity applies both to vectors and operators. For $0 \leqslant a < \pi/4$, consider the sector $S_a := S_{-a,a} = \{z = \rho e^{i\varphi} : -a < \varphi < a\}$. Then the set of S_a-dilation analytic vectors is defined as

$$D(S_a) = \{f \in L^2 : U(\rho)f \text{ has an analytic continuation to } U(\eta)f, \ \forall \eta = \rho e^{i\varphi} \in S_a\}. \tag{7.33}$$

Similarly, the potential V is called S_a-dilation analytic if $\mathsf{V}(\rho) = U(\rho)\mathsf{V}U(\rho)^{-1}$ admits an analytic continuation to $\mathsf{V}(\eta)$, $\forall \eta \in S_a$. For such an interaction, one considers the spectrum of the (non-self-adjoint) dilated Hamiltonian

$H(\eta) = U(\eta)HU(\eta)^{-1}$. For $\eta = \rho e^{i\varphi}$, it turns out that the bound states remain on the real axis, independently of φ, the absolutely continuous spectrum σ_{ac} is rotated into the lower half-plane by $-\varphi$, and the resonances appear as isolated points in the sector $-\varphi < \arg z < 0$, and they are in fact eigenvalues of $H(\eta)$. Thus the structure of $\sigma(H)$ becomes transparent, and the study of resonances much easier.

The key to the comparison between WVW and CSM is the definition of dilation analytic vectors in terms of the family of Hilbert spaces $D(a, b)$, defined in (4.50). We recall that $D(a, b)$ is a Hilbert space of functions analytic in the sector $T_{ab} = \{z \in \mathbb{C} : -b < \arg z < \pi - a\}$. Then it turns out that one may define the sets (7.33) of dilation analytic vectors in terms of the spaces $D(a, b)$:

$$D(S_a) = \bigcap_{0 \leqslant c < a} D(-c, c).$$

Furthermore, one can exploit the unitary equivalences between the three Hilbert spaces $G(a, b)$, $\widetilde{G}(a, b)$ and $D(2a, 2b)$ described in Figure 4.5. Using these three spaces, one may then show that any van Winter interaction that belongs to $G(-a, a)$ (local case) or to $\mathfrak{A}[G(-a, a)]$ (nonlocal case) is necessarily S_a-dilation analytic. In other words, the WVW formalism is a particular case of CSM, a result hitherto unknown.

7.2.3 A RHS Approach: Time-Asymmetric Quantum Mechanics

There is another approach to RHS quantum theory, the so-called *time-asymmetric* quantum mechanics. While this theory is still controversial on physical grounds, it offers some interesting mathematical aspects that justify its inclusion in the present survey. Of course, a thorough description is beyond the scope of the present volume, so we refer to the original literature quoted in the Notes.

If we concentrate on energy (assumed to be positive) and forget for a while the other observables making a CSCO, as explained in Section 7.1, we expand a given vector $\phi \in \mathcal{H}$ as

$$\phi = \int_0^\infty dE |E\rangle\langle E|\phi\rangle$$

and the inner product takes the form

$$\langle \psi | \phi \rangle = \int_0^\infty dE \, \langle \psi | E \rangle \langle E | \phi \rangle.$$

Now in a RHS $\Phi \hookrightarrow \mathcal{H} \hookrightarrow \Phi^\times$ as given in (7.4), the allowed wave functions $\phi(E) = \langle E|\phi \rangle$ which are admissible in the Dirac formalism, as a function of E, are Schwartz functions. This suggests to identify the space Φ with the Schwartz function space $\mathcal{S}$.

However, this Schwartz RHS is not sufficient for a quantum theory of scattering and decay, where one analytically continues the S-matrix into the complex energy plane. In the empirical description of resonance phenomena, one uses energy (or, in the relativistic case, invariant mass) values belonging to the complex plane and works with Gamow vectors which are associated with the complex eigenvalues of the Hamiltonian. One also uses Lippmann-Schwinger kets with $\pm i\epsilon$ in the energy denominator. The Schwartz RHS accommodates neither Lippmann-Schwinger kets nor exponentially decaying Gamow kets and thus cannot provide a relation between the lifetime τ of a decaying state and the width Γ (or the complex pole position) of a resonance.

To obtain a mathematical theory unifying quantum resonance and decay phenomena, one needs to take a step beyond the confines of Dirac's formalism. What is remarkable is that this step beyond the Schwartz space theory can be taken within the general mathematical framework of RHSs. Specifically, this extension requires a careful mathematical distinction between the set of *prepared in-states* and the set of *observed out-states* (more precisely, out-observables). Quantum theory is grounded in the notions of states ϕ, which are prepared by a preparation apparatus, and observables $\Lambda = |\psi\rangle\langle\psi|$, which are registered by a detector. The theory predicts the Born probabilities $|\langle\phi|\Lambda(t)\phi\rangle|^2$ which represent the normalized detector counts of events $\frac{N(t)}{N}$. In scattering theory, the distinction between in-states ϕ^+ and out-states ψ^- appears in separate basis vector expansions

$$\phi^+ = \int_0^\infty dE\, |E^+\rangle\langle^+E|\phi^+\rangle \quad \text{and} \quad \psi^- = \int_0^\infty dE\, |E^-\rangle\langle^-E|\psi^-\rangle \quad (7.34)$$

However, in the usual mathematical formulation, the set of state vectors $\{\phi\}$ is identified with the set of observable vectors $\{\psi\}$, usually by associating both with the same Hilbert space $\mathcal{H}$. Similarly, the kets $|E^\pm\rangle$ of expansions (7.34) are thought of as two sets of basis vectors for the same vector space. In contrast, the RHS theory of scattering and decay phenomena generalizes the Schwartz RHS formulation of the Dirac formalism to a theory with two RHSs, one for the prepared in-states $\{\phi^+\}$,

$$\{\phi^+\} = \Phi_- \hookrightarrow \mathcal{H} \hookrightarrow \Phi_-^\times, \quad \text{with } |E^+\rangle \in \Phi_-^\times, \quad (7.35)$$

and the other for the detected observable vectors ψ^-

$$\{\psi^-\} = \Phi_+ \hookrightarrow \mathcal{H} \hookrightarrow \Phi_+^\times, \quad \text{with } |E^-\rangle \in \Phi_+^\times, \quad (7.36)$$

where $\mathcal{H}$ is the same Hilbert space. Thus, this RHS theory elevates the physical content of the notions of state and observable vectors into a mathematical principle.

From this pair of RHSs which distinguish states and observable vectors, a mathematically consistent theory of resonance scattering and decay phenomena can be obtained by letting the spaces Φ_- and Φ_+ to be defined in their energy representation by Hardy spaces $\mathfrak{H}^2_\pm$ on the lower and upper complex half-planes, respectively (the Hardy spaces $\mathfrak{H}^p_\pm$ are defined in Section 4.6.5, Eq.(4.65)). In particular, the energy wavefunctions $\langle {}^+E|\phi^+\rangle = \phi^+(E)$ and $\langle {}^-E|\psi^-\rangle = \psi^-(E)$ in (7.34) are smooth, rapidly decreasing Hardy functions on the upper and lower complex half-planes. The basis kets $|E^\pm\rangle$ can now be defined correctly as elements of the dual spaces $\Phi^\times_\mp$, and therewith Dirac-type basis vector expansions (7.34) of ϕ^+ and ψ^- can be rigorously obtained in terms of $|E^\pm\rangle$ by way of the nuclear spectral theorem. The theory based on RHSs (7.35)-(7.36) also contains exponentially decaying Gamow vectors and Breit-Wigner resonance amplitude as well-defined mathematical concepts. This Hardy space theory has been subsequently extended to relativistic resonances and decaying states. One of the important outcomes of the relativistic extension is the unique and unambiguous definitions it provides for the mass and width of a relativistic resonance, a much debated problem since the early 1990s.

One particularly important aspect in which the Hardy-type RHSs differ from the Schwartz-type RHSs consists in the class of allowed representations of symmetry groups, including noncompact space-time symmetry groups. In the Schwartz-type construction, described in Section 7.1.2, the unitary representations of such groups in the Hilbert space $\mathcal{H}$ can be restricted to Φ and extended to $\Phi^\times$ to obtain differentiable representations in these spaces. Thus, quantum mechanical symmetry transformations can be well accommodated in the Schwartz-type RHSs, and many of the elements of the algebra of observables arise as the derivatives of these representations in Φ and $\Phi^\times$, as we have explained (see also Section 7.4 below). In contrast, Hardy-type RHSs do not furnish representations of the space-time symmetry groups. In particular, in the nonrelativistic version, the time evolution in $\Phi_\pm$ is given by the one parameter *semigroups* $U_\pm(t)$ with $t \geqslant 0$. In the relativistic version, the space-time evolution in $\Phi_\pm$ is given by semigroups $U_\pm(I,a)$ with $a_0 \geqslant 0$ and $a^2 \geqslant 0$, i.e., by representations of the Poincaré semigroup into the forward lightcone. According to the authors of the theory, these semigroup representations encode the fundamental causal structure of physics and provide the foundation for the unified theory of resonance and decay phenomena.

7.3 Quantum Field Theory

All sorts of PIP-spaces have been used in quantum field theory, from distribution spaces (RHS), to scales or lattices of Hilbert spaces (LHS). This section is devoted to a brief survey of several of these applications.

7.3.1 Quantum Field Theory: The Axiomatic Wightman Approach

Axiomatic quantum field theory, in the Wightman version, makes an essential use of distributions, since a Wightman field is precisely defined as an operator valued tempered distribution. As we know, distribution theory is best formulated in a RHS approach and from there the transition to PIP-spaces is easy.

For simplicity, we restrict ourselves here to the case of a single scalar field. The Wightman axioms may be formulated as follows.

(W_0) *Assumptions of a relativistic quantum theory:*

- The (pure) states of the theory are described by unit rays in a separable Hilbert space $\mathcal{H}$.
- The relativistic transformation law of the states is given by a strongly continuous unitary representation $U(a, \Lambda)$ of the Poincaré group $\mathcal{P}(1, 3)$.
- *Spectral condition:* For translations, in particular, one may write $U(a, I) = \exp(iP^\mu a_\mu)$, where P^μ is the energy-momentum operator. Then the eigenvalues of P^μ must lie in the closed future light cone.
- *Existence and uniqueness of vacuum:* There exists a unique (up to a constant phase factor) vector $\Psi_0 \in \mathcal{H}$ such that

$$U(a, \Lambda)\Psi_0 = \Psi_0, \quad \forall (a, \Lambda) \in \mathcal{P}(1, 3).$$

(W_1) *(Wightman) fields:* To each test function $f \in \mathcal{S}(\mathbb{R}^4)$, there is associated an operator $A(f)$ defined, together with its adjoint $A(f)^*$, on a dense subspace $\mathcal{D} \subset \mathcal{H}$. The domain $\mathcal{D}$ contains the vacuum state Ψ_0 and is invariant under all $U(a, \Lambda), A(f)$ and $A(f)^*$. For any $\xi, \eta \in \mathcal{D}$, $f \mapsto \langle \xi | A(f)\eta \rangle$ is a tempered distribution.

(W_2) *Relativistic covariance:* For any $(a, \Lambda) \in \mathcal{P}(1, 3)$, one has, on $\mathcal{D}$,

$$U(a, \Lambda)A(f)U(a, \Lambda)^{-1} = A(f_{(a,\Lambda)}),$$

where $f_{(a,\Lambda)} := f\big(\Lambda^{-1}(x - a)\big)$.

(W_3) *Cyclicity of the vacuum:* when applied to the vacuum, the polynomials in the field operators, $\{P(A(f))\Psi_0, f \in \mathcal{S}(\mathbb{R}^4), P(\cdot)$ a polynomial$\}$, generate a dense subspace in $\mathcal{H}$.

(W_4) *Locality:* if $\operatorname{supp} f$, $\operatorname{supp} g$ are space-like separated, then $[A(f), A(g)]\xi = 0, \forall \xi \in \mathcal{D}$, where $[A(f), A(g)] = A(f)A(g) - A(g)A(f)$.

Now, since the field $A(f)$ is an operator valued tempered distribution, it is customary to write it in terms of an unsmeared field (field at a point) $A(x)$, as

$$A(f) = \int_{\mathbb{R}^4} A(x)f(x)dx, \quad f \in \mathcal{S}(\mathbb{R}^4). \tag{7.37}$$

In terms of $A(x)$, the axioms (W$_2$) and (W$_4$) take the simple form

- covariance: $U(a, \Lambda) A(x) U(a, \Lambda)^{-1} = A(\Lambda x + a),$ (7.38)
- locality: $[A(x), A(y)] = 0$ whenever $(x - y)^2 < 0.$ (7.39)

The question, of course, is how to define the unsmeared field $A(x)$ in a rigorous way. Traditionally, it is considered as a sesquilinear form on the domain $\mathcal{D}$. Now, if one endows $\mathcal{D}$ with a suitable topology, one builds the familiar RHS $\mathcal{D} \hookrightarrow \mathcal{H} \hookrightarrow \mathcal{D}^\times$, as in Section 5.4.1. Then, as discussed in Section 6.3, one considers the space $\mathcal{L}(\mathcal{D}, \mathcal{D}^\times)$ of all continuous linear operators from $\mathcal{D}$ into $\mathcal{D}^\times$, or, equivalently, of all separately continuous bilinear forms over $\mathcal{D} \times \mathcal{D}$. If the topology of $\mathcal{D}$ is regular enough, this space is a quasi *-algebra, with distinguished subalgebra $\mathcal{L}^\dagger(\mathcal{D})$. Then one may consider $\mathcal{L}(\mathcal{D}, \mathcal{D}^\times)$-valued fields.

For instance, a natural choice is $\mathcal{D} = \mathcal{D}^\infty(H)$ where $H = P^0$ is the energy operator. In that case, one works with the scale generated by $H = P^0$ and, as we shall see below, more precise results may be obtained. Before that, we discuss in more details the RHS approach itself.

7.3.2 Quantum Field Theory: The RHS Approach

In the specific case of quantum mechanics, we have discussed in Section 7.1.1 its formulation in a RHS $\Phi \hookrightarrow \mathcal{H} \hookrightarrow \Phi^\times$, including the physical interpretation of the two extreme spaces. In this scheme, the space Φ is assumed to be reflexive, complete and nuclear. For a nonrelativistic particle, either free or submitted to a nice potential, it turns out that Φ may be taken as the Schwartz space $\mathcal{S}(\mathbb{R}^3)$, being essentially built on the powers of the operators of position and momentum, or, equivalently, on the powers of creation and annihilation operators, as used in the theory of canonical coherent states.

Then the same construction may be extended to quantum field theory, simply by taking the Fock construction over the RHS, as described in Section 1.1.3, Example (iv). More precisely, we start from the triplet

$$\mathcal{S}(\mathcal{V}_m^+) \hookrightarrow L^2(\mathcal{V}_m^+, d\mu) \hookrightarrow \mathcal{S}^\times(\mathcal{V}_m^+),$$

where $\mathcal{V}_m^+$ denotes the forward mass shell $\mathcal{V}_m^+ = \{p \in \mathbb{R}^4 : p^2 = m^2, p_0 > 0\}$ and $d\mu$ the Lorentz invariant measure d^3p/p_0 on it (one can also replace $\mathcal{V}_m^+$ by $\mathbb{R}^3$). Write $\Phi_1 = \mathcal{S}(\mathcal{V}_m^+)$, the space of "good" one-particle states. As usual, this space may be described via the scale generated by the powers of a single operator, namely, the one-particle Hamiltonian $h^{(1)} = 1 + b_1^* b_1 + b_2^* b_2 + b_3^* b_3$, where $b_j^\#$ are the creation/annihilation operators. Then define

$$\Phi_n = \Phi_1^{\otimes_s n},$$

where the right-hand side denotes the symmetrized tensor product of n copies of Φ_1, corresponding to n-boson states. Again Φ_n is reflexive, complete and nuclear with respect to its natural topology, and it can be described as the end space of a scale of Hilbert spaces. Finally, define

$$\Phi = \bigoplus_{n=0}^{\infty} \Phi_n, \tag{7.40}$$

that is, the topological direct sum of the component spaces. Elements of Φ are finite sequences $f = \{f_0, f_1, \ldots, f_n, \ldots\}$, $f_0 \in \mathbb{C}$, $f_n \in \Phi_n$, that is, totally symmetric functions of Schwartz type. The space Φ is reflexive, complete and nuclear with respect to the direct sum topology. Its dual is the topological product

$$\Phi^{\times} = \prod_{n=0}^{\infty} \Phi_n^{\times}.$$

Thus we get a suitable RHS, in which the central Hilbert space $\mathcal{H}$ is Fock space $\Gamma(\Phi_1)$.

Another possibility is to consider directly the scale built on the powers of the operator $H = \Gamma(h^{(1)})$. Here, as before, $\Gamma(\cdot)$ denotes the second quantization functor.

Another approach yet, which allows one to make contact with the Wightman theory, is to start from the so-called Borchers or field algebra. This object is an example of tensor algebra, as discussed in Section 6.4.3.1. Since it is very much in the spirit of PIP-space theory, it is worthwhile to give some details. We start with the following space:

$$\underline{S} := \bigoplus_{i=0}^{\infty} S(\mathbb{R}^{4i}). \tag{7.41}$$

Elements of $\underline{S}$ are finite sequences $f = \{f_0, f_1, \ldots, f_i, \ldots\}$, $f_0 \in \mathbb{C}$, $f_i \in S(\mathbb{R}^{4i})$. The space $\underline{S}$ is reflexive, complete and nuclear for its natural topology. In addition, it can be made into a *-algebra with the following operations:

- Product of two elements:

$$f \times g := \{f_0 g_0, f_0 g_1 + f_1 g_0, \ldots, \sum_{i+j=l} f_i \times g_j, \ldots\},$$

where

$$(f_i \times g_j)(x_1, \ldots, x_{i+j}) := f_i(x_1, \ldots, x_i) g_j(x_{i+1}, \ldots, x_{i+j}), \ x_k \in \mathbb{R}^4.$$

- Involution: $f^+ := \{f_0^+, f_1^+, \ldots, f_n^+, \ldots\}$, where $f_0^+ = \overline{f_0}$ and

$$f_i^+(x_1, \ldots, x_i) = \overline{f_i(x_i, x_{i-1}, \ldots, x_1)}, \ x_k \in \mathbb{R}^4.$$

The unit element is $1 = \{1, 0, 0, \ldots\}$ and the Poincaré group $\mathcal{P}(1,3)$ acts on $\underline{S}$ as

$$(a, \Lambda)f = f_{(a,\Lambda)}, \ f \in \underline{S}, \ (a, \Lambda) \in \mathcal{P}(1,3),$$

where

$$f_{(a,\Lambda)} := \{f_0, f_1[\Lambda^{-1}(x-a)], \ldots, f_n[\Lambda^{-1}(x_1-a), \ldots, \Lambda^{-1}(x_n-a)], \ldots\}.$$

The two operations being continuous, $\underline{S}$ is a topological nuclear *-algebra. Since $\underline{S}$ is a topological direct sum, its dual is the topological direct product of the duals, namely $\underline{S}^\times = \prod_{i=0}^\infty \mathcal{S}^\times(\mathbb{R}^{4i})$.

Next define a Wightman functional as a positive, normalized, continuous, conjugate linear functional over $\underline{S}$, say $W \in \underline{S}^\times$, satisfying in addition, conditions of Poincaré invariance, namely, $\langle W | f_{(a,\Lambda)} \rangle = \langle W | f \rangle$, $\forall (a, \Lambda) \in \mathcal{P}(1,3)$, and locality, that we will not explain. Positivity here means $\langle W | f^+ \times f \rangle \geqslant 0$, for every $f \in \underline{S}$. Given such a Wightman functional W, it is easy to show that the set $J_W := \{f \in \underline{S} : \langle W | f^+ \times f \rangle = 0\}$ is a left ideal of $\underline{S}$. The rest is standard. First the space $\Phi := \underline{S}/J_W$ is reflexive, complete and nuclear. From the functional W, one generates a Hilbert space $\mathcal{H}$ by a GNS construction:

(i) The inner product is defined on Φ by

$$\langle [\phi] | [\psi] \rangle := \langle W | \phi^+ \times \psi \rangle, \ [\phi], [\psi] \in \Phi.$$

(ii) $\mathcal{H}$ is the completion of Φ with respect to this inner product, so that one gets the RHS $\Phi \hookrightarrow \mathcal{H} \hookrightarrow \Phi^\times$, as usual.
(iii) Field operators are defined by the representation of $\underline{S}$ on Φ : $A(\phi)[\psi] = [\phi \times \psi]$, $\phi \in \underline{S}$. In particular, the usual field operators are given by $A(\phi_{(1)})$, where $\phi_1 = \{0, \phi_{(1)}, 0, 0, \ldots\}$.
(iv) Poincaré covariance: $U(a, \Lambda)[\psi] = [\psi_{(a,\Lambda)}]$ and $U(a, \Lambda)A(\phi)U(a, \Lambda^{-1}) = A(\phi_{(a,\Lambda)})$.

In conclusion, the RHS can be completely recovered from the given Wightman functional, which itself is in fact a full set of Wightman functions. We refer to the literature for further details.

7.3.3 Euclidean Field Theory

A great step forward in the development of QFT was the discovery in the 1970s that field theory could be formulated in Euclidean space (formally, replacing time t by it) and then recovered by analytic continuation. The advantage is that a Euclidean field theory is a theory of random processes, so that the full arsenal of probabilistic methods becomes available.

Particularly interesting for us is Nelson's formulation of Euclidean field theory, in which he performs the transition Euclidean $\to$ Minkowskian in what amounts essentially to a PIP-space language. Indeed, a quantum field is precisely defined as an operator (in the PIP-space sense) in the scale generated by the powers of the Hamiltonian. So the latter provides the natural framework for discussing regularity and locality properties of such fields. Let us sketch this method.

One starts with the Sobolev space $H^{-1}(\mathbb{R}^d) = W^{-1,2}(\mathbb{R}^d)$, that is, the completion of $\mathcal{D}(\mathbb{R}^d)$ or $\mathcal{S}(\mathbb{R}^d)$ with respect to the inner product $\langle g|f\rangle_{-1} = \langle g|(1-\Delta)^{-1}f\rangle$ and consider a random field $\phi(f)$ over $H^{-1}(\mathbb{R}^d)$, that is, a stochastic process indexed by $f \in H^{-1}(\mathbb{R}^d)$. Then a Euclidean field over $H^{-1}(\mathbb{R}^d)$ is a random field that verifies a Markov-type condition ('one-step memory') and is covariant under the Euclidean group. The field $\phi(f)$ is called Hermitian if $\phi(\bar{f}) = \overline{\phi(f)}$.

Consider the underlying probability space $(\Omega, \mathcal{B}, \mu)$, where Ω is a set, $\mathcal{B}$ a σ-algebra of subsets, and μ a normalized positive measure on Ω. Then take the corresponding Hilbert space $\mathcal{H} = L^2(\Omega, \mathcal{B}, \mu)$, the closed subspace $\mathcal{K} = \{u \in \mathcal{H} : \operatorname{supp} u \subset \mathbb{R}^{d-1}\}$ (that is, vectors at fixed time $t = 0$), and the orthogonal projection $E_0 : \mathcal{H} \to \mathcal{K}$. Note that the Euclidean field $\phi(f)$ lives in $\mathcal{H}$, whereas the corresponding quantum field $\theta(f)$ will live in $\mathcal{K}$. For $0 \leqslant t < \infty$ and $u \in \mathcal{H}$, one defines the operator

$$P^t u := E_0 T_t u,$$

where T_t is the operator of translation by t. Then one shows that there is a unique positive self-adjoint operator (Hamiltonian) H on $\mathcal{H}$ such that $P^t = e^{-tH}$, $0 \leqslant t < \infty$.

Next, consider in $\mathcal{H}$ the scale built on the powers of H, that is, $\mathcal{H}_k = Q(H^k)$ with norm $\|u\|_k = \|(1 + H)^{k/2}u\|$, $-\infty < k < \infty$, $\mathcal{H}_\infty = \cap_k \mathcal{H}_k$, and $\mathcal{H}_{\overline{\infty}} = \cup_k \mathcal{H}_k$. For $f \in \mathcal{S}(\mathbb{R}^{d-1})$, define $\phi_0(f) = \phi(f \otimes \delta)$ (field at time 0). Then the main assumption is that $\phi(f)$ be a continuous operator in the scale, in the PIP-space sense, that is, there exist finite k, l such that, for all $f \in \mathcal{S}(\mathbb{R}^{d-1})$, $\phi_0(f)$ is a bounded operator from $\mathcal{H}_k$ to $\mathcal{H}_l$ and $f \mapsto \phi_0(f)$ is continuous.

For $f \in \mathcal{S}(\mathbb{R}^{d-1})$, define the time-zero quantum field as $\theta_0(f) = \phi_0(f)$. Since, for real τ, the operator $e^{i\tau H}$ leaves every $\mathcal{H}_k$ invariant (that is, it is totally regular) and is unitary on it, the field

$$\theta_\tau(f) := e^{i\tau H}\theta_0(f)e^{-i\tau H}$$

also maps $\mathcal{H}_k$ into $\mathcal{H}_l$. For $f \in \mathcal{S}(\mathbb{R}^d)$, let $f_\tau(\mathbf{x}) = f(\mathbf{x}, \tau)$ and define

$$\theta(f) = \int_{-\infty}^{\infty} \theta_\tau(f_\tau)d\tau = \int_{-\infty}^{\infty} e^{i\tau H}\phi_0(f_\tau)e^{-i\tau H}d\tau.$$

This a Wightman field and one shows that $\theta(f)$ maps $\mathcal{H}_\infty$ continuously into itself (thus it is a regular operator). Since the vacuum Ψ_0, corresponding to the function identically equal to 1 on Ω, belongs to $\mathcal{H}_\infty$, the vacuum expectation values $\langle \Psi_0 | \theta(f_1) \cdots \theta(f_n) \Psi_0 \rangle$ are well-defined and separately continuous in $f_1, \ldots, f_n \in \mathcal{S}(\mathbb{R}^d)$. Hence, the Schwartz kernel theorem implies the existence of a unique tempered distribution $\mathcal{W}^{(n)}$ on $\mathbb{R}^{nd}$ such that

$$\int \cdots \int \mathcal{W}^{(n)}(x_1, \ldots, x_n) f_1(x_1) \cdots f_n(x_n) \, dx_1 \cdots dx_n = \langle \Psi_0 | \theta(f_1) \cdots \theta(f_n) \Psi_0 \rangle.$$

Then it remains to prove that the Wightman distributions $\mathcal{W}^{(n)}$ satisfy all the properties required for a relativistic QFT.

Examples of such fields are the free Euclidean field of mass m (then one uses the operator $(-\Delta + m^2)$ instead of $(-\Delta + 1)$ in the definition of the Sobolev space $H^{-1}(\mathbb{R}^d)$) and elements of the Borchers class of the latter.

The construction described so far is fine for QFT in two dimensions. For higher dimensions, one should replace Euclidean fields over the Sobolev space $H^{-1}(\mathbb{R}^d)$ by Euclidean fields over $\mathcal{D}(\mathbb{R}^d)$ and perform the same construction, replacing the Hamiltonian H by the quadratic Casimir operator of the translation group, that is, the operator

$$K = \left(\sum_{j=1}^d X_j^* X_j \right)^{1/2},$$

where $X_1, \ldots, X_d$ are the generators of translations in the representation of the Euclidean group implied by the covariance condition. The rest is the same, a Euclidean field being defined as an operator, in the PIP-space sense, in the scale built on the powers of K.

7.3.4 Fields at a Point

As we said above, the unsmeared field $A(x)$ requires a mathematically rigorous definition. Actually four different possibilities have been proposed in the literature:

(f$_1$) $\widetilde{\mathcal{L}^\dagger(\mathcal{D})}$-valued fields, i.e., fields that take their values in the weak sequential completion of $\mathcal{L}^\dagger(\mathcal{D})$;

(f$_2$) $\mathcal{L}(\mathcal{D}, \mathcal{D}^\times)$-valued fields, i.e., fields that take their values in the quasi *-algebra $\mathcal{L}(\mathcal{D}, \mathcal{D}^\times)$;

(f$_3$) H-bounded fields, i.e., fields $A(x)$ satisfying the requirement that $R^k A(0) R^k$ be a bounded sesquilinear form for some $k \in \mathbb{N}$; here H is the energy operator and $R := (1 + H)^{-1}$.

(f$_4$) Op(V)-valued fields, i.e., fields that are operators in some PIP-space V.

However, if one takes $\mathcal{D} = \mathcal{D}^\infty(S)$ for some self-adjoint operator $S \geqslant 1$, then the first three definitions are equivalent, as results from the following easy lemma. This is the case, in particular, for $S = H$, which is a self-adjoint positive operator in virtue of the spectral condition.

Lemma 7.3.1. *Let* $\mathcal{D} = \mathcal{D}^\infty(S)$ *with* $S \geqslant 1$ *and* β *a sesquilinear form on* $\mathcal{D} \times \mathcal{D}$. *Then the following statements are equivalent:*

(i) *There exists a net* $(X_\alpha) \subset \mathcal{L}^\dagger(\mathcal{D})$ *such that*

$$\beta(\xi, \eta) = \lim_\alpha \langle \xi | X_\alpha \eta \rangle, \ \forall \xi, \eta \in \mathcal{D}.$$

(ii) $\beta = \beta_X$ *for some* $X \in \mathcal{L}(\mathcal{D}, \mathcal{D}^\times)$, *i.e.,*

$$\beta(\xi, \eta) = \langle \xi | X \eta \rangle, \ \forall \xi, \eta \in \mathcal{D}.$$

(iii) *For some* $k \in \mathbb{N}$, *the sesquilinear form* $S^{-k} \beta S^{-k}$ *is bounded on* $\mathcal{D}$, *where*

$$(S^{-k} \beta S^{-k})(\xi, \eta) \equiv \beta(S^{-k}\xi, S^{-k}\eta), \ \xi, \eta \in \mathcal{D}.$$

In particular, a $\mathcal{L}(\mathcal{D}, \mathcal{D}^\times)$-valued field is necessarily H-bounded:

Proposition 7.3.2. *Let* $x \mapsto A(x)$ *be an* $\mathcal{L}(\mathcal{D}, \mathcal{D}^\times)$-*valued field with* $\mathcal{D} = \mathcal{D}^\infty(H)$, *let* $R = (1 + H)^{-1}$; *then* $A(\cdot)$ *satisfies the following H-bound condition:*
    There exists $k \in \mathbb{N}$ *such that* $R^k A(x) R^k$ *is defined as a bounded sesquilinear form on* $\mathcal{D} \times \mathcal{D}$.

Suppose now $A(x)$ is a an $\mathcal{L}(\mathcal{D}, \mathcal{D}^\times)$-valued field. Does it generate a Wightman field upon smearing with a test function? As a first step, one shows that $A(f)$ belongs also to $\mathcal{L}(\mathcal{D}, \mathcal{D}^\times)$.

Proposition 7.3.3. *Let* $x \mapsto A(x)$ *be an* $\mathcal{L}(\mathcal{D}, \mathcal{D}^\times)$-*valued field with* $\mathcal{D} = \mathcal{D}^\infty(H)$. *Then the integral*

$$\langle \xi | A(f)\eta \rangle := \int_{\mathbb{R}^4} f(x)\langle \xi | A(x)\eta \rangle \, dx$$

converges for all $\xi, \eta \in \mathcal{D}$ *and defines for each* $f \in \mathcal{S}(\mathbb{R}^4)$ *an operator of* $\mathcal{L}(\mathcal{D}, \mathcal{D}^\times)$. *Moreover, for every* $\xi, \eta \in \mathcal{D}$, *the map* $f \mapsto \langle \xi | A(f)\eta \rangle$ *is a tempered distribution.*

Proof. First we notice that the translation invariance contained in (7.38) implies the continuity of the map $x \mapsto \langle \xi | A(x)\eta \rangle$, for any pair $\xi, \eta \in \mathcal{D}$. Then joint continuity of $A(0)$ as a sesquilinear form implies that there exist $\gamma > 0$ and $k \in \mathbb{N}$ such that

$$|\langle\xi|A(f)\eta\rangle| = \left|\int_{\mathbb{R}^4} f(x)\langle U(-x,I)\xi|A(0)U(-x,I)\eta\rangle\, dx\right|$$

$$\leqslant \gamma \int_{\mathbb{R}^4} |f(x)|\, \|R^{-k}U(-x,I)\xi\|\, \|R^{-k}U(-x,I)\eta\|\, dx$$

$$= \gamma\, \|R^{-k}\xi\|\, \|R^{-k}\eta\| \int_{\mathbb{R}^4} |f(x)|\, dx.$$

These estimates imply all the statements. ∎

Of course, one needs more for $A(f)$ to be a Wightman field, namely, one requires that $A(f) \in \mathcal{L}^\dagger(\mathcal{D})$, $\forall f \in \mathcal{S}(\mathbb{R}^4)$.

Proposition 7.3.4. *Let $A(x)$ be an $\mathfrak{L}(\mathcal{D},\mathcal{D}^\times)$-valued field. Let k be the minimal integer such that $R^k A(0)R^k$ is bounded. Then $R^{k-l} A(f)R^{k+l}$ is bounded for any $l \in \mathbb{N}$. Moreover $A(f) \in \mathcal{L}^\dagger(\mathcal{D})$, $\forall f \in \mathcal{S}(\mathbb{R}^4)$.*

Proof. The boundedness of $R^k A(0)R^k$ implies, as seen in Proposition 7.3.3, that also $R^k A(f)R^k$ is bounded for any $f \in \mathcal{S}(\mathbb{R}^4)$. By repeated use of the equality

$$RA(f) - A(f)R = RA(\partial_t f)R, \quad f \in \mathcal{S}(\mathbb{R}^4),$$

we get the statement for any integer l.

The boundedness of $R^{k-l} A(f)R^{k+l}$, for any $l \in \mathbb{N}$, easily implies that $A(f)$ maps $\mathcal{D}^\infty(H)$ into itself. Moreover $A(\overline{f}) = A(f)^\dagger$. Therefore $A(f) \in \mathcal{L}^\dagger(\mathcal{D})$, $\forall f \in \mathcal{S}(\mathbb{R}^4)$. ∎

As a matter of fact, the H-bound condition is not only sufficient, but also necessary.

Proposition 7.3.5. *Let $\mathcal{S}(\mathbb{R}^4) \ni f \mapsto A(f) \in \mathcal{L}^\dagger(\mathcal{D})$, with $\mathcal{D} = \mathcal{D}^\infty(H)$, be a Wightman field. Then there exists a $k \in \mathbb{N}$, independent of f, such that $A(f)R^k$ is a bounded operator.*

For a proof of this proposition, we refer to the original paper quoted in the Notes.

Conversely, a Wightman field $A(f)$ defines an $\mathfrak{L}(\mathcal{D},\mathcal{D}^\times)$-valued field $A(x)$ under a H-bound condition.

Proposition 7.3.6. *Let $\mathcal{D} = \mathcal{D}^\infty(H)$ and $f \mapsto A(f)$ a Hermitian (scalar) Wightman field. Then, in order that there exists an $\mathfrak{L}(\mathcal{D},\mathcal{D}^\times)$-valued field $A(x)$ such that, for every $f \in \mathcal{S}(\mathbb{R}^4)$,*

$$\langle\xi|A(f)\eta\rangle = \int_{\mathbb{R}^4} f(x)\langle\xi|A(x)\eta\rangle\, dx, \quad \forall\, \xi, \eta \in \mathcal{D},$$

it is necessary and sufficient that the sesquilinear form $R^k A(0)R^k$ be bounded for some k, where $A(0) = U(-x,I)A(x)U(x,I)$.

Proof. Necessity has been proved above. As for sufficiency, let $\{f_n\}$ be a sequence of functions in $\mathcal{S}(\mathbb{R}^4)$ converging to the Dirac measure δ in the weak topology of $\mathcal{S}^\times(\mathbb{R}^4)$. For any $\xi, \eta \in \mathcal{D}$, put

$$\beta(\xi, \eta) = \lim_{n \to \infty} \langle \xi | A(f_n) \eta \rangle.$$

With help of Proposition 7.3.5, it is easily seen that β is jointly continuous. We define $A(0)$ as the corresponding operator of $\mathfrak{L}(\mathcal{D}, \mathcal{D}^\times)$. The other properties of a field are easily proved.   ∎

This concludes our analysis of cases (f_1)-(f_3). As for the genuine PIP-space version (f_4), it takes place in the generalized Fock space $V = \Gamma(V^{(1)})$, where $V^{(1)} = L^1_{\mathrm{loc}}(X, d\mu)$, as defined in Section 1.1.3, Example (iv). Creation and annihilation operators $a^\dagger(f), a(f)$ on V have been defined abstractly in Section 3.1.4 (vii). Now we consider the case where X is the positive hyperboloid of mass m in Minkowski space, and $d\mu = d^3k/k_0$ the invariant measure on X.

First we consider Poincaré invariance. On the one-particle Hilbert space $\mathcal{H}_0$, we have the usual unitary representation

$$[U_{00}(a, \Lambda) f_0](k) = e^{-ik \cdot a} f_0(\Lambda^{-1} k), \quad (a, \Lambda) \in \mathcal{P}(1, 3).$$

Then the representation U_{00} extends to a unitary representation U in the one-particle PIP-space $V^{(1)}$, as already mentioned in Example 3.3.19 (2). Note that the operator $U(a, \Lambda)$ is a unitary isomorphism, but it is not regular. Next we extend the representation U to the Fock space $V = \Gamma(V^{(1)})$. Define $\Lambda^{(n)}$ and $U^{(n)}$ in the obvious fashion on $V^{(1)}$. Let $\underline{U}(a, \Lambda) := \bigoplus_{n=0}^\infty U^{(n)}(a, \Lambda)$. Then clearly $\underline{U}$ is a unitary representation of $\mathcal{P}(1, 3)$ in $\Gamma(V^{(1)})$.

Now we come back to the free field $A(f) = a^\dagger(f) + a(\bar{f})$. If we take in particular $f = f^{(z)}$, where $f^{(z)}(k) = (2\pi)^{-3/2} e^{ikz}$, $k \in X$, $z \in \mathbb{C}^4$, then the operator $A(z) := a^\dagger(f^{(z)}) + a(\overline{f^{(-z)}})$ is the (nonsmeared) free field at the (real or complex) point z. From here on, one can study Wick products, analyticity in z, etc. In particular, the asymmetry in the domains of $a^\dagger(f)$ and $a(f)$ shows that Wick-ordered products are better behaved (i.e., have a larger $\mathrm{j}(\cdot)$) than other products. For further details, we refer to the original paper.

7.4 Representations of Lie Groups

Let us return to the problem stated in Section 3.3.4. We start with a strongly continuous unitary representation U_{00} of a Lie group G in a Hilbert space $\mathcal{H}_0$ and seek to build a PIP-space V_I, with $\mathcal{H}_0$ its central Hilbert space, such that U_{00} extends to a unitary representation U into V_I.

The solution of this problem is well-known from the theory of analytic vectors. Let $\overline{\Delta}$ be the closure of the Nelson operator $\Delta := \sum_{j=1}^n X_j^2$, where

$\{X_j, j = 1, \ldots, n\}$ are the representatives under U_{00} of the elements of a basis of the Lie algebra $\mathfrak{g}$ of G. Δ is essentially self-adjoint on the Gårding domain $\mathcal{H}_0^G$ (Section 7.1.2), $\overline{\Delta}$ is self-adjoint and $\overline{\Delta} \geqslant 0$. Then define $V_I := \{\mathcal{H}_n, n \in \mathbb{Z}\}$ as the canonical scale of Hilbert spaces generated by the operator $\overline{\Delta} + 1$ (see Sections I.1 and 5.2.1). Thus, for each $n \in \mathbb{N}$, one has $\mathcal{H}_n = D(\overline{\Delta}^n)$ with norm $\|\psi\|_n = \|(\overline{\Delta} + 1)^n \psi\|$ and one gets the usual triplet upon completing the Hilbert scale V_I:

$$V^\# := \mathcal{D}^\infty(\overline{\Delta}) \hookrightarrow \mathcal{H}_0 \hookrightarrow V := \mathcal{D}_{\overline{\infty}}(\overline{\Delta}). \tag{7.42}$$

First, one has $\mathcal{D}^\infty(\overline{\Delta}) = \bigcap_{n=1}^\infty D(\overline{\Delta}^n) = \mathcal{H}_0^\infty$, the space of C^∞-vectors of U_{00}.

Next, for every $g \in G$, $U_{00}(g)$ leaves each $\mathcal{H}_n, n \in \mathbb{N}$, invariant and its restriction $U_{nn}(g) : \mathcal{H}_n \to \mathcal{H}_n$ is continuous; thus it can be transposed to a continuous map $U_{nn}(g^{-1}) : \mathcal{H}_{\overline{n}} \to \mathcal{H}_{\overline{n}}$. It follows that U_{00} extends to a unitary representation U in the LHS V_I. Corresponding to the triplet (7.42), we have three representations of G, namely U_{00}, its restriction $U_{\infty\infty}$ and the dual $U_{\overline{\infty}\overline{\infty}}$ of the latter, which is an extension of the first two. All three are continuous. Moreover, if one of the three is topologically irreducible (i.e., there is no proper invariant closed subspace), so are the other two.

In addition to the representations of the group G, the scale V_I is the natural tool for studying the properties of the operators representing elements of the Lie algebra $\mathfrak{g}$ or the universal enveloping algebra $\mathfrak{U}(\mathfrak{g})$ of G. For every element $x \in \mathfrak{g}$ or $L \in \mathfrak{U}(\mathfrak{g})$, the representative $U(x)$, resp. $U(L)$, originally defined on $\mathcal{H}_0^G$, extends to a regular operator on V_I. These regular operators have in general no $\{0, 0\}$-representative, since x and L are represented in $\mathcal{H}_0$ by unbounded operators. As in the case of the group G, one gets three *-representations of the enveloping algebra $\mathfrak{U}(\mathfrak{g})$, and in particular of the Lie algebra $\mathfrak{g}$, in the three spaces of the triplet (7.42). Namely, one has, for every $L, L_1, L_2 \in \mathfrak{U}(\mathfrak{g})$,

$$U(L_1)U(L_2) = U(L_1 L_2),$$
$$U(L^+) = U(L)^\times,$$

where $L \leftrightarrow L^+$ is the involution on $\mathfrak{U}(\mathfrak{g})$. These representations have the same irreducibility properties as the corresponding ones of the group.

The space $\mathcal{H}_0^\infty = \mathcal{D}^\infty(\overline{\Delta})$ is a natural domain of essential self-adjointness for several classes of operators. More generally, given an element $L \in \mathfrak{U}(\mathfrak{g})$, one may ask whether the restriction of $U(L)$ to $\mathcal{H}_0^\infty$ satisfies the relation $\overline{U(L^+)} = U(L)^*$ (Hilbert space adjoint). Besides the generators $\{X_j, j = 1, \ldots, n\}$ of $\mathfrak{g}$, the answer is positive in two cases:

(i) L is elliptic (i.e. it is elliptic as a partial differential operator on G), for instance the Nelson operator Δ of G or of any Lie subgroup H of G;

(ii) L^+L commutes with a symmetric elliptic element $M = M^+ \in \mathfrak{U}(\mathfrak{g})$, for instance Δ itself; this applies in particular to all symmetric central elements of $U(\mathfrak{U}(\mathfrak{g}))$ or $U(\mathfrak{U}(\mathfrak{h}))$ (the so-called Casimir operators).

Therefore, if L^+L commutes with Δ, $U(L^+L)$ is essentially self-adjoint on $\mathcal{H}_0^\infty$ and $\overline{U(L^+L)}$ commutes strongly with the operator $A := \overline{\Delta} + 1$, i.e., their spectral projections commute. Thus, $\overline{U(L^+L)}$ is affiliated to the von Neumann algebra $\mathfrak{A}' = \mathfrak{C}$ (see Proposition 5.5.15), and similarly for $\overline{U(L)}$ when L is symmetric ($L = L^+$). In other words, the Nelson scale V_I has many totally orthogonal projections, and thus orthocomplemented subspaces, namely the spectral projections of all those operators $\overline{U(L^+L)}$, resp. $\overline{U(L)}$.

As stated in Section 7.1.1 (i), the space $\mathcal{H}_0^\infty$ is nuclear if, for each n, there is an $m > n$ such that the embedding $\mathcal{H}_m \to \mathcal{H}_n$ is a Hilbert-Schmidt operator. This the key to spectral analysis, with appropriate operators having generalized eigenvectors in the large space $\mathcal{D}_{\overline{\infty}}(\overline{\Delta})$. The corresponding PIP-space version has been described above.

In addition to $\mathcal{H}_0^\infty$, an essential tool in the study of self-adjointness is the space $\mathcal{H}_0^\omega$ of analytic vectors for the representation U_{00}, defined in Section 7.1.2. Now it is easily seen that $\mathcal{H}_0^\omega$ equals the analyticity space $\mathcal{S}_{\mathcal{H}_0, \overline{\Delta}^{1/2}}$ (here we see the justification of the terminology). Then one can show that the operators U_{00} map $\mathcal{S}_{\mathcal{H}_0, \overline{\Delta}^{1/2}}$ continuously into itself, and so do the generators X_j of the Lie algebra $\mathfrak{g}$. Many more results on this topic are available in the literature, we refer the interested reader to the papers and textbooks quoted in the Notes.

Notes for Chapter 7

Section 7.1. The idea to formulate the Dirac formalism in a rigorous fashion was implemented independently by Böhm [47], Roberts [171, 172], and Antoine [8, 9]. See also the monograph of Böhm–Gadella [BG89], Prugovečki [169, 170], and Melsheimer [149]. The notion of 'labeled observables' is due to Roberts [171, 172]. The RHS formulation of quantum mechanics is developed systematically, both for the nonrelativistic and the relativistic cases, in the textbook of Bogolubov–Logunov–Todorov [BLT75, Chap.4]. It forms also the backbone of several textbooks on quantum mechanics, such as those of Ballentine [Bal90], Bohm [Boh93] or Dubin–Hennings [DH90].

- As already pointed out in Section 4.5, the idea of formulating a rigorous version of the Dirac formalism in terms of analyticity/trajectory spaces is due to van Eijndhoven–de Graaf [Eij83, EG85, EG86]. Indeed the mathematical machinery was explicitly designed for that purpose.
- For *-algebras of unbounded operators, see Powers [168], Lassner [138], or Schmüdgen's monograph [Sch90]. For the GNS construction and the notions of *-automorphism groups and derivations, also in the case of

(partial) algebras of unbounded operators, one may consult the monograph of Antoine–Inoue–Trapani [AIT02].

- For the nuclear spectral theorem of Gel'fand and Maurin, see Gel'fand's monograph [GS64]. The notion of tight rigging is due to Fredricks [99]. Sufficient conditions for a rigging to be tight have been given also by Napiórkowski [152] and Babbitt [38]. The notion of eigenoperator is due to Foiaş [95].

- The need to consider self-adjoint Hamiltonians with complex generalized eigenvalues in a RHS has been repeatedly advocated by Böhm and Gadella [48, 49, 102]. This is the basis of the so-called time-asymmetric quantum mechanics developed by these authors within a RHS context.

- The alternative formulation of the RHS approach using simultaneously $\Phi^\times$ and Φ' is due to de la Madrid [72], with the avowed goal of staying as close as possible to the Dirac formalism. Note the similarity with the analyticity/trajectory space formulation of van Eijndhoven–de Graaf [Eij83, EG85, EG86].

- The description of symmetries in quantum mechanics was pioneered by Wigner [187]. The extension to a symmetry (Lie) group is due to Bargmann [41]. An extensive analysis may be found in the monograph of Fonda–Ghirardi [FG70]. The generalization of this approach to the RHS approach is due to Antoine [8, 9]. For the PIP-space version of it, see [15]. The continuity property for the contragredient representation is a result of Bruhat [56]. For the SO(4) invariance of the Coulomb problem, see for instance Bohm [Boh93, Sec.VI.3] or Fonda–Ghirardi [FG70, §4.2.2].

- The whole discussion of Section 7.1.3 on singular interactions in a PIP-space language is taken from the papers of Grossmann–Hoegh-Krohn–Mebkhout [118,119]. See also the textbooks of Albeverio et al. [AGHH88, App.G] and Albeverio–Kurasov [AK00, Chap.1].

Section 7.2. We refer to Reed–Simon's treatise [RS79] for further details on the phase space analysis and specific applications in quantum scattering theory.

- An interesting generalization of this construction was used by Hunziker [126] for a Schrödinger Hamiltonian $H = p^2 + V(x)$.

- The Hermite representation of S was introduced by Simon [179]. Hilbert spaces of type S, introduced by Grossmann [115], have been described in Section 5.4.3.

- The scattering theory of Weinberg-van Winter may be found in [188,189]. The complex scaling method (CSM) wad introduced by Aguilar–Balslev–Combes[2, 39].

- The formulation of the WVW scattering theory in terms of the LHS language is due to Gollier [Gol82]. The definition of dilation analytic vectors given in (7.33) is due to Balslev–Grossmann–Paul [40]. The essential result is that the WVW formalism is a particular case of CSM, and this is due to Klein [Kle87].

- For the notion of S-matrix, see any textbook on quantum scattering theory, for example, that of Goldberger–Watson [GW64]. Gamow vectors were introduced long ago by Gamow [103]. The extension of Hardy space methods to relativistic resonances and decaying states is due to Böhm–Kaldass–Wickramasekara [50].

Section 7.3. For the Wightman version of axiomatic quantum field theory, see Streater–Wightman [SW64]. A complete RHS formulation of QFT may be found in the textbook of Bogolubov–Logunov–Todorov [BLT75]. The second quantization functor has been introduced by Nelson [155,156]. The Borchers or field algebra has been introduced in [51, 52]. An essentially equivalent construction is given by Kristensen–Mejlbo–Thue Poulsen [133].

- Nelson's formulation of Euclidean field theory may be found in [155]. The example of the free Euclidean field is given by Nelson [156] and that of the Borchers class of the free Euclidean field is due to Karwowski [129].
- The results presented here mostly come from Epifanio–Trapani [75]. The idea of considering a point-like field as a map from the Minkowski space-time into the weak-sequential completion $\overline{\mathcal{L}^{\dagger}(\mathcal{D})}$ of $\mathcal{L}^{\dagger}(\mathcal{D})$ was proposed by Ascoli–Epifanio–Restivo in [34, 35]. It was further explored by Shabani in [177]. Quasi *-algebras valued fields were considered in Epifanio–Trapani [75]. The regularity condition (high energy bound) was proposed by Fredenhagen–Hertel in [98].
- The formulation of QFT in terms of operators on a PIP-space is due to Grossmann [117]. This was actually one of the motivations for introducing nested Hilbert spaces.

Section 7.4. For the theory of C^{∞} and analytic vectors, we refer to the work of Nelson [153], Nelson–Stinespring [158], Goodman [108,109] and Nagel [150]. A systematic treatment may be found in the monograph of Barut - Rączka [BR77, Chap.11]. The connection between analytic vectors and the analyticity space $\mathcal{S}_{\mathcal{H}_0, \overline{\Delta}^{1/2}}$ may be found in the work of van Eijndhoven-de Graaf [Eij83, EG85, EG86], together with several concrete examples (the Heisenberg group, the Euclidean group E(2), the rotation group SO(3)).

Chapter 8
PIP-Spaces and Signal Processing

Contemporary signal processing makes an extensive use of function spaces, always with the aim of getting a precise control on smoothness and decay properties of functions. In this chapter, we will discuss several classes of such function spaces that have found interesting applications, namely, mixed-norm spaces, amalgam spaces, modulation spaces, or Besov spaces. It turns out that all those spaces come in families indexed by one or more parameters, that specify, for instance, the local behavior or the asymptotic properties. In general, a single space, taken alone, does not have an intrinsic meaning, it is the family as a whole that does, which brings us to the very topic of this volume. In addition, several rigged Hilbert spaces (also called Gel'fand triplets) have a particular interest, notably the one generated by the so-called Feichtinger algebra. This too deserves a detailed discussion in the sequel.

Note that, unlike the previous chapter, we will treat each class with the corresponding applications. Also, we will merely state the relevant results/propositions, referring the interested reader to the vast literature quoted in the Notes.

8.1 Mixed-Norm Lebesgue Spaces

The first type of function space is the family of mixed-norm Lebesgue spaces, already described briefly in Section 4.4, Example 4.4.5. For the commodity of the reader, we repeat the general definition.

Let (X, μ) and (Y, ν) be two σ-finite measure spaces and $1 \leqslant p, q \leqslant \infty$. Then, a function $f(x, y)$ measurable on the product space $X \times Y$ is said to belong to $L^{(p,q)}(X \times Y)$ if the number obtained by taking successively the p-norm in x and the q-norm in y, in that order, is finite. If $p, q < \infty$, the norm is given in (4.19). The analogous norm for p or $q = \infty$ is obvious.

The case $X = Y = \mathbb{R}^d$ with Lebesgue measure is the important one for signal processing. More generally, one can add a weight function m and obtain the spaces $L^{p,q}_m(\mathbb{R}^d)$ (we switch to a notation more suitable for the applications):

J.-P. Antoine and C. Trapani, *Partial Inner Product Spaces:*
Theory and Applications, Lecture Notes in Mathematics 1986,
DOI 10.1007/978-3-642-05136-4_8, © Springer-Verlag Berlin Heidelberg 2009

$$L_m^{p,q}(\mathbb{R}^2) = \{f \text{ Lebesgue measurable on } \mathbb{R}^{2d} : \|f\|_{L_m^{p,q}} < \infty\}, \ 1 \leqslant p, q < \infty,$$

where

$$\|f\|_{L_m^{p,q}} := \left(\int_{\mathbb{R}^d} \left(\int_{\mathbb{R}^d} |f(x,\omega)|^p \, m(x,\omega)^p dx \right)^{q/p} d\omega \right)^{1/q}.$$

Here m is a weight function, that is, a non-negative locally integrable function on $\mathbb{R}^{2d}$. In addition, m is assumed to be v-moderate, i.e., $m(z_1 + z_2) \leqslant v(z_1)m(z_2)$, for all $z_1, z_2 \in \mathbb{R}^{2d}$, with v a submultiplicative weight function, that is, $v(z_1 + z_2) \leqslant v(z_1)v(z_2)$, for all $z_1, z_2 \in \mathbb{R}^{2d}$. The typical weights are of polynomial growth: $v_s(z) = (1 + |z|)^s$, $s \geqslant 0$.

Once again, things simplify when $p = q$: $L_m^{p,p}(\mathbb{R}^{2d}) = L_m^p(\mathbb{R}^{2d})$, a weighted L^p space.

The spaces $L_m^{p,q}(\mathbb{R}^{2d})$ have the properties inherited from the general case $L^{(p,q)}$, namely:

(i) *Completeness:* $L_m^{p,q}(\mathbb{R}^{2d})$ is a Banach space for the norm $\|\cdot\|_{L_m^{p,q}}$.

(ii) *Hölder's inequality:* If $f \in L_m^{p,q}(\mathbb{R}^{2d})$ and $h \in L_{1/m}^{\overline{p},\overline{q}}(\mathbb{R}^{2d})$, with $1/p + 1/\overline{p} = 1$, $1/q + 1/\overline{q} = 1$, then $f\,h \in L^1(\mathbb{R}^{2d})$ and

$$\left| \int_{\mathbb{R}^{2d}} \overline{f(z)}\, h(z)\, dz \right| \leqslant \|f\|_{L_m^{p,q}} \|h\|_{L_{1/m}^{\overline{p},\overline{q}}}.$$

(iii) *Duality:* If $p, q < \infty$, then $(L_m^{p,q})^\times = L_{1/m}^{\overline{p},\overline{q}}$.

(iv) *Translation invariance:* $L_m^{p,q}(\mathbb{R}^{2d})$ is invariant under translations $(T_z g)(w) = g(w - z)$, $z, w \in \mathbb{R}^{2d}$, if, and only if, m is v-moderate; then one has

$$\|T_z f\|_{L_m^{p,q}} \leqslant Cv(z)\|f\|_{L_m^{p,q}}, \text{ for all } f \in L_m^{p,q}.$$

(v) *Convolution:* If m is v-moderate, $f \in L_v^1(\mathbb{R}^{2d})$, and $g \in L_m^{p,q}(\mathbb{R}^{2d})$, then

$$\|f * g\|_{L_m^{p,q}} \leqslant C\|f\|_{L_v^1} \|g\|_{L_m^{p,q}},$$

that is, $L_v^1 * L_m^{p,q} \subseteq L_m^{p,q}$. This property generalizes the usual one of L^p spaces, $L^1 * L^p \subseteq L^p$.

Concerning lattice properties of the family of $L_m^{p,q}$ spaces, we cannot expect more than for the L^p spaces. Two $L_m^{p,q}$ spaces are never comparable, even for the same weight m, so one has to take the lattice generated by intersection and duality. Nevertheless, properties (iv) and (v) mean that translation and convolution by L_v^1 are totally regular operators in whatever PIP-space is constructed out of the $L_m^{p,q}$ spaces.

A different type of mixed-norm spaces is obtained if one takes $X = Y = \mathbb{Z}^d$, with the counting measure. Thus one gets the space $\ell_m^{p,q}(\mathbb{Z}^{2d})$, which consists of all sequences $a = (a_{kn})$, $k, n \in \mathbb{Z}^d$, for which the following norm is finite:

$$\|a\|_{\ell_m^{p,q}} := \left(\sum_{n \in \mathbb{Z}^d} \left(\sum_{k \in \mathbb{Z}^d} |a_{kn}|^p \, m(k,n)^p \right)^{q/p} \right)^{1/q} .$$

Contrary to the continuous case, here we do have inclusion relations:

Lemma 8.1.1. *If $p_1 \leqslant p_2, q_1 \leqslant q_2$ and $m_2 \leqslant Cm_1$, then $\ell_{m_1}^{p_1,q_1} \subseteq \ell_{m_2}^{p_2,q_2}$.*

Proof. First take $m_1 = m_2 \equiv 1$. Then the inclusion results from the obvious inequality $\|a\|_{\ell^{p_2,q_2}} \leqslant \|a\|_{\ell^{p_1,q_1}}$. Then, if the weights satisfy $m_2 \leqslant Cm_1$, the result follows from the relations

$$\|a\|_{\ell_{m_2}^{p_2,q_2}} = \|am_2\|_{\ell^{p_2,q_2}} \leqslant \|am_2\|_{\ell^{p_1,q_1}} \leqslant C\|am_1\|_{\ell^{p_1,q_1}} = C\|a\|_{\ell_{m_1}^{p_1,q_1}} . \qquad \blacksquare$$

As for the lattice properties, we have (for a fixed weight m)

$$\ell_m^{\min(p_1,p_2),\min(q_1,q_2)} \subset \ell_m^{p_1,q_1} \cap \ell_m^{p_2,q_2},$$

$$\ell_m^{p_1,q_1} + \ell_m^{p_2,q_2} \subset \ell_m^{\max(p_1,p_2),\max(q_1,q_2)},$$

but we conjecture that the equality is not obtained in general. Thus the set of spaces $\ell_m^{p,q}$, $1 \leqslant p, q \leqslant \infty$ is *not* a lattice, and one has to consider again the lattice it generates. For fixed m, however, one gets chains by varying either q or p, but not both.

Discrete mixed-norm spaces have been used extensively in functional analysis and signal processing. For instance, they are key to the proof that certain operators are bounded between two given function spaces, such as modulation spaces (see Section 8.3.1) or ℓ^p spaces. For instance, if $\{\psi_{j,k}, (j,k) \in I\}$, is a wavelet basis or frame, mixed-norm spaces may be used to prove boundedness of the analysis operator $D : L^2(\mathbb{R}) \to \ell^2(I)$ given by $Df = \left(\langle \psi_{j,k} | f \rangle \right)_{(j,k) \in I}$.

In general, a mixed-norm space will prove useful whenever one has a signal consisting of sequences labeled by two indices that play different roles. An obvious example is time-frequency or time-scale analysis: a Gabor or wavelet basis (or frame) is written as $\{\psi_{j,k}, j, k \in \mathbb{Z}\}$, where j indexes the scale or frequency and k the time. More generally, this applies whenever signals are expanded with respect to a dictionary with two indices. An example is provided by multichannel signals, where a first index labels dictionary elements and a second one labels channels. A variant is the case where indices are hierarchized. Coefficients are split into independent groups and coefficients within the same group are dependent. Thus one index labels groups and the other one elements within a group, and of course, the two not interchangeable. Interesting applications of this procedure are found in the papers quoted in the Notes.

8.2 Amalgam Spaces

A situation intermediate between the mixed-norm spaces (for $m \equiv 1$) $L^{p,q}(\mathbb{R}^{2d})$ and the spaces $\ell^{p,q}(\mathbb{Z}^{2d})$ is that of the so-called *amalgam spaces*. They were introduced specifically to overcome the inability of the L^p norms to distinguish between the local and the global behavior of functions.

The simplest ones are the spaces $W(L^p, \ell^q)$ (sometimes denoted (L^p, ℓ^q) or $W(L^p, L^q)$) consisting of functions on $\mathbb{R}$ which are locally in L^p and have ℓ^q behavior at infinity, in the sense that the L^p norms over the intervals $(n, n+1)$ form an ℓ^q sequence. For $1 \leqslant p, q < \infty$, the norm

$$\|f\|_{p,q} = \left(\sum_{n=-\infty}^{\infty} \left[\int_n^{n+1} |f(x)|^p dx \right]^{q/p} \right)^{1/q}$$

makes $W(L^p, \ell^q)$ into a Banach space. The same is true for the obvious extensions to p and/or q equal to ∞. Notice that $W(L^p, \ell^p) = L^p$. Also it is easy to see that these spaces $W(L^p, \ell^q)$ are a particular case of the mixed-norm spaces $L^{(p,q)}(X \times Y)$. Taking indeed $X = [0,1]$ with Lebesgue measure, $Y = \mathbb{Z}$ with the counting measure, and $g(x, n) = f(x+n)$, one gets

$$\|g\|_{(p,q)} = \left(\sum_{n=-\infty}^{\infty} \left[\int_0^1 |f(x+n)|^p dx \right]^{q/p} \right)^{1/q} = \|f\|_{p,q}.$$

Actually, an equivalent definition of the space $W(L^p, \ell^q)$ is obtained if one replaces the covering of $\mathbb{R}$ given by $\cup_{n=-\infty}^{\infty}[n, n+1] = \mathbb{R}$ by a so-called *bounded uniform partition of unity (BUPU)*. This means a family of functions $(\psi_i)_{i \in I}$, with I a countable index set, such that:

(1) $0 \leqslant \psi_i(x) \leqslant 1$, for all $i \in I$;
(2) There is a compact neighborhood W of 0 and a countable set of points $(x_i)_{i \in I}$ such that supp $\psi_i \subseteq x_i + W$ for all $i \in I$;
(3) Every point $x \in \mathbb{R}$ belongs to a finite number of subsets $x_k + W$;
(4) $\sum_{i \in I} \psi_i(x) \equiv 1$, so that $\bigcup_{i \in I}(x_i + W) = \mathbb{R}$.

For instance, in $\mathbb{R}$, one may take $W = [0,1]$, $x_i = i \in \mathbb{Z}$, and $\psi_i = T_i \chi_W = \chi_{i+W}$ (which gives back the original partition)[1] or $W_\alpha := [0, \alpha]$ and $\psi_i = T_{\alpha i} \chi_{W_\alpha}$. Alternatively, one may replace χ_W by a nicer function with compact support, such as a triangular ('tent') function or a spline function, satisfying condition (4). Similar considerations apply, of course, to $\mathbb{R}^d$.

The corresponding norm then reads as

$$\|f\|'_{p,q} := \left(\sum_{i \in I} \left[\int_{\mathbb{R}} |f(x)\psi_i(x)|^p dx \right]^{q/p} \right)^{1/q},$$

---

[1] As usual, χ_W is the characteristic function of the set W.

and it is equivalent to the original one $\| \cdot \|_{p,q}$, which we denote henceforth by $\| \cdot \|'_{p,q} \asymp \| \cdot \|_{p,q}$. It should be noted that the choice of a particular BUPU is irrelevant: two different ones will give the same space $W(L^p, \ell^q)$, with equivalent norms.

From condition (4) in the definition of a BUPU, we see that any function $f \in W(L^p, \ell^q)$ may be written as $f = \sum_{i \in I} f \psi_i$, where each component $f \psi_i$ has a compact support centered on x_i. This is the key step in the *localization* of functions or distributions, a fundamental tool in signal processing.

The spaces $W(L^p, \ell^q)$ obey the following (immediate) inclusion relations, with all embeddings continuous:

- If $q_1 \leqslant q_2$, then $W(L^p, \ell^{q_1}) \hookrightarrow W(L^p, \ell^{q_2})$.
- If $p_1 \leqslant p_2$, then $W(L^{p_2}, \ell^q) \hookrightarrow W(L^{p_1}, \ell^q)$.

From this it follows that the smallest space is $W(L^\infty, \ell^1)$ and the largest one is $W(L^1, \ell^\infty)$, and therefore

- If $p \geqslant q$, then $W(L^p, \ell^q) \subset L^p \cap L^q \subset L^s, \forall \, q < s < p$.
- If $p \leqslant q$, then $W(L^p, \ell^q) \supset L^p + L^q$.

Once again, Hölder's inequality is satisfied. Whenever $f \in W(L^p, \ell^q)$ and $g \in W(L^{\overline{p}}, \ell^{\overline{q}})$, with $1/p + 1/\overline{p} = 1$, $1/q + 1/\overline{q} = 1$, then $fg \in L^1$ and one has

$$\|fg\|_1 \leqslant \|f\|_{p,q} \, \|g\|_{\overline{p},\overline{q}}.$$

Therefore, one has the expected duality relation:

$$W(L^p, \ell^q)^\times = W(L^{\overline{p}}, \ell^{\overline{q}}), \quad \text{for } 1 \leqslant q, p < \infty.$$

The interesting fact is that, for $1 \leqslant p, q \leqslant \infty$, the set $\mathcal{J}$ of all amalgam spaces $\{W(L^p, \ell^q)\}$ may be represented by the points (p, q) of the *same* unit square J as in the example of the L^p spaces (Section 4.1.2), with the *same* order structure. However, $\mathcal{J}$ is not a lattice with respect to the order (4.3). One has indeed

$$W(L^p, \ell^q) \wedge W(L^{p'}, \ell^{q'}) \supset W(L^{p \vee p'}, \ell^{q \wedge q'}),$$
$$W(L^p, \ell^q) \vee W(L^{p'}, \ell^{q'}) \subset W(L^{p \wedge p'}, \ell^{q \vee q'}),$$

where again $\wedge$ means intersection with projective norm and $\vee$ means vector sum with inductive norm, but equality is not obtained. Thus, as in the previous case, one gets chains by varying either p or q, but not both.

Standard properties hold here too.

(i) *Convolution:* If $f \in W(L^{p_1}, \ell^{p_2})$ and $g \in W(L^{q_1}, \ell^{q_2})$, where $1/p_i + 1/q_i \geqslant 1$, $i = 1, 2$, then $f * g \in W(L^{r_1}, \ell^{r_2})$, where $1/r_i = 1/p_i + 1/q_i - 1$, and

$$\|f * g\|_{r_1,r_2} \leqslant C \|f\|_{p_1,p_2} \, \|g\|_{q_1,q_2}.$$

(ii) *Fourier transform:* If $f \in W(L^p, \ell^q)$, with $1 \leqslant p, q \leqslant 2$, then $\mathcal{F}f \in W(L^{\bar{q}}, \ell^{\bar{p}})$ and there is a constant $C_{p,q}$ such that

$$\|\mathcal{F}f\|_{\bar{q},\bar{p}} \leqslant C_{p,q} \|f\|_{p,q}.$$

(iii) *Multiplication:* Let $1 \leqslant p_i, q_i \leqslant \infty$ and $1/r_i = \max(0, 1/q_i - 1/p_i)$, $i = 1, 2,$. Then $fg \in W(L^{q_1}, \ell^{q_2})$ whenever $g \in W(L^{p_1}, \ell^{p_2})$ if, and only if, $f \in W(L^{r_1}, \ell^{r_2})$.

(iv) *Translation invariance:* $W(L^p, \ell^q)$ is invariant under translations $T_x, x \in \mathbb{R}^d$, and one has

$$\|T_x F\|_{p,q} \leqslant C_{pq} \|F\|_{p,q}, \quad \text{where } C_{pq} = \max\{2^{1/p-1/q}, 2^{1/q-1/p}\}.$$

Thus, here again, translation is a totally regular operator in any PIP-space constructed out of the $W(L^p, \ell^q)$ spaces.

The spaces $W(L^p, \ell^q)$ may be generalized considerably. The first obvious modification is to replace ℓ^q by a weighted space ℓ^q_m, with a suitable (v-moderate) weight m, thus getting the space $W(L^p, \ell^q_m)$. Actually, there is an alternative definition for that space. Let χ_Q denote the characteristic function of $Q := [0, 1]$ (or any compact interval). Then consider the following norm:

$$\|f\|''_{p,q} = \| \|f \cdot T_x \chi_Q\|_{L^p} \|_{L^q_m}$$

$$= \left(\int_{\mathbb{R}} \left(\int_{\mathbb{R}} |f(t)|^p \chi_Q(t - x) \, dt \right)^{q/p} m(x)^q \, dx \right)^{1/q}$$

$$= \left(\int_{\mathbb{R}} \left(\int_{\mathbb{R}} |f(t)|^p \chi_Q(t - x) \, m(x)^p \, dt \, dx \right)^{q/p} \right)^{1/q}.$$

It turns out that this ('continuous') norm is equivalent to the ('discrete') norm $\| \cdot \|_{p,q}$, thus it defines the same space $W(L^p, \ell^q_m) \equiv W(L^p, L^q_m)$, but, in addition, this definition justifies the latter notation.

In such a setting, the properties listed above generalize in an obvious way. For instance, translation invariance of $W(L^p, \ell^q_m)$ becomes

$$\|T_x f\|_{W(L^p, \ell^q_m)} \leqslant C\, v(x) \|F\|_{W(L^p, \ell^q_m)}, \quad \text{for all } x \in \mathbb{R}^d.$$

The next step consists in replacing the "local" space L^p by a suitable Banach space B. Consider first the weighted spaces L^p_m. First we note that L^1_m is a Banach algebra with respect to convolution (called the *Beurling algebra*). Moreover, as mentioned already, L^1_v acts on L^p_m through convolution, $L^1_v * L^p_m \subseteq L^p_m$.

Next, consider the space $\mathcal{F}L^1_m$ of Fourier transforms of functions $f \in L^1_m$, with norm $\|\mathcal{F}f\| = \|f\|^1_m$ and a symmetric v-moderate weight, $m(-x) =$

$m(x)$. This guarantees that $\mathcal{F}^{-1}L^1_m = \mathcal{F}L^1_m$. Applying a Fourier transform yields $\mathcal{F}L^1_v \cdot \mathcal{F}L^p_m = \mathcal{F}(L^1_v * L^p_m) \subseteq \mathcal{F}L^p_m$. The space $\mathcal{F}L^1_m$ possesses a BUPU $(\psi_i)_{i\in I}$, so that the construction above applies.

Let now $(B, \|\cdot\|_B)$ be a Banach space of functions (or distributions), invariant under translations $(T_x g)(y) = g(y-x)$ and modulations $(M_\omega g)(y) = e^{2\pi i y \omega} g(y)$, and such that $\mathcal{S}(\mathbb{R}) \hookrightarrow B \hookrightarrow \mathcal{S}^\times(\mathbb{R})$. In addition, assume that $\mathcal{F}L^1_m$ acts on B by pointwise multiplication. Then the (Wiener) generalized amalgam space $W(B, \ell^q_m)$ consists of all functions or distributions $f \in B_{\mathrm{loc}}$ such that

$$\|f\|_{B, \ell^q_m} := \left(\sum_{i\in I} \|f\psi_i\|^q_B \, m(x_i)^q \right)^{1/q} = \left\| (\|f\psi_i\|_B) \right\|_{\ell^q_m} < \infty. \qquad (8.1)$$

Here we have introduced the space $B_{\mathrm{loc}} := \{ f \in \mathcal{S}^\times(\mathbb{R}) : hf \in B$ for any $h \in \mathcal{D}(\mathbb{R})\}$. As usual, different BUPUs $(\psi_i)_{i\in I}$ and sets of points $(x_i)_{i\in I}$ yield equivalent norms, hence the same space. For $B \equiv L^p$, one recovers the usual space $W(L^p, \ell^q_m)$. An interesting case is $B = \mathcal{F}L^p_u$, which indeed satisfies all the conditions stated above, but many other spaces may be chosen, such as Besov spaces, Bessel potential spaces, etc.

The spaces $W(\mathcal{F}L^p_u, \ell^q_m)$ have many interesting properties. Given submultiplicative weights v and w, assume m is a v-moderate weight and u is a w-moderate weight. For this case, there is also an equivalent, 'continuous' norm, namely,

$$\|f\|'_{W(\mathcal{F}L^p_u, \ell^q_m)} = \left(\int_{\mathbb{R}} (\|f \cdot T_x h\|_{\mathcal{F}L^p_u})^q \, m(x)^q \, dx \right)^{1/q},$$

where h can be any nonzero element of $W(\mathcal{F}L^1_v, \ell^1_w)$. Then the following holds:

(i) *Completeness:* $W(\mathcal{F}L^p_u, \ell^q_m)$ is a Banach space for the norm $\|\cdot\|_{W(\mathcal{F}L^p_u, \ell^q_m)}$.

(ii) *Invariance:* $W(\mathcal{F}L^p_u, \ell^q_m)$ is invariant under translation and modulation and the corresponding operators are bounded.

(iii) *Duality:* $W(\mathcal{F}L^p_u, \ell^q_m)^\times = W(\mathcal{F}L^{\bar{p}}_{1/u}, \ell^{\bar{q}}_{1/m})$, with the usual notation.

(iv) *Convolution, multiplication:* one has, with all embeddings continuous,

$$L^1_v * W(\mathcal{F}L^p_u, \ell^q_m) \hookrightarrow W(\mathcal{F}L^p_u, \ell^q_m),$$

$$\mathcal{F}L^1_w \cdot W(\mathcal{F}L^p_u, \ell^q_m) \hookrightarrow W(\mathcal{F}L^p_u, \ell^q_m).$$

(v) *The Fourier transform* $\mathcal{F}$ is an isomorphism between the spaces $W(\mathcal{F}L^1_w, \ell^1_v)$ and $W(\mathcal{F}L^1_v, \ell^1_w)$. More generally, for weight functions v, w at most of polynomial growth, and for given $\alpha, \beta \in \mathbb{R}$, $\mathcal{F}$ extends to an isomorphism between the spaces $W(\mathcal{F}L^p_u, \ell^p_m)$ and $W(\mathcal{F}L^p_m, \ell^p_u)$, where $u := w^\alpha$ and $m := v^\beta$. For proving this, one starts from the case

$p = q = 1$. Then, since v, w are assumed to be at most of polynomial growth, one transposes the relation to the dual spaces $W(\mathcal{F}L_{1/w}^\infty, \ell_{1/v}^\infty)$ and $W(\mathcal{F}L_{1/v}^\infty, \ell_{1/w}^\infty)$, and finally one gets the case (p, p) by interpolation.

(vi) *Inclusion relations :* the spaces $W(\mathcal{F}L_u^p, \ell_m^q)$ are ordered by inclusion as follows. If $1 \leqslant p_1 \leqslant p_2 \leqslant \infty$ and $1 \leqslant q_1 \leqslant q_2 \leqslant \infty$, then $W(\mathcal{F}L_u^{p_1}, \ell_m^{q_1}) \hookrightarrow W(\mathcal{F}L_u^{p_2}, \ell_m^{q_2})$. Furthermore,

(a) if $p_1 < p_2$ and $u_2 \leqslant C u_1$, then $W(\mathcal{F}L_{u_1}^{p_1}, \ell_m^q) \hookrightarrow W(\mathcal{F}L_{u_2}^{p_2}, \ell_m^q)$;

(b) if $q_1 < q_2$ and $m_2 \leqslant C m_1$, then $W(\mathcal{F}L_u^p, \ell_{m_1}^{q_1}) \hookrightarrow W(\mathcal{F}L_u^p, \ell_{m_2}^{q_2})$; the same result holds if $q_1 < q_2$ and $m_2/m_1 \in L^r$, for $1/r = 1/q_1 - 1/q_1$.

The two statements (a) and (b) are proven essentially as in Lemma 8.1.1. The second part of (b) follows from the Hölder inequality.

We proceed now with further generalizations. The first step is to replace the global space ℓ_m^q by a solid, translation invariant Banach space Y. By this, we mean a Banach space of locally integrable functions, continuously embedded in L_{loc}^1, and such that $f \in Y$, $g \in L_{\mathrm{loc}}^1$ and $|g(x)| \leqslant |f(x)|$ (a.e.) imply $g \in Y$ and $\|g\|_Y \leqslant \|f\|_Y$. Translation invariance here means that $\|T_x f\|_Y \leqslant c w_\gamma(x) \|f\|_Y$, where $w_\gamma(x) := (1 + |x|)^\gamma$. We also assume that $f * g \in Y$ for $f \in Y, g \in L_{w_\gamma}^1$ and $\|f * g\|_Y \leqslant \|g\|_{L_{w_\gamma}^1} \|f\|_Y$. Typical examples are the weighted spaces $L_{w_s}^p$ with $0 \leqslant |s| \leqslant \gamma$.

Fixing a window function $h \in \mathcal{D}(\mathbb{R})$ (that is, a positive C^∞ 'bump' function with compact support), we consider the *control function* $F_h(x) := \|T_x h \cdot f\|_B$. Then, given B, Y as above, we define the *generalized Wiener amalgam space*

$$W(B, Y) := \{f \in B_{\mathrm{loc}} : F_h \in Y\}$$

with norm $\|f\|_{B,Y} := \|F_h\|_Y$. As usual, there is an equivalent (discrete) formulation in terms of a BUPU $\{\psi_i, x_i\}$, as in (8.1). Namely, considering the function $F_W \in Y$ defined by

$$F_W(x) := \sum_{i\,:\,x \in x_i + W} \|f \psi_i\|_B \cdot \chi_{x_i + W}(x),$$

the norm $\|F_W\|_Y$ of the function F_W in Y defines an equivalent norm on $W(B, Y)$. If Y is a sequence space, this norm reads simply

$$\|f\|'_{B,Y} = \big\| (\|f \psi_i\|_B) \big\|_Y.$$

Taking $Y = \ell_m^q$, we recover the previous class of amalgam spaces.

Among the (very general) results about these spaces (see the literature), an interesting one concerns convolution between amalgam spaces. Assume (B_1, B_2, B_3) and (Y_1, Y_2, Y_3) are Banach convolution triples, i.e., that

$$\|f * g\|_{B_3} \leqslant C_1 \|g\|_{B_1} \|f\|_{B_2}, \quad \text{for all } g \in B_1, f \in B_2,$$

and

$$\|F \cdot G\|_{Y_3} \leqslant C_2 \|G\|_{Y_1} \|F\|_{Y_2}, \text{ for all } G \in Y_1, F \in Y_2.$$

Then $(W(B_1, Y_1), W(B_2, Y_2), W(B_3, Y_3))$ is a Banach convolution triple, that is, one has, for some constant $C_3 > 0$,

$$\|f * g\|_{B_3, Y_3} \leqslant C_3 \|g\|_{B_1, Y_1} \|f\|_{B_2, Y_2}.$$

This is a far reaching generalization of Young's identity.

The last step in the generalization process is to replace $\mathbb{R}^d$ by a locally compact abelian group G with Haar measure dx and dual $\widehat{G}$ (which consists of all continuous unitary characters of G). In that case, the Fourier transform becomes

$$\mathcal{F}f(\chi) := \int_G \overline{\chi(x)} f(x) \, dx,$$

where $\chi \in \widehat{G}$ is a unitary character of G. Hence, if L^p is a space of functions on G, its Fourier transform $\mathcal{F}L^p$ is a space of functions on $\widehat{G}$. Translations are as usual, while modulation becomes multiplication by characters. With these modifications, the whole theory goes through.

Summarizing, we see that mixed-norm spaces and amalgam spaces consist of large families of (mostly) reflexive Banach spaces, indexed by one or several real indices. Among these, one finds plenty of Banach chains and lattices of Banach spaces. On the corresponding PIP-space structures, operations like translation and modulation may be seen as regular operators, Fourier transform and convolution may also be reinterpreted, etc. Thus the whole theory may be rewritten in PIP-space language. It remains to be seen to what extent this approach improves and/or simplifies it. We will see in the next sections that the natural framework for Gabor or time-frequency analysis is the family of so-called modulation spaces $M_m^{p,q}$, but it turns out that these may often be replaced by amalgam spaces $W(L^p, \ell_m^q)$. Thus, these too have a important role in signal processing.

8.3 Modulation Spaces

Among the function spaces that play a central role in signal processing, several classes are closely related to well-known integral transforms like the Gabor and the wavelet transforms, namely, the modulation spaces and the Besov spaces, respectively. We treat them successively.

8.3.1 General Modulation Spaces

Modulation spaces are closely linked to, and in fact defined in terms of, the Short-Time Fourier (or Gabor) Transform.

Given a C^∞ window function $g \neq 0$, the *Short-Time Fourier Transform* (STFT) of $f \in L^2(\mathbb{R}^d)$ is defined by

$$(V_g f)(x, \omega) = \langle M_\omega T_x g | f \rangle := \int_{\mathbb{R}^d} \overline{g(y - x)} \, f(y) \, e^{-2\pi i y \omega} \, dy, \quad x, \omega \in \mathbb{R}^d,$$

(8.2)

where, as usual, $(T_x g)(y) = g(y - x)$ (translation) and $(M_\omega h)(y) = e^{2\pi i y \omega} h(y)$ (modulation).

Notice that an equivalent definition is often used, namely $(\widetilde{V}_g f)(x, \omega) = \langle T_x M_\omega g | f \rangle$, the connection between the two resulting from the identity

$$T_x M_\omega = e^{-2\pi i x \omega} M_\omega T_x.$$

Then, given a v-moderate weight function $m(x, \omega)$, the modulation space $M_m^{p,q}$ is defined in terms of a mixed-norm of a STFT:

$$M_m^{p,q}(\mathbb{R}^d) = \{f \in S^\times(\mathbb{R}^d) : V_g f \in L_m^{p,q}(\mathbb{R}^{2d})\}, \quad 1 \leqslant p, q \leqslant \infty.$$

For $p = q$, one writes $M_m^p \equiv M_m^{p,p}$. The space $M_m^{p,q}$ is a Banach space for the norm

$$\|f\|_{M_m^{p,q}} := \|V_g f\|_{L_m^{p,q}}$$

Actually, the original definition was slightly more restrictive, in that it used the weight function $m_s(x, \omega) = w_s(\omega) = (1 + |\omega|)^s$, $s \geqslant 0$, (or, equivalently, $\widetilde{m}_s(x, \omega) = (1 + |\omega|^2)^{s/2}$), so that the norm reads

$$\|f\|_{M_{w_s}^{p,q}} = \left(\int_{\mathbb{R}^d} \left(\int_{\mathbb{R}^d} |\langle M_\omega T_x g | f \rangle|^p \, dx \right)^{q/p} (1 + |\omega|)^{sq} \, d\omega \right)^{1/q}.$$

Equivalently, one may define the modulation spaces as the inverse Fourier transform of a Wiener amalgam space:

$$M_{w_s}^{p,q} = \mathcal{F}^{-1}(W(L^p, \ell_{w_s}^q)).$$

This space is independent of the choice of window g, in the sense that different window functions define equivalent norms.

The class of modulation spaces $M_{w_s}^{p,q}$ contains several well-known spaces, such as:

(i) The Bessel potential spaces or fractional Sobolev spaces $H^s = M_{\widetilde{m}_s}^2$:

$$H^s(\mathbb{R}^d) = M_{\widetilde{m}_s}^2(\mathbb{R}^d) = \{f \in S^\times : \int_{\mathbb{R}^d} |\widehat{f}(t)|^2 \, (1 + |t|^2)^s \, dt < \infty\}, \quad s \in \mathbb{R}.$$

(ii) $L^2(\mathbb{R}^d) = M^2(\mathbb{R}^d)$.

(iii) The Feichtinger algebra $\mathcal{S}_0 = M^1$, that we shall describe in detail in Section 8.3.2.

The main properties of the modulation spaces $M_m^{p,q}(\mathbb{R}^{2d})$ follow from the similar ones of the spaces $L_m^{p,q}(\mathbb{R}^{2d})$.

(i) *Duality:* if $1 \leqslant p, q < \infty$, then $(M_m^{p,q})^\times = M_{1/m}^{\bar{p},\bar{q}}$, with the usual notation.

(ii) *Translation invariance:* $M_m^{p,q}$ is invariant under time-frequency shifts and
$$\|T_x M_\omega f\|_{M_m^{p,q}} \leqslant C v(x,\omega) \|f\|_{L_m^{p,q}}$$
(m is assumed to be v-moderate).

(iii) *Fourier transform:* If $p = q$ and $m(\omega, -x) \leqslant C m(x,\omega)$, then M_m^p is invariant under the Fourier transform.

(iv) *Density:* If $|m(z)| \leqslant (1 + |z|)^N$ and $1 \leqslant p, q < \infty$, then $\mathcal{S}(\mathbb{R}^d)$ is dense in $M_m^{p,q}(\mathbb{R}^d)$.

The lattice properties of the family $\{M_m^{p,q}, 1 \leqslant p, q \leqslant \infty\}$ are, of course, the same as those of the mixed-norm spaces $L_m^{p,q}$. Here also, statements (ii) and (iii) may be translated in PIP-space language, in terms of totally regular operators.

Similar inclusion relations hold:

Lemma 8.3.1. *If $p_1 \leqslant p_2$, $q_1 \leqslant q_2$, and $m_2 \leqslant C m_1$, for some constant $C > 0$, then $M_{m_1}^{p_1,q_1} \subseteq M_{m_2}^{p_2,q_2}$.*

In particular, one has

$$M_v^1 \subseteq M_m^{p,q} \subseteq M_{1/v}^\infty.$$

The proof follows immediately from Lemma 8.1.1.

By construction, modulation spaces are function spaces well-adapted to *Gabor analysis*. A wealth of information about the spaces and their application in Gabor analysis may be found in the monograph of Gröchenig [Grö01]. Here we just indicate a few relevant points.

Given a nonzero window function $g \in L^2(\mathbb{R}^d)$ and lattice parameters $\alpha, \beta > 0$, the set of vectors

$$\mathcal{G}(g, \alpha, \beta) = \{M_{n\beta} T_{k\alpha} g, \, k, n \in \mathbb{Z}^d\}$$

is called a *Gabor system*. The system $\mathcal{G}(g, \alpha, \beta)$ is a *Gabor frame* if there exist two constants $\mathsf{m} > 0$ and $\mathsf{M} < \infty$ such that

$$\mathsf{m}\|f\|^2 \leqslant \sum_{k,n \in \mathbb{Z}^d} |\langle M_{n\beta} T_{k\alpha} g | f \rangle|^2 \leqslant \mathsf{M}\|f\|^2, \text{ for all } f \in L^2(\mathbb{R}^d).$$

The associated Gabor frame operator $S_{g,g}$ is given by

$$S_{g,g}f := \sum_{k,n \in \mathbb{Z}^d} \langle M_{n\beta}T_{k\alpha}g|f\rangle M_{n\beta}T_{k\alpha}g. \tag{8.3}$$

The main results of the Gabor time-frequency analysis stem from the following proposition.

Proposition 8.3.2. *If $\mathcal{G}(g,\alpha,\beta)$ is a Gabor frame, there exists a dual window $\breve{g} = S^{-1}g$ such that $\mathcal{G}(\breve{g},\alpha,\beta)$ is a frame, called the dual frame. Then one has, for every $f \in L^2(\mathbb{R}^d)$,*

$$f = \sum_{k,n \in \mathbb{Z}^d} \langle M_{n\beta}T_{k\alpha}g|f\rangle M_{n\beta}T_{k\alpha}\breve{g} \tag{8.4}$$

$$= \sum_{k,n \in \mathbb{Z}^d} \langle M_{n\beta}T_{k\alpha}\breve{g}|f\rangle M_{n\beta}T_{k\alpha}g, \tag{8.5}$$

with unconditional convergence in $L^2(\mathbb{R}^d)$.

The two relations (8.4) and (8.5) mean that the function f may be reconstructed from suitable samples of its STFT. But this raises a number of questions:

(1) For which values of α, β is $\mathcal{G}(g,\alpha,\beta)$ a frame? For which class of windows g?

The answer is that $\mathcal{G}(g,\alpha,\beta)$ is a frame if $g \in W(L^\infty, \ell^1)$ and α, β are sufficiently small (for the technical meaning of 'sufficiently small ', see [Grö01, Sec.6.5]). In particular, $\alpha\beta \leqslant 1$ is a necessary condition.

(2) Can one replace the regular lattice $\mathbb{Z}^d$ by an irregular set of points in $\mathbb{R}^d$?

The answer is positive, but this is a difficult problem, related to irregular sampling and number theory.

(3) Under which conditions are the operators associated to Gabor frames (analysis, synthesis, frame operator) well-defined and bounded?

Here, the analysis operator $C_g : L^2(\mathbb{R}^{2d}) \to \ell^2(\mathbb{Z}^{2d})$ is defined by $(C_g f)_{kn} = \langle M_{n\beta}T_{k\alpha}g|f\rangle$ and the synthesis operator $D_g : \ell^2(\mathbb{Z}^{2d}) \to L^2(\mathbb{R}^{2d})$ by $D_g c = \sum_{k,n \in \mathbb{Z}^d} c_{kn} M_{n\beta}T_{k\alpha}g$. Then, one has $C_g^* = D_g$. As a slight generalization of (8.3), we still call Gabor frame operator the operator

$$S_{g,\breve{g}}f := D_{\breve{g}}C_g$$

$$= \sum_{k,n \in \mathbb{Z}^d} \langle M_{n\beta}T_{k\alpha}g|f\rangle M_{n\beta}T_{k\alpha}\breve{g}.$$

This is where the modulation spaces $M_m^{p,q}$ turn out to be the natural class of function spaces. First, the optimal space for window functions is M_v^1. Next one has (in a somewhat abbreviated form):

(i) If $g \in M_v^1$, then C_g is bounded from $M_m^{p,q}$ into $\ell_{\widetilde{m}}^{p,q}(\mathbb{Z}^{2d})$, for all v-moderate weights m, $1 \leqslant p, q \leqslant \infty$ and all lattice constants α, β. Here $\widetilde{m}(k, n) = m(k\alpha, n\beta)$.

(ii) If $g \in M_v^1$, then D_g is bounded from $\ell_{\widetilde{m}}^{p,q}(\mathbb{Z}^{2d})$ into $M_m^{p,q}$, for all p, q. If $p, q < \infty$, the series expressing D_g converges unconditionally in $M_m^{p,q}$.

(iii) If $g, \breve{g} \in W(L^\infty, \ell^1)$, then the Gabor frame operator $S_{g,\breve{g}}$ is bounded on every $L^p(\mathbb{R}^{2d})$, $1 \leqslant p \leqslant \infty$.

(iv) If $g, \breve{g} \in M_v^1$, then $S_{g,\breve{g}}$ is bounded on $M_m^{p,q}$ for all $1 \leqslant p, q \leqslant \infty$, all v-moderate weights m, and all α, β.

(v) If $\breve{g}$ is a dual window of g, that is, $S_{g,\breve{g}} = 1$ on L^2, then the two expansions (8.4) and (8.5) converge unconditionally in $M_m^{p,q}$ if $p, q < \infty$.

(vi) If $g \in \mathcal{S}$, then a tempered distribution $f \in \mathcal{S}^\times(\mathbb{R}^d)$ belongs to $M_m^{p,q}$ if, and only if, $C_g f \in \ell_{\widetilde{m}}^{p,q}$.

Notice once again that statements (iii) and (iv) can be translated into PIP-space language, by saying that $S_{g,\breve{g}}$ is a totally regular operator in the chain $\{L^p, 1 \leqslant p \leqslant \infty\}$, resp. any PIP-space built from modulation spaces.

Most of these results are highly nontrivial and their proof requires deep analysis. As for the result (v), it is a first example where membership of f in the modulation space $M_m^{p,q}$ is *characterized* by membership of the sequence of its Gabor coefficients $C_g f$ in $\ell_{\widetilde{m}}^{p,q}$. This type of result is quite strong and in general valid only for the pair $(L^2 \leftrightarrow \ell^2)$. Here, in fact, lies the power of Gabor analysis, and of wavelet analysis as well, as we shall see below.

These results should suffice to convince the reader that the modulation spaces $M_m^{p,q}$ are the 'natural' spaces for Gabor analysis. Actually, most of this remains true if one replaces modulation spaces by amalgam spaces $W(L^p, \ell_m^q)$. Second, it is obvious that most of the statements have a distinctly PIP-space flavor: it is not some individual space $M_m^{p,q}$ or $W(L^p, \ell_m^q)$ that counts, but the whole family, with many operators being regular in the sense of PIP-spaces.

8.3.2 The Feichtinger Algebra

A particularly interesting case of modulation space is the space M^1, the smallest of them, which consists of all functions with integrable Gabor transform. This space is also known as the *Feichtinger algebra*, denoted by $\mathcal{S}_0(\mathbb{R}^d)$, and it plays an important role in abstract harmonic analysis. As for general modulation spaces, $\mathcal{S}_0$ may also be defined as an amalgam space, namely $\mathcal{S}_0 = W(\mathcal{F}L^1, \ell^1)$.

By definition, $f \in \mathcal{S}_0$ if $V_{g_0} f$ is integrable, where g_0 is the Gaussian (which could be replaced by any function in $\mathcal{S}$). The space $\mathcal{S}_0$ has many interesting properties, for instance:

(i) $\mathcal{S}_0$ is a Banach space for the norm $\|f\|_{\mathcal{S}_0} = \|V_{g_0} f\|_1$, and $\mathcal{S} \hookrightarrow \mathcal{S}_0 \hookrightarrow L^2$, with all embeddings continuous with dense range.

(ii) $\mathcal{S}_0$ is a Banach algebra with respect to pointwise multiplication and convolution.

(iii) Time-frequency shifts $T_x M_\omega$ are isometric on $\mathcal{S}_0$: $\|T_x M_\omega f\|_{\mathcal{S}_0} = \|f\|_{\mathcal{S}_0}$. $\mathcal{S}_0$ is continuously embedded in any Banach space with the same property and containing g_0, thus it is the smallest Banach space with this property.

(iv) The Fourier transform is an isometry on $\mathcal{S}_0$: $\|\mathcal{F} f\|_{\mathcal{S}_0} = \|f\|_{\mathcal{S}_0}$.

Next we turn to the (conjugate) dual $\mathcal{S}_0^\times$ of $\mathcal{S}_0$. Since $\mathcal{S}_0 = M^1$, we have $\mathcal{S}_0^\times = M^\infty$, a Banach space with norm $\|f\|_{\mathcal{S}_0^\times} = \|V_g f\|_\infty$. The space $\mathcal{S}_0^\times$ contains both the δ function and the pure frequency $\chi_\omega(x) = e^{-2\pi i x \omega}$.

In virtue of (i) above, we have

$$\mathcal{S} \hookrightarrow \mathcal{S}_0 \hookrightarrow L^2 \hookrightarrow \mathcal{S}_0^\times \hookrightarrow \mathcal{S}^\times, \tag{8.6}$$

where all embeddings are continuous and have dense range. In the terminology of Section 5.2.1, $\mathcal{S}_0$ and $\mathcal{S}_0^\times$, are interspaces for the RHS $\mathcal{S} \hookrightarrow L^2 \hookrightarrow \mathcal{S}^\times$. In the quintuplet of spaces (8.6), the central triplet

$$\mathcal{S}_0(\mathbb{R}^d) \hookrightarrow L^2(\mathbb{R}^d) \hookrightarrow \mathcal{S}_0^\times(\mathbb{R}^d) \tag{8.7}$$

is the prototype of a *Banach Gel'fand triple*, that is a RHS (or LBS) in which the extreme spaces are (nonreflexive) Banach spaces. By (iii) and (iv) above, both time-frequency shifts and Fourier transform are isomorphisms of $\mathcal{S}_0^\times(\mathbb{R}^d)$, and indeed of the three spaces of the triple (8.7), and the Parseval formula holds:

$$\langle f|g \rangle = \langle \mathcal{F} f | \mathcal{F} g \rangle, \quad \text{for all } (f, g) \in \mathcal{S}_0(\mathbb{R}^d) \times \mathcal{S}_0^\times(\mathbb{R}^d).$$

Things becomes even more interesting in the discrete case. First there is the following striking result.

Theorem 8.3.3. *Let* $\mathcal{G}(g, \alpha, \beta)$ *be a Gabor frame with* $g \in \mathcal{S}_0(\mathbb{R}^d)$. *Then the dual window* $\breve{g} = S_{g,\breve{g}}^{-1} g$, *where* $S_{g,\breve{g}}$ *is the Gabor frame operator, belongs to* $\mathcal{S}_0(\mathbb{R}^d)$ *as well.*

Next, one can characterize membership of a function f in $\mathcal{S}_0$ or $\mathcal{S}_0^\times(\mathbb{R}^d)$ in terms of its Gabor coefficients. For simplicity, we put $\alpha = \beta = 1$ (so that we are in the so-called critical case $\alpha\beta = 1$).

Proposition 8.3.4. *Let $\mathcal{G}(g, 1, 1)$ be a Gabor frame with $g \in \mathcal{S}_0(\mathbb{R}^d)$. Then f belongs to $\mathcal{S}_0(\mathbb{R}^d)$ if and only if the sequence of Gabor coefficients $(\langle M_n T_k g | f \rangle)_{k,n \in \mathbb{Z}^d}$ belongs to $\ell^1(\mathbb{Z}^d)$. In addition, one has the equivalence of norms:*

$$C_1 \|f\|_{\mathcal{S}_0} \leqslant \sum_{k,n \in \mathbb{Z}^d} |\langle M_n T_k g | f \rangle| \leqslant C_2 \|f\|_{\mathcal{S}_0}, \quad \text{for all } f \in \mathcal{S}_0(\mathbb{R}^d).$$

Similarly, $f \in \mathcal{S}_0^\times(\mathbb{R}^d)$ if and only if the sequence of Gabor coefficients $(\langle M_n T_k g | f \rangle)_{k,n \in \mathbb{Z}^d}$ belongs to $\ell^\infty(\mathbb{Z}^d)$.

This result can be formalized in terms of the so-called *localization operators*. Let g be a window function and σ a bounded non-negative function on $\mathbb{R}^{2d}$. Then the localization operator associated to the symbol σ is the operator H_σ defined by

$$H_\sigma f := \int_{\mathbb{R}^{2d}} \sigma(x, \omega) \, V_g f(x, \omega) \, M_\omega T_x g \, dx \, d\omega.$$

If $\sigma \equiv 1$ and $\|g\|_2 = 1$, then $H_\sigma = 1$ and the relation above is nothing but the inversion formula of the STFT. If σ has compact support $\Omega \subset \mathbb{R}^{2d}$, then, intuitively, $H_\sigma f$ represents the part of f that lives in Ω, hence the name. This statement can be made precise as follows. Given a time-frequency shift $T_j, j \in \mathbb{Z}^{2d}$, consider the collection of localization operators $\{H_j := H_{T_j \sigma}, j \in \mathbb{Z}^{2d}\}$. Then the map $f \mapsto \{H_j f\}$ can be interpreted as the decomposition of f into (localized) components $H_j f$ living essentially on $\operatorname{supp} T_j \sigma = j + \operatorname{supp} \sigma$ in the time-frequency plane and the norm $\|H_j f\|_2^2$ is the energy of that component.

Using this concept, one has the following fundamental result.

Theorem 8.3.5. *Let $\sigma \in L^1(\mathbb{R}^{2d})$ be a non-negative symbol satisfying the condition*

$$A \leqslant \sum_{j \in \mathbb{Z}^{2d}} T_j \sigma \leqslant B, \quad a.e.,$$

for two constants $A, B > 0$ and assume that $g \in \mathcal{S}_0(\mathbb{R}^d)$. Then $f \in \mathcal{S}_0(\mathbb{R}^d)$ if and only if $\sum_{j \in \mathbb{Z}^{2d}} \|H_j f\|_2 < \infty$ and this quantity defines an equivalent norm on $\mathcal{S}_0(\mathbb{R}^d)$.

Similarly, the following norm equivalences characterize $\mathcal{S}_0^\times$ and L^2:

$$\|f\|_{\mathcal{S}_0^\times} \asymp \sup_{j \in \mathbb{Z}^{2d}} \|H_j f\|_2, \quad \|f\|_2^2 \asymp \sum_{j \in \mathbb{Z}^{2d}} \|H_j f\|_2^2.$$

Using the notion of Gel'fand triples, this result takes a simpler form.

Corollary 8.3.6. *Under the conditions of Theorem 8.3.5, the map $\iota : f \mapsto (\|H_j f\|_2)_{j \in \mathbb{Z}^{2d}}$ is an isomorphism between the Gel'fand triple $(\mathcal{S}_0, L^2, \mathcal{S}_0^\times)$ and a closed subspace of the triple $(\ell^1, \ell^2, \ell^\infty)$.*

Actually, one can go further. Since $\mathcal{S}_0 = M^1$ and $\mathcal{S}_0^\times = M^\infty$, all the modulation spaces $M^p, 1 \leqslant p \leqslant \infty$ may be obtained by interpolation between $\mathcal{S}_0$

and $\mathcal{S}_0^\times$, so that the statement of Corollary 8.3.6 extends to the whole chain. Thus the map ι is a monomorphism from the LHS $\{M^p\}$ into the LHS $\{\ell^p\}$.

The Feichtinger algebra $\mathcal{S}_0$ is often used in time-frequency analysis and it is considered by many authors as the natural space of test functions. Indeed, the Banach Gel'fand triple $(\mathcal{S}_0, L^2, \mathcal{S}_0^\times)$ often replaces advantageously Schwartz' space RHS $\mathcal{S} \hookrightarrow L^2 \hookrightarrow \mathcal{S}^\times$.

8.4 Besov Spaces

Besov spaces were introduced around 1960 for providing a precise control on the smoothness of solutions of certain partial differential equations. Later on, it was discovered that they are closely linked to wavelet analysis, exactly as the (much more recent) modulation spaces are structurally adapted to Gabor analysis. In fact, there are many equivalent definitions of Besov spaces. We begin by a 'discrete' formulation, based on a dyadic partition of unity.

Let us consider a weight function $\varphi \in \mathcal{S}(\mathbb{R})$ with the following properties:

- supp $\varphi = \{\xi : 2^{-1} \leqslant |\xi| \leqslant 2\}$,
- $\varphi(\xi) > 0$ for $2^{-1} < |\xi| < 2$,
- $\sum_{j=-\infty}^{\infty} \varphi(2^{-j}\xi) = 1$ $(\xi \neq 0)$.

Then one defines the following functions by their Fourier transform:

- $\widehat{\varphi_j}(\xi) = \varphi(2^{-j}\xi)$, $j \in \mathbb{Z}$: high "frequency" for $j > 0$, low "frequency" for $j < 0$,
- $\widehat{\psi}(\xi) = 1 - \sum_{j=1}^{\infty} \varphi(2^{-j}\xi)$: low "frequency" part.

Given the weight function φ, the inhomogeneous Besov space B_{pq}^s is defined as

$$B_{pq}^s = \{f \in \mathcal{S}^\times : \|f\|_{pq}^s < \infty\}, \tag{8.8}$$

where $\|\cdot\|_{pq}^s$ denotes the norm

$$\|f\|_{pq}^s := \|\psi * f\|_p + \left(\sum_{j=1}^{\infty} (2^{sj}\|\varphi_j * f\|_p)^q\right)^{1/q} , \quad s \in \mathbb{R}, 1 \leqslant p, q \leqslant \infty. \tag{8.9}$$

The space B_{pq}^s is a Banach space and it does not depend on the choice of the weight function φ, since a different choice defines an equivalent norm. Note that $B_{22}^s = H^s$, the (fractional) Sobolev space or Bessel potential space.

For $f \in B_{pq}^s$, one may write the following (weakly converging) expansion, known as a *dyadic Littlewood–Paley decomposition*:

$$f = \psi * f + \sum_{j=1}^{\infty} \varphi_j * f. \tag{8.10}$$

Clearly the first term represents the (relatively uninteresting) low "frequency" part of the function, whereas the second term analyzes in detail the high "frequency" component.

An equivalent, 'continuous', definition is based on the notion of *modulus of smoothness*. For $f \in L^p$ and $h > 0$, this is the quantity

$$\omega_p(f, h) := \|f(\cdot + h) - f(\cdot)\|_p .$$

Then, for $0 < s < 1$ and $q < \infty$, the space B_{pq}^s consists of all functions $f \in L^p$ for which the following norm is finite:

$$\|f\|_{B_{pq}^s} := \|f\|_p + \left(\int_0^\infty [h^{-s} \omega_p(f, h)]^q \, \frac{dh}{h} \right)^{1/q}.$$

This norm is equivalent to the norm (8.9). A similar norm may be defined for $s > 1$ and for $q = \infty$.

Another equivalent norm (again for $0 < s < 1$) yet is the following:

$$\|f\|_{B_{pq}^s} \asymp \|f\|_p + \left(\sum_{j=0}^\infty [2^{sj} \omega_p(f, 2^{-j})]^q \right)^{1/q}.$$

Besov spaces enjoy many familiar properties (for more details, we refer to the literature):

(i) *Inclusion relations:* The following relations hold, where all embeddings are continuous:

- $\mathcal{S} \hookrightarrow B_{pq}^s \hookrightarrow \mathcal{S}^\times$;
- $B_{pq}^s \hookrightarrow L^p$, if $1 \leqslant p, q \leqslant \infty$ and $s > 0$;
- for $s_1 < s_2$, $B_{pq}^{s_2} \hookrightarrow B_{pq}^{s_1}$ ($1 \leqslant q, p \leqslant \infty$);
- for $1 \leqslant q_1 < q_2 \leqslant \infty$, $B_{pq_1}^s \hookrightarrow B_{pq_2}^s$ ($s \in \mathbb{R}, 1 \leqslant p \leqslant \infty$);
- for $s - 1/p = s_1 - 1/p_1$, $B_{pq}^s \hookrightarrow B_{p_1 q_1}^{s_1}$ ($s, s_1 \in \mathbb{R}, 1 \leqslant p \leqslant p_1 \leqslant \infty, 1 \leqslant q \leqslant q_1 \leqslant \infty$).

In the terminology of Section 5.2.1, the first statement means that the spaces B_{pq}^s are interspaces for the RHS $\mathcal{S} \hookrightarrow L^2 \hookrightarrow \mathcal{S}^\times$. The inclusion relations above mean that the family of spaces B_{pq}^s contains again many chains of Banach spaces, but no more.

(ii) *Interpolation:* Besov spaces enjoy nice interpolation properties, in all three parameters s, p, q.

(iii) *Duality:* one has $(B_{pq}^s)^\times = B_{\bar{p}\bar{q}}^{-s}$ ($s \in \mathbb{R}$).

(iv) *Translation and dilation invariance:* every space B_{pq}^s is invariant under translation and dilation.

(v) *Regularity shift:* let $J^\sigma : \mathcal{S}^\times \to \mathcal{S}^\times$ denote the operator $J^s f = \mathcal{F}^{-1}\{(1 + |\cdot|^2)^{s/2} \mathcal{F} f\}$, $s \in \mathbb{R}$. Then J^σ is an isomorphism from B_{pq}^s onto $B_{pq}^{s-\sigma}$. Thus J^σ is totally regular for $\sigma \leqslant 0$, but not for $\sigma > 0$.

It is also useful to consider the homogeneous Besov space $\dot{B}_{pq}^s$, defined as the set of all $f \in \mathcal{S}^\times$ for which $\|f\|_{pq}^{\dot{s}} < \infty$, where the quasi-norm $\| \cdot \|_{pq}^{\dot{s}}$ is defined by

$$\|f\|_{pq}^{\dot{s}} := \left(\sum_{j=-\infty}^{\infty} (2^{sj} \|\varphi_j * f\|_p)^q \right)^{1/q}$$

(this is only a quasi-norm since $\|f\|_{pq}^{\dot{s}} = 0$ if and only if supp $\hat{f} = \{0\}$, i.e., f is a polynomial). Note that, if $0 \notin \text{supp} \hat{f}$, then $f \in \dot{B}_{pq}^s$ if and only if $f \in B_{pq}^s$.

The spaces $\dot{B}_{pq}^s$ have properties similar to the previous ones and, in addition, one has $B_{pq}^s = L^p \cap \dot{B}_{pq}^s$ for $s > 0$, $1 \leqslant p, q \leqslant \infty$. In particular, every space $\dot{B}_{pq}^s$ is invariant under translation and dilation, which is not surprising, since these spaces are in fact based on the $ax + b$ group, consisting precisely of dilations and translations of the real line, via the coorbit space construction (see Section 8.5(ii) below).

Besov spaces are well-adapted to *wavelet analysis*, because the definition (8.8) essentially relies on a dyadic partition (powers of 2). Historically, the connection was made with the *discrete* wavelet analysis, for that reason. Indeed, there exists an equivalent definition given in terms of decay of wavelet coefficients. More precisely, if a function f is expanded in a wavelet basis, the decay properties of the wavelet coefficients allow to characterize precisely to which Besov space the function f belongs. In addition, the Besov spaces may also be characterized in terms of the *continuous* wavelet transform. These properties will be discussed below.

In order to go into details, we have to some recall basic facts about the wavelet transform (for simplicity, we restrict ourselves to one dimension). Whereas the STFT is defined in terms of translation and modulation, the continuous wavelet transform is based on translations and dilations:[2]

$$(W_\psi s)(b, a) = a^{-1} \int_{-\infty}^{\infty} \overline{\psi(a^{-1}(x - b))} s(x) \, dx, \quad a > 0, b \in \mathbb{R}, s \in L^2(\mathbb{R}).$$

$$(8.11)$$

In this relation, the wavelet ψ is assumed to satisfy the admissibility condition

$$c_\psi := \int_{-\infty}^{\infty} d\omega \, |\omega|^{-1} |\hat{\psi}(\omega)|^2 < \infty,$$

which implies $\int_{-\infty}^{\infty} \psi(x) \, dx = 0$. In addition, the wavelet ψ is said to have N *vanishing moments* ($N \in \mathbb{N}$) if it verifies the conditions

---

[2] This is the so-called L^1 normalization. It is more frequent to use the L^2 normalization, in which the prefactor is $a^{-1/2}$ instead of a^{-1}.

$$\int_{\mathbb{R}} x^n \, \psi(x) \, dx = 0, \quad \text{for } n = 0, 1, \dots, N - 1.$$

This property improves the efficiency of ψ at detecting singularities in a signal, since the wavelet ψ is then blind to polynomials up to order $N - 1$, which constitute the smoothest part of the signal, i.e., the part which contains the smallest amount of information.

However, discretizing the two parameters a and b in (8.11) leads in general only to frames. In order to get orthogonal wavelet bases, one relies on the so-called *multiresolution analysis* of $L^2(\mathbb{R})$. This is defined as an increasing sequence of closed subspaces of $L^2(\mathbb{R})$:

$$\dots \subset \mathcal{V}_{-2} \subset \mathcal{V}_{-1} \subset \mathcal{V}_0 \subset \mathcal{V}_1 \subset \mathcal{V}_2 \subset \dots \tag{8.12}$$

with $\bigcap_{j \in \mathbb{Z}} \mathcal{V}_j = \{0\}$ and $\bigcup_{j \in \mathbb{Z}} \mathcal{V}_j$ dense in $L^2(\mathbb{R})$, and such that

(1) $f(x) \in \mathcal{V}_j \Leftrightarrow f(2x) \in \mathcal{V}_{j+1}$;
(2) There exists a function $\phi \in \mathcal{V}_0$, called a *scaling function*, such that the family $\{\phi(\cdot - k), \, k \in \mathbb{Z}\}$ is an orthonormal basis of $\mathcal{V}_0$.

Combining conditions (1) and (2), one sees that $\{\phi_{jk} \equiv 2^{j/2}\phi(2^j \cdot -k), \, k \in \mathbb{Z}\}$ is an orthonormal basis of $\mathcal{V}_j$. The space $\mathcal{V}_j$ can be interpreted as an *approximation* space at resolution 2^j. Defining $\mathcal{W}_j$ as the orthogonal complement of $\mathcal{V}_j$ in $\mathcal{V}_{j+1}$, i.e., $\mathcal{V}_j \oplus \mathcal{W}_j = \mathcal{V}_{j+1}$, we see that $\mathcal{W}_j$ contains the additional *details* needed to improve the resolution from 2^j to 2^{j+1}. Thus one gets the decomposition $L^2(\mathbb{R}) = \bigoplus_{j \in \mathbb{Z}} \mathcal{W}_j$. The crucial theorem then asserts the existence of a function ψ, called the *mother wavelet*, explicitly computable from ϕ, such that $\{\psi_{jk} \equiv 2^{j/2}\psi(2^j \cdot -k), \, k \in \mathbb{Z}\}$ constitutes an orthonormal basis of $\mathcal{W}_j$ and thus $\{\psi_{jk} \equiv 2^{j/2}\psi(2^j \cdot -k), \, j, k \in \mathbb{Z}\}$ is an orthonormal basis of $L^2(\mathbb{R})$: these are the *orthonormal wavelets*. Thus the expansion of an arbitrary function $f \in L^2$ into an orthogonal wavelet basis $\{\psi_{jk}, \, j, k \in \mathbb{Z}\}$ reads

$$f = \sum_{j,k \in \mathbb{Z}} c_{jk} \, \psi_{jk}, \quad \text{with } c_{jk} = \langle \psi_{jk} | f \rangle. \tag{8.13}$$

Additional regularity conditions can be imposed to the scaling function ϕ. Given $r \in \mathbb{N}$, the multiresolution analysis corresponding to ϕ is called *r-regular* if

$$\left| \frac{d^n \phi}{dx^n} \right| \leqslant c_m (1 + |x|^m), \quad \text{for all } n \leqslant r \text{ and all integers } m \in \mathbb{N}.$$

Well-known examples include the Haar wavelets, the B-splines, and the various Daubechies wavelets.

As a result of the 'dyadic' definition (8.8)-(8.9), it is natural that Besov spaces can be characterized in terms of an r-regular multiresolution analysis $\{\mathcal{V}_j\}$. Let $E_j : L^2 \to \mathcal{V}_j$ be the orthogonal projection on $\mathcal{V}_j$ and $D_j = E_{j+1} - E_j$ that on $\mathcal{W}_j$. Let $0 < s < r$ and $f \in L^p(\mathbb{R})$. Then, $f \in B^s_{pq}(\mathbb{R})$ if, and only if, $\|D_j f\|_p = 2^{-js} \delta_j$, where $(\delta_j) \in \ell^q(\mathbb{N})$, and one has

$$\|f\|^s_{pq} \asymp \|E_0 f\|_p + \Big(\sum_{j \in \mathbb{Z}} 2^{jsq} \|D_j f\|^q_p \Big)^{1/q}.$$

Specializing to $p = q = 2$, one gets a similar result for Sobolev spaces: given $f \in H^{-r}(\mathbb{R})$ and $|s| < r$, $f \in H^s(\mathbb{R})$ if, and only if, $E_0 f \in L^2(\mathbb{R})$ and $\|D_j f\|_2 = 2^{-js} \epsilon_j$, $j \in \mathbb{N}$, where $(\epsilon_j) \in \ell^2(\mathbb{N})$.

But there is more. Indeed, modulation spaces and Besov spaces admit decomposition of elements into wavelet bases and each space can be uniquely characterized by the decay properties of the wavelet coefficients. To be precise, let $\{\psi_{jk}, j, k \in \mathbb{Z}\}$ be an orthogonal wavelet basis coming from an r-regular multiresolution analysis based on the scaling function ϕ. Then the following results are typical:

(i) *Inhomogeneous Besov spaces:* $f \in B^s_{pq}(\mathbb{R})$ if it can be written as

$$f(x) = \sum_{k \in \mathbb{Z}} \beta_k \phi(x - k) + \sum_{j \geqslant 0, k \in \mathbb{Z}} c_{jk}\, \psi_{jk},$$

where $(\beta_k) \in \ell^p$ and $\big(\sum_{k \in \mathbb{Z}} |c_{jk}|^p \big)^{1/p} = 2^{-j(s+1/2-1/p)} \gamma_j$, with $(\gamma_j) \in \ell^q(\mathbb{Z})$.

(ii) *Homogeneous Besov spaces:* let $|s| < r$. Then, if $f \in \dot{B}^s_{pq}(\mathbb{R})$, its wavelet coefficients c_{jk} verify $\big(\sum_{k \in \mathbb{Z}} |c_{jk}|^p \big)^{1/p} = 2^{-j(s+1/2-1/p)} \gamma_j$, where $(\gamma_j) \in \ell^q(\mathbb{Z})$. Conversely, if this condition is satisfied, then $f = g + P$, where $g \in \dot{B}^s_{pq}$ and P is a polynomial.

We conclude this section with some examples of unconditional wavelet bases, as announced in Section 3.4.4. For precise definitions, we refer to the literature.

- The Haar wavelet basis is defined by the scaling function $\phi_H = \chi_{[0,1]}$ and the mother wavelet $\psi_H = \chi_{[0,1/2]} - \chi_{[1/2,1]}$. It is a standard result that the Haar system is an unconditional basis for every $L^p(\mathbb{R})$, $1 < p < \infty$.
- The Lemarié-Meyer wavelet basis is an unconditional basis for all L^p spaces, Sobolev spaces, homogeneous Besov spaces $\dot{B}^s_{pq}$ $(1 \leqslant p, q < \infty)$.
- There is a class of wavelet bases (Wilson bases of exponential decay) that are unconditional bases for every modulation space $M^{p,q}_m$, $1 \leqslant p, q < \infty$, but *not* for L^p, $1 < p < \infty$, $p \neq 2$.

The characterization of Besov spaces in terms of discrete wavelet coefficients is standard, but there exists also an interesting one in terms of the continuous wavelet transform (CWT). A preliminary step is to reformulate the CWT (8.11) in L^p. The result is that, for any admissible wavelet ψ, the CWT map $W_\psi : f(x) \mapsto (W_\psi f)(b, a)$ is a bounded linear operator from $L^p(\mathbb{R})$ into $L^p(\mathbb{R}) \times L^2(\mathbb{R}_+^*, \frac{da}{a})$ and one has

$$\|f\|_p \asymp \left(\int_{-\infty}^{+\infty} \left(\int_0^{+\infty} |(W_\psi f)(b, a)|^2 \frac{da}{a} \right)^{p/2} db \right)^{1/p}.$$

The familiar Parseval formula extends to this context too (for pairs of vectors belonging to L^p, resp. $L^{\bar{p}}$, with $1/p + 1/\bar{p} = 1$, as usual), and so does the reconstruction formula.

Now we come back to Besov spaces. For simplicity, we quote the results only in the simplest case, which is s non-integer, denoting by $[s]$ the integer part of s. Then one has:

Proposition 8.4.1. *(1) Let $f \in B_{pq}^s(\mathbb{R})$, $1 \leqslant p, q < \infty$, s non-integer. Let ψ be a wavelet such that $(x^{s-[s]}\psi) \in L^1(\mathbb{R})$, with $[s] + 1$ vanishing moments. Then the wavelet transform of f satisfies the condition*

$$\int_0^\infty \left(a^{-s} \|(W_\psi f)(\cdot, a)\| \right)^q \frac{da}{a} < \infty.$$

(2) Conversely, let $s > 0$, non-integer, and ψ a real-valued $C^{[s]+1}$ wavelet, with all derivatives rapidly decreasing. If $f, f', \ldots, f^{[s]} \in L^p(\mathbb{R})$, $1 < p < \infty$, and if $a^{-s} \|W_\psi f(\cdot, a)\|_p \in L^q(\mathbb{R}_+^*, \frac{da}{a})$, $1 \leqslant q \leqslant \infty$, then f belongs to $B_{pq}^s(\mathbb{R})$.

Thus, as expected, the behavior at small scales of the wavelet transform indeed characterizes Besov spaces.

8.4.1 α-Modulation Spaces

The α-modulation spaces ($\alpha \in [0, 1]$) are spaces intermediate between modulation and Besov spaces, to which they reduce for $\alpha = 0$ and $\alpha \to 1$, respectively. A possible definition of these spaces runs as follows. Whereas the modulation spaces are defined in terms of the Gabor transform, the α-modulation spaces rely on the so-called *flexible Gabor-wavelet transform*, that is,

$$(V_\psi^\alpha f)(x, \omega) = \langle T_x M_\omega D_{w-\alpha(\omega)} \psi | f \rangle, \tag{8.14}$$

where D_a is the unitary dilation operator:

$$D_a f(x) = a^{-d/2} f(a^{-1}x), \ a > 0, \ f \in L^2(\mathbb{R}^d), \tag{8.15}$$

and $w_{-\alpha}$ is, as usual, the weight function $w_{-\alpha}(\omega) = (1 + |\omega|)^{-\alpha}$, $\alpha \in [0, 1)$. Clearly, for $\alpha = 0$, this reduces to the Gabor transform, whereas, for $\alpha = 1$, one gets a simple variant of the wavelet transform. The intermediate case $\alpha = 1/2$ appears in the literature under the name of *FBI transform* (for Fourier-Bros-Iagolnitzer).

Then, for $s \in \mathbb{R}$, for all $1 \leqslant p, q \leqslant \infty$ and for $\alpha \in [0, 1]$, one can define the α-modulation space via the relation

$$M^{p,q}_{s+\alpha(1/q-1/2),\alpha} := \{f \in \mathcal{S}^\times(\mathbb{R}^d) : V^\alpha_\psi f \in L^{p,q}_{w_s}(\mathbb{R}^{2d})\} \qquad (8.16)$$

with the norm

$$\|f\|_{M^{p,q}_{s+\alpha(1/q-1/2),\alpha}} = \|V^\alpha_\psi f\|^{p,q}_{w_s}.$$

Here $L^{p,q}_{w_s}(\mathbb{R}^{2d})$ denotes, as usual, the weighted mixed-norm L^2 space with weight $w_s = (1 + |\omega|)^s$:

$$\|F\|_{L^{p,q}_{w_s}} := \left(\int_{\mathbb{R}^d} \left(\int_{\mathbb{R}^d} |F(x,\omega)|^p dx \right)^{q/p} (1 + |\omega|)^{sq} d\omega \right)^{1/q}, 1 \leqslant p, q < \infty.$$

The usual modifications apply when $p = \infty$ or $q = \infty$.

For $\alpha = 0$, the space $M^{p,q}_{s,0}$ coincides with the modulation space $M^{p,q}_{w_s}$. For $\alpha \to 1$, the space $M^{p,q}_{s,\alpha}$ tends to the inhomogeneous Besov space $B^s_{p,q}$. Thus we may write $M^{p,q}_{s,1} = \lim_{\alpha \to 1} M^{p,q}_{s,\alpha}$, where the limit is to be taken in a geometrical (but somewhat imprecise) sense. In order to appreciate the true signification of these facts in signal processing, one needs some group-theoretical technology that we will introduce in the next section.

8.5 Coorbit Spaces

Coorbit spaces constitute a far reaching generalization of the function spaces described above. They provide a unified description of a number of function spaces useful in signal processing, some examples of which will be detailed at the end of this section. The construction is based on integrable group representations and thus requires a substantial amount of new concepts. Therefore our treatment will be very sketchy here, since otherwise this would lead us too far from our main subject. In particular, propositions will be stated without proof.

The starting point is the notion of (square) integrable group representation. Let G be a locally compact group with left Haar measure dg and U a (strongly continuous) irreducible representation in a Hilbert space $\mathcal{H}$. For a fixed nonzero vector $\eta \in \mathcal{H}$, denote by $V_\eta \phi$ the representation coefficient (matrix element) $V_\eta \phi(g) := \langle U(g)\eta | \phi \rangle$, a continuous bounded function on G.

Then the representation U is said to be *square integrable*, resp. *integrable*, if there exists at least one nonzero vector $\eta \in \mathcal{H}$ (called admissible) such that $V_\eta \eta \in L^2(G, dg)$, resp. $V_\eta \eta \in L^1(G, dg)$. Every integrable representation is square integrable.

Let U be a square integrable representation in $\mathcal{H}$ and let $\mathcal{A}$ denote the set of all admissible vectors. The crucial fact is the existence of *orthogonality relations*. Namely, there exists a unique positive, self-adjoint, invertible operator[3] C in $\mathcal{H}$, with dense domain $\mathcal{D}(C)$ equal to $\mathcal{A}$, such that, for any two admissible vectors η and η' and arbitrary vectors $\phi, \phi' \in \mathcal{H}$, one has

$$\int_G \overline{V_{\eta'} \phi'(g)} \, V_\eta \phi(g) \, dg = \langle C\eta | C\eta' \rangle \, \langle \phi' | \phi \rangle. \tag{8.17}$$

Furthermore $C = \lambda I$, $\lambda > 0$, if, and only if, G is unimodular. As an important consequence of the relations (8.17), one has the convolution identity

$$V_\eta \phi * V_{\eta'} \phi' = \langle C\eta | C\phi' \rangle V_{\eta'} \phi, \ \forall \eta, \phi' \in \mathcal{D}(C), \ \eta', \phi \in \mathcal{H}. \tag{8.18}$$

Here the convolution on G is defined as

$$(\chi * \xi)(g) = \int_G \chi(g_1) \xi(g_1^{-1} g) \, dg_1.$$

In particular, normalizing the vector $\eta \in \mathcal{D}(C)$ by $\|C\eta\| = 1$, one gets the *reproduction formula*

$$V_\eta \phi = V_\eta \phi * V_\eta \eta. \tag{8.19}$$

In other words, the function $K(g, g_1) = \langle U(g_1^{-1} g)\eta | \eta \rangle$ is a reproducing kernel on G.

Given a fixed admissible vector $\eta \in \mathcal{H}$, the map $V_\eta : \phi \mapsto V_\eta \phi(g)$ is called the *coherent state transform* or *CS transform* on G (sometimes called abusively the wavelet transform). The map V_η is an isometry from $\mathcal{H}$ into $L^2(G, dg)$ and satisfies $V_\eta(U(g)\phi) = L_g V_\eta \phi$, where L_g is the left regular representation. In other words, V_η intertwines U and L, and U is equivalent to a subrepresentation of L, hence U belongs to the discrete series of G.

The orthogonal projection from $L^2(G)$ onto the range of V_η is given by the convolution operator $\xi \mapsto \xi * V_\eta \eta$. Thus a function $\xi \in L^2(G)$ belongs to the range of V_η, i.e., $\xi = V_\eta \phi$ for some $\phi \in \mathcal{H}$ if and only if $\xi * V_\eta \eta = \xi$.

Now we are ready for defining coorbit spaces. We start with a unitary, irreducible, integrable representation U of G in $\mathcal{H}$. Given a weight function w (i.e., a positive continuous submultiplicative function) on G, define the following set of analyzing vectors:

$$\mathcal{A}_w := \{\eta \in \mathcal{H} : V_\eta \eta \in L^1_w(G)\}.$$

---

[3] The operator C is called the Duflo-Moore operator.

Since U is irreducible, $\mathcal{A}_w$ is a dense subspace of $\mathcal{H}$, assumed to be nontrivial. Then, fixing a nonzero vector $\eta \in \mathcal{A}_w$, one defines the space

$$\mathcal{H}_w^1 := \{\phi \in \mathcal{H} : V_\eta \phi \in L_w^1(G)\}.$$

$\mathcal{H}_w^1$ is a U-invariant Banach space for the norm

$$\|\phi\|_{\mathcal{H}_w^1} := \|V_\eta \phi\|_w^1.$$

It is dense in $\mathcal{H}$ and independent of the choice of the vector $\eta \in \mathcal{A}_w$.

Next one considers the conjugate dual $(\mathcal{H}_w^1)^\times$ of $\mathcal{H}_w^1$ (called the *reservoir*) and thus one gets the triplet

$$\mathcal{H}_w^1 \hookrightarrow \mathcal{H} \hookrightarrow (\mathcal{H}_w^1)^\times. \tag{8.20}$$

In other words, we obtain a RHS whose extreme spaces are Banach spaces, thus a Banach Gel'fand triple and a PIP-space. The action of U on $\mathcal{H}_w^1$ can be extended to $(\mathcal{H}_w^1)^\times$ by duality:

$$\langle \phi, U(g)\psi \rangle := \langle U(g^{-1})\phi, \psi \rangle, \quad \text{for } \phi \in \mathcal{H}_w^1, \ \psi \in (\mathcal{H}_w^1)^\times.$$

Therefore the CS transform can also be extended as $V_\eta \psi(g) := \langle U(g)\eta, \psi \rangle$ for $\psi \in (\mathcal{H}_w^1)^\times$. This extension has the following properties, which clearly mean that we are in a PIP-space-setting.

Proposition 8.5.1. *(i) The inner product of $\mathcal{H}$ extends to a sesquilinear U-invariant pairing between $\mathcal{H}_w^1$ and $(\mathcal{H}_w^1)^\times$. For any $\psi \in (\mathcal{H}_w^1)^\times$, the CS transform $V_\eta \psi(g) := \langle U(g)\eta, \psi \rangle$ is a continuous function in $L_{1/w}^\infty(G)$.*

(ii) The map $V_\eta : (\mathcal{H}_w^1)^\times \to L_{1/w}^\infty(G)$ is one-to-one and intertwines U and L, i.e., one has $V_\eta(U(g)\psi) = L_g V_\eta \psi, \ \forall \psi \in (\mathcal{H}_w^1)^\times$.

(iii) If η is normalized by $\|C\eta\| = 1$, the reproducing formula holds true:

$$V_\eta \psi = V_\eta \psi * V_\eta \eta, \quad \text{for all } \psi \in (\mathcal{H}_w^1)^\times.$$

Let now Y be a solid Banach function space (actually, a Köthe function space) on G, that is, a Banach space of functions on G, continuously embedded in $L_{loc}^1(G)$, and satisfying the solidity condition (Section 4.4). Then the *coorbit space of Y under the representation U* is the space

$$\text{Co}Y := \{\psi \in (\mathcal{H}_w^1)^\times \text{ with } V_\eta \psi \in Y\}.$$

As natural norm, one takes $\|\psi\|_{\text{Co}Y} := \|V_\eta \psi\|_Y$. The basic properties of these spaces are as follows.

Theorem 8.5.2. *(i) CoY is a U-invariant Banach space, continuously embedded into $(\mathcal{H}_w^1)^\times$.*

(ii) $\mathcal{C}oY$ is independent of the choice of the analyzing vector $\eta \in \mathcal{A}_w$, i.e.,
different vectors define the same space with an equivalent norm.

(iii) $\mathcal{C}oY$ is independent of the reservoir $(\mathcal{H}_w^1)^\times$, i.e., if w_1 is another weight
with $w(g) \leqslant Cw_1(g)$ for all $g \in G$ and $\mathcal{A}_{w_1} \neq \{0\}$, then both weights
generate the same space $\mathcal{C}oY$.

The proof of this theorem relies on the following proposition, which is crucial
for the applications.

Proposition 8.5.3. (i) Given $\eta \in \mathcal{A}_w$, a function $\Psi \in Y$ is of the form
$V_\eta \psi$ for some $\psi \in \mathcal{C}oY$ if and only if Ψ satisfies the reproducing formula,
i.e., $\Psi = \Psi * V_\eta \eta$. It follows that

(ii) $V_\eta : \mathcal{C}oY \to Y$ establishes an isometric isomorphism between $\mathcal{C}oY$ and
the closed subspace $Y * V_\eta \eta$ of Y, whereas $\Psi \mapsto \Psi * V_\eta \eta$ defines a bounded
projection from Y onto that subspace.

(iii) Every function $\Psi = \Psi * V_\eta \eta$ is continuous and belongs to $L_{1/w}^\infty(G)$.

Further interesting properties of coorbit spaces are summarized in the
following

Proposition 8.5.4. (i) $\mathcal{C}oL_{1/w}^\infty = (\mathcal{H}_w^1)^\times$.

(ii) $\mathcal{C}oL^2 = \mathcal{H}$.

(iii) Assume that Y has an absolutely continuous norm (i.e., $Y^\times = Y^\alpha$, the
Köthe dual, see Section 4.4). Then

$$(\mathcal{C}oY)^\times = \mathcal{C}oY^\alpha = \mathcal{C}oY^\times.$$

As a consequence, $\mathcal{C}oY$ is reflexive if Y is reflexive.

Besides the fact that coorbit spaces provide a unified description of a number of useful function spaces, their advantage is that, for all these spaces, the
coorbit language yields interesting *atomic decompositions*. This means that
every element in a given space of functions or distributions can be represented as a sum of simpler functions, called *atoms*. Then many properties of
the space, such as duality, interpolation, operator theory, growth and smoothness properties, can be characterized in terms of such atoms. Furthermore,
the atoms are obtained in a unified way by the action of a group on the space,
this being, of course, the coherent state formalism. In turn, such atomic decompositions may be used as a discretization technique that allows to obtain
in a simpler way various types of *frames* in the spaces in question. This is of
crucial importance for the applications, in particular approximation theory.

The key to the atomic decompositions is that one can associate to each
Banach function space Y a sequence space Y_d that characterizes the properties of Y. One starts with a discrete set of points $X = (x_i)_{i \in I}$ in G such as
the one used in the definition of a BUPU in Section 8.2, that is,

(i) For a given neighborhood U_o of the identity in G, the family X is U_o-
dense, i.e., $(x_i U_o)_{i \in I}$ covers G.

(ii) The family X is relatively separated, that is, for any relatively compact set W with nonempty interior, $\sup_{i \in I} \operatorname{card}\{k : x_k W \cap x_i W \neq \emptyset\} < \infty$ (here card stands for the cardinality of the set).

Then, given a Banach function space Y as before and a discrete family $X = (x_i)_{i \in I}$, one defines the *associated discrete Banach space* as

$$Y_d := \{\Lambda = (\lambda_i)_{i \in I} \text{ with } \sum_{i \in I} \lambda_i \chi_{x_i W} \in Y\}$$

with the natural norm $\|\Lambda\|_{Y_d} := \|\sum_{i \in I} |\lambda_i| \chi_{x_i W}\|_Y$. The space Y_d does not depend of the choice of W, different sets yield the same space Y_d with equivalent norms. If the functions of compact support are dense in Y, then the finite sequences form a dense subspace of Y_d. To give an example, if $Y = L_m^p$, then $Y_d = \ell_m^p$, with the weights $m(i) = m(x_i)$.

Using this tool, the central result is the atomic decomposition in $\mathcal{C}oY$. Let $X = (x_i)_{i \in I}$ be a discrete family as above and Y_d the discrete Banach space associated to Y, with $\Lambda = (\lambda_i)_{i \in I}$. Then, roughly speaking, one has:

(i) *Analysis:* There exists a bounded operator $A : \mathcal{C}oY \to Y_d$, thus

$$\|Af\|_{Y_d} \leqslant C_0 \|f\|_{\mathcal{C}oY},$$

such that every $f \in \mathcal{C}oY$ can be represented as $f = \sum_{i \in I} \lambda_i U(x_i)\eta$, where $Af = \Lambda = (\lambda_i)_{i \in I}$.

(ii) *Synthesis:* Conversely, every element $\Lambda \in Y_d$ defines an element $f = \sum_{i \in I} \lambda_i U(x_i)\eta$ in $\mathcal{C}oY$ with

$$\|f\|_{\mathcal{C}oY} \leqslant C_1 \|\Lambda\|_{Y_d}.$$

In both cases, convergence is in the sense of the norm of $\mathcal{C}oY$, if the finite sequences are dense in Y_d, in the weak*-sense of $(\mathcal{H}_w^1)^\times$ otherwise.

Associated discrete Banach spaces are the key to a number of interesting results about coorbit spaces. In fact, the Banach space structure of $\mathcal{C}oY$ is closely related to that of Y_d, although it is not known whether the two are always isomorphic. For instance:

(1) $\mathcal{C}oY \subseteq \mathcal{C}oZ$ if, and only if, $Y_d \subseteq Z_d$. In particular, $\mathcal{C}oY = \mathcal{C}oZ$ if and only if $Y_d = Z_d$.
(2) $\mathcal{C}oY$ shares with Y_d all properties which are inherited by closed subspaces and finite direct sums of Banach spaces.
(3) $\mathcal{C}oY$ is reflexive if and only if Y_d is reflexive.
(4) Whenever $\mathcal{C}oY \subseteq \mathcal{C}oZ$, the inclusion $J : \mathcal{C}oY \to \mathcal{C}oZ$ is automatically continuous. The same is true for $J_d : Y_d \to Z_d$.
(5) Moreover, J is compact (resp. Hilbert-Schmidt, nuclear) if and only if J_d is compact (resp. Hilbert-Schmidt, nuclear)

In order to go further into the properties of the class of coorbit spaces, in the PIP-space spirit, we need one more qualification. A coorbit space CoY is called *minimal* if $\mathcal{H}_w^1$ is norm dense in CoY. It is called *maximal* if it is not properly contained in another coorbit space defining the same norm on $\mathcal{H}_w^1$. Then one has:

Proposition 8.5.5. (i) CoY is a minimal coorbit space if and only if the finite sequences are dense in Y_d, if and only if $(Y_d)^\times = (Y_d)^a$.
(ii) CoY is a maximal coorbit space if and only if $CoY = CoZ^\alpha$ for some Banach function space Z, if and only if $Y_d = Z_d^\alpha$ for some appropriate sequence space Z.

Finally, there is a result that pertains to lattice properties of the coorbit spaces (there are further results concerning the hereditary properties of interpolation methods).

Proposition 8.5.6. (i) The family of all minimal coorbit spaces is closed with respect to finite intersections and sums.
(ii) The family of all maximal coorbit spaces is closed with respect to intersections and sums.
(iii) The family of all reflexive coorbit spaces is closed with respect to duality, intersections and sums.

For instance, for two minimal coorbit spaces, one has:

$$(CoY^1 \cap CoY^2)^\times = (CoY^1)^\times + (CoY^2)^\times$$
$$(CoY^1 + CoY^2)^\times = (CoY^1)^\times \cap (CoY^2)^\times,$$

and all four spaces are minimal coorbit spaces. Similarly for reflexive spaces.

It is clear that further PIP-space-type results could be obtained by combining the coorbit space methodology with the theory of Köthe sequence spaces developed in Section 4.3.

We are going now to indicate very briefly a number of examples of coorbit spaces of interest for signal processing.

(i) The Weyl-Heisenberg group and modulation spaces

The (reduced) Weyl-Heisenberg group is $\mathbb{H}_d = \mathbb{R}^d \times \mathbb{R}^d \times \mathbb{T}$, with elements $h = (x, y, \tau)$ and group law $h_1 h_2 = (x_1, y_1, \tau_1)(x_2, y_2, \tau_2) := (x_1 + x_2, y_1 + y_2, \tau_1 \tau_2 e^{iy_1 x_2})$. The group $\mathbb{H}_d$ is unimodular, with Haar measure $dh = dx\, dy\, d\tau$. The relevant representation is the so-called *Schrödinger representation*, which forms the basis of nonrelativistic quantum mechanics, namely,

$$\big(U(x, y, \tau)\phi\big)(z) := \tau\big(M_y T_x \phi\big)(z) = \tau e^{2\pi i x y}\phi(z - x), \ z \in \mathbb{R}^d, \ \phi \in L^2(\mathbb{R}^d, dz).$$
$$(8.21)$$

Notice there are many different normalizations in the literature, both for the group and the representation. It follows that $V_\eta \phi$ is simply the Gabor transform, multiplied by the innocuous factor $\tau \in \mathbb{T}$, and one shows that the Schrödinger representation is integrable. To that effect, one proves that

$$\int_{\mathbb{H}_d} |V_\eta \phi(h)| \, dh < \infty,$$

whenever both functions η, ϕ have compactly supported Fourier transforms. Choose now the weight function $w_s(x, y, \tau) := (1 + |y|)^s$, $s \geqslant 0$, and the weighted L^p spaces $L^p_{w_s}$. Then it turns out that the corresponding coorbit spaces are $CoL^p_{w_s} = M^{p,p}_{w_s}(\mathbb{R}^{2d})$, belonging to the family of modulation spaces. For $p = 1$, in particular, one gets the Feichtinger algebra, $CoL^1 = S_0$. For $p = 2$, one recovers the fractional Sobolev or Bessel potential spaces H^s. Finally, applying the discretization procedure mentioned above, one gets for the atomic decomposition simply the familiar Gabor frames.

In addition to the representation (8.21), the Weyl-Heisenberg group admits other, nonequivalent, unitary irreducible representations (UIRs), namely, $U^k(x, y, \tau) := \tau^k M_y T_x$, $k \in \mathbb{Z} \setminus \{0\}$. However, the Stone-von Neumann uniqueness theorem says that any unitary irreducible representation of $\mathbb{H}_d$ is equivalent to some U^k and, moreover, all these representations yield the same coorbit spaces.

(ii) The affine group and Besov spaces

The full affine group of the line is $G_{\mathrm{aff}} = \mathbb{R} \rtimes \mathbb{R}_* := \{(b, a) : b \in \mathbb{R}, a \neq 0, \}$, with the natural action $x \mapsto ax + b$ and group law $(b, a)(b', a') = (b + ab', aa')$. The group G_{aff} is non-unimodular, the left Haar measure is $d\mu(b, a) = |a|^{-2} da \, db$ and the right Haar measure is $d\mu_r(b, a) = |a|^{-1} da \, db$.

Up to unitary equivalence, G_{aff} has a unique UIR, acting in $L^2(\mathbb{R}, dx)$, namely,

$$(U(b, a)\psi)(x) := (T_b D_a \psi)(x) = |a|^{-1/2} \psi\left(\frac{x - b}{a}\right), \quad \psi \in L^2(\mathbb{R}, dx) \quad (8.22)$$

where D_a is the unitary dilation operator defined in (8.15). We may also write

$$(V_\eta \phi)(b, a) = \langle T_b D_a \eta | \phi \rangle = (D_a \eta^\nabla * \phi)(b),$$

where $\eta^\nabla(x) := \overline{\eta(-x)}$. This representation is square integrable, even integrable. This is shown as in the Weyl-Heisenberg case, starting with a function $\eta \in L^2(\mathbb{R})$ such that $\operatorname{supp} \hat{\eta}$ is compact and bounded away from 0. Then $V_\eta \eta \in L^1_w$ for many weights w, in particular $r_s(b, a) := |a|^{-s}$, $s \in \mathbb{R}$. Then it follows that $\phi \in CoL^p_{r_s}$ if and only if

$$\int_{\mathbb{R}_*} \|D_a \eta^\nabla * \phi\|_p \, |a|^{-sp} \frac{da}{|a|^2} < \infty.$$

This means that $\mathcal{C}oL^p_{r_s} = \dot{B}^{s-1/2-1/p}_{pp}$, a homogeneous Besov space. Of course, the resulting atomic decompositions are simply wavelet expansions.

The extension to the multidimensional case is easy. One starts with the similitude group of $\mathbb{R}^d$, consisting of translations, rotations and dilations, namely, $\mathrm{SIM}(d) = \mathbb{R}^d \rtimes (\mathbb{R}^+_* \times \mathrm{SO}(d))$. Again this group has, up to unitary equivalence, a unique UIR, acting in $L^2(\mathbb{R}^d)$:

$$\big(U(b, a, R)\psi\big)(x) = a^{-d/2}\psi(a^{-1}R^{-1}(x - b)), \ a > 0, \ b \in \mathbb{R}^d, \ R \in \mathrm{SO}(d).$$

Then the analysis is the same as for $d = 1$ and leads to multidimensional wavelet expansions.

(iii) SL(2,ℝ) and Bergman spaces

The group $\mathrm{SL}(2, \mathbb{R})$ is the group of all real 2×2 matrices of determinant equal to 1 and it is unimodular. It has a family of square integrable representations (the discrete series), acting in Hilbert spaces of functions analytic in the upper half-plane $\mathbb{C}^+ := \{z = x + iy \in \mathbb{C}, y > 0\}$. The representation spaces are special cases of the so-called *Bergman spaces* $A^{p,\beta}, 1 \leqslant p < \infty, \beta > 1$, defined as follows:

$$A^{p,\beta} := \{f \text{ analytic in } \mathbb{C}^+ : \|f\|^p_{p,\beta} = \iint_{\mathbb{C}^+} |f(z)|^p \, y^\beta \frac{dxdy}{y^2} < \infty\}. \quad (8.23)$$

For any integer $m \geqslant 2$, the discrete series representation U_m is defined on $A^{2,m}$ by

$$\left(U_m \begin{pmatrix} a & b \\ c & d \end{pmatrix} f\right)(z) := f\left(\frac{dz - b}{-cz + a}\right)(-cz + a)^{-m}. \quad (8.24)$$

Consider now the simpler functions $f_m(z) := (z + i)^{-m}$. The following properties are known:

(i) $f_m \in A^{2,m}$ for all $m \geqslant 2$.
(ii) $V_{f_m} f_m = \langle U_m(\cdot) f_m | f_m \rangle \in L^1(\mathrm{SL}(2, \mathbb{R})), \forall m \geqslant 3$, that is, $f_m \in \mathcal{H}^1(U_m)$, but $\mathcal{H}^1(U_2) = \{0\}$.
(iii) For $m \geqslant 3$, one has $f_m \in A^{p,pm/2}, 1 \leqslant p < \infty$, and U_m acts isometrically on $A^{p,pm/2}$.

As a consequence, the Bergman spaces are coorbit spaces of $L^p(\mathrm{SL}(2, \mathbb{R}))$ under the representation U_m, namely, $A^{p,pm/2} = \mathcal{C}o(L^p, U_m)$.

(iv) The Weyl-Heisenberg group and Fock-Bargmann spaces

The Fock-Bargmann space $\mathfrak{F} \equiv \mathfrak{F}^0$, introduced in Section 1.1.3, Example (v), may be generalized as follows:

$$\mathfrak{F}^{(p)} = \{f(z) \text{ entire on } \mathbb{C}^d : \|f\|^p_{\mathfrak{F}^{(p)}} := \int_{\mathbb{C}^d} |f(z)|^p \, d\mu(z) < \infty\}, \qquad (8.25)$$

where $d\mu(z) = \pi^{-d} e^{-|z|^2} d\nu(z)$ is the Gaussian measure on $\mathbb{C}^d$. Thus $\mathfrak{F}^{(2)} = \mathfrak{F}$ is a Hilbert space, on which the Weyl-Heisenberg group acts via the following representation:

$$\left(U(x,y,\tau)f\right)(z) = \tau e^{-i\pi xy} e^{-|w|^2/2} e^{wz} f(z - \overline{w}), \quad w = x + iy \in \mathbb{C}^d, \ f \in \mathfrak{F}^{(2)}. \tag{8.26}$$

Choosing the function $g(z) \equiv 1$, one sees that

$$|(V_g f)(x,y,\tau)| = |f(\overline{w})| e^{-|w|^2/2},$$

so that $V_g f \in L^p(\mathbb{H}_d)$ if and only if $f \in \mathfrak{F}^{(p)}$. In other words, $\mathcal{C}o(L^p(\mathbb{H}_d)) = \mathfrak{F}^{(p)}$. As for atomic decompositions, one gets all sorts of sampling theorems for entire functions.

As a last remark, it should be mentioned that the whole theory of coorbit spaces may be generalized to the case of a representation of a locally compact group G which is only integrable modulo a subgroup H. In that case, the analysis takes place not on the group G itself, but on the quotient manifold $X = G/H$. A good example is the two-sphere $S^2 = \mathrm{SO}(3)/\mathrm{SO}(2)$.

Notes for Chapter 8

Section 8.1. Mixed norm spaces are described in detail in Benedek–Panzone [44], Bertrandias–Datry–Dupuis [45, 46]. For their applications in functional analysis and, in particular, the Schur tests of boundedness, we refer to Samarah et al. [174]. A nice application in signal processing, in the context of sparse representations of signals, may be found in Kowalski–Torrésani [130]. Here the authors consider hierarchized indices, as described in the text, and use (1,2)- and (2,1)-norms. The idea is that ℓ^1 norms favor sparsity, whereas ℓ^2 norms do not.

- Amalgam spaces (L^p, ℓ^q) are discussed in detail in the review papers by Fournier–Stewart [97] and Holland [123]. Some information may also be found in the monograph of Gröchenig [Grö01, Sec.11.1]. In the notation $W(L^p, \ell^q)$ or $W(L^p, L^q)$), W stands for Wiener, since this author was the first to consider a space of this type. Indeed he introduced the spaces $W(L^1, \ell^2)$ and $W(L^2, \ell^1)$ in [190], then $W(L^1, \ell^\infty)$ and $W(L^\infty, \ell^1)$ in [191] and the textbook [Wie33]. Weighted Wiener amalgams are reviewed by Heil [122]. The general theory was developed by Feichtinger [79–81], using the notion of bounded uniform partition of unity (BUPU). It is often applied to Gabor analysis, see for instance in Gröchenig–Heil–Okoudjou [113] and in Feichtinger–Weisz [92].

Section 8.3. Time-frequency analysis, more precisely Gabor analysis, and modulation spaces are studied in the monograph of Gröchenig [Grö01, Sec.11.1], that we follow closely. A good review paper on modulation spaces is Feichtinger [82].

- The Feichtinger algebra $\mathcal{S}_0$ was introduced by Feichtinger [78]. A comprehensive study may be found in Feichtinger–Zimmermann [93]. Banach Gel'fand triples and their application in Gabor analysis, in particular to localization operators, are studied by Dörfler–Feichtinger–Gröchenig [74] and Feichtinger–Luef–Cordero [91]. The concept extends naturally to locally compact abelian groups, see for instance, Feichtinger–Kozek [90] or the monograph of Reiter–Stegeman [RS00]. As a replacement of the standard Schwartz' space RHS $(\mathcal{S}, L^2, \mathcal{S}^\times)$, the Banach Gel'fand triple $(\mathcal{S}_0, L^2, \mathcal{S}_0^\times)$ has found a natural role in the connection between Gabor analysis and noncommutative geometry [145, 146].

Section 8.4. Besov spaces are described in the monographs of Bergh–Löfström [BL76, Sec.6.2] and Triebel [Tri78b, vol.II, Chap.2]. For Littlewood-Paley dyadic decompositions, see Stein [Ste70, Sec. IV.5].

- A standard reference for wavelet analysis is the textbook of Daubechies [Dau92]. The characterization of Besov spaces in terms of the decay of discrete wavelet coefficients is analyzed in the monograph of Meyer [Mey90, Chap.II.9 and Chap.VI.10]. The analogous result in terms of the continuous wavelet transform is due to Perrier–Basdevant [164].
- Several examples of unconditional wavelet bases are given by Gröchenig [112]. The case of Wilson bases is due to Feichtinger–Gröchenig–Walnut [87].
- The flexible Gabor-wavelet transform was introduced by Feichtinger–Fornasier [89] as a transform intermediate between the Gabor ($\alpha = 0$) and the wavelet transforms ($\alpha = 1$). The case ($\alpha = 1/2$) is called the Fourier–Bros–Iagolnitzer or FBI transform. For the latter, we refer to the monograph of Delort [Del92]. The α-modulation spaces based on this transform have been introduced independently by P. Gröbner [83, Grö92] and Päivärinta–Somersalo [161], and further analyzed by Dahlke *et al.* [64] and Fornasier *et al.* [96]. Actually they are a particular case of the family of function spaces intermediate between modulation and Besov spaces introduced by Nazaret–Holschneider [151]. There is a considerable literature about the α-modulation spaces, mostly in the context of pseudodifferential operators. Typical examples are the papers by Borup [53] and Borup–Nielsen [54, 55]. As a general reference for pseudodifferential operators, we may mention the classical text of Shubin [Shu01] or Folland's monograph [Fol89]. On the other hand, α-modulation spaces provide an intrinsic adaptivity which is useful for the analysis of very complex signals, containing both stationary components and transients. A nice example is their use for disentangling car crash signals [159].

Section 8.5. Coorbit spaces were introduced by Feichtinger–Gröchenig [84–86, 88]. Here we follow closely [85] and [86], in particular Proposition 8.5.1, Theorem 8.5.2 and Proposition 8.5.3 are taken from the former paper.

- For information about square integrable representations, see the review paper of Ali–Antoine–Gazeau–Mueller [6], the textbook of Ali–Antoine–Gazeau [AAG00], especially Chapter 8, and the papers by Grossmann–Morlet–Paul [120, 121]. A deeper analysis, including integrable representations, may be found in Warner's treatise [War72, Sec. 4.5.9].
- Coorbit spaces on quotient manifolds have been constructed by Dahlke [63, 64], using the theory of square integrable representations modulo a subgroup developed by Ali–Antoine–Gazeau [3]-[6]. See also the textbook of Ali–Antoine–Gazeau [AAG00].
- The Weyl-Heisenberg group $\mathbb{H}_d$ is often denoted G_{WH} in the physics literature, in particular in Ali *et al.* [6] and in the textbook [AAG00]. For instance, there one writes $\tau = e^{i\theta} \in \mathbb{T}$ and uses a different normalization, namely,

$$g = (\theta, q, p), \ \theta \in \mathbb{R}, \ (q, p) \in \mathbb{R}^{2d},$$

with multiplication law

$$g_1 g_2 = \big(\theta_1 + \theta_2 + \xi\big((q_1, p_1); (q_2, p_2)\big), \ q_1 + q_2, \ p_1 + p_2\big),$$

where ξ is the multiplier function

$$\xi\big((q_1, p_1); (q_2, p_2)\big) = \frac{1}{2}(p_1 q_2 - p_2 q_1).$$

There the Schrödinger representation takes the form

$$\big(U(\theta, q, p)\phi\big)(x) = e^{i\theta} e^{ip(x - \frac{q}{2})} \phi(x - q), \ \phi \in L^2(\mathbb{R}, dx).$$

A standard reference for the Weyl-Heisenberg group is Folland's monograph [Fol89].
- For representations of SL(2,$\mathbb{R}$), see, for instance, the monograph of Lang [Lan75, chap. IX]. Actually, there are two versions of the discrete series representations mentioned here. In the SL(2,$\mathbb{R}$) presentation, the standard representation space is the Bergman space $A^{2,m}$ of functions analytic in the upper half-plane. But, since SL(2,$\mathbb{R}$) is isomorphic to SU(1,1), there is an equivalent version for which the representation space is the Bergman space $\mathcal{K}_{m-2}$ of functions analytic in the unit disk. The spaces $\mathcal{K}_\alpha$ are defined in (4.55).

Appendix A
Galois Connections

Let $\mathcal{M}$ be a partially ordered set, with order relation $\subseteq$. A *closure* on $\mathcal{M}$ is a map $X \mapsto \overline{X}$ from $\mathcal{M}$ to $\mathcal{M}$ such that: (i) $X \subseteq \overline{X}$; (ii) $\overline{\overline{X}} = \overline{X}$; (iii) $X \subseteq Y$ implies $\overline{X} \subseteq \overline{Y}$. An element $X \in \mathcal{M}$ is said to be *closed* whenever $X = \overline{X}$. Let $\mathcal{C}(\mathcal{M})$ be the set of all closed elements of $\mathcal{M}$, with the induced order. Then if $\mathcal{M}$ is a complete lattice (see Section 1.1.2), so is $\mathcal{C}(\mathcal{M})$ with respect to the lattice operations ($\mathcal{N} \subseteq \mathcal{C}(\mathcal{M})$)

$$
\bigwedge_{X \in \mathcal{N}} X \Big|_{\mathcal{C}(\mathcal{M})} = \bigwedge_{X \in \mathcal{N}} X \Big|_{\mathcal{M}} ,
$$

$$
\bigvee_{X \in \mathcal{N}} X \Big|_{\mathcal{C}(\mathcal{M})} = \overline{\left(\bigvee_{X \in \mathcal{N}} X \Big|_{\mathcal{M}} \right)}.
$$

Let now $\mathcal{L}$ and $\mathcal{M}$ be two partially ordered sets. A *Galois connection* [160, Bir66] between $\mathcal{L}$ and $\mathcal{M}$ is a pair of maps $\alpha : \mathcal{L} \to \mathcal{M}$ and $\beta : \mathcal{M} \to \mathcal{L}$, such that (we write $X^\alpha := \alpha(X)$ and $Y^\beta := \beta(Y)$:

(i) both α and β reverse order,
(ii) $S \subseteq S^{\alpha\beta}$ for each $S \in \mathcal{L}$ and $T \subseteq T^{\beta\alpha}$ for each $T \in \mathcal{M}$.

It follows from the definition that $S \mapsto S^{\alpha\beta}$ (resp. $T \mapsto T^{\beta\alpha}$) is a closure on $\mathcal{L}$ (resp. $\mathcal{M}$).

From now on, we will assume that both $\mathcal{L}$ and $\mathcal{M}$ are complete lattices. So are then the two sets of closed elements $\mathcal{C}(\mathcal{L})$ and $\mathcal{C}(\mathcal{M})$. Furthermore, α (resp. β) is a lattice anti-isomorphism of $\mathcal{C}(\mathcal{L})$ onto $\mathcal{C}(\mathcal{M})$ [resp. $\mathcal{C}(\mathcal{M})$ onto $\mathcal{C}(\mathcal{L})$], that is, for every subset $\mathcal{N} \subseteq \mathcal{C}(\mathcal{L})$, one has

$$
\left(\bigvee_{X \in \mathcal{N}} X \right)^\alpha = \bigwedge_{X \in \mathcal{N}} X^\alpha, \tag{A.1}
$$

$$
\left(\bigwedge_{X \in \mathcal{N}} X \right)^\alpha = \bigvee_{X \in \mathcal{N}} X^\alpha, \tag{A.2}
$$

and similarly for β. Actually [165], in the case where both $\mathcal{L}$ and $\mathcal{M}$ are complete lattices, the two maps α and β are not independent: α generates a Galois connection if, and only if, it satisfies the single condition (A.1); β is then uniquely determined and given by

$$T^\beta = \bigvee_{X^\alpha \geqslant T} X.$$

Further insight into the structure of Galois connections can be found in the paper of Shmuely [178]. Two points are of interest for us.

(1) Galois correspondences between the complete lattices $\mathcal{L}$ and $\mathcal{M}$ are in 1-1 correspondence with certain subsets of $\mathcal{L} \times \mathcal{M}$, called G-ideals. Since the latter form a complete lattice with the natural order inherited $\mathcal{L} \times \mathcal{M}$, it follows that the set of all Galois maps $\alpha : \mathcal{L} \to \mathcal{M}$ that generate a Galois connection also form a complete lattice.

(2) Every Galois connection between $\mathcal{L}$ and $\mathcal{M}$ can be generated by a *binary relation*, that is a subset $\varUpsilon \subseteq \mathcal{L} \times \mathcal{M}$; for instance $\varUpsilon = \{(S,T) : T \leqslant S^\alpha\}$. Conversely, every binary relation $\varUpsilon \subseteq \mathcal{L} \times \mathcal{M}$ generates a Galois connection, namely the one that corresponds to the minimal G-ideal generated (by lattice operations) by $\varUpsilon, (0_{\mathcal{L}}, 1_{\mathcal{M}})$ and $(1_{\mathcal{L}}, 0_{\mathcal{M}})$ where 0, resp. 1, denotes the smallest, resp. largest, element of the lattice indicated.

Next we specialize the discussion to the case $\mathcal{L} = \mathcal{M} = \mathcal{P}(S)$, the complete lattice of all subsets of a given set S. If we assume furthermore that $\alpha = \beta$ (such an α is called an *involution*), the resulting self-dual Galois connection on $\mathcal{P}(S)$ is exactly what was called *compatibility on S* in Definition 1.1.4. Indeed $\alpha = \beta$ is equivalent to the corresponding binary relation $\varUpsilon$ being symmetric: $(X, Y) \in \varUpsilon$ if and only if $(Y, X) \in \varUpsilon$, which we can write, as usual, $X \# Y$ (with $\# \equiv \alpha = \beta$). The closed elements of $\mathcal{P}(S)$ are precisely the *assaying subspaces*, which constitute the complete lattice $\mathcal{F}(S, \#)$. The map $\#$ of $\mathcal{F}(S, \#)$ onto itself is an involution and a lattice anti-isomorphism. Property (1) above means that the set $\mathrm{Comp}(S)$ of all compatibilities on S is in a 1-1 correspondence with the set of all symmetric G-ideals of $\mathcal{P}(S) \times \mathcal{P}(S)$ and the latter is a complete lattice with respect to the order inherited from $\mathcal{P}(S) \times \mathcal{P}(S)$. That order is exactly the notion of the weak comparability ("weakly finer", etc.) introduced in Section 1.5. Property (2) yields the notion of generating subset for a Galois connection. These are exactly our *generating subsets*, discussed in Section 1.4.

Finally, we come back to the linear case. Let V be a vector space and $\#$ a linear compatibility on V. By the very definition, the relation $f \# g$ ($f, g \in V$) is equivalent to $[f] \# [g]$, where $[f]$ is the one-dimensional subspace generated by f. Thus we may take as complete lattice $\mathcal{L}(V)$ the set of all vector subspaces of V. A linear compatibility on V is the same thing as a self-dual (or involutive) Galois map on $\mathcal{L}(V)$. The whole discussion above then goes through.

Note. For lattice theory, we refer to Birkhoff [Bir66]. For Galois connections, see also Ore [160], Pickert [165] or Shmuely [178]. The consideration of Galois connections in the PIP-space context was first made in Antoine–Gustafson [20].

Appendix B
Some Facts About Locally Convex Spaces

In this appendix, we will recall some basic definitions and facts concerning locally convex topological vector spaces (LCS), i.e., topological vector spaces (TVS) which have a base of neighborhoods of zero consisting of convex sets, or equivalently, spaces with a topology that can be defined in terms of a family of seminorms. Our reference is the textbook of Köthe [Köt69], except for the notation of the different topologies, where we follow Schaefer [Sch71].

B.1 Completeness

A LCS $E[t]$ is *complete* if every Cauchy net has a limit in E; it is *quasi-complete* if every closed bounded set in $E[t]$ is complete; it is sequentially complete if every Cauchy sequence has a limit in E. Of course, completeness $\Rightarrow$ quasi-completeness $\Rightarrow$ sequential completeness, and the three notions coincide for metrizable spaces, i.e., Banach or Fréchet spaces (a Fréchet space is a complete metrizable LCS).

B.2 Dual Pairs and Canonical Topologies

Two vector spaces E, F form a dual pair $\langle E, F \rangle$ if there is a bilinear form $\langle \cdot | \cdot \rangle$ on $E \times F$, separating in both arguments: $\langle e|f \rangle = 0$, $\forall f \in F$, implies $e = 0$, $\langle e|f \rangle = 0$, $\forall e \in E$, implies $f = 0$. For any LCS E, with dual E', that is, the space of continuous linear functionals on E, $\langle E, E' \rangle$ is a dual pair. So is $\langle E, E^\times \rangle$, where $E^\times$ denotes the space of continuous conjugate linear functionals on E.

Given the dual pair $\langle E, F \rangle$, the weak topology $\sigma(E, F)$ is the coarsest topology on E for which the linear forms $e \mapsto \langle e|f \rangle$, $f \in F$, are continuous, and in fact, for which the dual of E is F. It is locally convex and Hausdorff. A basis of neighborhoods of zero consists of the sets S^o, where S runs over all finite subsets of F, and $S^o := \{e \in E : |\langle e|f \rangle| \leqslant 1, \forall f \in F\}$ is the (absolute) polar of $S \subset F$. The weak topology on F, $\sigma(F, E)$, is defined similarly.

The Mackey topology $\tau(E, F)$ can be defined as the finest topology on E such that the dual is F (its existence is the content of the Mackey-Arens theorem); a basis of neighborhoods of zero is given by the sets T^o, where T runs over all absolutely convex, $\sigma(F, E)$-compact subsets of F.

The strong topology $\beta(F, E)$ is defined by the basis of neighborhoods of zero $\{U^o\}$ where U runs over all absolutely convex $\sigma(F, E)$-closed and bounded subsets of F. We recall that a subset B of a LCS E is *bounded* if, for every neighborhood of zero U, there is a $\rho > 0$ such that $B \subset \rho U$.

A topology $t(E)$ on E is called a *topology of the dual pair* $\langle E, F \rangle$ if the dual of $E[t(E)]$ is F. Then one has, for any topology $t(E)$ of the dual pair:

$$\sigma(E, F) \prec t(E) \prec \tau(E, F) \prec \beta(E, F).$$

Thus, by definition, the Mackey topology is a topology of the dual pair, while the strong one is not, in general. If we start with a given topology $t(E)$ on E we have the same inclusions with $F = E'$ (equivalently, with $F = E^\times$).[1]

In a dual pair $\langle E, F \rangle$, several classes of subsets of E depend only on the dual pair and *not* on the topology of E, i.e., they coincide for all topologies of the dual pair. Such are: closed subspaces, convex closed subsets, dense and total subsets, bounded subsets.

A LCS $E[t]$ is *barreled* if $t(E) = \tau(E, E') = \beta(E, E')$. A metrizable LCS always carries its Mackey topology, $t(E) = \tau(E, E')$, but need not be barreled. A complete metrizable LCS, i.e., a Banach or a Fréchet space, is always barreled. The strong dual of a metrizable LCS is a complete (DF)-space. This class, whose definition is quite technical, contains also all normed spaces (see [Köt69, Sec. 29.3]).

A LCS E is *bornological* if every seminorm that is bounded on bounded sets is continuous. Every metrizable space is bornological, but a bornological space need not be barreled.

A barreled LCS $E[t]$ is called a *Montel space* if every bounded subset of E is relatively compact. A Montel space is necessarily quasi-complete and reflexive (see below), and its strong dual is also a Montel space. An infinite dimensional Banach space cannot be Montel. Typical examples are $\omega, \varphi, \mathcal{S}(\mathbb{R}^n), \mathcal{S}^\times(\mathbb{R}^n)$.

B.3 Linear Maps

Let $E[t(E)]$, $F[t(F)]$ be locally convex spaces with topologies $t(E)$ and $t(F)$ and duals E', F', respectively. Let $\alpha : E \to F$ be a linear map. Consider the following statements

---

[1] In fact, all the statements that follow remain valid if one replaces the dual E' by the conjugate dual $E^\times$.

(i) α is continuous from $E[\mathsf{t}(E)]$ into $F[\mathsf{t}(F)]$.
(ii) α is continuous from $E[\tau(E, E')]$ into $F[\tau(F, F')]$.
(iii) α is continuous from $E[\sigma(E, E')]$ into $F[\sigma(F, F')]$.
(iv) There exists a linear map $\alpha' : F' \to E'$ such that

$$\langle \alpha(f)|g \rangle = \langle f|\alpha'(g) \rangle, \quad \forall f \in E, g \in F'.$$

(v) The sesquilinear form $b(f, g) = \langle f|\alpha'(g) \rangle$ is separately continuous in each of its arguments $f \in E, g \in F'$.

Then (ii)-(v) are equivalent and (i) implies all of them. The map α', which is automatically Mackey and weakly continuous, is called the *adjoint* or *transposed* map of α. For Banach or Fréchet spaces, the five statements are obviously equivalent. This also happens if the spaces under consideration are *reflexive* (see below).

B.4 Reflexivity

Given an LCS E, the canonical topologies $\sigma(E', E), \tau(E', E), \beta(E', E)$ are defined in the same way; thus, with the notation of Section 2.3:

$$E'|_\beta \hookrightarrow E'|_\tau \hookrightarrow E'|_\sigma.$$

By definition $\left(E'|_\tau \right)' = E$, but the dual of the strong dual, called the *bidual*, $E'' = (E'|_\beta)'$ may be strictly larger than E. A LCS E is called *semi-reflexive* if E'' coincides with E as a vector space. E is called *reflexive* if, in addition, the strong bidual $E''[\beta(E'', E')] := (E'|_\beta)'|_\beta$ coincides with E as a TVS. The two notions are different in general, but they coincide for Fréchet spaces. In fact, for a Fréchet or a Banach space E, the following properties are equivalent: (i) E is reflexive, (ii) E is semi-reflexive; (iii) E is weakly quasi-complete; and (iv) the strong dual $E'|_\beta$ is reflexive. Notice that an incomplete normed or metrizable space can never be reflexive.

A dual pair $\langle E, F \rangle$ is *reflexive* if each space is the strong dual of the other: $E|'_\beta = F, F|'_\beta = E$. Equivalently, if $E|_\tau$ and $F|_\tau$ are both semi-reflexive, or if they are both barreled: $\tau(E, F) = \beta(E, F)$ and $\tau(F, E) = \beta(F, E)$. In a reflexive pair, both spaces are reflexive and quasi-complete for their weak and their Mackey (= strong) topology.

B.5 Projective Limits

Let be given a vector space E, a family $\{E_\alpha\}$ of LCS and maps $i_\alpha : E \to E_\alpha$ such that, for every nonzero $x \in E$, there is some α with $i_\alpha(x) \neq 0$. Then there is a coarsest topology on E that makes all the maps i_α continuous; it is called the *projective topology* and E with this topology, $E_{\text{proj}} := \varprojlim_\alpha E_\alpha$, is called the *projective limit* of $\{E_\alpha\}$ with respect to the maps i_α. The projective limit is said to be *reduced* if $i_\alpha(E)$ is dense in E_α for each α (this can always be achieved without restriction of generality). The following properties are useful:

- E_{proj} is complete (resp. quasi-complete, sequentially complete) if every E_α is.
- Given any LCS Y, a linear map $t : Y \to E_{\text{proj}}$ is continuous if, and only if, each composed map $t_\alpha = i_\alpha \circ t : Y \to E_\alpha$ is continuous.

The following examples are important:

(i) Let E be a LCS, H a subspace of E. The *subspace topology* on H is the projective topology with respect to the embedding $i : H \to E$.

(ii) Let $\{E_\alpha\}$ be as above and $E = \prod_\alpha E_\alpha$ the product vector space, i.e., the set-theoretic product $\{x = (x_\alpha)\}$, $x_\alpha \in E_\alpha$, with addition and scalar multiplication defined componentwise. The *product topology* on E is the projective topology with respect to the projection maps $p_\alpha : E \to E_\alpha$.

(iii) In a general projective limit, E_{proj} is isomorphic to a closed subspace of the product $\prod_\alpha E_\alpha$. In the case considered here, $\{E_\alpha\}$ is a family of vector subspaces of a given vector space V, each of which is itself a LCS. Then $E_{\text{proj}} = \varprojlim_\alpha E_\alpha$ is the subspace $\cap_\alpha E_\alpha$ with the projective topology. E_{proj} is metrizable if, and only if, the family $\{E_\alpha\}$ contains a cofinal countable subfamily of metrizable spaces (this makes sense since the subspaces are partially ordered by inclusion).

(iv) In particular, if the family $\{E_\alpha\}$ consists of a countable family $\{\mathcal{H}_n, n \in \mathbb{N}\}$ of Hilbert spaces, with mutually consistent norms (see Section 2.2), then $E_{\text{proj}} = \varprojlim_n \mathcal{H}_n$ is called a *countably Hilbert space*. Such spaces and their relation to PIP-spaces have been studied by Antoine–Karwowski [22].

(v) A countably Hilbert space $E_{\text{proj}} = \varprojlim_n \mathcal{H}_n$ is said to be *nuclear* if, for every $\mathcal{H}_n$, there is a larger $\mathcal{H}_m$ such that the embedding $E_{mn} : \mathcal{H}_n \to \mathcal{H}_m$ is a nuclear (i.e., trace class) operator or, equivalently, a Hilbert-Schmidt operator. Such nuclear spaces are precisely those in which the nuclear spectral theorem holds true (see Section 7.1.1).

B.6 Inductive Limits

Let be given a vector space F, a family $\{F_\kappa\}$ of LCS, with $\{\kappa\}$ a directed set, and maps $j_\kappa : F_\kappa \to F$. Then there is a finest topology on F that makes all the maps j_κ continuous; it is called the *inductive topology* and F with this topology, denoted $F_{\text{ind}} := \varinjlim_\kappa F_\kappa$, is called the *inductive limit* of $\{F_\kappa\}$ with respect to the maps j_κ. The inductive limit is called *strict* if $\nu \leqslant \kappa$ implies that $F_\nu \hookrightarrow F_\kappa$ and F_κ induces on F_ν its original topology. Given any LCS, Y, a linear map $t : F_{\mathrm{ind}} \to Y$ is continuous if, and only if, each composed map $t_\kappa = t \circ j_\kappa : F_\kappa \to Y$ is continuous. Again three cases are worth mentioning.

(i) If E is a LCS and H a closed subspace, the *quotient topology* on E/H is the inductive topology with respect to the canonical surjection $\pi : E \to E/H$.

(ii) For any family $\{F_\kappa\}$, let $F = \sum_\kappa F_\kappa$ be the direct sum, that is, the subspace of $\prod_\kappa F_\kappa$ consisting of elements with finitely many nonzero coordinates;[2] the *direct sum topology* on F is the inductive topology with respect to the embeddings $j_\kappa : F_\kappa \to F$.

(iii) In particular, a LCS is called an *(LF)-space* if it can be represented as a strict inductive limit of an increasing sequence of Fréchet spaces.

(iv) For a general inductive limit, F_{ind} is isomorphic to a quotient of $\sum_\kappa F_\kappa$.

B.7 Duality and Hereditary Properties

Let E be a LCS, H a closed subspace. The orthogonal space of H is $H^\perp := \{f \in E' : \langle f | h \rangle = 0, \forall h \in H\}$. $H^\perp$ is a closed subspace of E'. Then the dual of H is $E'/H^\perp$, and the dual of E/H is $H^\perp$.

As for canonical topologies, the hereditary properties are the following:

- The *Mackey topology* is inherited by quotients, but not by closed subspaces in general:

$$\tau(E, E')|_{E/H} = \tau(E/H, H^\perp), \qquad \tau(E, E')|_H \prec \tau(H, E'/H^\perp).$$

We do get equality for subspaces in two cases: if $\tau(E, E')|_H$ is metrizable, or if H is a dense subspace (hence not closed).

- The *weak topology* is inherited both by quotients and closed subspaces, whereas the *strong topology* is inherited by neither of them, in general.

Direct sums and products are dual to each other:

$$\left(\prod_\alpha E_\alpha\right)'_{\mathrm{proj}} = \sum_\alpha E'_\alpha,$$
$$\left(\sum_\kappa F_\kappa\right)'_{\mathrm{ind}} = \prod_\kappa F'_\kappa,$$

and Mackey topologies go through:

$$\left(\prod_\alpha E_\alpha\right)\Big|_\tau = \left(\prod_\alpha E_\alpha|_\tau\right)_{\mathrm{proj}} \quad \text{and} \quad \left(\sum_\kappa F_\kappa\right)\Big|_\tau = \left(\sum_\kappa F_\kappa|_\tau\right)_{\mathrm{ind}}.$$

---

[2] Some authors denote the direct sum as $\bigoplus_\kappa F_\kappa$

Reduced projective limits and inductive limits are also dual to each other:

$$\begin{aligned} (\varprojlim_\alpha E_\alpha)' &= \varinjlim_\alpha E'_\alpha \\ (\varinjlim_\kappa F_\kappa)' &= \varprojlim_\kappa F'_\kappa \end{aligned} \qquad \text{(if the left-hand side is reduced).}$$

Combining all the above results, we get finally that the Mackey topology on a projective, resp. inductive, limit is finer than the projective, resp. inductive, limit of the respective Mackey topologies, with equality if the former is metrizable:

$$\begin{aligned} \varprojlim_\alpha (E_\alpha|_\tau) &\prec \left.\left(\varprojlim_\alpha E_\alpha\right)\right|_\tau \\ \varinjlim_\kappa (F_\kappa|_\tau) &\prec \left.\left(\varinjlim_\kappa F_\kappa\right)\right|_\tau \end{aligned} \qquad \text{(equality if the left-hand side is metrizable)}$$

where, with a slight abuse of language, the symbol $\prec$ means that the topology of the space on the left-hand side is coarser than that of the space on the right-hand side.

Weak topologies go through projective limits only, and there is no general result for strong topologies.

Epilogue

Now it is time to draw some conclusions for this volume. As we explained in the Prologue, PIP-spaces emerged as a common backbone underlying a number of different structures, such as Rigged Hilbert spaces, Nested Hilbert spaces, scales of Banach or Hilbert spaces, etc. The aim of all these constructions is to bypass the inconvenients of the Hilbert space, and notably the L^2 spaces which are the natural environment of quantum mechanics. Indeed, Hilbert space is too "narrow", in the sense that it cannot accommodate many useful objects (δ functions, distributions, plane waves, ...). At the same time, it is too "large", because it contains very singular functions, which result from the requirement of completeness in the Lebesgue norm topology. Of course, the way out of the dilemma is to go beyond Hilbert space, essentially towards distribution theory. But this creates another problem, by creating two different kinds of mathematical objects, namely, nice (test) functions and continuous linear functionals on them. By contrast, the PIP-space point of view stays closer to the original Hilbert space, in the sense that there is a unique vector space, with an inner product, defined for certain pairs only. This is in fact the central point: the existence of an inner product between two vectors is *not* a property of each of them separately, but a property of the pair. Once this fact has been formalized in the form of a compatibility relation on the vector space at hand, all the rest follows naturally.

As we have seen, the development of the subject, while indeed rooted in the mathematical formulation of quantum mechanics, gets its own momentum. Many concrete examples have been discussed, showing the versatility of the concept of PIP-space. Actually the latter brings in a clear mathematical advantage. Indeed, as for any procedure of mathematical abstraction, the theory of PIP-spaces puts at our disposal a series of techniques and mathematical tools which are independent of the specific "example" we are dealing with. We have emphasized the word "example", since the family of spaces that constitute any one of them has been in most cases the subject of an extensive literature in functional analysis. All examples discussed in Chapter 4 and in Chapter 8 enter in this category.

Yet many problems remain open, and we have occasionally stated some of them. More fundamentally, the task ahead is to apply these ideas to concrete situations, in particular in the realm of signal processing. It is there that ever more sophisticated types of function spaces emerge, generated by specific applications. They always come in families, indexed by one or several parameters controlling smoothness, local or asymptotic behavior, etc. And systematically the relevant objects are not the individual spaces, but the family as a whole. This is precisely the idea of PIP-spaces, and this is why we thought that the time has come to review the formalism.

Bibliography

A. Books and Theses

[Ada75] R.A. Adams, *Sobolev Spaces*, Academic Press, New York, 1975.

[AGHH88] S. Albeverio, F. Gesztesy, R. Hoegh-Krohn, and H. Holden, *Solvable Models in Quantum Mechanics*, Springer-Verlag, New York, Berlin, Heidelberg, 1988.

[AK00] S. Albeverio and P. Kurasov, *Singular Perturbations of Differential Operators*, London Math. Soc. Lect. Notes Series, Vol. 271, Cambridge Univ. Press, Cambridge (UK), 2000.

[AAG00] S.T. Ali, J-P. Antoine, and J-P. Gazeau, *Coherent States, Wavelets and Their Generalizations*, Springer-Verlag, New York, Berlin, Heidelberg, 2000.

[AIT02] J-P. Antoine, A. Inoue, and C. Trapani, *Partial *-Algebras and Their Operator Realizations*, Kluwer, Dordrecht, 2002.

[Bal90] L.E. Ballentine, *Quantum Mechanics*, Prentice-Hall, Englewood Cliffs, NJ, 1990.

[BR77] A.O. Barut and R. Rączka, *Theory of Group Representations and Applications*, PWN – Polish Scientific Publishers, Warszawa, 1977.

[Ber68] Yu.M. Berezanskii, *Expansions in Eigenfunctions of Self-Adjoint Operators*, Amer. Math. Soc., Providence, RI, 1968.

[BL76] J. Bergh and J. Löfström, *Interpolation Spaces*, Springer-Verlag, Berlin, 1976.

[Bir66] G. Birkhoff, *Lattice Theory*, 3rd ed., Amer. Math. Soc., Coll. Publ., Providence, RI., 1966.

[Bog74] J. Bognar, *Indefinite Inner Product Spaces*, Springer-Verlag, Berlin-New York, 1974.

[BLT75] N.N. Bogolubov, A.A. Logunov, and I.T. Todorov, *Introduction to Axiomatic Quantum Field Theory*, Benjamin, Reading, MA, 1975.

[BG89] A. Böhm and M. Gadella, *Dirac Kets, Gamow Vectors and Gel'fand Triplets*, Lecture Notes in Physics, Vol. 348, Springer-Verlag, Berlin, 1989.

[Boh93] A. Bohm, *Quantum Mechanics, Foundations and Applications*, 3rd ed., Springer-Verlag, Berlin, 1993.

[BR79] O. Bratteli and D.W. Robinson, *Operator Algebras and Quantum Statistical Mechanics I, II*, Springer-Verlag, Berlin, 1979.

[Bre86] H. Brézis, *Analisi Funzionale – Teoria e Applicazioni*, Liguori, Napoli, 1986.

[Chr03] O. Christensen, *An Introduction to Frames and Riesz Bases*, Birkhäuser, Boston, MA, 2003.

[Dau92] I. Daubechies, *Ten Lectures on Wavelets*, SIAM, Philadelphia, 1992.

[Del92] J-M. Delort, *F.B.I. transformation. Second Microlocalization, and Semilinear Caustics*, Lecture Notes in Mathematics, Vol. 1522, Springer-Verlag, Berlin, Heidelberg, New York, 1992.

[Dir30] P.A.M. Dirac, *The Principles of Quantum Mechanics*, (1st ed. 1930) 4th ed. Oxford University Press, Oxford, 1958.

[DH90] D.A. Dubin and M.A. Hennings, *Quantum Mechanics, Algebras and Distributions*, Pitman Research Notes in Math. Series, Longman, Harlow, 1990.

[Dur70] P.L. Duren, *Theory of H^p Spaces*, Academic Press, New York, San Francisco, and London, 1970.

[Eij83] S.J.L. van Eijndhoven, *Analyticity Spaces, Trajectory Spaces, and Linear Mappings Between Them*, Ph.D. Dissertation, Technical University of Eindhoven, 1983.

[EG85] S.J.L. van Eijndhoven and J. de Graaf, *Trajectory Spaces, Generalized Functions and Unbounded Operators*, Lecture Notes in Mathematics, Vol. 1162, Springer-Verlag, Berlin, Heidelberg, New York, Tokyo, 1985.

[EG86] S.J.L. van Eijndhoven and J. de Graaf, *A Mathematical Introduction to Dirac's Formalism*, North-Holland, Amsterdam, New York, Oxford, Tokyo, 1986.

[FW68] K. Floret and J. Wloka, *Einführung in die Theorie der lokalkonvexen Räume*, Lecture Notes in Mathematics, Vol. 56, Springer-Verlag, Berlin, Heidelberg, New York, 1968 (esp. §9).

[Fol89] G.B. Folland, *Harmonic Analysis in Phase Space*, Princeton Univ. Press, Princeton, NJ, 1989.

[FG70] L. Fonda and G.C. Ghirardi, *Symmetry Principles in Quantum Physics*, M.Dekker, New York, 1970.

[GS64] I.M. Gel'fand and G.E. Shilov, *Generalized Functions*, Vols. I-III, Academic Press, New York and London, 1964–1968.

[GV64] I.M. Gel'fand and N.Ya. Vilenkin, *Generalized Functions*, Vol. IV, Academic Press, New York and London, 1964.

[GK69] I.C. Gohberg and M.G. Krein, *Introduction to the Theory of Linear Non-selfadjoint Operators*, Math. Monogr. Amer.Math. Soc., Providence, RI., 1969; French trad. *Introduction à la théorie des opérateurs linéaires non auto-adjoints dans un espace hilbertien*, Dunod, Paris, 1971.

[GW64] M.L. Goldberger and K.M. Watson, *Collision Theory*, Wiley, New York, 1964.

[Gol82] A. Gollier, *Espaces de fonctions analytiques et théorie de la diffusion*, Mémoire de licence, UCL, 1982 (unpublished).

[GR65] I.S. Gradshteyn and I.M. Ryzhik, *Table of Integrals, Series and Products*, Academic, New York and London, 1965 (cf. 0.224).

[Grö92] P. Gröbner, *Banachräume glatter Funktionen and Zerlegungs-Methoden*, Ph.D. Dissertation, University of Vienna, 1992.

[Grö01] K. Gröchenig, *Foundations of Time-Frequency Analysis*, Birkhäuser, Boston, MA, 2001.

[Gro66] A. Grothendieck, *Produits tensoriels topologiques et espaces nucléaires*, Memoirs Amer. Math. Soc. Nr.16, Providence, RI, 1966.

[Hör63] L. Hörmander, *Linear Partial Differential Operators*, Springer-Verlag, Berlin, Heidelberg, New York, 1963.

[Kat76] T. Kato, *Perturbation Theory for Linear Operators*, Springer-Verlag, Berlin, 1976.

[Kle87] M. Klein, *Phénomènes de résonance et espaces de fonctions analytiques*, Mémoire de licence, UCL, 1987 (unpublished).

[Koo80] P. Koosis, *Introduction to H^p Spaces*, London Math. Soc. Lect. Notes Series, vol. 40, Cambridge Univ. Press, Cambridge (UK), 1980.

[Köt69] G. Köthe, *Topological Vector Spaces, Vols. I, II*, Springer-Verlag, Berlin, 1969, 1979.

[KZPS66] M.A. Krasnoselskii, P.P. Zabreiko, E.I. Pustylnik, and P.E. Sbolevskii, *Integral Operators in Spaces of Summable Functions*, Nauka, Moscow, 1966; Engl. Transl., Noordhoff, Leyden, 1976.

[Lan75] S. Lang, $SL_2(\mathbb{R})$, Addison-Wesley, Reading, MA, 1975.

[LM68] J-L. Lions and E. Magenes, *Problèmes aux limites non homogènes et applications* (3 vols.), Dunod, Paris, 1968.

[Mat75] F. Mathot, *Opérateurs dans les espaces à produit interne partiel*. Thèse de doctorat, Univ. Cath. Louvain, 1975.

[Mey90] Y. Meyer, *Ondelettes et Opérateurs. I. Ondelettes.* Hermann, Paris, 1990.

[Mit65] B. Mitchell, *Theory of Categories.* Academic Press, New York, 1965.

[Nac72] A.-M. Nachin, *Une échelle d'espaces de Hilbert dans la représentation de Bargmann,* Thèse de 3e cycle, Université d'Aix-Marseille, 1972.

[Obe92] M. Oberguggenberger, *flexplication of distributions and applications to partial differential equations,* Pitman Research Notes in Mathematics Series, n. 259, Longman, Harlow, 1992.

[New72] R.G. Newton, *Scattering Theory of Waves and Particles,* 2nd ed., Springer-Verlag, Heidelberg, 1982.

[Oos73] W. Oostenbrink, *Normed ideals,* Ph.D. Dissertation, University of Groningen, 1973.

[Pie80] A. Pietsch, *Operator Ideals,* North-Holland, Amsterdam, New York and Oxford, 1980.

[Pru71] E. Prugovečki, *Quantum Mechanics in Hilbert Space,* Academic Press, New York and London, 1971.

[RS72] M. Reed and B. Simon, *Methods of Modern Mathematical Physics. I. Functional Analysis,* Academic Press, New York and London, 1972, 1980.

[RS75] M. Reed and B. Simon, *Methods of Modern Mathematical Physics. II. Fourier Analysis, Self-Adjointness,* Academic Press, New York, San Francisco and London, 1975.

[RS79] M. Reed and B. Simon, *Methods of Modern Mathematical Physics. III. Scattering Theory,* Academic Press, New York, San Francisco and London, 1979.

[RS00] H. Reiter and J.D. Stegeman, *Classical Harmonic Analysis and Locally Compact Groups.* 2nd ed., Clarendon Press, Oxford, 2000.

[RN56] F. Riesz and B. Sz.-Nagy, *Vorlesungen über Funktionalanalysis,* DVW, Berlin, 1956; Ed. fr. *Leçons d'Analyse Fonctionnelle,* Gauthier-Villars, Paris, et Akadémiai Kiadó, Budapest, 1968.

[Sch71] H.H. Schaefer, *Topological Vector Spaces,* Springer-Verlag, Berlin, 1971.

[Sch70] R. Schatten, *Norm Ideals of Completely Continuous Operators,* Springer-Verlag, Berlin-New York, 1970.

[Sch90] K. Schmüdgen, *Unbounded Operator Algebras and Representation Theory,* Akademie-Verlag, Berlin, 1990.

[Sch57] L. Schwartz, *Théorie des distributions,* I, II, Hermann, Paris, 1957–1959.

[Sim71] B. Simon, *Quantum Mechanics for Hamiltonians Defined as Quadratic Forms,* Princeton Univ. Press, Princeton, NJ, 1971.

[Sim79] B. Simon, *Trace Ideals and Their Applications,* London Math. Soc. Lect. Notes Series, Vol. 35, Cambridge Univ. Press, Cambridge (UK), 1979.

[Ste70] E.M. Stein, *Singular integrals and Differentiability Properties of Functions,* Princeton Univ. Press, Princeton, NJ, 1970.

[Sch62] S.S. Schweber, *An Introduction to Relativistic Quantum Field Theory,* Harper & Row, New York, 1962.

[Shu01] M.A. Shubin, *Pseudodifferential Operators and Spectral Theory,* Springer-Verlag, Berlin, 2001.

[SZ79] S. Strătilă and L. Zsidó, *Lectures on von Neumann Algebras,* Editura Academiei, Bucharest and Abacus Press, Tunbridge Wells, Kent, 1979.

[SW64] R.F. Streater and A.S. Wightman, *PCT, Spin and Statistics, and All That,* Benjamin, New York, 1964.

[Tsc02] F. Tschinke, **-Algebre parziali di distribuzioni,* Ph.D. Thesis, U. Palermo, 2002.

[Tre67] F. Trêves, *Topological Vector Spaces, Distributions and Kernels,* Academic Press, New York and London, 1967.

[Tri78a] H. Triebel, *Interpolation Theory, Function Spaces, Differential Operators,* North-Holland, Amsterdam, 1978.

[Tri78b] H. Triebel, *Function Spaces I–III,* Birkhäuser, Basel, Boston, 1978, 1992, 2006.

[vNe55] J. von Neumann, *Mathematical Foundations of Quantum Mechanics*, English transl. of the original German edition (1933) by R.T. Byer, Princeton University Press, Princeton, NJ, 1955.

[Wag97] E. Wagner, *Eine Verallgemeinerung der GNS-Konstruktion für partielle *-Algebren auf der Basis von 𝔅-Wichten*, Diplomarbeit, Univ. Leipzig, 1997.

[War72] G. Warner, *Harmonic Analysis on Semi-Simple Lie Groups. I*, Springer-Verlag, Berlin-Heidelberg-New York, 1972.

[Wie33] N. Wiener, *The Fourier Integral and Certain of its Applications*, Cambridge Univ. Press, Cambridge (UK), 1933; reprinted in Dover, New York, 1958.

[Zaa61] A.C. Zaanen, *Integration*, 2nd. ed., Chap. 15; North-Holland, Amsterdam, 1961.

B. Articles

[1] S. Agmon and L. Hörmander, Asymptotic properties of solutions of differential equations with simple characteristics, *J. Anal. Math. (Jerusalem)* **30** (1976), 1–38.

[2] J. Aguilar and J-M. Combes, A class of analytic perturbations for one-body Schrödinger Hamiltonians, *Commun. Math. Phys.* **22** (1971), 269–279.

[3] S.T. Ali, J-P. Antoine, and J-P. Gazeau, Square integrability of group representations on homogeneous spaces I. Reproducing triples and frames, *Ann. Inst. H. Poincaré* **55** (1991), 829–856.

[4] S.T. Ali, J-P. Antoine, and J-P. Gazeau, Square integrability of group representations on homogeneous spaces II. Coherent and quasi-coherent states – The case of the Poincaré group, *Ann. Inst. H. Poincaré* **55** (1991), 857–890.

[5] S.T. Ali, J-P. Antoine, and J-P. Gazeau, Continuous frames in Hilbert space, *Ann. Phys. (NY)* **222** (1993), 1–37.

[6] S.T. Ali, J-P. Antoine, J-P. Gazeau, and U.A. Mueller, Coherent states and their generalizations: A mathematical overview, *Reviews Math. Phys.* **7** (1995), 1013–1104.

[7] I. Amemiya and H. Araki, A remark on Piron's paper, *Publ. Res. Inst. Math. Sci., Kyoto Univ.* **A 2** (1967), 423–427.

[8] J-P. Antoine, Dirac formalism and symmetry problems in Quantum Mechanics. I. General Dirac formalism, *J. Math. Phys.* **10** (1969), 53–69.

[9] J-P. Antoine, Dirac formalism and symmetry problems in Quantum Mechanics. II. Dirac formalism and symmetry problems in Quantum Mechanics, *J. Math. Phys.* **10** (1969), 2276–2290.

[10] J-P. Antoine, Super-Hilbert spaces and quantum field theory, in *Recent Developments in Relativistic Quantum Field Theory and its Applications*, J.Lopuszanski, (ed.), Vol. II, pp.139–173, Acta Univ. Wratislaviensis No. 207, Wrocław, 1974.

[11] J-P. Antoine, Partial inner product spaces: Another approach to unbounded operators, *Proc. Intern. Conf. on Operator Algebras, Ideals and their Applications in Theoretical Physics*, pp.279–289; H. Baumgärtel *et al.* (eds.), Teubner, Leipzig, 1978.

[12] J-P. Antoine, Partial inner product spaces. III. Compatibility relations revisited, *J. Math. Phys.* **21** (1980), 268–279.

[13] J-P. Antoine, Partial inner product spaces. IV. Topological considerations, *J. Math. Phys.* **21** (1980), 2067–2079.

[14] J-P. Antoine, Quantum mechanics beyond Hilbert space. Applications to scattering theory, in *Quantum Theory in Rigged Hilbert Spaces – Semigroups, Irreversibility and Causality*, pp.3–33; A. Böhm, H. D. Doebner, and P. Kielanowski (eds.), Springer-Verlag, Berlin, 1998.

[15] J-P. Antoine, Symmetries of quantum systems: A partial inner product space approach, *J. Phys. A: Math. Theor.* **40** (2007), 1–13.

[16] J-P. Antoine, Partial *-algebras, a tool for the mathematical description of physical systems? in *Contributions in Mathematical Physics: A Tribute to Gerard G. Emch*, pp.37–68; S. Twareque Ali and Kalyan B. Sinha (eds), Hindustan Book Agency, New Delhi, 2007.

[17] J-P. Antoine and A. Grossmann, Partial inner product spaces. I. General properties, *J. Funct. Analysis*, **23** (1976), 369–378.

[18] J-P. Antoine and A. Grossmann, Partial inner product spaces. II. Operators, *J. Funct. Analysis*, **23** (1976,) 379–391.

[19] J-P. Antoine and A. Grossmann, Orthocomplemented subspaces of nondegenerate partial inner product spaces. *J. Math. Phys.* **19** (1978), 329–335.

[20] J-P. Antoine and K. Gustafson, Partial inner product spaces and semi-inner product spaces, *Adv. in Math.* **41** (1981), 281–300.

[21] J-P. Antoine, A. Inoue, and C. Trapani, Partial *-algebras of closable operators. II. States and representations of partial *-algebras, *Publ. Res. Inst. Math. Sci., Kyoto Univ.* **27** (1991), 399–430.

[22] J-P. Antoine and W. Karwowski, Countably Hilbert spaces and partial inner product spaces, *Bull. Acad. Pol. Sci., Série Sc. Phys. et Astr.* **27** (1979), 223–232.

[23] J-P. Antoine and W. Karwowski, Interpolation theory and refinement of nested Hilbert spaces, *J. Math. Phys.* **22** (1981), 2489–2496.

[24] J-P. Antoine and W. Karwowski, Partial *-algebras of closed operators, in *Quantum Theory of Particles and Fields*, pp.13–30; B. Jancewicz and J. Lukierski (eds.) World Scientific, Singapore, 1983.

[25] J-P. Antoine and W. Karwowski, Partial *-algebras of closed linear operators in Hilbert space, *Publ. Res. Inst. Math. Sci., Kyoto Univ.* **21** (1985), 205–236 ; Add./Err. *ibid.* **22** (1986), 507–511.

[26] J-P. Antoine and F. Mathot, Partial inner product spaces generated by algebras of unbounded operators, preprint UCL-IPT-81-10 (unpublished).

[27] J-P. Antoine and F. Mathot, Regular operators on partial inner product spaces, *Ann. Inst. H. Poincaré* **37** (1982), 29–50.

[28] J-P. Antoine and F. Mathot, The canonical nested Hilbert space associated to a partial inner product space, (preprint UCL-IPT-82-06, unpublished).

[29] J-P. Antoine and F. Mathot, Partial *-algebras of closed operators and their commutants. I. General structure, *Ann. Inst. H. Poincaré* **46** (1987), 299–324.

[30] J-P. Antoine, Y. Soulet, and C. Trapani, Weights on partial *-algebras, *J. Math. Anal. Appl.* **192** (1995), 920–941.

[31] J-P. Antoine and M. Vause, Partial inner product spaces of entire functions, *Ann. Inst. H. Poincaré* **35** (1981), 195–224.

[32] H. Araki and E.J. Woods, Representations of the canonical commutation relations describing a nonrelativistic infinite free Bose gas, *J. Math. Phys.* **4** (1963), 637–662.

[33] R. Arens, The space L^ω and convex topological rings, *Bull. Amer. Math. Soc.* **52** (1946), 931–935.

[34] R. Ascoli, G. Epifanio, and A. Restivo, On the mathematical description of quantized fields, *Commun. Math. Phys.* **18** (1970), 291–300.

[35] R. Ascoli, G. Epifanio, and A. Restivo, *-Algebrae of unbounded operators in scalar-product spaces, *Rivista Mat. Univ. Parma* **3** (1974), 21–32.

[36] R. Ascoli, P. Gabbi, and G. Palleschi, On the representation of linear operators in L^2 spaces by means of "generalized matrices", *J. Math. Phys.* **19** (1978), 1023–1027.

[37] J. Avron, A. Grossmann, R. Rodriguez, and J. Zak, Spectral properties of Bloch Hamiltonians, *Ann. Phys. (NY)*, **103** (1977), 47–63.

[38] D. Babbitt, Rigged Hilbert spaces and one-particle Schrödinger operators *Rep. Math. Phys.* **3** (1972), 37–42.

[39] E. Balslev and J-M. Combes, Spectral properties of many-body Schrödinder operators with dilatation-analytic interactions, *Commun. Math. Phys.* **22** (1971), 280–294.

[40] E. Balslev, A. Grossmann, and Th. Paul, A characterization of dilation-analytic operators, *Ann. Inst. H. Poincaré* **45** (1986), 277–292.

[41] V. Bargmann, On unitary ray representations of continuous groups, *Ann. Math.* **59** (1954), 1–46.

[42] V. Bargmann, On a Hilbert space of analytic functions and an associated integral transform. Part I, *Commun. Pure Appl. Math.* **14** (1961), 187–214.

[43] V. Bargmann, On a Hilbert space of analytic functions and an associated integral transform. Part II. A family of related function spaces; Application to distribution theory, *Commun. Pure Appl. Math.* **20** (1967), 1–101.

[44] A. Benedek and R. Panzone, The spaces L^P, with mixed norm, *Duke Math. J.* **28** (1961), 301–324.

[45] J.-P. Bertrandias, C. Datry et C. Dupuis, Unions et intersections d'espaces L^p invariantes par translation ou convolution, *Ann. Inst. Fourier, Grenoble* **28** (1978), 53–324.

[46] J.-P. Bertrandias et C. Dupuis, Transformation de Fourier sur les espaces $\ell^p(L^{p'})$ invariantes par translation ou convolution, *Ann. Inst. Fourier, Grenoble* **29** (1979), 189–206.

[47] A. Böhm, The Rigged Hilbert Space in quantum mechanics, in *Lectures in Theoretical Physics*, Vol. IX A, W.A Brittin *et al.* (eds.), pp. 255–315; Gordon & Breach, New York, 1967.

[48] A. Bohm, Gamow state vectors as functionals over subspaces of the nuclear space, *Lett. Math. Phys.* **3** (1979), 455–461.

[49] A. Bohm, Resonance poles and Gamow vectors in the rigged Hilbert space formulation of quantum mechanics, *J. Math. Phys.* **22** (1981), 2813–2823.

[50] A. Bohm, H. Kaldass, and S. Wickramasekara, Time asymmetric quantum theory. II. Relativistic resonances from S-matrix poles, *Fortschr. Phys.* **51** (2003), 569–603; id. III. Decaying states and the causal Poincaré group, *Fortschr. Phys.* **51**(2003), 604–634.

[51] H.J. Borchers, On the structure of the algebra of field operators, *Nuovo Cimento* **24** (1962), 214–236.

[52] H.J. Borchers, Algebraic aspects of Wightman Field Theory, in *Statistical Mechanics and Field Theory*, pp.31–79; R.N. Sen and C. Weil (eds.), Halsted Press, New York, 1972.

[53] L. Borup, Pseudodifferential operators on α-modulation spaces, *J. Funct. Spaces Appl.* **2** (2004), 107–123.

[54] L. Borup and M. Nielsen, Nonlinear approximation in α-modulation spaces, *Math. Nachr.* **279** (2006), 101–120.

[55] L. Borup and M. Nielsen, Boundedness for pseudodifferential operators on multivariate α-modulation spaces, *Ark. Mat* **44** (2006), 241–259.

[56] F. Bruhat, Sur les représentations induites des groupes de Lie, *Bull. Soc. Math. France* **84** (1956), 97–205.

[57] J.M. Chaiken, Number operators for representations of the commutation relations, *Commun. Math. Phys.* **8** (1968), 164–184.

[58] J. Cigler, Normed ideals in $L^1(G)$, *Nederl. Akad. Wetensch., Indag. Math.* **31** (1969), 273–282.

[59] J-M. Combes, Recent Developments in Quantum Scattering Theory, in *Mathematical Problems in Theoretical Physics*, Lecture Notes in Physics, Vol. 116, K. Osterwalder (ed.), Springer-Verlag, Berlin, 1980.

[60] J-M. Combes, in *Rigorous Atomic and Molecular Physics (Proceedings, Erice, 1980)*, G. Velo and A.S. Wightman (eds.), Reidel, Dordrecht, 1981.

[61] E. Cordero and L. Rodino, Short-Time Fourier Transform analysis of localization operators, in *Frames and Operator Theory in Analysis and Signal Processing*, D.R. Larson *et al.* (eds.), pp.47–68; Contemporary Mathematics, Vol. 451, Amer. Math. Soc., Providence, RI, 2008.

[62] L. Crone, A characterization of matrix operators in l^2, *Math. Z.* **123** (1971), 315–317.

[63] S. Dahlke, G. Steidl, and G. Teschke, Coorbit spaces and Banach frames on homogeneous spaces with applications to the sphere, *Adv. in Comput. Math.* **21** (2004), 147–180.

[64] S. Dahlke, M. Fornasier, H. Rauhut, G. Steidl, and G. Teschke, Generalized coorbit theory, Banach frames, and the relation to α-modulation spaces, *Proc. London Math. Soc.* **96** (2008), 464–506.

[65] I. Daubechies, On the distributions corresponding to bounded operators in the Weyl quantization, *Commun. Math. Phys.* **75** (1980), 229–238.

[66] I. Daubechies and A. Grossmann, An integral transform related to quantization. I. *J. Math. Phys.* **21** (1980), 2080–2090.

[67] I. Daubechies, A. Grossmann, and J. Reignier, An integral transform related to quantization. II. *J. Math. Phys.* **24** (1983), 239–254.

[68] H.W. Davis, F.J. Murray, and J.K. Weber, Families of L_p spaces with inductive and projective topologies, *Pacific J. Math.* **34** (1970), 619–638.

[69] H.W. Davis, F.J. Murray, and J.K. Weber, Inductive and projective limits of L_p spaces, *Portug. Math.* **31** (1972), 21–29.

[70] F. Debacker-Mathot, Some operator algebras on Nested Hilbert Spaces, *Commun. Math. Phys.* **42** (1975), 183–193.

[71] R. Debacker-Mathot, Integral decomposition of unbounded operator families, *Commun. Math. Phys.* **71** (1980), 47–58.

[72] R. de la Madrid, The role of the rigged Hilbert space in Quantum Mechanics, *Eur. J. Phys.* **26** (2005), 287–312.

[73] J. Dieudonné, Sur les espaces de Köthe, *J. Anal. Math.* (Jerusalem) **I** (1951), 81–115.

[74] M. Dörfler, H.G. Feichtinger, and K. Gröchenig, Time-frequency partitions for the Gelfand triple $(\mathcal{S}_0, L^2, \mathcal{S}_0^{\times})$, *Math. Scand.* **98** (2006), 81–96.

[75] G. Epifanio and C. Trapani, Quasi *-algebras valued quantized fields, *Ann. Inst. H. Poincaré* **46** (1987), 175–185.

[76] G. Epifanio and C. Trapani, Some topics in pre-Hilbert space, *J. Math. Phys.* **23** (1982), 39–42.

[77] G. Epifanio and C. Trapani, Partial *-algebras and Volterra convolution of distribution kernels, *J. Math. Phys.* **32** (1991), 1096–1101.

[78] H.G. Feichtinger, On a new Segal algebra, *Monatsh. Math.* **92** (1981), 269–289.

[79] H.G. Feichtinger, Banach convolution algebras of Wiener type, in *Proc. Conf. on Functions, Series, Operators, Budapest 1980*, pp.509–524; Colloq. Math. Soc. Janos Bolyai, Vol.35, North-Holland, Amsterdam, 1983.

[80] H.G. Feichtinger, Generalized amalgams, with applications to Fourier transform, *Can. J. Math.* **42** (1990), 395–409.

[81] H.G. Feichtinger, Wiener amalgams over Euclidean spaces and some of their applications, in *Function spaces, Proc. Conf. Edwardsville, IL (USA) 1990*, (K.Jarosz, ed.), pp.123–137, Lect. Notes Pure Appl. Math., Vol.136, Marcel Dekker, New York, 1992.

[82] H.G. Feichtinger, Modulation spaces: looking back and ahead, *Sampl. Theory Sig. Image Proc.* **5** (2006), 109–140.

[83] H.G. Feichtinger and P. Gröbner, Banach spaces of distributions defined by decomposition methods. I, *Math. Nachr.* **123** (1985), 97–120.

[84] H.G. Feichtinger and K. Gröchenig, A unified approach to atomic decompositions via integrable group representations, in *Function Spaces and Applications (Proc. Lund 1986)*, M. Cwikel *et al.* (eds.) Lecture Notes in Mathematics, Vol. 1302, 52–73; Springer-Verlag, Berlin, Heidelberg, New York, 1988.

[85] H.G. Feichtinger and K. Gröchenig, Banach spaces related to integrable group representations and their atomic decompositions. I, *J. Funct. Anal.* **86** (1989), 307–340.

[86] H.G. Feichtinger and K. Gröchenig, Banach spaces related to integrable group representations and their atomic decompositions. Part II. *Monatsh. Math.* **108** (1989), 129–148.

[87] H.G. Feichtinger, K. Gröchenig, and D. Walnut, Wilson bases and modulation spaces, *Math. Nachr.* **155** (1992), 7–17.

[88] H.G. Feichtinger and K. Gröchenig, Non-orthogonal wavelet and Gabor expansions, and group representations, in *Wavelets and their Applications*, M.B. Ruskai *et al.* (eds.), pp.353–375; Jones and Barlett, Boston and London, 1992.

[89] H.G. Feichtinger and M. Fornasier, Flexible Gabor-wavelet atomic decompositions for L^2-Sobolev spaces, *Ann. Matem.* **185** (2006), 105–131.

[90] H.G. Feichtinger and W. Kozek, Quantization of TF lattice-invariant operators on elementary LCA groups, in *Gabor Analysis and Algorithms — Theory and Applications*, H.G. Feichtinger and T. Strohmer (eds.), pp.233–266; Birkhäuser, Boston, 1998.

[91] H.G. Feichtinger, F. Luef, and E. Cordero, Banach Gel'fand triples for Gabor analysis, in *Pseudo-Differential Operators*, R. Rodino and M.W. Wong (eds.), pp.1–33; Lecture Notes in Mathematics, Vol. 1949, Springer-Verlag, Berlin, Heidelberg, New York, 2008.

[92] H.G. Feichtinger and F. Weisz, Gabor analysis on Wiener amalgams, *Sampl. Theory Signal Image Proc.* **6** (2007), 129–150.

[93] H.G. Feichtinger and G. Zimmermann, A Banach space of test functions for Gabor analysis, in *Gabor Analysis and Algorithms — Theory and Applications*, H.G. Feichtinger and T. Strohmer (eds.), pp.123–170; Birkhäuser, Boston, 1998.

[94] P.A. Fillmore and J.P. Williams, On operator ranges, *Adv. in Math.* **7** (1971), 254–281.

[95] C. Foiaş, Décompositions en opérateurs et vecteurs propres. I. Etude de ces décompositions et leurs relations avec les prolongements de ces opérateurs, *Rev. Math. Pures Appl.* **7** (1962), 241–282; II. Eléments de théorie spectrale dans les espaces nucléaires, *ibid.* **7** (1962), 571–602.

[96] M. Fornasier, Banach frames for α-modulation spaces, *Appl. Comput. Harmon. Anal.* **22** (1999), 157–175.

[97] J.J.F. Fournier and J. Stewart, Amalgams of L^p and ℓ^q, *Bull. Amer. Math. Soc.* **13** (1985), 1–21.

[98] K. Fredenhagen and J. Hertel, Local algebras of observables and pointlike localized fields, *Commun. Math. Phys.* **80** (1981), 555–561.

[99] D. Fredricks, Tight riggings for a complete set of commuting observables, *Rep. Math. Phys.* **8** (1975) 277–293.

[100] M. Friedrich and G. Lassner, Angereicherte Hilberträume, die zu Operatorenalgebren assoziiert sind, *Wiss. Z. KMU, Leipzig, Math.-Nat.R.* **27** (1978), 245–251.

[101] M. Friedrich and G. Lassner, Rigged Hilbert spaces and topologies on operator algebras, preprint E5-12420, JINR, Dubna (1979).

[102] M. Gadella, A rigged Hilbert space of Hardy-class functions: Applications to resonances, *J. Math. Phys.* **24**(1983), 1462–1469.

[103] G. Gamow, Zur Quantentheorie des Atomkernes, *Z. Phys.* **51** (1928), 204–212.

[104] D.J.H. Garling , The β- and γ-duality of sequence spaces, *Proc Camb. Phil. Soc.* **63** (1967), 963–981.

[105] J. Ginibre and G. Velo, The classical limit of nonrelativistic bosons. I. Borel summability for bounded potentials, *Ann. Phys. (NY)* **128** (1980), 243–285.

[106] J.P. Girardeau, Sur l'interpolation entre un espace localement convexe et son dual, *Rev. Fac. Ciencias Univ. Lisboa* **11** (1964–1965), 165–186.

[107] S. Goes and R. Welland, Some remarks on Köthe spaces, *Math. Annalen* **175** (1968), 127–131.

[108] R. Goodman, Analytic and entire vectors for representations of Lie groups, *Trans. Amer. math. Soc.* **143** (1969), 55–76.

[109] R. Goodman, One parameter groups generated by operators in an enveloping algebra, *J. Funct. Anal.* **6** (1970), 218–236.

[110] C. Goulaouic, Prolongements de foncteurs d'interpolation et applications, *Ann. Inst. Fourier (Grenoble)* **18** (1968), 1–98.

[111] G.G. Gould, On a class of integration spaces, *J. London Math. Soc.* **34** (1959), 161–172.

[112] K. Gröchenig, Unconditional bases in translation and dilation invariant function spaces on $\mathbb{R}^n$, *Constructive theory of functions (Varna 1987)*, pp.174–183; Bulg. Acad. Sc., Sofia, 1988.

[113] K. Gröchenig, C. Heil, and K. Okoudjou, Gabor analysis in weighted amalgam spaces, *Sampl. Theory Signal Image Proc.* **1** (2002), 225–259.

[114] A. Grossmann, Elementary properties of nested Hilbert spaces, *Commun. Math. Phys.* **2** (1965), 1–30.

[115] A. Grossmann, Hilbert spaces of type S, *J. Math. Phys.* **6** (1965), 54–67.

[116] A. Grossmann, Homomorphisms and direct sums of nested Hilbert spaces, *Commun. Math. Phys.* **4** (1967), 190–202.

[117] A. Grossmann, Fields at a point, *Commun. Math. Phys.* **4** (1967), 203–216.

[118] A. Grossmann, R. Hoegh-Krohn, and M. Mebkhout, A class of explicitly soluble, local, many center, Hamiltonians for one-particle Quantum Mechanics in two and three dimensions. I, *J. Math. Phys.* **21** (1980), 2376–2385.

[119] A. Grossmann, R. Hoegh-Krohn, and M. Mebkhout, The one particle theory of periodic point interactions, *Commun. Math. Phys.* **77** (1980), 87–110.

[120] A. Grossmann, J. Morlet, and Th. Paul, Integral transforms associated to square integrable representations. I. General results, *J. Math. Phys.* **26** (1985), 2473–2479.

[121] A. Grossmann, J. Morlet, and Th. Paul, Integral transforms associated to square integrable representations. II. Examples, *Ann. Inst. H. Poincaré* **45** (1986), 293–309.

[122] C. Heil, An introduction to weighted Wiener amalgams, in *Wavelets and their Applications (Chennai 2002)*, M. Krishna *et al.* (eds.), pp.183–216; Allied Publ., New Delhi, 2003.

[123] F. Holland, Harmonic analysis on amalgams of L^p and ℓ^q, *J. London Math. Soc.* (2) **10** (1975), 295–305.

[124] R.A. Hirschfeld, Projective limits of unitary representations, *Math. Ann.* **194** (1971), 180–196.

[125] L.P. Horwitz and E. Katznelson, A partial inner product space of analytic functions for resonances, *J. Math. Phys.* **24** (1983), 848–859.

[126] W. Hunziker, On the space-time behavior of Schrödinger wavefunctions, *J. Math. Phys.* **7** (1966), 300–304.

[127] A. Inoue, On a class of unbounded operator algebras. II. *Pacific J. Math.* **66** (1976), 411–431.

[128] A. Inoue, L^p-spaces and maximal unbounded Hilbert algebras, *J. Math. Soc. Japan* **30** (1978), 667–686.

[129] W. Karwowski, On Borchers class of Markoff fields, *Proc. Camb. Phil. Soc.* **76** (1974), 457–463.

[130] M. Kowalski and B. Torrésani, Sparsity and persistence: mixed norms provide simple signal models with dependent coefficients, *Signal, Image, Video Proc.* **3** (2009), 251–264.

[131] G. Köthe and O. Toeplitz, Lineare Räume mit unendlichvielen Koordinaten und Ringe unendlicher Matrizen, *J. Reine ang. Math.* **171** (1934), 193–226.

[132] S.G. Krein and Ju.I. Petunin, Scales of Banach spaces, *Usp. Math. Nauk* **21** (1966), 89–168; Engl. transl. *Russian Math. Surveys* **21** (1966), 85–159. This review paper contains references to earlier work.

[133] P. Kristensen, L. Mejlbo, and E. Thue Poulsen, Tempered distributions in infinitely many dimensions. I. Canonical field operators, *Commun. Math. Phys.* **1** (1965), 175–214; id. II. Displacement operators, *Math. Scand.* **14** (1964), 129–150; id. III. *Commun. Math. Phys.* **6** (1967), 29–48.

[134] K-D. Kürsten, Ein Gegenbeispiel zum Reflexivitätsproblem fur gemeinsame Definitionsbereiche von Operatorenalgebren im separablen Hilbertraum, *Wiss. Z. KMU-Leipzig, Math.-Naturwiss. R.* **31** (1982), 49–54.

[135] K-D. Kürsten, The completion of the maximal O_p*-algebra on a Fréchet domain, *Publ. Res. Inst. Math. Sci., Kyoto Univ.* **22** (1986), 151–175.

[136] K-D. Kürsten, On algebraic properties of partial algebras, *Rend. Circ. Mat. Palermo, Ser.II*, Suppl. **56** (1998), 111–122.

[137] K-D. Kürsten and M. Läuter, An extreme example concerning factorization products on the Schwartz space $\mathfrak{S}(\mathbb{R}^n)$ *Note Mat.* **25** (2005/06), 31–38.

[138] G. Lassner, Topological algebras of operators, *Rep. Math. Phys.* **3** (1972), 279–293.

[139] G. Lassner, Topologien auf Op*-Algebren, *Wiss. Z. KMU-Leipzig, Math.-Naturwiss. R.* **24** (1975), 465–471.

[140] G. Lassner, G.A. Lassner, and C. Trapani, Canonical commutation relations on the interval, *J. Math. Phys.* **28** (1987), 174–177.

[141] G. Lassner and W. Timmermann, Classification of domains of closed operators, *Rep. Math. Phys.* **9** (1976), 157–170.

[142] G. Lassner and W. Timmermann, Classification of domains of operator algebras, *Rep. Math. Phys.* **9** (1976), 205–217.

[143] J. Lindenstrauss and L. Tzafriri, On the complemented subspaces problem, *Israel J. Math.* **9** (1971), 263–269.

[144] D.H. Luecking, Representation and duality in weighted spaces of analytic functions, *Ind. Univ. Math. J.* **34** (1985), 319–336.

[145] F. Luef, Gabor analysis, noncommutative tori and Feichtinger's algebra, in *Gabor and wavelet frames*, S.S. Goh *et al.* (eds.), pp.77–106; Lecture Notes Series, Institute for Mathematical Sciences, National University of Singapore, vol. 10, World Scientific, Hackensack, NJ, 2007.

[146] F. Luef and Yu.I. Manin, Quantum theta functions and Gabor frames for modulation spaces, *Lett. Math. Phys.* **88** (2009), 131–161.

[147] W.A.J. Luxemburg and A.C. Zaanen, Some examples of normed Köthe spaces, *Math. Ann.* **162** (1966), 337–350.

[148] F. Mathot, Topological properties of unbounded bicommutants, *J. Math. Phys.* **26** (1985), 1118–1124.

[149] O. Melsheimer, Rigged Hilbert space formalism as an extended mathematical formalism for quantum systems. I. General theory, *J. Math. Phys.* **15** (1974), 902–916; id. II. Transformation theory in nonrelativistic quantum mechanics, *J. Math. Phys.* **15** (1974), 917–925.

[150] B. Nagel, Generalized eigenvectors in group representations, in *Studies in Mathematical Physics*, pp.135–154; A.O. Barut (ed.), Reidel, Dordrecht, 1973.

[151] B. Nazaret and M. Holschneider, An interpolation family between Gabor and wavelet transformations. Application to differential calculus and construction of anisotropic Banach spaces, in *Nonlinear Hyperbolic Equations, Spectral Theory, and Wavelet Transformations*, S. Albeverio *et al.*, (eds.), pp.363–394; Operator Theory: Advances and Applications, vol. 145, Birkhäuser, Basel, 2003.

[152] K. Napiórkowski, Good and bad generalized eigenvectors. I, II, *Bull. Acad. Pol. Sc.* **22** (1974) 1215–1218; **23** (1975), 251–252.

[153] E. Nelson, Analytic vectors, *Ann. Math.* **70** (1959), 572–615.

[154] E. Nelson, Interaction of nonrelativistic particles with a quantized scalar field, *J. Funct. Anal.* **11** (1972), 211–219.

[155] E. Nelson, Time-ordered operator products of sharp-time quadratic forms, *J. Math. Phys.* **5** (1964), 1190–1197.

[156] E. Nelson, Construction of quantum fields from Markoff fields, *J. Funct. Anal.* **12** (1973), 97–112.

[157] E. Nelson, The free Markoff field, *J. Funct. Anal.* **12** (1973), 211–227.

[158] E. Nelson and W.F. Stinespring, Representation of elliptic operators in an enveloping algebra, *Amer. J. Math.* **81** (1959), 547–560.

[159] D.M. Onchiş and E.M. Súarez Sánchez, The flexible Gabor-wavelet transform for car crash analysis, *Int. J. Wavelets Multiresolut. Inf. Process.* **7** (2009), 481–490.

[160] O. Ore, Galois connexions, *Trans. Amer. Math. Soc.* **55** (1944), 493–515.

[161] L. Päivärinta and E. Somersalo, A generalization of the Calderon-Vaillancourt theorem to L^p and ℓ^p, *Math. Nachr.* **138** (1988), 145–156.

[162] R. Palais, Chains of Hilbertian spaces, in *Seminar on the Atiyah-Singer Index Theorem*, Chap. VIII, pp.125–146; Tensor products, *ibid.* Chap. XIV, pp.197–214, *Annals of Math. Studies, # 57*, Princeton University, Princeton, N.J., 1965.

[163] Th. Paul, Functions analytic on the half-plane as quantum mechanical states, *J. Math. Phys.* **25** (1984), 3252–3263.

[164] V. Perrier and C. Basdevant, Besov norms in terms of the continuous wavelet transform. Application to structure functions, *Math. Models Meth. Appl. Sci.* **6** (1996), 649–664.

[165] G. Pickert, Bemerkungen über Galois-Verbindungen, *Archiv Math.* **3** (1952), 385–389.

[166] C. Piron, Axiomatique quantique, *Helv. Phys. Acta* **37** (1964), 439–468.

[167] Z. Popowicz, Remarks on dual structures in a Dirac space, *Bull. Acad. Pol. Sci. Ser. Sci. Math.* **23** (1975), 1119–1124.

[168] R.T. Powers, Self-adjoint algebras of unbounded operators, *Commun. Math. Phys.* **2** (1971), 85–124.

[169] E. Prugovečki, Topologies on generalized inner product spaces, *Canad. J. Math.* **21** (1969), 158–169.

[170] E. Prugovečki, The bra and ket formalism in extended Hilbert space, *J. Math. Phys.* **14**, 1410–1422 (1973).

[171] J.E. Roberts, The Dirac bra and ket formalism, *J. Math. Phys.* **7** (1966), 1097–1104.

[172] J.E. Roberts, Rigged Hilbert spaces in quantum mechanics, *Commun. Math. Phys.* **3** (1966), 98–119.

[173] A. Russo and C. Trapani, Quasi *-algebras and multiplication of distributions, *J. Math. Anal. Appl.* **215** (1997), 423–442.

[174] S. Samarah, S. Obeidat, and R. Salman, A Schur test for weighted mixed-norm spaces, *Anal. Math.* **31** (2005), 277–289.

[175] L. Schwartz, Sous-espaces hilbertiens d'espaces vectoriels topologiques et noyaux associés (noyaux reproduisants), *J. Anal. Math. (Jerusalem)* **13** (1964), 115–256.

[176] I.E. Segal, A noncommutative extension of abstract integration, *Ann. Math.* **57** (1953), 401–457.

[177] J. Shabani, Quantized fields and operators on a partial inner product space, *Ann. Inst. H. Poincaré* **48** (1988), 97–104.

[178] Z. Shmuely, The structure of Galois connections, *Pacific J. Math.* **54** (1974), 209–225.

[179] B. Simon, Distributions and their Hermite expansions, *J. Math. Phys.* **12** (1971), 140–148.

[180] E. Skibsted, Resonances eigenfunctions of a dilation-analytic Schrödinger operator, based on the Mellin transform, *J. Math. Anal. Appl.* **117** (1986), 198–219.

[181] G. Svetlichny, On the domains of generalized operators, *Commun. Math. Phys.* **33** (1973), 243–251.

[182] C. Trapani, Some results on CQ*-algebras of operators, *Suppl. Rend. Circolo Mat. Palermo* **56** (1998), 199–205.

[183] C. Trapani and F. Tschinke, Partial multiplication of operators in rigged Hilbert spaces, *Integral Equations Operator Theory* **51** (2005), 583–600.

[184] C. Trapani and F. Tschinke, Partial *-algebras of distributions, *Publ. Res. Inst. Math. Sci., Kyoto Univ.* **41** (2005), 259–279.

[185] F. Tschinke, Some results about operators in nested Hilbert spaces, *Rend. Circ. Mat. Palermo, Ser.II*, **54** (2005), 81–92.

[186] P. Werner, A distribution-theoretical approach to certain Lebesgue and Sobolev spaces, *J. Math. Anal. Appl.* **29** (1970), 18–78.

[187] E.P. Wigner, Unitary representations of the inhomogeneous Lorentz group, *Ann. Math.* **40** (1939), 149–204.

[188] C. van Winter, Fredholm equations on a Hilbert space of analytic functions, *Trans. Amer. Math. Soc.* **162** (1971), 103–139.

[189] C. van Winter, Complex dynamical variables for multiparticle systems with analytic interactions. I, II, *J. Math. Anal. Appl.* **47** (1974), 633–670, **48** (1974), 368–399.

[190] N. Wiener, On the representation of functions by trigonometric integrals, *Math. Z.* **24** (1926), 575–616.

[191] N. Wiener, Tauberian theorems, *Annals of Math.* **33** (1932), 1–100.

Index

Lecture Notes in Mathematics

For information about earlier volumes
please contact your bookseller or Springer
LNM Online archive: springerlink.com

Vol. 1897: R. Doney, Fluctuation Theory for Lévy Processes, Ecole d'Été de Probabilités de Saint-Flour XXXV-2005. Editor: J. Picard (2007)

Vol. 1898: H.R. Beyer, Beyond Partial Differential Equations, On linear and Quasi-Linear Abstract Hyperbolic Evolution Equations (2007)

Vol. 1899: Séminaire de Probabilités XL. Editors: C. Donati-Martin, M. Émery, A. Rouault, C. Stricker (2007)

Vol. 1900: E. Bolthausen, A. Bovier (Eds.), Spin Glasses (2007)

Vol. 1901: O. Wittenberg, Intersections de deux quadriques et pinceaux de courbes de genre 1, Intersections of Two Quadrics and Pencils of Curves of Genus 1 (2007)

Vol. 1902: A. Isaev, Lectures on the Automorphism Groups of Kobayashi-Hyperbolic Manifolds (2007)

Vol. 1903: G. Kresin, V. Maz'ya, Sharp Real-Part Theorems (2007)

Vol. 1904: P. Giesl, Construction of Global Lyapunov Functions Using Radial Basis Functions (2007)

Vol. 1905: C. Prévôt, M. Röckner, A Concise Course on Stochastic Partial Differential Equations (2007)

Vol. 1906: T. Schuster, The Method of Approximate Inverse: Theory and Applications (2007)

Vol. 1907: M. Rasmussen, Attractivity and Bifurcation for Nonautonomous Dynamical Systems (2007)

Vol. 1908: T.J. Lyons, M. Caruana, T. Lévy, Differential Equations Driven by Rough Paths, Ecole d'Été de Probabilités de Saint-Flour XXXIV-2004 (2007)

Vol. 1909: H. Akiyoshi, M. Sakuma, M. Wada, Y. Yamashita, Punctured Torus Groups and 2-Bridge Knot Groups (I) (2007)

Vol. 1910: V.D. Milman, G. Schechtman (Eds.), Geometric Aspects of Functional Analysis. Israel Seminar 2004-2005 (2007)

Vol. 1911: A. Bressan, D. Serre, M. Williams, K. Zumbrun, Hyperbolic Systems of Balance Laws. Cetraro, Italy 2003. Editor: P. Marcati (2007)

Vol. 1912: V. Berinde, Iterative Approximation of Fixed Points (2007)

Vol. 1913: J.E. Marsden, G. Misiołek, J.-P. Ortega, M. Perlmutter, T.S. Ratiu, Hamiltonian Reduction by Stages (2007)

Vol. 1914: G. Kutyniok, Affine Density in Wavelet Analysis (2007)

Vol. 1915: T. Bıyıkoğlu, J. Leydold, P.F. Stadler, Laplacian Eigenvectors of Graphs. Perron-Frobenius and Faber-Krahn Type Theorems (2007)

Vol. 1916: C. Villani, F. Rezakhanlou, Entropy Methods for the Boltzmann Equation. Editors: F. Golse, S. Olla (2008)

Vol. 1917: I. Veselić, Existence and Regularity Properties of the Integrated Density of States of Random Schrödinger (2008)

Vol. 1918: B. Roberts, R. Schmidt, Local Newforms for GSp(4) (2007)

Vol. 1919: R.A. Carmona, I. Ekeland, A. Kohatsu-Higa, J.-M. Lasry, P.-L. Lions, H. Pham, E. Taflin, Paris-Princeton Lectures on Mathematical Finance 2004. Editors: R.A. Carmona, E. Çinlar, I. Ekeland, E. Jouini, J.A. Scheinkman, N. Touzi (2007)

Vol. 1920: S.N. Evans, Probability and Real Trees. Ecole d'Été de Probabilités de Saint-Flour XXXV-2005 (2008)

Vol. 1921: J.P. Tian, Evolution Algebras and their Applications (2008)

Vol. 1922: A. Friedman (Ed.), Tutorials in Mathematical BioSciences IV. Evolution and Ecology (2008)

Vol. 1923: J.P.N. Bishwal, Parameter Estimation in Stochastic Differential Equations (2008)

Vol. 1924: M. Wilson, Littlewood-Paley Theory and Exponential-Square Integrability (2008)

Vol. 1925: M. du Sautoy, L. Woodward, Zeta Functions of Groups and Rings (2008)

Vol. 1926: L. Barreira, V. Claudia, Stability of Nonautonomous Differential Equations (2008)

Vol. 1927: L. Ambrosio, L. Caffarelli, M.G. Crandall, L.C. Evans, N. Fusco, Calculus of Variations and Non-Linear Partial Differential Equations. Cetraro, Italy 2005. Editors: B. Dacorogna, P. Marcellini (2008)

Vol. 1928: J. Jonsson, Simplicial Complexes of Graphs (2008)

Vol. 1929: Y. Mishura, Stochastic Calculus for Fractional Brownian Motion and Related Processes (2008)

Vol. 1930: J.M. Urbano, The Method of Intrinsic Scaling. A Systematic Approach to Regularity for Degenerate and Singular PDEs (2008)

Vol. 1931: M. Cowling, E. Frenkel, M. Kashiwara, A. Valette, D.A. Vogan, Jr., N.R. Wallach, Representation Theory and Complex Analysis. Venice, Italy 2004. Editors: E.C. Tarabusi, A. D'Agnolo, M. Picardello (2008)

Vol. 1932: A.A. Agrachev, A.S. Morse, E.D. Sontag, H.J. Sussmann, V.I. Utkin, Nonlinear and Optimal Control Theory. Cetraro, Italy 2004. Editors: P. Nistri, G. Stefani (2008)

Vol. 1933: M. Petkovic, Point Estimation of Root Finding Methods (2008)

Vol. 1934: C. Donati-Martin, M. Émery, A. Rouault, C. Stricker (Eds.), Séminaire de Probabilités XLI (2008)

Vol. 1935: A. Unterberger, Alternative Pseudodifferential Analysis (2008)

Vol. 1936: P. Magal, S. Ruan (Eds.), Structured Population Models in Biology and Epidemiology (2008)

Vol. 1937: G. Capriz, P. Giovine, P.M. Mariano (Eds.), Mathematical Models of Granular Matter (2008)

Vol. 1938: D. Auroux, F. Catanese, M. Manetti, P. Seidel, B. Siebert, I. Smith, G. Tian, Symplectic 4-Manifolds and Algebraic Surfaces. Cetraro, Italy 2003. Editors: F. Catanese, G. Tian (2008)

Vol. 1939: D. Boffi, F. Brezzi, L. Demkowicz, R.G. Durán, R.S. Falk, M. Fortin, Mixed Finite Elements, Compatibility Conditions, and Applications. Cetraro, Italy 2006. Editors: D. Boffi, L. Gastaldi (2008)

Vol. 1940: J. Banasiak, V. Capasso, M.A.J. Chaplain, M. Lachowicz, J. Miękisz, Multiscale Problems in the Life Sciences. From Microscopic to Macroscopic. Będlewo, Poland 2006. Editors: V. Capasso, M. Lachowicz (2008)

Vol. 1941: S.M.J. Haran, Arithmetical Investigations. Representation Theory, Orthogonal Polynomials, and Quantum Interpolations (2008)

Vol. 1942: S. Albeverio, F. Flandoli, Y.G. Sinai, SPDE in Hydrodynamic. Recent Progress and Prospects. Cetraro, Italy 2005. Editors: G. Da Prato, M. Röckner (2008)

Vol. 1943: L.L. Bonilla (Ed.), Inverse Problems and Imaging. Martina Franca, Italy 2002 (2008)

Vol. 1944: A. Di Bartolo, G. Falcone, P. Plaumann, K. Strambach, Algebraic Groups and Lie Groups with Few Factors (2008)

Vol. 1945: F. Brauer, P. van den Driessche, J. Wu (Eds.), Mathematical Epidemiology (2008)

Vol. 1946: G. Allaire, A. Arnold, P. Degond, T.Y. Hou, Quantum Transport. Modelling, Analysis and Asymptotics. Cetraro, Italy 2006. Editors: N.B. Abdallah, G. Frosali (2008)

Vol. 1947: D. Abramovich, M. Mariño, M. Thaddeus, R. Vakil, Enumerative Invariants in Algebraic Geometry and String Theory. Cetraro, Italy 2005. Editors: K. Behrend, M. Manetti (2008)

Vol. 1948: F. Cao, J-L. Lisani, J-M. Morel, P. Musé, F. Sur, A Theory of Shape Identification (2008)

Vol. 1949: H.G. Feichtinger, B. Helffer, M.P. Lamoureux, N. Lerner, J. Toft, Pseudo-Differential Operators. Quantization and Signals. Cetraro, Italy 2006. Editors: L. Rodino, M.W. Wong (2008)

Vol. 1950: M. Bramson, Stability of Queueing Networks, Ecole d'Eté de Probabilités de Saint-Flour XXXVI-2006 (2008)

Vol. 1951: A. Moltó, J. Orihuela, S. Troyanski, M. Valdivia, A Non Linear Transfer Technique for Renorming (2009)

Vol. 1952: R. Mikhailov, I.B.S. Passi, Lower Central and Dimension Series of Groups (2009)

Vol. 1953: K. Arwini, C.T.J. Dodson, Information Geometry (2008)

Vol. 1954: P. Biane, L. Bouten, F. Cipriani, N. Konno, N. Privault, Q. Xu, Quantum Potential Theory. Editors: U. Franz, M. Schuermann (2008)

Vol. 1955: M. Bernot, V. Caselles, J.-M. Morel, Optimal Transportation Networks (2008)

Vol. 1956: C.H. Chu, Matrix Convolution Operators on Groups (2008)

Vol. 1957: A. Guionnet, On Random Matrices: Macroscopic Asymptotics, Ecole d'Eté de Probabilités de Saint-Flour XXXVI-2006 (2009)

Vol. 1958: M.C. Olsson, Compactifying Moduli Spaces for Abelian Varieties (2008)

Vol. 1959: Y. Nakkajima, A. Shiho, Weight Filtrations on Log Crystalline Cohomologies of Families of Open Smooth Varieties (2008)

Vol. 1960: J. Lipman, M. Hashimoto, Foundations of Grothendieck Duality for Diagrams of Schemes (2009)

Vol. 1961: G. Buttazzo, A. Pratelli, S. Solimini, E. Stepanov, Optimal Urban Networks via Mass Transportation (2009)

Vol. 1962: R. Dalang, D. Khoshnevisan, C. Mueller, D. Nualart, Y. Xiao, A Minicourse on Stochastic Partial Differential Equations (2009)

Vol. 1963: W. Siegert, Local Lyapunov Exponents (2009)

Vol. 1964: W. Roth, Operator-valued Measures and Integrals for Cone-valued Functions and Integrals for Cone-valued Functions (2009)

Vol. 1965: C. Chidume, Geometric Properties of Banach Spaces and Nonlinear Iterations (2009)

Vol. 1966: D. Deng, Y. Han, Harmonic Analysis on Spaces of Homogeneous Type (2009)

Vol. 1967: B. Fresse, Modules over Operads and Functors (2009)

Vol. 1968: R. Weissauer, Endoscopy for GSP(4) and the Cohomology of Siegel Modular Threefolds (2009)

Vol. 1969: B. Roynette, M. Yor, Penalising Brownian Paths (2009)

Vol. 1970: M. Biskup, A. Bovier, F. den Hollander, D. Ioffe, F. Martinelli, K. Netočný, F. Toninelli, Methods of Contemporary Mathematical Statistical Physics. Editor: R. Kotecký (2009)

Vol. 1971: L. Saint-Raymond, Hydrodynamic Limits of the Boltzmann Equation (2009)

Vol. 1972: T. Mochizuki, Donaldson Type Invariants for Algebraic Surfaces (2009)

Vol. 1973: M.A. Berger, L.H. Kauffmann, B. Khesin, H.K. Moffatt, R.L. Ricca, De W. Sumners, Lectures on Topological Fluid Mechanics. Cetraro, Italy 2001. Editor: R.L. Ricca (2009)

Vol. 1974: F. den Hollander, Random Polymers: École d'Été de Probabilités de Saint-Flour XXXVII – 2007 (2009)

Vol. 1975: J.C. Rohde, Cyclic Coverings, Calabi-Yau Manifolds and Complex Multiplication (2009)

Vol. 1976: N. Ginoux, The Dirac Spectrum (2009)

Vol. 1977: M.J. Gursky, E. Lanconelli, A. Malchiodi, G. Tarantello, X.-J. Wang, P.C. Yang, Geometric Analysis and PDEs. Cetraro, Italy 2001. Editors: A. Ambrosetti, S.-Y.A. Chang, A. Malchiodi (2009)

Vol. 1978: M. Qian, J.-S. Xie, S. Zhu, Smooth Ergodic Theory for Endomorphisms (2009)

Vol. 1979: C. Donati-Martin, M. Émery, A. Rouault, C. Stricker (Eds.), Séminaire de Probablitiés XLII (2009)

Vol. 1980: P. Graczyk, A. Stos (Eds.), Potential Analysis of Stable Processes and its Extensions (2009)

Vol. 1981: M. Chlouveraki, Blocks and Families for Cyclotomic Hecke Algebras (2009)

Vol. 1982: N. Privault, Stochastic Analysis in Discrete and Continuous Settings. With Normal Martingales (2009)

Vol. 1983: H. Ammari (Ed.), Mathematical Modeling in Biomedical Imaging I. Electrical and Ultrasound Tomographies, Anomaly Detection, and Brain Imaging (2009)

Vol. 1984: V. Caselles, P. Monasse, Geometric Description of Images as Topographic Maps (2010)

Vol. 1985: T. Linß, Layer-Adapted Meshes for Reaction-Convection-Diffusion Problems (2010)

Vol. 1986: J.-P. Antoine, C. Trapani, Partial Inner Product Spaces. Theory and Applications (2009)

Vol. 1987: J.-P. Brasslet, J. Seade, T. Suwa, Vector Fields on Singular Varieties (2010)

Recent Reprints and New Editions

Vol. 1702: J. Ma, J. Yong, Forward-Backward Stochastic Differential Equations and their Applications. 1999 – Corr. 3rd printing (2007)

Vol. 830: J.A. Green, Polynomial Representations of GL_n, with an Appendix on Schensted Correspondence and Littelmann Paths by K. Erdmann, J.A. Green and M. Schoker 1980 – 2nd corr. and augmented edition (2007)

Vol. 1693: S. Simons, From Hahn-Banach to Monotonicity (Minimax and Monotonicity 1998) – 2nd exp. edition (2008)

Vol. 470: R.E. Bowen, Equilibrium States and the Ergodic Theory of Anosov Diffeomorphisms. With a preface by D. Ruelle. Edited by J.-R. Chazottes. 1975 – 2nd rev. edition (2008)

Vol. 523: S.A. Albeverio, R.J. Høegh-Krohn, S. Mazzucchi, Mathematical Theory of Feynman Path Integral. 1976 – 2nd corr. and enlarged edition (2008)

Vol. 1764: A. Cannas da Silva, Lectures on Symplectic Geometry 2001 – Corr. 2nd printing (2008)

LECTURE NOTES IN MATHEMATICS  Springer

Edited by J.-M. Morel, F. Takens, B. Teissier, P.K. Maini

Editorial Policy (for the publication of monographs)

1. Lecture Notes aim to report new developments in all areas of mathematics and their applications - quickly, informally and at a high level. Mathematical texts analysing new developments in modelling and numerical simulation are welcome.

    Monograph manuscripts should be reasonably self-contained and rounded off. Thus they may, and often will, present not only results of the author but also related work by other people. They may be based on specialised lecture courses. Furthermore, the manuscripts should provide sufficient motivation, examples and applications. This clearly distinguishes Lecture Notes from journal articles or technical reports which normally are very concise. Articles intended for a journal but too long to be accepted by most journals, usually do not have this "lecture notes" character. For similar reasons it is unusual for doctoral theses to be accepted for the Lecture Notes series, though habilitation theses may be appropriate.

2. Manuscripts should be submitted either online at www.editorialmanager.com/lnm to Springer's mathematics editorial in Heidelberg, or to one of the series editors. In general, manuscripts will be sent out to 2 external referees for evaluation. If a decision cannot yet be reached on the basis of the first 2 reports, further referees may be contacted: The author will be informed of this. A final decision to publish can be made only on the basis of the complete manuscript, however a refereeing process leading to a preliminary decision can be based on a pre-final or incomplete manuscript. The strict minimum amount of material that will be considered should include a detailed outline describing the planned contents of each chapter, a bibliography and several sample chapters.

    Authors should be aware that incomplete or insufficiently close to final manuscripts almost always result in longer refereeing times and nevertheless unclear referees' recommendations, making further refereeing of a final draft necessary.

    Authors should also be aware that parallel submission of their manuscript to another publisher while under consideration for LNM will in general lead to immediate rejection.

3. Manuscripts should in general be submitted in English. Final manuscripts should contain at least 100 pages of mathematical text and should always include

    – a table of contents;
    – an informative introduction, with adequate motivation and perhaps some historical remarks: it should be accessible to a reader not intimately familiar with the topic treated;
    – a subject index: as a rule this is genuinely helpful for the reader.

    For evaluation purposes, manuscripts may be submitted in print or electronic form (print form is still preferred by most referees), in the latter case preferably as pdf- or zipped ps-files. Lecture Notes volumes are, as a rule, printed digitally from the authors' files. To ensure best results, authors are asked to use the LaTeX2e style files available from Springer's web-server at:

    ftp://ftp.springer.de/pub/tex/latex/svmonot1/ (for monographs) and
    ftp://ftp.springer.de/pub/tex/latex/svmultt1/ (for summer schools/tutorials).

Additional technical instructions, if necessary, are available on request from: lnm@springer.com.

4. Careful preparation of the manuscripts will help keep production time short besides ensuring satisfactory appearance of the finished book in print and online. After acceptance of the manuscript authors will be asked to prepare the final LaTeX source files and also the corresponding dvi-, pdf- or zipped ps-file. The LaTeX source files are essential for producing the full-text online version of the book (see http://www.springerlink.com/openurl.asp?genre=journal&issn=0075-8434 for the existing online volumes of LNM).

   The actual production of a Lecture Notes volume takes approximately 12 weeks.

5. Authors receive a total of 50 free copies of their volume, but no royalties. They are entitled to a discount of 33.3% on the price of Springer books purchased for their personal use, if ordering directly from Springer.

6. Commitment to publish is made by letter of intent rather than by signing a formal contract. Springer-Verlag secures the copyright for each volume. Authors are free to reuse material contained in their LNM volumes in later publications: a brief written (or e-mail) request for formal permission is sufficient.

Addresses:

Professor J.-M. Morel, CMLA,
École Normale Supérieure de Cachan,
61 Avenue du Président Wilson, 94235 Cachan Cedex, France
E-mail: Jean-Michel.Morel@cmla.ens-cachan.fr

Professor F. Takens, Mathematisch Instituut,
Rijksuniversiteit Groningen, Postbus 800,
9700 AV Groningen, The Netherlands
E-mail: F.Takens@rug.nl

Professor B. Teissier, Institut Mathématique de Jussieu,
UMR 7586 du CNRS, Équipe "Géométrie et Dynamique",
175 rue du Chevaleret,
75013 Paris, France
E-mail: teissier@math.jussieu.fr

For the "Mathematical Biosciences Subseries" of LNM:

Professor P.K. Maini, Center for Mathematical Biology,
Mathematical Institute, 24-29 St Giles,
Oxford OX1 3LP, UK
E-mail: maini@maths.ox.ac.uk

Springer, Mathematics Editorial, Tiergartenstr. 17,
69121 Heidelberg, Germany,
Tel.: +49 (6221) 487-259
Fax: +49 (6221) 4876-8259
E-mail: lnm@springer.com